Lecture Notes in Bioinformatics 4955

Subseries of Lecture Notes in Computer Science

Martin Vingron   Limsoon Wong (Eds.)

Research in Computational Molecular Biology

12th Annual International Conference, RECOMB 2008
Singapore, March 30 – April 2, 2008
Proceedings

 Springer

Series Editors

Sorin Istrail, Brown University, Providence, RI, USA
Pavel Pevzner, University of California, San Diego, CA, USA
Michael Waterman, University of Southern California, Los Angeles, CA, USA

Volume Editors

Martin Vingron
Max Planck Institute for Molecular Genetics
Ihnestr. 63-73, 14195 Berlin, Germany
E-mail: vingron@molgen.mpg.de

Limsoon Wong
National University of Singapore, School of Computing
Law Link, Singapore 117590
E-mail: wongls@comp.nus.edu.sg

Library of Congress Control Number: 2008922908

CR Subject Classification (1998): F.2.2, F.2, E.1, G.2, H.2.8, G.3, I.2, J.3

LNCS Sublibrary: SL 8 – Bioinformatics

ISSN   0302-9743
ISBN-10   3-540-78838-7 Springer Berlin Heidelberg New York
ISBN-13   978-3-540-78838-6 Springer Berlin Heidelberg New York

Springer is a part of Springer Science+Business Media

springer.com

© Springer-Verlag Berlin Heidelberg 2008

Typesetting: Camera-ready by author, data conversion by Scientific Publishing Services, Chennai, India
Printed on acid-free paper   SPIN: 12247264   06/3180   5 4 3 2 1 0

Preface

This volume contains the papers presented at the 12th Annual International Conference on REsearch in COmputational Molecular Biology (RECOMB 2008) held in Singapore, March 30-April 2, 2008. The RECOMB conference series was founded in 1997 by Sorin Istrail, Pavel Pevzner and Michael Waterman. Its history is summarized in the table herein. The general development of the RECOMB conference series is guided by a Steering Committee. RECOMB 2008 was hosted by the National University of Singapore and held at the University Cultural Centre. The Conference Chair was Limsoon Wong, who was supported by a 16-member-strong Organizing Committee.

Out of 193 papers submitted to RECOMB 2008, the Program Committee (PC) selected 34 papers for presentation at the conference. The PC consisted of 46 members, who could further draw on the help of subreviewers. Three PC members with possible help of the external reviewers first reviewed each paper, subsequently a Web-based discussion led to the final selection of papers.

This year, RECOMB and the journal *Genome Research* teamed up to offer authors *Genome Research* as an alternative forum for publication. Approximately 55 of the RECOMB submissions, selected by a subset of the PC headed by Serafim Batzoglou, were forwarded to *Genome Research* for consideration. The reviewing procedures by RECOMB and *Genome Research* were conducted independently, leading to the acceptance of a small number of papers to both RECOMB and *Genome Research*. Other papers were accepted to *Genome Research* but not to RECOMB. The papers accepted by both were presented at the conference and are also summarized in this volume. Their final versions appear in *Genome Research*.

In addition to the contributed presentations, RECOMB 2008 featured seven keynote lectures, held by Howard Cedar (Hebrew University, Jerusalem), Vivian G. Cheung (University of Pennsylvania, Philadelphia), Suzanne Cory (Walter and Eliza Hall Institute of Medical Research, Melbourne), Sang Yup Lee (Korea Advanced Institute of Science and Technology, Daejeon), Edison Liu (Genome Institute of Singapore, Singapore), Andrei Lupas (Max Planck Institute for Developmental Biology, Tübingen), and Temple F. Smith (Boston University, Boston). The Ulam Lecture was delivered by Temple Smith, the Distinguished Technology Lecture by Edison Liu and the Distinguished Biology Lecture by Suzanne Cory.

In addition to the oral presentations, RECOMB 2008 also provided a forum for presenting approximately 150 posters. The poster sessions are an integral part of the conference and foster communication among participants.

A conference like RECOMB lives from the dedication of the people helping and supporting it. Thanks go to the RECOMB Steering Committee for their advice and guidance, to the PC and the external reviewers for their efforts, to the Poster Committee for going through a large number of posters in short time, and to the local organizers for their hard work. The reviewing process profited greatly from the use of the EasyChair conference management system. Patricia Marquardt's help in putting together this proceedings volume is greatly appreciated. We are further indebted to the National University of Singapore, the Institute for Infocomm Research (A*STAR), and the Bioinformatics Institute (A*STAR) for their support of RECOMB 2008. We also appreciate the generous sponsorships from the Lilly Singapore Center for Drug Discovery, IBM, the International Society for Computational Biology, World Scientific Publishing Company, and Taylor & Francis Asia Pacific. Special thanks go to the participants, to those who contributed posters or oral presentations, and to the keynote speakers for their input. It is this community that makes RECOMB such an inspiring, lively, and enjoyable conference.

February 2008   Martin Vingron

Organization

Program Committee

Tatsuya Akutsu	Kyoto University, Japan
Serafim Batzoglou	Stanford University, USA
Bonnie Berger	MIT, Cambridge, USA
Mathieu Blanchette	McGill University, Montreal, Quebec, Canada
Joachim Buhmann	ETH Zürich, Switzerland
Kun-Mao Chao	National Taiwan University, Taipei, Taiwan
Dannie Durand	Carnegie Mellon University, Pittsburgh, USA
Eleazar Eskin	University of California, Los Angeles, USA
Nir Friedman	Hebrew University, Jerusalem, Israel
Dan Geiger	Technion, Haifa, Israel
Gaston Gonnet	ETH Zürich, Switzerland
Dan Gusfield	University of California, Davis, USA
Arndt von Haeseler	Max F. Perutz Laboratories, Vienna, Austria
Alexander Hartemink	Duke University, Durham, USA
Trey Ideker	University of California, San Diego, USA
Sorin Istrail	Brown University, Providence, USA
Tao Jiang	University of California, Riverside, USA
John Kececioglu	University of Arizona, Tucson, USA
Manolis Kellis	Massachusetts Institute of Technology, Cambridge, USA
Anders Krogh	University of Copenhagen, Denmark
Jens Lagergren	Royal Institute of Technology, Stockholm, Sweden
Thomas Lengauer	Max Planck Institute for Computer Science, Saarbrücken, Germany
Arthur Lesk	Pennsylvania State University, University Park, USA
Ming Li	University of Waterloo, Canada
Michal Linial	Hebrew University, Jerusalem, Israel
Satoru Miyano	Tokyo University, Japan
William Noble	University of Washington, USA
Pavel Pevzner	University of California, San Diego, USA
Knut Reinert	Free University of Berlin, Germany
Stephan Robin	AgroParisTech, Paris, France
Ron Shamir	Tel Aviv University, Israel
Roded Sharan	Tel Aviv University, Israel
Adam Siepel	Cornell University, Ithaca, USA
Terence Speed	University of California, Berkeley, USA
Peter Stadler	University of Leipzig, Germany
Jens Stoye	University of Bielefeld, Germany
Fengzhu Sun	University of Southern California, Los Angeles, USA

Wing-Kin Sung	National University of Singapore, Singapore
Simon Tavaré	Cornell University, Ithaca, USA
Jerzy Tiuryn	University of Warsaw, Poland
Martin Vingron (Chair)	Max Planck Institute for Molecular Genetics, Berlin, Germany
Lusheng Wang	City University, Hong Kong
Michael Waterman	University of Southern California, Los Angeles, USA
Limsoon Wong	National University of Singapore, Singapore
Wing Hung Wong	Stanford University, USA
Eric Xing	Carnegie Mellon University, Pittsburgh, USA
Zohar Yakhini	Agilent Laboratories, Santa Clara, USA

Poster Committee

Hon-Nian Chua	Institute for Infocomm Research, Singapore
Neil Clarke	Genome Institute of Singapore, Singapore
Frank Eisenhaber	Bioinformatics Institute, Singapore
Wing-Kin Sung	National University of Singapore, Singapore
Greg Tucker-Kellogg	Lilly Singapore Centre for Drug Discovery, Singapore
Martin Vingron (Co-chair)	Max Planck Institute for Molecular Genetics, Berlin, Germany
Limsoon Wong (Co-chair)	National University of Singapore, Singapore

Steering Committee

Serafim Batzoglou	Stanford University, USA
Sorin Istrail	Brown University, Providence, USA
Thomas Lengauer	Max Planck Institute for Computer Science, Saarbrücken, Germany
Michal Linial	Hebrew University, Jerusalem, Israel
Pavel Pevzner	University of California, San Diego, USA
Ron Shamir	Tel Aviv University, Israel
Terrence Speed	University of California, Berkeley, USA
Martin Vingron	Max Planck Institute for Molecular Genetics, Berlin, Germany

Organizing Committee

Agnes Ang	National University of Singapore, Singapore
Lay Khim Chng	National University of Singapore, Singapore
Kwok Pui Choi	National University of Singapore, Singapore
Alexia Leong	National University of Singapore, Singapore
Hon Wai Leong	National University of Singapore, Singapore

Wai Kin Leong	National University of Singapore, Singapore
Stefanie Ng	National University of Singapore, Singapore
Wing-Kin Sung	National University of Singapore, Singapore
Martti Tammi	National University of Singapore, Singapore
Limsoon Wong (chair)	National University of Singapore, Singapore
Siang Yong Yap	National University of Singapore, Singapore
Xin Chen	Nanyang Technological University, Singapore
Chee Keong Kwoh	Nanyang Technological University, Singapore
Guillaume Bourque	Genome Institute of Singapore, Singapore
Alan Christoffels	Temasek Lifesciences Laboratory, Singapore
Dong-Yup Lee	Bioprocessing Technology Institute, Singapore
Gunaretnam Rajagopal	BioInformatics Institute, Singapore

Previous RECOMB Meetings

Date / Location	Hosting Institution	Program Chair	Conference Chair
January 20-23, 1997 Santa Fé, NM, USA	Sandia National Lab	Michael Waterman	Sorin Istrail
March 22-25, 1998 New York, NY, USA	Mt. Sinai School of Medicine	Pavel Pevzner	Gary Benson
April 22-25, 1999 Lyon, France	INRIA	Sorin Istrail	Mireille Regnier
April 8-11, 2000 Tokyo, Japan	University of Tokyo	Ron Shamir	Satoru Miyano
April 22-25, 2001 Montréal, Canada	Université de Montréal	Thomas Lengauer	David Sankoff
April 18-21. 2002 Washington, DC, USA	Celera	Gene Myers	Sridhar Hannenhalli
April 10-13, 2003 Berlin, Germany	Max Planck Institute for Molecular Genetics	Webb Miller	Martin Vingron
March 27-31, 2004 San Diego, CA, USA	UC San Diego	Dan Gusfield	Philip E. Bourne
May 14-18, 2005 Boston, MA, USA	Broad Institute of MIT and Harvard	Satoru Miyano	Jill P. Mesirov and Simon Kasif
April 2-5, 2006 Venice, Italy	University of Padova	Alberto Apostolico	Concettina Guerra
April 21-25, 2007 Oakland, CA, USA	California Institute for Quantitative Biomedical Research	Terry Speed	Sandrine Dudoit

The RECOMB 2008 Program Committee gratefully acknowledges the valuable input received from the following external reviewers

Sarah Aerni
Edoardo Airoldi
Frank Alber
Andrey Alexeyenko
Adrian Altenhoff
Peter Arndt
Lars Arvestad
George Asimenos
Manor Askenazi
Chris Bailey-Kellogg
Sourav Bandyopadhyay
Hideo Bannai
Vikas Bansal
Avner Bar-Hen
Nuno Barbosa-Morais
Markus Bauer
Jan Baumbach
Amir Ben-Dor
Sivan Bercovici
Doron Betel
Chris Bielow
Alexander Bockmayr
Wouter Boomsma
Laurent Brehelin
Brona Brejova
Dan Brown
Brian Browning
Jocelyne Bruand
Sebastian Böcker
Gal Chechik
Ming-Chiang Chen
Kuan-Yu Chen
Xiaoyu Chen
Benny Chor
Han-Yu Chuang
Trees-Juen Chuang
Inbar Cohen-Gihon
Miklos Csuros
Tim Danford
Ofir Davidovich
Minghua Deng
Zhihong Ding

Chuong Do
Norbert Dojer
Francisco Domingues
Banu Dost
Robin Dowell
Janusz Dutkowski
Eran Eden
Nathan Edwards
Tal El-Hay
Ran Elkon
Arne Elofsson
Olof Emanuelsson
Nikolas Fechner
Bernd Fischer
Jason Flannick
Geoffrey Fox
Ari Frank
Eugene Fratkin
Shiri Freilich
Weijie Fu
Mickael Guedj
Anna Gambin
Irit Gat-Viks
Guy Geva
Manuel Gil
Uri Gophna
Pawel Gorecki
Assaf Gottlieb
Ewing Greg
Frauke Gräter
Clemens Gröpl
Karlebach Guy
Naomi Habib
Eran Halperin
Yonit Halperin
Thomas Hamelryck
Buhm Han
Greg Hannum
Morihiro Hayashida
Jacques van Helden
Torsten Herrmann
Andreas Hildebrandt

Ivo Hofacker
Barbara Holland
Liisa Holm
Judie Howrylak
Yao-Ting Huang
Xiaoqiu Huang
Katharina Huber
Man Hung
Daniel Huson
Seiya Imoto
Anders Jacobsen
Katharina Jahn
Ariel Jaimovich
Rui Jiang
Inge Jonassen
Maxim Kalaev
Lukas Kall
Hans-Michael
Kaltenbach
Hetunandan Kamisetty
Hyun Min Kang
Tommy Kaplan
Yuki Kato
Uri Keich
Orgad Keller
Pouya Kheradpour
Seyoung Kim
Gad Kimmel
Bonnie Kirkpatrick
Aaron Klammer
Gunnar W. Klau
Boguslaw Kluge
Oliver Kohlbacher
Rachel Kolodny
Carolin Kosiol
Dennis Kostka
Roland Krause
David Kreil
Michael Lappe
Berthold Lausen
Chris Lee
Hans-Peter Lenhof

Ulf Leser
Mike Lin
Stinus Lindgreen
Chaim Linhart
Simone Linz
Hsiao-Fei Liu
Yaniv Loewenstein
Chin Lung Lu
Xiaotu Ma
Craig Mak
John Marioni
Paul Marjoram
Tobias Marschall
Troels Marstrand
Marie-Laure
Martin-Magniette
Osamu Maruyama
Tristan Mary-Huard
Jochen Maydt
Aurelien Mazurio
Paul Medvedev
Dirk Metzler
Martin Middendorf
Istvan Miklos
Axel Mosig
Simon Myers
Cedric Notredame
Noa Novershtern
Sanne Nygaard
Charles O'Donnell
Sean O'Rourke
Peter Orbanz
Ivan Ovcharenko
Rod Page
Utz Pape
Dana Pe'er
Jakob Skou Pedersen
Peter Pfaffelhuber
Franck Picard
Ron Pinter
Robert Preissner
Teresa Przytycka

Jian Qiu
Predrag Radivojac
Sven Rahmann
Jörg Rahnenführer
Ben Raphael
Matt Rasmussen
Tobias Rausch
Pradipta Ray
Sheila Reynolds
Markus Ringnér
Peter Robinson
Ingo Roeder
Volker Roth
Bashir Sadjad
Marie-France Sagot
Nikolaos Sahinidis
Albin Sandelin
Oliver Sander
Marc Schaub
Maya Schuldiner
Russell Schwartz
Ariel Schwartz
Bengt Sennblad
Oliver Serang
Itai Sharon
Maxim Shatsky
Tetsuo Shibuya
Teppei Shimamura
Tomer Shlomi
Suyash Shringarpure
Alexandra
Shulman-Peleg
Sagi Snir
Kyung-Ah Sohn
Nan Song
Yun Song
Rainer Spang
Rohith Srivas
Alex Stark
Israel Steinfeld
Christine Steinhoff
Jörg Stelling

Andreas Sundquist
Silpa Suthram
Alain Trubuil
Takeyuki Tamura
Kai Tan
Eric Tang
Haixu Tang
Willie Taylor
Robert Thurman
Ali Tofigh
Martin Tompa
Zhidong Tu
Eivind Valen
Roy Varshavsky
Balaji Venkatachalam
Stéphane Vialette
Antoine Vigneron
Jens Vilstrup Johansson
Tomas Vinar
Le Sy Vinh
Jerome Waldispuhl
Lin Wan
Li Wang
Hung-Lung Wang
Ilan Wapinski
David Weese
Hugo Willy
Ole Winther
Roland Wittler
Yufeng Wu
Xiting Yan
Zizhen Yao
Yuzhen Ye
Chun Ye
Daniel Yekutieli
Nir Yosef
Noah Zaitlen
Alex Zelikovsky
Tomasz Zemojtel
Shaojie Zhang
Xianghong Jasmine Zhou

Organizers

Supporting Organizers

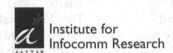

Platinum Sponsors

Gold Sponsors

Table of Contents

Computational Biology:
Its Challenges Past, Present, and Future

Temple F. Smith

Boston University, Boston
tsmith@darwin.bu.edu

Dedication: To a friend and inspiration, Stanislaw Ulam.

I was lucky not only to have known Stan (Dr. Ulam), but to have been be-friended by him. I would note that it was with his support that my first "computational" biology paper was published and that I was then a physics graduate student who had not taken or read biology since the sixth grade! I will outline just a few Ulam anecdotes in his memory.

Abstract. The recognition of the role of mathematics and computer science in modern biology has led to new terminology, as did chemistry with biochemistry, and physics with biophysics. We need to think only of bioinformatics, computational biology, and even system biology and genomics for example. These terms seem to strongly suggest that this is all rather new. Yet a short review of the work of those such as J.B.S. Haldane, Sewell Wright, DArcy Thompson and R.A. Fisher, to say nothing of scientists like Luria and Delbrueck or Hodgkin and Huxley or Thomas Hunt Morgan, is useful. Their work and foresight set the stage for modern applications of mathematical modeling and statistics in the biological sciences.

It has often been said that the only difference between now and then is the increase in data—a lot more data. This is clearly not the full story. In addition, we have computational power unimaginable to these earlier researchers, as well as to anyone only forty years ago. So what are our challenges? Some are clear, including the modeling and analysis of biologys complex systems such as a cells signaling, metabolic and differentiation. Also needed are analysis and models of complex neural systems and ecological structures. The latter, for example, will require a nearly full revamping of the early field of population genetics and evolution in order to exploit both modern genomics and new field studies of multiple species and environmental interactions. And there will be more, much of which will only become apparent as new data and questions arise. One example would be RNAi and micro-arrays inducing the development of new analysis tools.

About the keynote speaker. Dr. Temple Smith graduated with a Ph.D. in Nuclear Physics from University of Colorado. He did a joint postdoctoral fellowship under the direction of the mathematician, Stanislaw Ulam and the molecular biologist, John Sadler. He was one of the founders of GenBank at Los Alamos. Dr. Smith has been the Director of the BioMolecular Engineering Research Center in the College of

M. Vingron and L. Wong (Eds.): RECOMB 2008, LNBI 4955, pp. 1–2, 2008.

Engineering at Boston University since 1991. He is a professor in the Department of Biomedical Engineering and co-founder of the company, Modular Genetics, Inc.

Dr. Smith is a co-developer of the Smith-Waterman sequence alignment algorithm, the standard tool used in most DNA and protein sequence comparison. His research is centered on the application of various computer science and mathematical methods to the discovery of the syntactic and semantic patterns in nucleic acid and amino acid sequences. These include the development of new sequence pattern extraction tools, multidomain dissection methods, and protein inverse folding prediction algorithms. In addition, Dr. Smith has carried out research in the application of many such methods ranging from the time calibration of HIV viral evolution analysis and modeling of the WD repeat family of proteins, to ribosomal protein evolution.

Bootstrapping the Interactome: Unsupervised Identification of Protein Complexes in Yeast

Caroline C. Friedel*, Jan Krumsiek, and Ralf Zimmer

Institut für Informatik, Ludwig-Maximilians-Universität München, Amalienstraße 17,
80333 München, Germany
Caroline.Friedel@bio.ifi.lmu.de

Abstract. Protein interactions and complexes are major components of biological systems. Recent genome-wide applications of tandem affinity purification (TAP) in yeast have increased significantly the available information on such interactions. From these experiments, protein complexes were predicted with different approaches first from the individual experiments only and later from their combination. The resulting predictions showed surprisingly little agreement and all of the corresponding methods rely on additional training data. In this article, we present an unsupervised algorithm for the identification of protein complexes which is independent of the availability of additional complex information. Based on a bootstrap approach, we calculated intuitive confidence scores for interactions which are more accurate than previous scoring metrics. The complexes determined from this confidence network are of similar quality as the complexes identified by the best supervised approaches. Despite the similar quality of the latest predictions and our predictions, considerable differences are still observed between all of them. Nevertheless, the set of consistently identified complexes is more than four times as large as for the first two studies. Our results illustrate that meaningful and reliable complexes can be determined from the purification experiments alone. As a consequence, the approach presented in this article is easily applicable to large-scale TAP experiments for any organism.

1 Introduction

Cellular processes are shaped by proteins physically associated in complexes. Accordingly, significant efforts are put into the experimental identification of such protein interactions. Commonly used techniques are yeast two-hybrid (Y2H) [1,2] and affinity purification followed by mass spectrometry (e.g. Co-immuno-precipitation (Co-IP)[3] or tandem affinity purification (TAP)[4,5,6]). Since new methods are generally first applied to the yeast *Saccharomyces cerevisiae*, its interactome is the most thoroughly studied one.

While Y2H identifies only direct physical interactions, affinity purification can also identify indirect interactions in protein complexes. In the so far only

---

* Corresponding author.

M. Vingron and L. Wong (Eds.): RECOMB 2008, LNBI 4955, pp. 3–16, 2008.
© Springer-Verlag Berlin Heidelberg 2008

genome-scale studies on complexes, the more effective TAP system [7] was applied to yeast separately by Gavin et al. [5] and Krogan et al. [6]. In the first experiment, 1,993 distinct TAP-tagged proteins (baits) were purified successfully and 2,760 distinct proteins (preys) identified in these purifications. In the second experiment, 2,357 baits were purified and 4,087 preys identified.

Because of the large size of these data sets, sophisticated methods were developed in both studies to infer individual protein complexes from the purification data. However, the resulting complexes showed surprisingly little overlap (see results). After the publication of the original results, improved prediction methods were developed by Pu et al. [8] based on the scoring method of Collins et al. [9] and by Hart et al. [10]. These methods used the data from both purification experiments.

A comparison of the different approaches outlines the important steps in complex determination (see Figure 1 **A**). First, purification experiments have to be combined for higher prediction quality. Second, propensities for individual protein interactions have to be determined from the purifications. Third, these propensities should be converted to confidence values assessing the likelihood of each protein interaction. In a fourth step, the network has to be clustered to obtain individual complexes after restricting it to the most confident interactions. If the corresponding clustering method only produces disjoint complexes, a final step has to be included to identify proteins shared between complexes. This is necessary because in biological systems proteins can be involved in more than one complex. Apart from the combined method by Collins et al. [9] and Pu et al. [8], all previous approaches leave out at least one of these steps. Figure 1 **A** gives an overview on which methods implement which steps in a supervised or unsupervised way.

All previous approaches rely more or less heavily on the availability of additional training data for at least one step. For yeast, hand-curated complexes can be taken from MIPS [11] and the study of Aloy et al. [12]. Furthermore, complexes can be automatically extracted from GO annotations [13]. Unfortunately, the resulting complexes are of lower quality than the hand-curated ones (see results). Recently, hand-curated complexes have become available for human and other mammals which cover about 12% of the protein-coding genes in human [14]. For other organisms, such complex information is not available which limits the applicability of the supervised approaches significantly.

Even if reference sets are available as for yeast, a large fraction of them have to be set aside as independent test sets to evaluate the quality of predicted complexes. Although most of the above mentioned approaches distinguished between test and training set by choosing one of the yeast reference sets for training and a different one for testing, these sets are not sufficiently disjoint to guarantee a reliable performance estimate. Indeed, generally more than half of the complexes in each reference set have a significant overlap to at least one complex in each of the other sets.

In this article, we present an unsupervised algorithm for the identification of protein complexes from the purification data alone which implements all steps

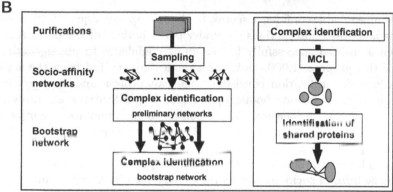

A

	GA	KR	CP	HA	BT
Experiment combination	-	-	S	U	U
Propensity calculation	U	S	S	U	U
Confidence values	-	S	S	-	U
Clustering	S	S	S	S	U
Shared Proteins	S	-	S	-	U

B

Purifications
Sampling
Socio-affinity networks
Complex identification
preliminary networks
Bootstrap network
Complex identification
bootstrap network

Complex identification
MCL
Identification of shared proteins

Fig. 1. Table **A** lists the major steps involved in complex identification and if they are realized by the approaches of Gavin et al. [5] (GA), Krogan et al. [6] (KR), Collins et al. and Pu et al. [9,8] (CP), Hart et al. [10] (HA) and the approach described in this article (BT). Furthermore, we indicate if a supervised approach based on additional training data is used (S, red) or an unsupervised approach (U, green). Figure **B** illustrates the major steps of the bootstrap approach described in this article. See methods for more details.

described above. Since no additional information on protein complexes is required, our approach is not limited to yeast but can be applied easily to large-scale purification experiments for other organisms.

Our results show that this approach is equivalent to the best supervised methods both with regard to functional and localization similarity observed in the resulting complexes as well as predictive performance with regard to known yeast complexes. We find that our predictions and the Pu and Hart predictions are much more consistent than the original Gavin and Krogan complexes. However, although the latest predictions are of similar quality, significant divergences between the predicted complexes are still observed. This clearly illustrates the need for further investigations of the individual protein complexes.

2 Methods

The algorithm we propose for the unsupervised identification of protein complexes implements all five necessary steps (see Figure 1 **A**). Purification experiments are combined by pooling them. Interaction confidences are determined by first identifying preliminary complexes for bootstrap samples from the set of purifications [15,16]. The resulting confidence network is then clustered with the MCL algorithm [17] and proteins shared between complexes are identified in a post-processing step. Here, all necessary parameters are tuned based on intrinsic measures calculated from the networks and complexes alone. The last two steps are also used to determine the preliminary complexes for the bootstrap samples (see Figure 1 **B**). Details of the algorithm are described in the following.

2.1 Bootstrap Sampling

To determine reliable confidence scores, the bootstrap technique [15,16] is used to estimate how stable interactions are under perturbations of the data. A similar approach is utilized successfully for assigning confidence to phylogenetic trees [18]. For this purpose, 1,000 bootstrap samples are created from the set of N purifications. A purification consists of one bait protein and its preys in this purification. To create one bootstrap sample, N purifications are drawn with replacement. Each purification can be contained in a bootstrap sample once, multiple times or not at all. Multiple copies of the same purification are treated as separate purifications.

For each bootstrap sample, we then calculate preliminary interaction propensities in the form of socio-affinity scores as described by Gavin et al. [5]. These scores compare the number of co-occurrences of two proteins against the random expectation using a combination of spoke and matrix model [19]. No additional training data is required to compute them from a set of purifications. From these preliminary networks, we then determine preliminary complex predictions for each individual bootstrap sample with the algorithm described in the following section. To allow for reasonably fast computation, only edges are included in the preliminary networks whose weight exceeds a certain threshold τ.

2.2 Identification of Protein Complexes

The algorithm for the prediction of complexes from a network consists of two steps: clustering of the network and subsequent identification of shared proteins.

Clustering. Networks are clustered using the Markov clustering algorithm (MCL) developed by van Dongen [17]. In a recent study [20], this algorithm was found to be superior to a selection of other graph clustering algorithms for the identification of protein complexes. Apart from the approach of Gavin et al., all previous approaches to complex identification from purification data use this method. The running-time of the MCL procedure is in $O(Nk^2)$ for a network with N nodes and a maximum degree of k. Thus, it is reasonably fast for sparse networks. Its most important parameter is the *inflation parameter* which influences the granularity of the identified clusters, i.e. their size and number. The higher the inflation parameter, the smaller are the resulting clusters and the more clusters are identified.

All previous approaches used additional training data in the form of known complexes to chose the optimal inflation parameter. In this article, we suggest to use an intrinsic measure which compares the resulting clusters against the original network. For this purpose, we use a performance measure for graph clustering proposed by van Dongen, the so-called *efficiency*. Details for the calculation of efficiency can be found in [17]. Basically, a clustering is highly efficient if proteins in the same cluster are connected by edges with high weights and proteins in different clusters have no or only low weight connections.

To determine the optimal inflation parameter, we clustered the socio-affinity networks for each sample using several, gradually increasing inflation parameters. For each inflation parameter we then calculated the average efficiency over all samples. We found that the efficiency always reaches a maximum for a certain inflation parameter and decreases on either side of this value if the socio-affinity networks were reasonably sparse. Accordingly, the optimal inflation parameter is chosen as the one with the highest average efficiency across all samples. This inflation parameter is then used to cluster the preliminary networks for the bootstrap samples.

Identification of shared proteins. The MCL algorithm, as most clustering methods, identifies only disjoint clusters. In real biological systems, however, proteins can be contained in more than one complex. If a protein has such strong associations with two complexes, the MCL procedure will either cluster those two complexes together or, if further associations between the complexes are missing, cluster this protein with only one of these complexes. We address this problem in a similar way as Pu et al. [8] by post-processing the clusters obtained from the MCL algorithm. Contrary to them, we do not optimize this step based on proteins shared between known yeast complexes, but again use an intrinsic measure based on the original network.

The following criteria are used for adding shared proteins. First, a protein is only added to another complex if it has sufficiently strong interactions to this complex. Second, the tighter the associations within this complex, the stronger have to be the interactions of the protein to this complex. Third, for large complexes strong interactions are only required to some of the complex proteins or, alternatively, weaker interactions to most of them. As a consequence, a protein p_i can be added to a complex C if

$$s(p_i, C) \geq \alpha \cdot s(C) \cdot \left(|C|^{-\gamma} / 2^{-\gamma} \right) \tag{1}$$

with $s(p_i, C)$ the average interaction score of p_i to proteins of C and $s(C)$ the average interaction score within the complex. Interactions not contained in the network are given a weight of zero.

The threshold for adding a protein to a complex decreases both with complex size and with complex confidence. To control the influence of complex size, a power function was chosen since it decreases steeply at first but then levels off for larger values. The power function is normalized to yield 1 for complexes of size 2. In this case the threshold depends only on the strength of the interaction between the two proteins.

This threshold definition has two parameters, α and γ. Here, α defines how much weaker than $s(C)$ the connections to the complex are allowed to be and γ controls to which fraction of the complex sufficiently strong interactions have to exist. Both parameters are set such that the weighted average score over all complexes after the post-processing is at least as high as a fraction λ of the original average score. For this purpose, α is set to λ and γ to the largest possible value such that this requirement is still met. Here, λ was set to 0.95 to add proteins only to complexes to which they have a strong association. Note that proteins are added to complexes in parallel. Accordingly, the complex memberships and the average complex score are not updated until all proteins have been processed and the result does not depend on the order of the proteins.

2.3 Calculation of Confidence Scores and Final Complexes

The final confidence scores are then determined by calculating the so-called bootstrap network from the complexes identified for each bootstrap sample. In the bootstrap network, two proteins are connected by an edge if they are clustered together in at least one sample. The fraction of samples for which they are contained in the same complex provides the weight for the corresponding edge and the confidence for this association (between 0 and 1).

Final complexes are then obtained by applying the complex identification algorithm on this bootstrap network. For this purpose, the optimal inflation parameter determined in the previous step is chosen. No threshold is applied to the network before complex identification but the size of the network is limited beforehand by choosing stringent cut-offs for the preliminary socio-affinity networks.

More confident predictions can be obtained from the original complexes in the following way. First, all edges are removed from the network with weight lower than a specific threshold and then connected components are calculated for each complex separately in this restricted network. As a consequence, complexes can either shrink, be subdivided or be removed completely. This approach is more efficient than the alternative approach of first restricting the network and then repeating complex identification but yields almost identical complexes (see results).

2.4 Criteria for the Evaluation of Complex Quality

Functional co-annotation within complexes. Since protein complexes are formed to carry out a specific function, the functions of proteins in the same complex should be relatively homogeneous. We evaluate the functional similarity between proteins predicted to be in the same complex by using the protein annotations of the Gene Ontology (GO) [13]. The functional similarity of two proteins is quantified in terms of the *semantic similarity* of GO terms annotated to these proteins. Several variations of semantic similarity have been described [21,22,23,24]. Here, we use the relevance similarity proposed recently by Schlicker at al. [24]. This measure is based both on the closeness of two GO terms to their common ancestors as well as the level of detail of these ancestors.

The GO score of a complex is the average relevance similarity of all protein pairs in this complex. The GO score for a set of complexes is the weighted mean over all complex scores and determined separately for the "biological process" and "molecular function" taxonomies. The final co-annotation score is then calculated as the geometric mean of the two values.

Co-localization within complexes. Since complexes can only be formed if the corresponding proteins are actually located together in the cell, a second quality measure is based on the similarity of protein localizations within a complex. For this purpose, we used the localization assignments and categories determined experimentally in yeast by Hu et al. [25].

The co-localization score for a complex is defined as the maximum fraction of proteins in this complex which are found at the same localization. The average co-localization score is calculated as the weighted average over all complexes and is defined as

$$L = \frac{\sum_j \max_i l_{i,j}}{\sum_j |C_j|} \tag{2}$$

Here, l_{ij} is the number of proteins of complex C_j assigned to the localization group i and $|C_j|$ is the number of proteins in the complex C_j with localization assignments.

Sensitivity and positive predictive value. To evaluate the accuracy of the predictions, *sensitivity* (Sn) and *positive predictive value* (PPV) were calculated with regard to the following reference sets: (a) hand-curated complexes from MIPS [11] (214 complexes after removing redundant complexes) and Aloy et al. [12] (101 complexes) as well as (b) complexes extracted from the SGD database [26] based on GO annotations (189 complexes). To compile the SGD set, GO-slim complex annotations to all yeast genes were taken from the SGD ftp site.

We used the definition of sensitivity and PPV for protein complexes by Brohee and van Helden [20]. Both measures are calculated from the number $T_{i,j}$ of proteins shared between a reference complex R_i and a predicted complex C_j:

$$Sn = \frac{\sum_i \max_j T_{i,j}}{\sum_i |R_i|} \quad \text{and} \quad PPV = \frac{\sum_j \max_i T_{i,j}}{\sum_j \sum_i T_{i,j}}. \tag{3}$$

3   Results

Bootstrap confidence values were calculated from the combined Krogan and Gavin purification experiments. This combined set contains 6498 purifications with 2995 distinct baits, more than half of which (1617) were purified more than once. On average, separate purifications of the same bait agree in about 27% of the retrieved preys between the two experiments. This is similar to the agreement between purifications of the same bait within the Krogan data set, but significantly smaller than within the Gavin set.

From these purifications the final bootstrap network was calculated. Only socio-affinity scores of at least 8 were included in the preliminary networks for the bootstrap samples. We chose a more stringent threshold than the one recommended by Gavin et al. for two reasons. First, the preliminary networks are much denser for the combined data than for the Gavin data alone and, as a consequence, the runtime of the MCL algorithm is increased considerably. Second, the final bootstrap network contains many more interactions than each individual preliminary network (in our case 20 time as many). Thus, the more stringent threshold both reduces runtime of the bootstrapping step and at the same time limits the size of the resulting bootstrap network.

The final bootstrap network contains 62,876 interactions between 5195 distinct proteins. Protein complexes were then determined from the bootstrap network with our method, resulting in 893 complexes (denoted as BT-893) which contain 5187 distinct proteins (397 of those shared between complexes).

To compare our results against the smaller Pu and Hart predictions, more confident complexes were extracted from the original set at a threshold of 0.32. This set contains 409 complexes with 1692 distinct proteins (101 shared between complexes) and will be referred to as BT-409 in the following. It is comparable in size with the Pu predictions of 400 complexes with 1914 distinct proteins (141 shared) and the Hart predictions of 390 complexes with 1689 distinct proteins (none shared). We also extracted a second set of 217 complexes (BT-217) at a threshold of 0.69 which has a similar size as the MIPS complexes. We compared our selection approach against the less efficient method of first restricting the network and clustering afterwards and found that the differences observed were negligible with mutual sensitivities of 0.97 on average.

3.1   Evaluation of Interaction Networks

The quality of the bootstrap network in comparison to previously suggested interaction propensities was evaluated using a *receiver operating characteristic* (ROC) curve [27]. In a ROC curve, true positive rates are plotted against false positive rates for gradually decreasing thresholds. True positive interactions were defined as interactions between proteins in the same MIPS complex. The large and small ribosomal subunits were excluded since they would otherwise make up 44% of the true positive interactions. A set of true negative interactions was defined by randomly selecting interactions between proteins assigned to different MIPS complexes and cellular localizations by Hu et al. [25].

Figure 2 **A** shows the resulting ROC curves for the Gavin, Krogan, Collins, Hart and bootstrap scores as well as socio-affinity scores calculated from the combined experiments. We see that for all networks calculated from the combined data, the curve is steeper and reaches a higher level than for the propensities calculated from each experiment alone. Furthermore, the bootstrap scores calculated with our method performed best at separating true interactions from noise. Among the scoring methods proposed after the publication of the original purification studies, the Collins scores performed worst. Nevertheless, they

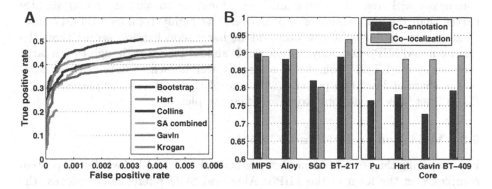

Fig. 2. ROC curves (**A**) are given for the Bootstrap, Hart and Collins scores, the socio-affinity scores for the combined data (SA combined) and the Gavin and Krogan scores (from top to bottom). True positive rates on the y coordinate are plotted against false positive rates on the x coordinate for gradually decreasing thresholds. Figure **B** illustrates co-annotation (dark blue) and co-localization (light blue) scores for the MIPS, Aloy and SGD complexes and the highly confident BT-217 complexes on the left hand side and the Pu, Hart, Gavin core and BT-409 predictions on the right hand side.

still perform slightly better for the given range than the socio-affinity scores computed from the combined experiments.

3.2   Co-annotation and Co-localization within Complexes

To assess the quality of the predicted complexes, we calculated the co-annotation and co-localization scores for the MIPS, Aloy and SGD complexes as well as for the Pu, Hart, Gavin and Krogan predictions and the BT-409 and BT-217 complexes (see Figure 2 **B**). Furthermore, the Gavin core set was evaluated which contains only the core components defined by Gavin et al. [5].

The lowest functional and localization similarity is observed for the Gavin and Krogan complexes (data not shown). By restricting to the more confident core components in the Gavin predictions, both co-annotation and co-localization can be increased significantly by 17% and 25%, respectively. Among all previous approaches, the highest co-annotation scores are obtained by the Pu and Hart predictions and the highest co-localization scores by the Hart predictions and the Gavin core set.

Functional and localization similarity is highest in the MIPS, Aloy and BT-217 complexes. Among the reference complexes, the SGD complexes perform worst. While co-annotation is still higher than in the best predictions, co-localization is significantly lower. This suggests that these automatically derived complexes are considerably less reliable than hand-curated ones.

When evaluating the complexes identified by our approach, we find that the BT-409 complexes perform significantly better than the Pu and Gavin core complexes with regard to functional and localization similarity and slightly better than the Hart complexes. Moreover, the highly confident BT-217 complexes show similar co-annotation and higher co-localization scores than the hand-curated MIPS and Aloy complexes. It should be noted though that a large fraction of the BT-217 complexes is already well-known as 64% of these complexes share at least two proteins with one of the reference complexes. In the BT-409 set, this applies only to 43% of the complexes.

3.3   Validation on Reference Complexes

By comparing the predicted complexes against the current knowledge on protein complexes in the form of the MIPS, Aloy and SGD reference complexes, the sensitivity and and the PPV of the corresponding methods can be estimated. One should keep in mind, though, that these estimates may be unprecise due to the incompleteness of current knowledge.

Results for the comparison against the MIPS complexes are shown in Figure 3 **A**. Similar trends are observed for the comparison against all reference sets. The worst results are obtained by the original Gavin complexes, followed by the Krogan complexes. Here, the Gavin complexes are generally more sensitive but less accurate in their predictions than the Krogan complexes. By restricting the Gavin complexes to the core components, the PPV can be increased beyond that of any other prediction. However, this improvement comes at the cost of a very low sensitivity.

When comparing the performance of the BT-409, Pu and Hart complexes, we observe that none of the predictions is clearly superior to the other two. Although the sensitivity of the Pu complexes is slightly higher than for the other two approaches, the corresponding PPV is in return lower. Thus, it appears that all predictions cover the reference complexes with similar quality. The PPV of the BT-409 complexes can be increased slightly by restricting to the more confident BT-217 complexes, however the loss of sensitivity again is considerable.

3.4   Towards a Consensus of Complex Predictions

Although functional and localization similarity within complexes is slightly higher for the BT-409 predictions than for the Pu and Hart predictions, the comparison against the reference complexes yielded very similar results for all three sets. In order to appreciate how much these predictions agree or diverge, we compared them at the level of the individual complexes.

First, we calculated pairwise PPV and sensitivity values by taking either set of complexes once as prediction and once as reference for each possible pairwise combination. Here, we observed an average PPV of 0.85 and sensitivity of 0.72. This suggests that the agreement between each pair of these new predictions is much higher than between the Gavin and Krogan complexes for which we observe a mutual PPV of 0.26 and a sensitivity of 0.29.

In a second step, we calculated for each pair of predictions the number of complexes with (a) no significant overlap (at least 2 proteins) to the other set,

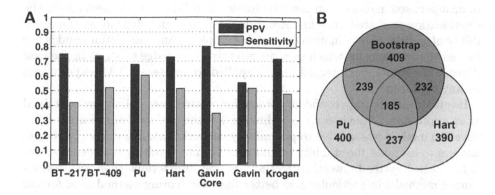

Fig. 3. PPV (red) and sensitivity (green) are shown in **A** for the BT-217, BT-409, Pu, Hart, Gavin core, Gavin and Krogan complexes. **B** illustrates the number of complexes which can be assigned consistently between the BT-409, Pu and Hart complexes. Results are shown for each pairwise comparison as well as the comparison of all three sets.

(b) a significant overlap to exactly one complex in the other set which again has no other overlaps and (c) significant overlaps but without an one-to-one correspondence as in (b). In the second case, we also distinguished between complexes with an exactly matching counterpart and complexes which contain additional proteins in at least one of the predictions. The same analysis was also performed for all three sets together.

For the second group of complexes, results are shown in Figure 3 **B**. For more than half of complexes in this group the predictions agree exactly. For the remaining complexes, each prediction adds on average 28-34% proteins to the proteins common to all two or three predictions. As can be seen, the consensus of each pair of predictions is much higher than for all predictions taken together. Nevertheless, even in the latter case the consensus is still larger at 46% (185 complexes) than between the Gavin and Krogan complexes where less than ten percent (45 complexes) have a clear one-to-one correspondence between the predictions.

Furthermore, 42% of the Gavin complexes and 64% of the Krogan complexes have no significant overlap to the respective other set. Contrary to that, only about 25% of complexes in the pairwise comparisons of the BT-409, Pu and Hart complexes and 16% in the comparison of all three predictions are unique to the corresponding prediction. These complexes are also characterized by considerably lower scores in the respective network and by smaller sizes than complexes in the other groups. Highest confidence values and medium complex sizes are observed for the second group while the last group generally contains the largest complexes.

4 Discussion

In this article, we presented an algorithm for the prediction of protein complexes from purification experiments alone. It implements all necessary steps in an unsupervised manner from the combination of different experiments up to the identification of shared proteins. Accordingly, it does not depend on the availability of additional information on protein complexes and interactions and is not limited to organisms for which such an extensive knowledge exists as in the case of yeast. Therefore, our method can be applied to large-scale TAP experiments for any organism.

Intuitive and accurate confidence scores for protein interactions were obtained by an application of the bootstrap technique. For this purpose, our complex identification method was applied to preliminary networks calculated from bootstrap samples to estimate the stability of interactions. The resulting confidence scores distinguished better between correct and wrong interactions than all previous scoring methods, in particular also better than the scoring method used for the preliminary networks.

The same complex identification method was then applied to the complete bootstrap network to yield a large set of complex predictions. From this large set, we extracted approximately the same number of high-confidence complexes (BT-409) as the so far best methods by Pu et al. and Hart et al. The comparison of functional and localization similarity within complexes showed slightly better results for the BT-409 complexes compared to the Pu and Hart predictions. Furthermore, the predictive performance with regard to known reference complexes proved to be equivalent. This suggests that meaningful complexes can be derived from the purification experiments without additional training data.

When analyzing the individual BT-409, Pu and Hart complexes, we found that about 60% of the complexes have a one-to-one correspondence in pairwise comparisons. Here, each prediction shows approximately the same agreement with each of the other predictions. When combining all three predictions, the fraction of complexes identified consistently drops to 46%. This shows, that the consensus between each pair of predictions is larger than between all three of them. Nevertheless, the degree of agreement is still significantly higher than observed between the original Gavin and Krogan predictions.

In general, complexes in the consensus set are assigned higher confidence values by each method than more diverging complexes. A possible explanation is that these complexes are most stable and strongly connected and thus are detected more consistently by all methods than more transient or weakly connected complexes. Furthermore, low confidence scores may be an indication for a higher degree of uncertainty. Thus, the confidence of complexes should be taken into account for any analysis based on these protein complexes. However, since the more confident complexes tend to be already covered to a large degree by existing biological knowledge, new information may be found preferentially in the less confident ones. Thus, these should not be discounted per se but validated in additional experiments. Here, the original large set of complexes (BT-893) identified in this study can be a valuable resource for biological hypothesis generation and testing.

5   Availability

Bootstrap scores for the combined purification experiments, as well as the BT-893, BT-409 and BT-217 complexes can be downloaded from
http://www.bio.ifi.lmu.de/Complexes.

References

1. Uetz, P., et al.: A comprehensive analysis of protein-protein interactions in Saccharomyces cerevisiae. Nature 403, 623–627 (2000)
2. Ito, T., et al.: A comprehensive two-hybrid analysis to explore the yeast protein interactome. Proc. Natl. Acad. Sci. USA 98, 4569–4574 (2001)
3. Ho, Y., et al.: Systematic identification of protein complexes in Saccharomyces cerevisiae by mass spectrometry. Nature 415, 180–183 (2002)
4. Gavin, A.-C., et al.: Functional organization of the yeast proteome by systematic analysis of protein complexes. Nature 415, 141–147 (2002)
5. Gavin, A.-C., et al.: Proteome survey reveals modularity of the yeast cell machinery. Nature 440, 631–636 (2006)
6. Krogan, N.J., et al.: Global landscape of protein complexes in the yeast Saccharomyces cerevisiae. Nature 440, 637–643 (2006)
7. von Mering, C., Krause, R., Snel, B., Cornell, M., Oliver, S.G., Fields, S., Bork, P.: Comparative assessment of large-scale data sets of protein-protein interactions. Nature 417, 399–403 (2002)
8. Pu, S., Vlasblom, J., Emili, A., Greenblatt, J., Wodak, S.J.: Identifying functional modules in the physical interactome of Saccharomyces cerevisiae. Proteomics 7, 944–960 (2007)
9. Collins, S.R., et al.: Toward a comprehensive atlas of the physical interactome of Saccharomyces cerevisiae. Mol. Cell. Proteomics 6, 439–450 (2007)
10. Hart, G.T., Lee, I., Marcotte, E.: A high-accuracy consensus map of yeast protein complexes reveals modular nature of gene essentiality. BMC Bioinformatics 8, 236 (2007)
11. Mewes, H.W., et al.: MIPS: analysis and annotation of proteins from whole genomes. Nucleic Acids Res. 32, 41–44 (2004)
12. Aloy, P., et al.: Structure-based assembly of protein complexes in yeast. Science 303, 2026–2029 (2004)
13. Ashburner, M., et al.: Gene ontology: tool for the unification of biology. The Gene Ontology Consortium. Nat. Genet. 25, 25–29 (2000)
14. Ruepp, A., Brauner, B., Dunger-Kaltenbach, I., Frishman, G., Montrone, C., Stransky, M., Waegele, B., Schmidt, T., Doudieu, O.N., Stümpflen, V., Mewes, H.W.: CORUM: the comprehensive resource of mammalian protein complexes. Nucleic Acids Res. 36, 646–650 (2008)
15. Efron, B.: Bootstrap methods: Another look at the jackknife. The Annals of Statistics 7, 1–26 (1979)
16. Efron, B., Tibshirani, R.J.: An Introduction to the Bootstrap. Chapman & Hall, Boca Raton (1994)
17. van Dongen, S.: Graph Clustering by Flow Simulation. Ph.D. thesis, University of Utrecht (2000)
18. Felsenstein, J.: Confidence limits on phylogenies: an approach using the bootstrap. Evolution 39, 783–791 (1985)

19. Bader, G.D., Hogue, C.W.V.: Analyzing yeast protein-protein interaction data obtained from different sources. Nat. Biotechnol. 20, 991–997 (2002)
20. Brohee, S., van Helden, J.: Evaluation of clustering algorithms for protein-protein interaction networks. BMC Bioinformatics 7, 488 (2006)
21. Resnik, P.: Semantic similarity in a taxonomy: an information-based measure and its application to problems of ambiguity in natural language. Journal of Artificial Intelligence Research 11, 95–130 (1999)
22. Lin, D.: An information-theoretic definition of similarity. In: Proc. 15th International Conf. on Machine Learning, pp. 296–304. Morgan Kaufmann, San Francisco (1998)
23. Lord, P.W., Stevens, R.D., Brass, A., Goble, C.A.: Investigating semantic similarity measures across the Gene Ontology: the relationship between sequence and annotation. Bioinformatics 19, 1275–1283 (2003)
24. Schlicker, A., Domingues, F.S., Rahnenführer, J., Lengauer, T.: A new measure for functional similarity of gene products based on Gene Ontology. BMC Bioinformatics 7, 302 (2006)
25. Huh, W.-K., et al.: Global analysis of protein localization in budding yeast. Nature 425, 686–691 (2003)
26. Dwight, S.S., et al.: Saccharomyces Genome Database (SGD) provides secondary gene annotation using the Gene Ontology (GO). Nucleic Acids Res. 30, 69–72 (2002)
27. Fawcett, T.: An introduction to ROC analysis. Pattern Recognition 27, 861–874 (2006)

CompostBin: A DNA Composition-Based Algorithm for Binning Environmental Shotgun Reads

Sourav Chatterji[1], Ichitaro Yamazaki[2], Zhaojun Bai[2], and Jonathan A. Eisen[1,3]

[1] Genome Center, U C Davis, Davis CA 95616, USA
schatterji@ucdavis.edu,jaeisen@ucdavis.edu
[2] Computer Science Department, U C Davis, Davis CA 95616, USA
yamazaki@cs.ucdavis.edu,bai@cs.ucdavis.edu
[3] The Joint Genome Institute, Walnut Creek CA 94598, USA

Abstract. A major hindrance to studies of microbial diversity has been that the vast majority of microbes cannot be cultured in the laboratory and thus are not amenable to traditional methods of characterization. Environmental shotgun sequencing (ESS) overcomes this hurdle by sequencing the DNA from the organisms present in a microbial community. The interpretation of this metagenomic data can be greatly facilitated by associating every sequence read with its source organism. We report the development of CompostBin, a DNA composition-based algorithm for analyzing metagenomic sequence reads and distributing them into taxon-specific bins. Unlike previous methods that seek to bin assembled contigs and often require training on known reference genomes, CompostBin has the ability to accurately bin raw sequence reads without need for assembly or training. CompostBin uses a novel weighted PCA algorithm to project the high dimensional DNA composition data into an informative lower-dimensional space, and then uses the normalized cut clustering algorithm on this filtered data set to classify sequences into taxon-specific bins. We demonstrate the algorithm's accuracy on a variety of low to medium complexity data sets.

Keywords: Metagenomics, Binning, Feature Extraction, Normalized Cut, weighted PCA, DNA composition metrics, Genome Signatures.

1 Introduction

Microbes are ubiquitous organisms that play pivotal roles in the earth's biogeochemical cycles. Their most visible effects on human well-being arise through their roles as mutualistic symbionts and hazardous pathogens. The study of microbes is crucial to our understanding of the earth's life processes and human health. Most of our knowledge about microbes has been obtained through the study of organisms cultured in artificial media in the laboratory. Although this approach has provided profound biological insights, it is inadequate for studying

M. Vingron and L. Wong (Eds.): RECOMB 2008, LNBI 4955, pp. 17–28, 2008.

the structure and function of many microbial communities. One obstacle has been that the vast majority of microbes have not been cultured and may not be culturable [1]. Even though culture independent methods such as 16S rRNA surveys [2] have been developed, they are unable to simultaneously answer two fundamental questions: Who is out there? and What are they doing? The application of genome sequencing methods is revolutionizing this field by enabling us for the first time to address those two questions for unculturable microbial communities [3,4,5]. These techniques, called environmental genomics or metagenomics, study microbial communities by analyzing the pooled genomes of all the organisms present in the community.

In one specific metagenomic method, *environmental shotgun sequencing* (ESS), DNA pooled from a microbial community is sampled randomly using whole genome shotgun sequencing. Thus, ESS data is made up of sequence reads from multiple species. This adds an additional layer of complexity compared to single-species genome sequencing, as it requires analysis of the metagenomic data in order to associate each sequence read with its source organism. Therefore, a critical first step in many metagenomic analyses is the distribution of reads into taxon-specific bins.

The difficulty of accurately binning ESS reads from whole genome data remains a significant hurdle in metagenomics. The taxonomic resolution achievable by the analysis depends on both the binning method and the complexity of the community. For instance, binning into species-specific bins can be achieved in low-complexity microbial communities (e.g., the dual-bacterial symbiosis of sharpshooters [6]). However, the problem becomes more difficult in high-complexity communities with hundreds of species, such as ocean microbial communities [7] and the human distal gut [5]. Because of these difficulties, many metagenomic studies (e.g., [8]) have resorted to analyzing at the level of the metagenome, essentially treating a microbial community as a bag of genes. This is not a satisfactory solution. Identifying and characterizing individual genomes can provide deeper insight into the structure of the community [6].

A variety of approaches have been developed for binning: assembly, phylogenetic analysis [9], database search [10], alignment with reference genome [7] and DNA composition metrics [11,12,13] Most current binning methods suffer from two major limitations: they require closely related reference genomes for training/alignment and they perform poorly on short sequences. To overcome the second difficulty, almost all current binning methods are applied to assembled contigs. However, most of the current generation assemblers can be confounded by metagenomic data since they implicitly assume that the shotgun data is from a single individual or clone. Therefore, we believe that assembly is risky when binning and that it is necessary to analyze raw sequence reads to get an unbiased look at the data.

To overcome the above-mentioned disadvantages of other binning methods, we have developed CompostBin, a binning algorithm based on DNA composition. CompostBin can bin raw sequence reads into taxon-specific bins with high accuracy and does not require training on currently available genomes.

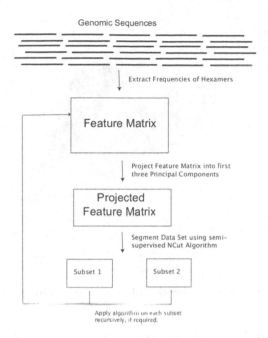

Fig. 1. High-level overview of the CompostBin algorithm. Principal Component Analysis is used to project the data into a lower-dimensional space. A semi-supervised normalized cut algorithm is used to segment the data set into two subsets. The algorithm is applied iteratively on the subsets to obtain the desired number of bins.

Like other composition-based methods, it seeks to distinguish different genomes based on their characteristic DNA compositional patterns, termed "signatures." For example, one of the most commonly used metrics measure the frequency of occurrence of Kmers (oligonucleotides of length K) in a sequence. Biases in $Kmer$ frequencies were analyzed extensively by Karlin and colleagues (e.g. [14]).

These biases have been extensively used for binning metagenomic sequences. For instance, TETRA[11] uses z-scores from tetramer frequencies to classify metagenomic sequences. A related program, MetaClust uses a combination of $Kmer$ frequency metrics to score metagenomic sequences and was used to classify sequences from the endosymbionts of a gutless worm [15]. However, the final assignment of sequences to bins in both these programs involve a significant manual component. Another class of methods [12,13] train their classifier using existing whole genome sequences and these classifiers can be even used to classify sequences from closely related novel genomes. However, as we discuss later, a serious drawback of these methods is that the pool of available genomes is very small and biased. Finally, the interpolated Markov model of the genefinder *Glimmer* can be used for binning in specific cases and has been used to distinguish symbiont sequences from the host sequences [16]. Unfortunately, these composition-based binning algorithms do not perform well on short fragments. Poor performance in shorter fragments is caused by the noise associated with the high dimensionality of the feature space and the associated curse of dimensionality.

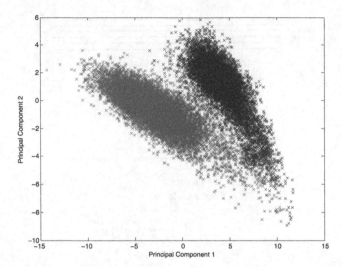

Fig. 2. Figure illustrating the separation of sequences according to species by using PCA. This data set contains sequences from two alphaproteobacteria, *Gluconobacter oxydans* (in red) and *Rhodospirillum rubrum* (in blue), which have GC content of 0.65 and 0.61, respectively. The data set is projected into the first two principal components.

When measuring the frequency of Kmers, the feature vector has 4^K dimensions (associated with measuring the frequencies of 4^K possible oligonucleotides of length K). Thus, for instance, if one looks at the frequency of hexamers in 2kb fragments, the dimensionality of the feature space is twice the length of the sequenced fragments.

CompostBin employs a new approach to deal with the noise arising from the high dimensionality of the feature vector (Figure 1). Instead of treating all components of the noisy feature space equally, we extract the most "important" directions and use these components for distinguishing between taxa. We use a weighted version of the standard Principal Component Analysis technique[17] to extract a "meaningful" lower dimensional sub-space. As shown in Figure 2, the algorithm can distinguish sequences from various species using just these first three principal components. The normalized cut clustering algorithm used to classify sequences into taxon-specific bins works on the lower dimensional sub-space and is guided by information from phylogenetic markers. We tested CompostBin on a wide variety of data sets and demonstrated that it is highly accurate in separating sequences into taxon-specific bins, even when processing raw reads of short sequences.

2   Methods

2.1   The CompostBin Algorithm

The input to CompostBin consists of raw sequence reads, along with mate pair information and the taxonomic assignment of reads containing phylogenetic markers. Either the number of abundant species or the number of taxonomic groups in the data set is provided to help the algorithm determine the number of bins in the output. This information can be obtained by analyzing the reads containing genes for ribosomal RNA or other marker genes [7]. In the simulation experiments, the number of bins is set to the number of species in the simulation. An overview of the algorithm is provided in Figure 1.

Feature Extraction by weighted PCA: Mate pairs are joined together and treated as a single sequence because they are highly likely to have originated from the same organism. Each sequence being analyzed is initially represented as a 4,096-dimensional feature vector, with each component denoting the frequency of one of the 4,096 hexamers. A weighted version of Principal Component Analysis (PCA) is then used to decrease the noise inherent in this high-dimensional data set by identifying the principal components of the feature matrix A.

In the standard form of Principal Component Analysis, the principal components are the orthogonal directions with highest variance and correspond to eigenvectors of the covariance matrix. If the *relative abundance* of various species is skewed, standard PCA might not be suitable for distinguishing the species. This is because the *within species variance* in the more abundant species might be overwhelming compared to *between species variance* and therefore the principal components cannot be used to distinguish between species. Therefore, we try to take the relative abundance into account in a weighting scheme that is used to normalize the effect of skewed relative abundance. We use a generalized variant of Principal Component Analysis that assigns a weight to each sequence and uses these weights to calculate the *weighted* covariance matrix of the data set. The principal components are the eigenvectors of this weighted covariance matrix. Further details about this generalization of PCA can be obtained from Chapter 14 of the book by Jolliffe [17].

We use a novel weighting scheme where the weight of each sequence is calculated by measuring its overlap with other reads in the data set. For each sequence, BLAT [18] was used to find overlaps with other sequences in the data set. This overlap information is then used to calculate the number of times a particular base in the sequence has been sequenced and thus estimate the coverage of the sequence (as defined in [19]). The weight of each sequence is set to the inverse of its *coverage*. The rationale behind this weighting strategy is that the sequences from the more abundant species will have higher coverage and thus will be weighed down. In fact, if there are sufficient number of sequences and the genome sizes of all species in the sample are equal, the average weight of the sequences from a particular species will be inversely proportional to the relative abundance of that species.

Determining the number of principal components required for analysis is crucial to the success of the algorithm. In our case, use of just the first three principal components is adequate to separate sequences from different species. For example, Figure 2 shows that for Data Set S5 which contains two alphaproteobacteria with similar GC content, almost complete separation is achieved by using only the first two principal components.

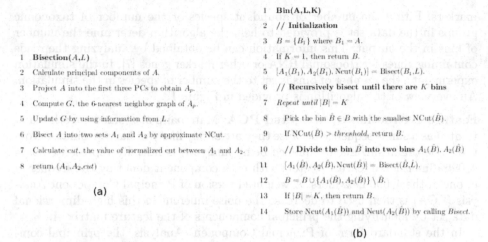

1 Bisection(*A,L*)

2 Calculate principal components of *A*.

3 Project *A* into the first three PCs to obtain A_p.

4 Compute *G*, the 6-nearest neighbor graph of A_p.

5 Update *G* by using information from *L*.

6 Bisect *A* into two sets A_1 and A_2 by approximate NCut.

7 Calculate *cut*, the value of normalized cut between A_1 and A_2.

8 return (A_1, A_2, cut)

(a)

1 Bin(A,L,K)

2 // Initialization

3 $B = \{B_1\}$ where $B_1 = A$.

4 If $K = 1$, then return *B*.

5 $[A_1(B_1), A_2(B_1), \text{Ncut}(B_1)] = \text{Bisect}(B_1, L)$.

6 // Recursively bisect until there are *K* bins

7 *Repeat until* $|B| = K$

8   Pick the bin $\hat{B} \in B$ with the smallest NCut($\hat{B}$).

9   If NCut($\hat{B}$) > *threshold*, return *B*.

10   // **Divide the bin** $\hat{B}$ **into two bins** $A_1(\hat{B}), A_2(\hat{B})$

11    $[A_1(\hat{B}), A_2(\hat{B}), \text{Ncut}(\hat{B})] = \text{Bisect}(\hat{B}, L)$.

12    $B = B \cup \{A_1(\hat{B}), A_2(\hat{B})\} \setminus \hat{B}$.

13   If $|B| = K$, then return *B*.

14   Store Ncut($A_1(\hat{B})$) and Ncut($A_2(\hat{B})$) by calling *Bisect*.

(b)

Fig. 3. Pseudocode describing the bisection and binning algorithm. *A* is the $N \times 4,096$ feature matrix, with each $4,096$-length feature vector representing a sequence. *L* contains labeling information obtained from phylogenetic markers, and *K* is the the desired number of bins.

Bisection by Normalized Cuts: The projection of the data matrix *A* into the first three principal components produces an $N \times 3$ data matrix A_p. A clustering algorithm is then applied to A_p to separate the N points into taxon-specific bins. A bisection algorithm is used to bisect a data set into two bins as detailed below. If the data set is to be divided into more than two bins, this algorithm is used recursively. Figure 3(a) shows pseudocode for the bisection algorithm. Given the projected matrix and phylogenetic markers as inputs, the procedure first computes the weighted graph over the sequences where the edge weights measure the similarity between corresponding sequences. Then, the normalized cut clustering algorithm [22] is employed to bisect the graph such that sequences from the same taxonomic group stay together.

Computation of Similarity Measure: As described earlier, the $4,096$-dimensional feature vector is projected into the first three principal components, and each sequence is represented as a point in 3-dimensional space. The clustering algorithm initially creates a 6-nearest neighbor graph $G(V, E, W)$ to capture the structure of the data set. The vertices in *V* correspond to the sequences, and an edge $(v_1, v_2) \in E$ between two sequences v_1 and v_2 exists only if one of the sequences is a 6-nearest neighbor of the other in Euclidean space. The nearest-neighbor graph reveals the global relation of the data set through this easily-computable local metric [20]. Each edge between two neighboring sequences v_1 and v_2 is weighted by their similarity $w(v_1, v_2)$, which is defined as the exponential inverse of their normalized Euclidean distance:

$$w(v_1, v_2) = \begin{cases} e^{-\frac{d(v_1, v_2)}{\alpha}} & \text{if } (v_1, v_2) \in E, \\ 0 & \text{otherwise,} \end{cases}$$

where $d(v_1, v_2)$ is the Euclidean distance between v_1 and v_2, and $\alpha = \max_{(v,u) \in E} d(v, u)$.

Semi-supervision Using Phylogenetic Markers: Marker genes, such as the genes that code for ribosomal proteins, are one of the most reliable tools for phylogenetically assigning reads to bins. Since these marker genes appear in only a small fraction of the reads, we used taxonomic information from 31 phylogenetic markers [21] to improve the clustering algorithm. This taxonomic information is provided to the binning algorithm as a label for each sequence, with each label corresponding to a single taxonomic group. Sequences without a taxonomic assignment are assigned the label "unknown."

A semi-supervised approach is then employed to incorporate this information into the clustering algorithm. Two vertices v_1 and v_2 are connected with the maximum edge weight (i.e., $w(v_1, v_2) = 1$) if the corresponding sequences are from the same taxonomic group, and the edge between v_1 and v_2 is removed (i.e., $w(v_1, v_2) = 0$) if they are from different groups.

Normalized Cut and its approximation: Given a weighted graph $G(V, E, W)$, the association between two subsets X and Y of V $W(X, Y)$ is defined as the total weight of the edges connecting X and Y: $W(X, Y) = \sum_{x \in X, y \in Y} w(x, y)$. The normalized cut algorithm bisects V into two disjoint subsets U and $\bar{U}$ such that the association within each cluster is large while the association between clusters is small, i.e., the normalized cut value $NCut$ is minimized, where

$$\text{NCut} = \frac{W(U, \bar{U})}{W(U, V)} + \frac{W(U, \bar{U})}{W(\bar{U}, V)}.$$

Since finding the exact solution to minimize $NCut$ is an NP-hard problem, an approximate solution is computed using a spectral analysis of the Laplacian matrix of the graph [22].

Generalization to Multiple Bins: If the data set needs to be divided into more than two bins, an iterative algorithm is used, where the bins are bisected recursively until the required number of bins is obtained. Figure 3(b) shows the pseudocode describing the algorithm. A set of bins, B is kept, where each element of B is a set of data points belonging to the same bin. The set B is initialized to be the singleton set $\{A\}$, where A contains all points in the data set. At each subsequent step of the algorithm, the bin with the lowest normalized cut value is bisected. The bisection continues until either B has the required number of bins or we no longer have a good bisection as measured by the normalized cut value.

2.2 Generation of Test Sets

In our experiments, we simulated the sequencing of low- to medium-complexity communities in which the number of species ranged from two to six and their

Table 1. Test Data Sets and Binning Accuracy

ID	Species	Ratio	Taxonomic Differences	Error
S1	*Bacillus halodurans* [0.44] & *Bacillus subtilis* [0.44]	1:1	Species	6.48%
S2	*Gluconobacter oxydans* [0.61] & *Granulobacter bethesdensis* [0.59]	1:1	Genus	3.39%
S3	*Escherichia coli* [0.51] & *Yersinia pestis* [0.48]	1:1	Genus	10.0%
S4	*Rhodopirellula baltica* [0.55] & *Blastopirellula marina* [0.57]	1:1	Genus	2.05%
S5	*Bacillus anthracis* [0.35] & *Listeria monocytogenes* [0.38]	1:2	Family	5.49%
S6	*Methanocaldococcus jannaschii* [0.31] & *Methanococcus mariplaudis* [0.33]	1:1	Family	0.51%
S7	*Thermofilum pendens* [0.58] & *Pyrobaculum aerophilum[0.51]*	1:1	Family	0.28%
S8	*Gluconobacter oxydans* [0.61] & *Rhodospirillum rubrum* [0.65]	1:1	Order	0.98%
S9	*Gluconobacter oxydans* [0.61], *Granulobacter bethesdensis* [0.59], & *Nitrobacter hamburgensis* [0.62]	1:1:8	Family Order	7.7%
S10	*Escherichia coli* [0.51], *Pseudomonas putida* [0.62], & *Bacillus anthracis* [0.35]	1:1:8	Order Phylum	1.96%
S11	*Gluconobacter oxydans* [0.61], *Granulobacter bethesdensis* [0.59], *Nitrobacter hamburgensis* [0.62], & *Rhodospirillum rubrum* [0.65]	1:1:4:4	Family Order	4.44%
S12	*Escherichia coli* [0.51], *Pseudomonas putida* [0.62], *Thermofilum pendens* [0.58], *Pyrobaculum aerophilum* [0.51], *Bacillus anthracis* [0.35], & *Bacillus subtilis* [0.44]	1:1: 1:1: 2:14	Species, Order Family, Phylum Kingdom	4.52%
R1	Glassy-winged sharpshooter endosymbionts	-	-	9.04%

relative abundance ranged from 1:1 to 1:14. ReadSim [23] was used to simulate paired-end Sanger sequencing from isolate genomes with an average read length of $1,000$ bp. The reads from various isolates were then combined in ratios corresponding to their relative species abundance in the data set to yield a simulated metagenomic data set of known composition. The 12 simulated data sets are described in Table 1. The GC content of each species' genome is listed in squared-brackets and can be used for assessing the diversity of DNA composition. The taxonomic levels are obtained from IMG[24] and can be used for assessing the phylogenetic diversity.

In addition, we tested the algorithm on a metagenomic data set containing reads obtained from gut bacteriocytes of the glassy-winged sharpshooter. The original study [7] had used phylogenetic markers to classify the sequence reads into three bins: reads from *Baumannia cicadellinicola* in Bin 1, reads from *Sulcia muelleri* in Bin 2, and reads from the host and miscellaneous unclassified reads in Bin 3. Due to the heterogeneity of Bin 3, the accuracy of the algorithm was tested only on its ability to distinguish between reads from Bin 1 and Bin 2.

3   Results

CompostBin was coded in C and Matlab. It is publicly available for download from   `http://bobcat.genomecenter.ucdavis.edu/souravc/compostbin/`. CompostBin was tested on a variety of low-to-medium complexity data-sets. Details of the test data sets and CompostBin's performance are provided in the next two sections.

3.1 Test Data Sets

Metagenomics being a relatively new field, very few standard data sets for testing binning algorithms have been developed [25]. One obstacle to their development has been that the "true" solution is still unknown for the sequence data generated by most metagenomic studies. To test the accuracy of a binning algorithm, one can instead simulate the shotgun sequences that would be obtained from a combination of organisms of known genome sequences. Simulated sequence reads from multiple genomes were pooled to simulate the challenges of metagenomic sequencing. When designing our simulated data sets, we took into account several variables that affect the difficulty of binning: the number of species in the sample, their relative abundance, their phylogenetic diversity, and the differences in GC content between genomes.

We also tested CompostBin on a publicly available metagenomic data set whose solution is well accepted. This data Set (R1) contains sequence reads obtained from gut bacteriocytes of the glassy-winged sharpshooter, *Homalodisca coagulata*. The data sets used for testing CompostBin are described in Table 1, and experimental details are provided in Methods.

3.2 Performance

The most self-evident way of measuring error rates would be to report the percentage of reads misclassified by the algorithm. However, this method can artificially decrease the error-rates of data sets with skewed relative abundance of species. For example, consider a data set consisting of 90 sequences from species 1 and 10 sequences from species 2. If we classify 5 sequences of species 2 inaccurately, the error rate would be just 5%, even though 50% of the sequences have been misclassified. Therefore, we report a normalized error rate, where we compute the error rate for each bin and the error rate for the whole data set is the mean of these error rates.

CompostBin's accuracy in classifying reads from the test data sets is reported in Table 1. The normalized error rates is bounded by 10% in all the 13 data sets. The error rates are correlated mostly with the phylogenetic distances between the species and the relative abundance of species. For example, the highest error rates measured was 10% for Data Set S3 (sequences from *E. coli* and *Y. pestis*), where the phylogenetic distances between the genomes is small. Similarly, the error rates are comparatively high in S9 because there are very few sequences from the less abundant *Gluconobacter oxydans* and *Granulobacter bethesdensis*, which are also phylogenetically very close.

4 Discussion

In this paper, we report the development of CompopstBin, a new algorithm for the taxonomic binning problem associated with the analysis of metagenomic data. The principal novel aspect of our method is the observation that the high-dimensional Kmer frequency data for short sequences is noisy, and that one can deal with the noise by projecting the data into a carefully chosen lower-dimensional space. We illustrate that CompostBin can accurately classify sequences from low to medium complexity data sets into taxon-specific bins.

Unlike previous methods, CompostBin doesn't require training of the algorithm with data from sequenced genomes. This is critical for success when binning environmental shotgun data because more than 99.9% of microbes are currently unculturable and unlikely to be represented in the training data set. Even closely related organisms living in different environments may have divergent genome signatures. For example, Bacillus anthracis and Bacillus subtilis have widely differing GC content and genome signatures. One should also keep in mind that the currently available genomes are not a phylogenetically random sample, but rather are a highly biased collection of biomedically interesting genomes combined with an overabundance of strains of model organisms such as Escherichia coli.

We used the frequencies of hexamers (oligonucleotides of length 6) as the metric for our analysis of short sequences. The choice of hexamers was motivated by both computational and biological rationale. Since the length of the feature vector for analyzing Kmers is $O(4^K)$, both the memory and the CPU requirements of the algorithm become infeasible for large data sets when K is greater than six. Using hexamers is biologically advantageous in that, being the length of two codons, their frequencies can capture biases in codon usage. Similarly, hexamer frequencies can detect genomic biases resulting from the observed avoidance of specific palindromic words of lengths 4 and 6 from genomes due to the presence of restriction enzymes [26]. It should be noted that the frequencies of lower-length words are linear combinations of hexamer frequencies. For example: $f(AAAAA) = f(AAAAAA) + f(AAAAAC) + f(AAAAAG) + f(AAAAAT)$. Thus, our PCA-based method implicitly takes into account any biases in the frequencies of lower length words.

CompostBin is a work in progress, with several refinements of the algorithm planned for the future. Our method of analysis is based primarily on DNA composition metrics and, like all such methods, it cannot distinguish between organisms unless their DNA compositions are sufficiently divergent. Thus, our method would probably be unable to distinguish between strains of the same species. We believe that an ideal binning algorithm would also utilize additional types of information, such as assembly (depth of coverage and overlap information) and population genetics parameters. We have taken an initial step in this direction by using taxonomic information from phylogenetic markers to guide the clustering algorithm. We intend to develop other hybrid methods in the future that can tackle the very formidable problem of classifying sequences in complex metagenomic communities.

Acknowledgments

We thank Ambuj Tewari, Martin Wu, Lior Pachter, Jonathan Dushoff, Joshua Weitz, Dongying Wu, Amber Hartman, and Jenna Morgan for their helpful suggestions and comments. This work was supported in part by a Laboratory Directed Research and Development Program (Lawrence Berkeley Lab) Grant to Jonathan A. Eisen. S.C. and J.E. were partially supported by the Defense Advanced Research Projects Agency under grants HR0011-05-1-0057 and FA9550-06-1-0478. I.Y. and Z.B. were supported in part by the NSF under grant 0313390.

References

1. Rappe, M.S., Giovannoni, S.J.: The uncultured microbial majority. Annu Rev Microbiol 57, 369–394 (2003)
2. Lane, D.J., Pace, B., Olsen, G.J., Stahl, D.A., Sogin, M.L., Pace, N.R.: Rapid determination of 16s ribosomal rna sequences for phylogenetic analyses. Proc. Natl Acad. Sci. USA 82(20), 6955–6959 (1985)
3. Venter, J.C., Remington, K., Heidelberg, J.F., Halpern, A.L., Rusch, D., Eisen, J.A., Wu, D., Paulsen, I., Nelson, K.E., Nelson, W., Fouts, D.E., Levy, S., Knap, A.H., Lomas, M.W., Nealson, K., White, O., Peterson, J., Hoffman, J., Parsons, R., Baden-Tillson, H., Pfannkoch, C., Rogers, Y.H., Smith, H.O.: Environmental genome shotgun sequencing of the sargasso sea. Science 304(5667), 66–74 (2004)
4. Tyson, G.W., Chapman, J., Hugenholtz, P., Allen, E.E., Ram, R.J., Richardson, P.M., Solovyev, V.V., Rubin, E.M., Rokhsar, D.S., Banfield, J.F.: Community structure and metabolism through reconstruction of microbial genomes from the environment. Nature 428(6978), 37–43 (2004)
5. Gill, S.R., Pop, M., Deboy, R.T., Eckburg, P.B., Turnbaugh, P.J., Samuel, B.S., Gordon, J.I., Relman, D.A., Fraser-Liggett, C.M., Nelson, K.E.: Metagenomic analysis of the human distal gut microbiome. Science 312(5778), 1355–1359 (2006)
6. Wu, D., Daugherty, S.C., Van Aken, S.E., Pai, G.H., Watkins, K.L., Khouri, H., Tallon, L.J., Zaborsky, J.M., Dunbar, H.E., Tran, P.L., Moran, N.A., Eisen, J.A.: Metabolic complementarity and genomics of the dual bacterial symbiosis of sharpshooters. PLoS Biol. 4(6), 188 (2006)
7. Rusch, D.B., Halpern, A.L., Sutton, G., Heidelberg, K.B., Williamson, S., Yooseph, S., Wu, D., Eisen, J.A., Hoffman, J.M., Remington, K., Beeson, K., Tran, B., Smith, H., Baden-Tillson, H., Stewart, C., Thorpe, J., Freeman, J., Andrews-Pfannkoch, C., Venter, J.E., Li, K., Kravitz, S., Heidelberg, J.F., Utterback, T., Rogers, Y.H., Falcon, L.I., Souza, V., Bonilla-Rosso, G., Eguiarte, L.E., Karl, D.M., Sathyendranath, S., Platt, T., Bermingham, E., Gallardo, V., Tamayo-Castillo, G., Ferrari, M.R., Strausberg, R.L., Nealson, K., Friedman, R., Frazier, M., Venter, J.C.: The sorcerer ii global ocean sampling expedition: Northwest atlantic through eastern tropical pacific. PLoS Biol. 5(3), e77 (2007)
8. Tringe, S.G., von Mering, C., Kobayashi, A., Salamov, A.A., Chen, K., Chang, H.W., Podar, M., Short, J.M., Mathur, E.J., Detter, J.C., Bork, P., Hugenholtz, P., Rubin, E.M.: Comparative metagenomics of microbial communities. Science 308(5721), 554–557 (2005)
9. von Mering, C., Hugenholtz, P., Raes, J., Tringe, S., Doerks, T., Jensen, L., Ward, N., Bork, P.: Quantitative phylogenetic assessment of microbial communities in diverse environments. Science 315(5815), 1126–1130 (2007)
10. Huson, D.H., Auch, A.F., Qi, J., Schuster, S.C.: Megan analysis of metagenomic data. Genome Research (in press, 2007)

11. Teeling, H., Waldmann, J., Lombardot, T., Bauer, M., Glockner, F.O.: Tetra: a web-service and a stand-alone program for the analysis and comparison of tetranucleotide usage patterns in dna sequences. BMC Bioinformatics 5(1471–2105 (Electronic)) (2004)

12. Abe, T., Sugawara, H., Kinouchi, M., Kanaya, S., Ikemura, T.: Novel phylogenetic studies of genomic sequence fragments derived from uncultured microbe mixtures in environmental and clinical samples. DNA Res 12(5), 281–290 (2005)

13. McHardy, A.C., Martin, H.G., Tsirigos, A., Hugenholtz, P., Rigoutsos, I.: Accurate phylogenetic classification of variable-length dna fragments. Nat Methods 4(1), 63–72 (2007)

14. Karlin, S., Burge, C.: Dinucleotide relative abundance extremes: a genomic signature. Trends Genet 11(7), 283–290 (1995)

15. Woyke, T., Teeling, H., Ivanova, N.N., Huntemann, M., Richter, M., Gloeckner, F.O., Boffelli, D., Anderson, I.J., Barry, K.W., Shapiro, H.J., Szeto, E., Kyrpides, N.C., Mussmann, M., Amann, R., Bergin, C., Ruehland, C., Rubin, E.M., Dubilier, N.: Symbiosis insights through metagenomic analysis of a microbial consortium. Nature 443(7114), 950–955 (2006)

16. Delcher, A.L., Bratke, K.A., Powers, E.C., Salzberg, S.L.: Identifying bacterial genes and endosymbiont dna with glimmer. Bioinformatics 23(6), 673–679 (2007)

17. Jolliffe, I.T.: Principal Component Analysis. Springer, Heidelberg (2002)

18. Kent, W.J.: Blat-the blast-like alignment tool. Genome Res 12(4), 656–664 (2002)

19. Lander, E.S., Waterman, M.S.: Genomic mapping by fingerprinting random clones: a mathematical analysis. Genomics 2(3), 231–239 (1988)

20. Tenebaum, J.B., Silva, V.D., Langford, J.C.: A global geometric framework for nonlinear dimensionality reduction. Science 190(5500), 2319–2323 (2000)

21. Wu, M., Eisen, J.: A simple, fast and accurate method for phylogenenomics inference approach (submitted, 2007)

22. Shi, J., Malik, J.: Normalized cuts and image segmentation. IEEE Transactions on Pattern Analysis and Machine Intelligence 22(8), 888–905 (2000)

23. Schmid, R., Schuster, S.C., Steel, M.A., Huson, D.H.: Readsim- a simulator for sanger and 454 sequencing (in press, 2007)

24. Markowitz, V.M., Korzeniewski, F., Palaniappan, K., Szeto, E., Werner, G., Padki, A., Zhao, X., Dubchak, I., Hugenholtz, P., Anderson, I., Lykidis, A., Mavromatis, K., Ivanova, N., Kyrpides, N.C.: The integrated microbial genomes (img) system. Nucleic Acids Res. 34(Database issue), D344–348 (2006)

25. Mavromatis, K., Ivanova, N., Barry, K., Shapiro, H., Goltsman, E., McHardy, A.C., Rigoutsos, I., Salamov, A., Korzeniewski, F., Land, M., Lapidus, A., Grigoriev, I., Richardson, P., Hugenholtz, P., Kyrpides, N.C.: Use of simulated data sets to evaluate the fidelity of metagenomic processing methods. Nat. Methods 4(6), 495–500 (2007)

26. Gelfand, M.S., Koonin, E.V.: Avoidance of palindromic words in bacterial and archaeal genomes: a close connection with restriction enzymes. Nucleic Acids Res. 25(12), 2430–2439 (1997)

Reconstructing the Evolutionary History of Complex Human Gene Clusters

Yu Zhang[1,2], Giltae Song[1], Tomáš Vinař[3], Eric D. Green[4], Adam Siepel[3], and Webb Miller[1]

[1] Center for Comparative Genomics and Bioinformatics, 506B Wartik Lab, Penn State University, University Park, PA 16802, USA
[2] Department of Statistics, Penn State University, University Park, PA 16802, USA
[3] Department of Biological Statistics and Computational Biology, Cornell University, Ithaca, NY 14853, USA
[4] Genome Technology Branch and NIH Intramural Sequencing Center, National Human Genome Research Institute, National Institutes of Health, Bethesda, Maryland 20892, USA

Abstract. Clusters of genes that evolved from single progenitors via repeated segmental duplications present significant challenges to the generation of a truly complete human genome sequence. Such clusters can confound both accurate sequence assembly and downstream computational analysis, yet they represent a hotbed of functional innovation, making them of extreme interest. We have developed an algorithm for reconstructing the evolutionary history of gene clusters using only human genomic sequence data. This method allows the tempo of large-scale evolutionary events in human gene clusters to be estimated, which in turn will facilitate primate comparative sequencing studies that will aim to reconstruct their evolutionary history more fully.

1 Introduction

Gene clusters in a genome provide substrates for genomic innovation, as gene duplication is often followed by functional diversification [1]. Also, genomic deletions associated with nearby segmental duplications cause several human genetic diseases [2]. One surprising discovery emerging from the sequencing of the human genome was the large extent of recent duplication in the human lineage. Analysis of the human genome sequence revealed that 5% consists of recent duplications [3]; subsequent studies have further found extensive copy-number variation among individuals [4].

Recently duplicated genomic segments are exceedingly difficult to sequence accurately and completely. Even the "finished" human genome sequence [5] contains about 300 gaps, many of which reflect regions harboring nearly identical tandemly duplicated segments. The situation with mammalian genomes sequenced by a whole-genome shotgun sequencing strategy [6] is typically much worse, with recently duplicated segments often grossly misassembled. The development of computational methods for analyzing gene clusters has therefore

M. Vingron and L. Wong (Eds.): RECOMB 2008, LNBI 4955, pp. 29–49, 2008.
© Springer-Verlag Berlin Heidelberg 2008

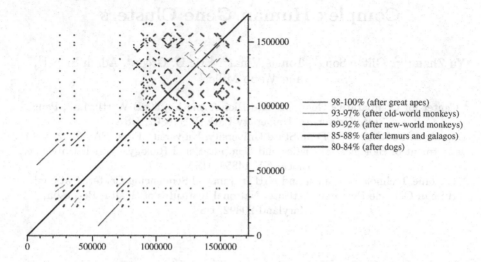

Fig. 1. Dot-plots of self-alignments of the human UGT2 cluster exceeding thresholds of percent identity chosen to roughly correspond to the divergence of the human lineage from great apes (98%), old-world monkeys (93%), new-world monkeys (89%), prosimians (85%) and dogs and other laurasiatherians (80%). We estimate that 2, 27, 51, 59, and 82 duplications respectively are needed to produce the current configuration from a duplication-free sequence (no deletions were predicted), suggesting a sustained growth of the cluster along the human lineage, with a burst of activity around the time that humans and apes diverged from old-world monkeys. The sequence alignments were computed using blastz [11] and post-processed as described in the text.

lagged far behind that for analyzing single-copy regions, due in part to the lack of accurate sequence data. Even the basic problem of formally defining what is meant by a multi-species sequence "alignment" of a region harboring a gene cluster (much less actually generating an accurate alignment of such a region) has only recently been addressed [7,8]. While the recent testing of several alignment methods with comparative sequence data representing 1% of the human genome [9] suggested adequate performance, a closer examination of the resulting alignments for those regions containing tandem gene clusters (e.g., both globin clusters) showed significant imperfections [10].

Here, we describe an algorithm for producing a theoretical ancestral sequence and a parsimonious set of duplication and deletion events explaining the observed state of a gene cluster. We start by setting a lower bound for the percent identity in self-alignments of a gene cluster (e.g., 93%; Fig.1). This defines the set of duplications that have occurred in a given time interval (such as the last 25 million years) and that have not subsequently been deleted. The ancestral configuration of each gene cluster is then deduced at several evolutionary points, and predictions are made about the parsimonious sets of duplications and deletions that converted the ancestral configuration into the extant one.

Similar problems have been studied before. Elemento *et al.* [12] and Lajoie *et al.* [13] developed algorithms for reconstruction of evolutionary histories of gene families allowing tandem duplications and inversions. Their basic assumption is that a gene is always duplicated as a whole unit and duplicated copies are always immediately adjacent to their sources. These assumptions are routinely violated in the real data, and thus their methods have limited applicability in genome-wide studies. In addition, Elemento *et al.* do not consider inversions, while Lajoie *et al.* only consider single gene duplications. Jiang *et al.* [14] recently used methods developed for repeat identification to infer ancestral "core duplicated elements". Their results provide useful insights about duplication histories, but without detailed reconstructions. In this paper, we aim to provide event-by-event reconstructions of duplication and deletion histories using local sequence alignments, allowing both tandem and interspersed duplications (potentially with inversions).

We have applied our algorithm to 25 human gene clusters, in each case predicting the evolutionary scenarios corresponding to five major divergence points along the lineage leading to human.[1] Our results provide distributions of the predicted sizes of rearranged segments. Also, using percent-identity thresholds associated with large increases in the estimated number of duplications and deletions, we can estimate dates of rapid cluster expansion.

In future work, we plan to use such estimates to examine a large number of human gene clusters in conjunction with experimental data on gene-family size in various primates, as generated by array comparative genome hybridization (aCGH) [15,16]. Our aim is to design a larger primate comparative sequencing project that will more deeply examine the evolutionary history of a set of human gene clusters. In turn, the availability of such comparative sequence data should provide important insights about primate genome evolution and catalyze the development of computational methods for analyzing gene clusters.

2   Problem Statement and Data Preparation

Our goal is to reconstruct the evolutionary history that has generated a gene cluster in the human genome. Given the cluster's DNA sequence in a single species, we first identify all local self-alignments in both forward and reverse-complement orientations using blastz [11]. We can visualize the identified alignments using a dot-plot, and our goal is equivalent to providing a set of instructions for generating the observed dot-plot from a duplication-free sequence using a series of evolutionary events (duplications and deletions).

We preprocess the initial dot-plot to satisfy the *transitive closure property*. That is, if the dot-plot contains local alignments for region A and B, and for region B and C, then the dot-plot must also contain a local alignment for region A and C. We also *maximize each alignment*, i.e., we ensure that the alignments cannot be extended at either end. Finally, a local alignment can be broken into smaller pieces by mutations and interspersed repeats. We have developed an

---

[1] We have also extended this analysis to 165 biomedically interesting clusters and the results are presented in Appendix C.

accurate algorithm to determine the transitive closure of a dot-plot and to *chain alignments* together if they are broken by these events.

Since after preprocessing the alignments are maximized and have the transitive closure property, we can represent the original sequence by a sequence of *atomic segments* that are separated by boundaries of the alignment (*atomic boundaries*). We will denote the atomic segments by letters $a, b, c, \ldots$, and their reverse complements by $\bar{a}, \bar{b}, \bar{c}, \ldots$. The atomic segments that are aligned to each other will have the same letter with different subscripts (e.g., $x a_1 y b_1 c_1 z \overline{c_2} a_2 \overline{b_2} w$ has 10 atomic segments, two of which are reverse complements; a_1 and a_2 are aligned, and so are b_1 and b_2, and c_1 and c_2).

We say that the two adjacent atomic segments xy can be *collapsed* into a single atomic segment z, if y is always immediately preceded by x, and x is always immediately followed by y (we also consider $\bar{x}$ and $\bar{y}$ in the reverse orientation). In such case, we can replace all occurrences of xy with z, and all occurrences of $\overline{yx}$ with $\bar{z}$. Since initially all alignments are maximized, our initial representation will have no collapsible atomic segments.

We will be looking at sequences of duplication events in reversed order of time, i.e., starting from the latest duplication. A duplication event copies region P of the sequence (which can consists of several consecutive atomic segments) to another location (possibly with reversal). Thus, we can always identify the latest duplication by a pair of regions (P, D), where D is a region identical to P except for atomic segment subscripts and perhaps orientation (e.g., $(a_1 b_1, \overline{b_2 a_4})$). If correctly identified, we can *unwind* a duplication (P, D) by removing segment D from the sequence, then collapsing all collapsible atomic segments. By unwinding all duplications, we obtain an atomic segment representation of the ancestral sequence. We are now ready to state our problem formally.

Definition 1 (Parsimonious reconstruction of duplication events). *Given a representation of the present-day DNA sequence by atomic segments, find the shortest sequence of duplication events* $(P_1, D_1), (P_2, D_2), \ldots, (P_k, D_k)$ *such that if we unwind these duplications, we obtain a sequence containing only a single atomic segment.*

3 Basic Combinatorial Algorithm

We first present a simple combinatorial algorithm that can correctly reconstruct all the duplication events (except for their order and orientation) under the following assumptions:

(1) A duplication event copies (possibly with reversal) a region of the sequence to any location except inside the originating region.
(2) The sequence evolves only by duplications (including duplications with reversal and tandem duplications). In particular, there are no deletions.
(3) No atomic boundaries are reused as duplication boundaries, except in tandem duplications. Here, boundaries of two aligned atomic segments (e.g. a_1 and a_2) are considered to be the same atomic boundary.

These assumptions are much more permissible than those of Elemento et al. [12], yet they are still often violated in the real data. Therefore, we also offer a more practical solution based on the sequential importance sampling in the next section. Note that assumption (3) is a stronger version of the commonly used no-breakpoint-reuse assumption [17] and can be justified by the usual arguments.

Definition 2 (Candidate alignments). *We call a pair of regions (P, D) a candidate alignment if P and D are identical except for subscripts and orientation, and if, after removing D, the atomic segment pair flanking D and the two pairs flanking each boundary of P can be collapsed.*

For example, for $xa_1yb_1c_1z\overline{c_2}a_2\overline{b_2}w$, the alignment (a_1, a_2) is a candidate alignment. This is because after removing a_2, the flanking atomic segment pair, $\overline{c_2}\overline{b_2}$ can be collapsed into a single atomic segment. Additionally, the atomic segment pairs flanking boundaries of a_1 (xa_1 and a_1y) can also be collapsed.

Lemma 1. *In a sequence of atomic segments that arose by the process satisfying the assumptions (1)-(3), the latest duplication is always among the candidate alignments.*

Lemma 1 suggests a simple and efficient *basic algorithm* for reconstructing a sequence of duplications:

1. Find a candidate alignment (P, D).
2. Output (P, D) as the latest duplication and unwind (P, D) by removing D from the sequence and collapsing all collapsible atomic segments.
3. Repeat until there is only a single atomic segment left.

Depending on the choice of candidate alignments in step 1, we can produce several duplication histories that could lead to the present-day sequence as represented by the sequence of atomic segments. Lemma 1 shows that one of those possible solutions is the real sequence of duplications. We can further show that all the other solutions produced by the basic algorithm are equally good solutions of the problem (proof relegated to Appendix A and B):

Theorem 1. *If assumptions (1)-(3) are met then the basic algorithm will always succesfully recover a sequence of duplications that will collapse the whole sequence into a single atomic segment, regardless of the order of choice of candidate alignments in step 1. Moreover, all of these solutions have the same number of events and they represent all parsimonious solutions of the duplication event reconstruction problem.*

For example, to apply the basic algoritm to $xa_1yb_1c_1z\overline{c_2}a_2\overline{b_2}w$, we note that alignment (a_1, a_2) is the only candidate alignment; $(b_1, \overline{b_2})$ and $(c_1, \overline{c_2})$ do not satisfy the definition of candidate alignment at this moment. We remove a_2 to obtain a new sequence $xa_1yb_1c_1z\overline{c_2}\overline{b_2}w$, and we remove the corresponding local alignment (a_1, a_2). We collapse the new sequence into a simpler form $ue_1z\overline{e_2}w$, where $u = xa_1y, e_1 = b_1c_1$, $\overline{e_2} = \overline{c_2}\overline{b_2}$. Now only one local alignment remains, which can be resolved by repeating the above procedure. Since both e_1 and $\overline{e_2}$ can be deleted, deleting either of them leads to a duplication-free sequence with different configurations.

4   Sequential Importance Sampling

The assumptions required for the basic algorithm to work are often violated in practice. In particular, large scale deletions in the gene clusters violating assumption (2) are likely to occur, and atomic boundary reuses violating assumption (3) are not uncommon. Once a boundary reuse occurs, regardless of its causes, we can no longer reconstruct the correct evolution history or even predict the true number of events. Even if assumptions (1)-(3) are satisfied, there are always multiple ways of reconstructing the history of a gene cluster. The number of the events will be the same, but the order of the events and the ancestral duplication-free sequence will be different among solutions. To make inference about the evolution history of a gene cluster, we need to summarize the feature of interest from all possible histories. However, enumerating all possible histories would be computationally expensive.

To address the atomic boundary reuse and to model deletions, we propose a stochastic algorithm that first samples many possible histories of a gene cluster from a target distribution, and then makes inference of evolutionary features from the collected samples. We use the target distribution to define the scope of histories and their relative contributions. For example, to make inference exclusively from histories that have no atomic boundary reuse, the target distribution can be uniform on all such histories and 0 otherwise. In practice, we will use more flexible target distributions to accommodate practical complications. To reconstruct a possible history from the target distribution, we use sequential importance sampling (SIS) [18]. SIS sequentially samples one event at a time from a pool of possible events until all local alignments in a dot-plot are resolved. We represent a history of the gene cluster by a series of T events $\mathcal{O}_T = (O_1, \ldots, O_T)$ reconstructed by SIS in reverse order of time. Here, both $\mathcal{O}_T$ and T are unknown. The basic algorithm is a special case in which every reconstructed event O_i corresponds to a *candidate alignment*. By repeating the SIS procedure, we obtain many possible histories. We then summarize the desired features by taking a weighted average, with weights calculated as the difference between the target distribution and the actual sampling distribution.

Given a gene cluster X, we specify the target distribution of histories to be $\pi(\mathcal{O}_T \mid X) \propto e^{aT+br}$, where T is the number of events, r is the number of reused atomic boundaries, and a, b are two penalty parameters. We chose $a = b = -5$; thus histories with fewer evolutionary events and boundary reuses will contribute more to the inference. The penalty (-5) was chosen to allow suboptimal solutions. When the penalty approaches $-\infty$, only the most parsimonious solutions with the least boundary reuse will influence the result. Note that we only need to specify the target distribution up to a normalizing constant.

Directly sampling histories from the target distribution is often intractable, and thus SIS is used. Suppose we already reconstructed t most recent events, we sample the next event O_{t+1} from a trial distribution $g_t(O_{t+1} \mid \mathcal{O}_t)$. Our goal in choosing the trial distribution is to allow easy sampling while resembling the target distribution as closely as possible. By sampling events until all alignments are resolved, we obtain a possible history $\mathcal{O}_T$, and by repeating this procedure we collect many possible histories. However, the collected histories will not follow the target distribution $\pi(\mathcal{O}_T \mid X)$, but instead $\prod_{t=0}^{T-1} g_t(O_{t+1} \mid \mathcal{O}_t)$. To correct this bias, we calculate weight $w = \pi(\mathcal{O}_T \mid X) / \prod_{t=0}^{T-1} g_t(O_{t+1} \mid \mathcal{O}_t)$ determining how much reliance we shall put on each reconstructed history. Finally, given m histories $\mathcal{O}_{T_1}^{(1)}, \mathcal{O}_{T_2}^{(2)}, \ldots, \mathcal{O}_{T_m}^{(m)}$ and their weights $w_1, \ldots, w_m$, we make a statistical inference about evolutionary features by approximating the expectation of any function $u(\mathcal{O}_T)$ of histories as $E[u(\mathcal{O}_T)] = \left(\sum_{i=1}^{m} w_i u(\mathcal{O}_{T_i}^{(i)}) \right) / \left(\sum_{i=1}^{m} w_i \right)$. For example, $u(\mathcal{O}_T) = T$ gives the number of events.

The choice of the trial distribution directly determines the efficiency of history reconstruction. For example, if assumptions (1)-(3) are met, we can let $g_t(O_{t+1} \mid \mathcal{O}_t)$ be uniform on all events O_{t+1} that involve a candidate alignment and 0 on all other events. As a result, the SIS algorithm will efficiently and precisely produce the same number of events as the basic algorithm.

We used simulations to choose a set of good trial distributions. In particular, we used $g_t(O_{t+1} \mid \mathcal{O}t) = (L-\ell)^{-k-2} f(s, \delta)/Z$ for duplication, and $g_t(O_{t+1} \mid \mathcal{O}t) = (L+\ell)^{-1} e^{-\ell/\lambda} f(s, \delta)/Z$ for deletion. For duplication $O_{t+1} = (P, D)$, $k \in \{0, 1, 2, 3\}$ denotes the number of reused atomic boundaries, i.e. the number of non-collapsible atomic segment pairs that flank D and the boundaries of P after removing D. Furthermore, L and ℓ denote the current sequence length and the duplication size, respectively. For deletion, ℓ and L denote the actual and the expected deletion size, respectively. We only consider deletions without atomic boundary reuse, and $\lambda = 10000$. Intuitively, we prefer to sample longer duplications and shorter deletions in each SIS step. We also prefer alignments with higher percent identity and those that resolve more local alignments, which is represented by function $f(s, \delta) = e^{(\delta - (100-s))/5}$ of the alignment percentage identity $s \in [0, 100]$ and the number δ of alignments resolved by O_{t+1}.

We only consider a deletion event if the atomic segment pair flanking a deletion site appears elsewhere in the sequence. Otherwise, no deletion information is available. For example, suppose $a_1 b_1$ flanks a deletion site, and we observe a_2 and b_2 elsewhere, then the region between a_2 and b_2 can be inserted in between $a_1 b_1$ to unwind a deletion. The relative orientation between a_1 and b_1 must match that between a_2 and b_2, and $a_1 b_1$ must not be located between a_2 and b_2. If all conditions are met, we calculate the percentage identity s from the flanking alignments (a_1, a_2) and (b_1, b_2), and the deletion event can be reconstructed. Finally, Z denotes the normalizing constant for the trial distribution. Compared with the normalizing constant for the target distribution, Z is much easier to calculate, because we can easily enumerate all possible events given $\mathcal{O}_t$.

5   Application to Human Gene Clusters

We have identified 457 duplicated regions in the human genome assembly hg18, based on alignments from UCSC browser self-chains [19] of length at least 500 bp, with at least 70% identity, and with both segments located within 500 Kbp of each other. The regions were defined by clustering overlapping duplications; only regions of substantial size (at least 50 Kbp) and non-trivial complexity (at least

Table 1. Estimated numbers of duplications and deletions in 25 human gene clusters following divergence from great apes (GA), old world monkeys (OWM), new world monkeys (NWM), prosimians (LG), and dogs and other laurasiatherians (DOG)

Name (possible disease association)	Location	GA	OWM	NWM	LG	DOG	gaps
PRAMEF	chr1p36.21	7	23	32	48	63	3
HIST2H (asthma; atrial fibrillation)	chr1q21.1-2	21	41	68	101	107	6
FCGR (systemic lupus erythematosus)	chr1q23.3	3	3	5	6	6	0
CFH (macular degeneration)	chr1q31.1	4	6	18	22	25	0
CCDC;CFC1 (left-right laterality defects)	chr2q21.1	3	5	12	12	15	0
UGT1A (neonatal hyperbilirubinemia)	chr2q37.1	0	2	13	17	23	0
UGT2 (prostate cancer)	chr4q13.2-3	2	27	51	59	82	1
SMA;SMN (motor neuron disease)	chr5q13.2	23	25	25	25	25	0
HIST1H;BTN (coronary heart disease)	chr6p22.2-1	0	1	9	19	35	0
HLA;TRIM (multiple sclerosis)	chr6p22.1-21.33	0	2	29	45	58	0
HLA;BAT (type 1 diabetes)	chr6p21.33	0	4	12	17	28	0
HLA-D (rheumatoid arthritis)	chr6p21.32	0	1	14	21	26	0
HLA-D;COL11A (acute lymphoblastic leukemia)	chr6p21.32	0	0	0	7	14	0
CCL;CTF2;PMS2 (rheumatoid arthritis)	chr7q11.23	21	31	38	40	45	1
IFN (cervical cancer)	chr9p21.3	0	11	15	20	41	0
SFTPA (tuberculosis)	chr10q22.3	6	7	8	10	12	1
OR5;HB;TRIM (thalassemia; sickle cell anemia)	chr11p15.4	4	6	10	10	27	0
KLR (immunological diseases)	chr12p13.2	0	1	1	2	3	0
CHRNA;KIAA (schizophrenia)	chr15q13.3-1	15	38	47	56	58	2
CYP1;DKFZ (lung cancer; macular degeneration)	chr15q24.1-3	2	14	23	26	28	0
LOC (rheumatoid arthritis)	chr16p11.2	3	6	6	6	8	0
NF1;EVI2 (intestinal neuronal dysplasia; autism)	chr17q11.2	3	9	10	10	10	0
CYP2 (lung cancer; esophageal cancer)	chr19q13.2	0	5	14	17	19	0
KIR;LILR (hepatitis C; liver cancer)	chr19q13.42	0	16	30	43	65	0
WFDC	chr20q13.12	0	0	0	1	2	0

two duplications) were retained. These regions cover ∼215 Mbp (7%) of the human genome. We targeted 165 biomedically interesting clusters (∼111 Mbp) that either overlap genes associated with a human disease (genetic association database [20]), or contain groups of similarly named genes [21].

Clusters were processed through a pipeline that included: (1) self-alignment by blastz; (2) production of subsets of the alignments roughly corresponding to duplications in the human lineage after divergence from great apes ($\geq$ 98% identity), old-world monkeys (93%), new-world monkeys (89%), lemurs (85%), and dogs (80%); (3) adjusting alignment endpoints to avoid predicting spurious tiny duplications; (4) chaining (i.e., local alignments of similar percent identity broken by small insertions/deletions or post-duplication insertion of interspersed repeats. For each of the resulting 825 combinations of gene cluster and divergence threshold, we estimated the number of duplications or deletions in the human lineage subsequent to the divergence. Selection of the results is shown in Table 1.

Table 1 reveals large differences in the evolutionary tempo among the gene clusters. For instance, the cluster of SMN genes appears to have been quiescent through almost all of primate evolution, then experienced an explosion of duplications in the last six million years. On the other hand, the cluster containing HLA-D appears to have changed little for 50 million years, while that containing UGT2 may have accumulated duplications fairly consistently throughout primate evolution, but with a surge of activity about 10-40 MYA.

We also estimated the size, spacing, and orientation of duplication events. Fig.2 shows estimated distributions of the size of the duplicated region and the

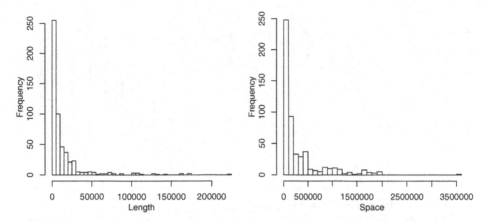

Fig. 2. Distribution of duplication lengths (left) and distances between the original and duplicate segments (right) for duplications with at least 93% sequence identity

spacing between the original and duplicated segments for duplications with at least 93% identity. For those duplication events, the copy was in the reverse orientation relative to the original segment in 39% of the cases.

We used these observed distributions and inversion rates to simulate the evolution of gene clusters, providing data to evaluate our pipeline. Starting from a 500 Kbp sequence, we simulated the formation of gene clusters via 10-100 duplications. For each event, we chose a random left end and length from the observed distribution. The procedure then chose an insertion point at a distance selected from the observed spacing distribution, and a copy of the "source" interval (or its reverse complement at a frequency of 0.39) was inserted. We also simulated deletions with frequency equal to 2% of the duplication rate (the observed frequency), using random left ends and length drawn from the empirical distribution. By simulating $N = 10, 20, 30, \ldots, 100$ events, we created 10 gene clusters for each N. The results of our pipeline were compared to the actual number of simulated events. Fig.3 shows that our algorithm accurately predicted the true number of events for the simulated gene clusters. The predicted numbers of events were slightly larger (4% on average) than the true number of events.

6   Discussion

We have designed and implemented a method to predict the duplication history of a gene cluster using sequence data from only one species. Our goal was to measure the tempo of cluster expansions throughout primate evolution for every human gene cluster, so as to help prioritize the selection of notably interesting gene clusters for more detailed comparative genomics studies. Our future plans include performing comparative sequence analysis of a series of human gene clusters, which will involve isolating and accurately sequencing the orthologous genomic regions in multiple primates.

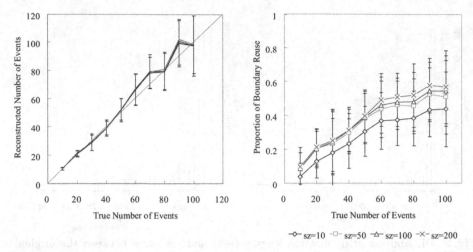

Fig. 3. Left: Actual versus reconstructed number of events with standard errors. Right: Proportion of breakpoint reuses within the reconstructed histories. For each simulated gene cluster, we used four minimum alignment length thresholds (sz): 10 bp, 50 bp, 100 bp, and 200 bp, as indicated (shorter alignments were omitted).

It will be fascinating to compare cluster dynamics in certain lineages to observed phenotypic differences among primates. For instance, Hurle et al. [22] look for correlations between differences in the WFDC cluster and several phenotypes, including female promiscuity. Note that Table 1 indicates a lack of recent WFDC expansions in the human lineage. Another potential use is illustrated by the PRAME cluster, where three gaps remain in the human assembly (Table 1). The rhesus cluster was straightforward to assemble because it lacks recent duplications [23], paving the way for evolutionary studies to help understand the cluster's function.

In addition, such sequence data should reveal differences among primate species of possible relevance for selecting species for further biomedical studies. Sequence data has already been gathered from primate orthologs of the HLA cluster, showing a large expansion in the macaque lineage [24,23], and effects of differences among the rhesus, cynomolgus, and pigtail macaque MHC clusters may be relevant for clinical studies of AIDS progression [25,26]. Similarly, the KLR cluster has been sequenced in marmoset by Averdam et al. [27] to help determine the value of that species as a primate model for immunological research. Our planned systematic project will provide a deeper understanding of primate genome evolution than would piecemeal studies of this sort.

The data should also fuel the development of computational methods for handling the complexities associated with comparative sequence data that include closely related duplicated segments. The approach described here is just one way of approaching this fascinating class of problems.

Acknowledgements. This project has been funded in part by NHGRI grant HG002238 to WM, and in part from support provided to EDG from the NHGRI

Intramural Program. Jian Ma produced a set of 141 clusters used at the start of this study, and Richard Burhans performed further analysis on those clusters.

References

1. Ohno, S.: Evolution by Gene Dupplication. Springer, Berlin (1970)
2. Lupski, J.R.: Genomic rearrangements and sporadic disease. Nat. Genet. 39(7 Suppl), 43–47 (2007)
3. Lander, E.S., et al.: Initial sequencing and analysis of the human genome. Nature 409(6822), 860–921 (2001)
4. Wong, K.K., de Leeuw, R.J., Dosanjh, N.S., Kimm, L.R., Cheng, Z., Horsman, D.E., MacAulay, C., Ng, R.T., Brown, C.J., Eichler, E.E., Lam, W.L.: A comprehensive analysis of common copy-number variations in the human genome. Am. J Hum. Genet. 80(1), 91–104 (2007)
5. International Human Genome Sequencing Consortium: Finishing the euchromatic sequence of the human genome. Nature 431(7011), 931–935 (2004)
6. Green, E.D.: Strategies for the systematic sequencing of complex genomes. Nat. Rev. Genet. 2(8), 573–573 (2001)
7. Blanchette, M., Kent, W.J., Riemer, C., Elnitski, L., Smit, A.F., Roskin, K.M., Baertsch, R., Rosenbloom, K., Clawson, H., Green, E.D., Haussler, D., Miller, W.: Aligning multiple genomic sequences with the threaded blockset aligner. Genome Res. 14(4), 708–715 (2004)
8. Raphael, B., Zhi, D., Tang, H., Pevzner, P.: A novel method for multiple alignment of sequences with repeated and shuffled elements. Genome Res. 14(11), 2336–2336 (2004)
9. Margulies, E.H., et al.: Analyses of deep mammalian sequence alignments and constraint predictions for 1% of the human genome. Genome Res. 17(6), 760–764 (2007)
10. Hou, M.: (unpublished data, 2007)
11. Schwartz, S., Kent, W.J., Smit, A., Zhang, Z., Baertsch, R., Hardison, R.C., Haussler, D., Miller, W.: Human-mouse alignments with BLASTZ. Genome Res. 13(1), 103–107 (2003)
12. Elemento, O., Gascuel, O., Lefranc, M.P.: Reconstructing the duplication history of tandemly repeated genes. Mol. Biol. Evol. 19(3), 278–278 (2002)
13. Lajoie, M., Bertrand, D., El-Mabrouk, N., Gascuel, O.: Duplication and inversion history of a tandemly repeated genes family. J Comput. Biol. 14(4), 462–468 (2007)
14. Jiang, Z., Tang, H., Ventura, M., Cardone, M.F., Marques-Bonet, T., She, X., Pevzner, P.A., Eichler, E.E.: Ancestral reconstruction of segmental duplications reveals punctuated cores of human genome evolution. Nat. Genet. 39(11), 1361–1368 (2007)
15. Wilson, G.M., Flibotte, S., Missirlis, P.I., Marra, M.A., Jones, S., Thornton, K., Clark, A.G., Holt, R.A.: Identification by full-coverage array CGH of human DNA copy number increases relative to chimpanzee and gorilla. Genome Res. 16(2), 173–181 (2006)
16. Dumas, L., Kim, Y.H., Karimpour-Fard, A., Cox, M., Hopkins, J., Pollack, J.R., Sikela, J.M.: Gene copy number variation spanning 60 million years of human and primate evolution. Genome Res. 17(9), 1266–1267 (2007)
17. Nadeau, J.H., Taylor, B.A.: Lengths of chromosomal segments conserved since divergence of man and mouse. Proc. Natl. Acad. Sci. USA 81(3), 814–818 (1984)

18. Liu, J.S.: Monte Carlo Strategies in Scientific Computing. Springer, New York (2001)
19. Kuhn, R.M., et al.: The UCSC genome browser database: update 2007. Nucleic Acids Res 35(Database issue), D668–D673 (2007)
20. Becker, K.G., Barnes, K.C., Bright, T.J., Wang, S.A.: The genetic association database. Nat. Genet. 36(5), 431–432 (2004)
21. Ma, J.: personal communication (2007)
22. Hurle, B., Swanson, W., Green, E.D.: Comparative sequence analyses reveal rapid and divergent evolutionary changes of the WFDC locus in the primate lineage. Genome Res. 17(3), 276–276 (2007)
23. The Rhesus Macaque Genome Sequencing and Analysis Consortium: Evolutionary and biomedical insights from the rhesus macaque genome. Science 316(5822), 222–224 (2007)
24. Daza-Vamenta, R., Glusman, G., Rowen, L., Guthrie, B., Geraghty, D.E.: Genetic divergence of the rhesus macaque major histocompatibility complex. Genome Res. 14(8), 1501–1505 (2004)
25. Krebs, K.C., Jin, Z., Rudersdorf, R., Hughes, A.L., O'Connor, D.H.: Unusually high frequency MHC class I alleles in Mauritian origin cynomolgus macaques. J Immunol. 175(8), 5230–5239 (2005)
26. Smith, M.Z., Fernandez, C.S., Chung, A., Dale, C.J., De Rose, R., Lin, J., Brooks, A.G., Krebs, K.C., Watkins, D.I., O'Connor, D.H., Davenport, M.P., Kent, S.J.: The pigtail macaque MHC class I allele Mane-A*10 presents an immundominant SIV Gag epitope: identification, tetramer development and implications of immune escape and reversion. J Med. Primatol 34(5-6), 282–283 (2005)
27. Averdam, A., Kuhl, H., Sontag, M., Becker, T., Hughes, A.L., Reinhardt, R., Walter, L.: Genomics and diversity of the common marmoset monkey NK complex. J Immunol. 178(11), 7151–7151 (2007)

A   Proof of the Basic Algorithm

Proof of Theorem 1: Denote the present day sequence of atomic segments S and the series of k duplications that created this sequence $O_1, O_2, \ldots, O_k$. To prove the claim, we will first show that for any candidate alignment (P, D), sequence S can also be created by a sequence of duplications $O'_1, O'_2, \ldots, O'_k$ of the same length (also satisfying assumptions (1)-(3)), where the last duplication O'_k is (P, D). All claims of the theorem are a direct consequence of this claim, proven simply by induction on the number of duplication events.

Now consider a candidate alignment (P, D) in sequence S. If we look at the duplication history in reverse, we can show that D will be always a D-segment of some candidate alignment until one of the following happens (see Lemma 2): (A) either D is deleted by unwinding a duplication (P', D), or (B) all the P-segments matching D are unwound, and the role D-segment is in fact gained by a duplication (D, P').

In case (A), we can find a segment P'' matching D such that there exist a sequence of k duplications that will create sequence S, where (P'', D) is the latest duplication (Lemma 3). Since both (P'', D) and (P, D) are candidate alignments in S, we can replace (P'', D) with (P, D) in the last duplication and still obtain the same sequence S with k duplications.

In case (B), the role of the D-segment has been gained by a duplication $O_i = (D, P')$ at time i. Immediately after this event, (D, P') must be a candidate alignment (Lemma 1). Since (P', D) is also a candidate alignment, we can replace $O_1, \ldots, O_i$ with some sequence of duplications $O'_1, \ldots, O'_i$ such that we obtain the same intermediate atomic segment sequence at time i, where $O'_i = (P', D)$ (Lemma 4). Using the sequence of duplications $O'_1, \ldots, O'_i, O_{i+1}, \ldots, O_k$, we reduce case (B) to case (A), for which we have already proven the claim.   □

We present the proofs of the following supporting lemmas in Appendix B.

Lemma 2. *If we consider duplication operations in reverse order, the D-segment of a candidate alignment will remain a D-segment of some (not necessarily the same) candidate alignment until either this D segment is removed from the sequence by unwinding a duplication (P, D), or all segments matching D are deleted, in which case the segment gains the role of D-segment by duplication (D, P).*

Lemma 3. *Let S be a sequence of atomic segments created by k duplications $O_1, \ldots, O_k$, and let $O_i = (P, D)$ for some i. If D is a D-segment of a candidate alignment in all intermediate sequences after duplication O_i, as well as in S (possibly with different P-segments, say P'), we can always find a sequence of duplications $O'_1, \ldots, O'_k$ leading to S such that $O'_k = (P', D)$.*

Lemma 4. *Let S be a sequence of atomic segments created by k duplications $O_1, \ldots, O_k$, where the last duplication is $O_k = (D, P)$. If (P, D) is also a candidate alignment, there exists a sequence of k duplications $O'_1, \ldots, O'_k$ such that the last operation is $O'_k = (P, D)$, and it creates the same sequence of atomic segments S.*

B   Proofs of Supporting Lemmas

Lemma 5. *For a candidate alignment (P, D), with $D = u|a_1 \cdots b_1|v$ and $P = x|a_2 \cdots b_2|y$, the D segment will not overlap with any other alignments unless (P, D) is a forward tandem duplication.*

Proof. Without loss of generality, we assume there is a copy of $u|a_1$ in the sequence, say $u_3|a_3$. If $u_3|a_3$ lies within or outside either $|a_1 \cdots b_1|$ or $|a_2 \cdots b_2|$, it will remain in the sequence after removing D. Since $x|a_2$ is collapsible after removing D, $u_3|a_3$ must equal to $x|a_2$, which means $u = u3 = x$, but this contradicts with the maximum alignment assumption.

Alternatively, either $u_3|a_3$ or $x|a_2$ is deleted when removing D. If $u_3|a_3$ is deleted by D, it must lie on the boundary $b_1|v$ of D, i.e., either $b_1|v \equiv u_3|a_3$ or $b_1|v \equiv \overline{a_3}|\overline{u_3}$; either way we will have the atomic pair flanking D non-collapsible after removing D. On the other hand, if $x|a_2$ is deleted by D, we must have either a forward tandem duplication $u|a_1 \cdots b_1|a_2 \cdots b_2|y$ or a backward tandem duplication $\overline{v}|\overline{b_1} \cdots \overline{a_1}|a_2 \cdots b_2|y$. The latter leads to a contradiction because $u = \overline{a_2}$ means $u_3|a_3 = \overline{a_2}|a$, and hence $\overline{v}|a_2$ is not collapsible after removing D.   □

Lemma 6. D_1 *of a candidate alignment* (P_1, D_1) *cannot lie within either* P_2 *or* D_2 *of another candidate alignment* (P_2, D_2), *but they can represent the same region, i.e.,* $D_1 \equiv D_2$.

Proof. By Lemma 5, the statement is true if (P_1, D_1) is not a forward tandem duplication. When (P_1, D_1) is a forward tandem duplication, without loss of generality, assume (P_1, D_1) has the form $D_1 | P_1 = u | a_1 \cdots b_1 | a_2 \cdots b_2 | y$. Suppose there is another candidate alignment (P_2, D_2), in which either P_2 or D_2 covers D_1. If D_1 completely lies within either P_2 or D_2 and shares no boundaries with them, then there is a second copy of $b_1 | a_2$, say $b_3 | a_3$ in the sequence. After removing D_1, we should have $u | a_2$ collapsible, which is impossible due to $b_3 | a_3$. On the other hand, suppose D_1 lies within either P_2 or D_2 and they share the boundary $u | a_1$; then the same arguments apply. Instead, if D_1 shares the boundary $b_1 | a_2$ with either P_2 or D_2, there are two situations:

Situation 1: P_2 covers D_1. In this case, after removing D_2, we should have $b_1 | a_2$ collapsible, which is impossible due to $b_2 | y$ in P_1.

Situation 2: D_2 covers D_1. In this case, we must have $D_2 = p | c1 \cdots u a_1 \cdots b_1 | a_2$, in which the segment $a_1 \cdots b_1$ is D_1, and $P_2 = w | c_2 \cdots u_4 a_4 \cdots b_4 | z$. After removing D_2, we have $p | a_2$ collapsible, which means $p = u$. After removing D_1, we should have $u | a_2$ collapsible, which means $(p | c_1) = (u | c_1) = (u | a_2)$, and thus $c_1 = a_2$. However, this means $w | c_2 = w | a_2$ in P_2 must also equal to $u | a_2$, and thus $w = u = p$, which contradicts with the maximum alignment assumption.   □

Definition 3 (Coupling). *Two candidate alignments* (P_1, D_1) *and* (P_2, D_2) *are coupled if* $P_1 \equiv D_2$ *and* $P_2 \equiv D_1$.

Lemma 7. D_1 *in a candidate alignment* $A = (P_1, D_1)$ *cannot share boundaries with* P_2 *in another candidate alignment* $B = (P_2, D_2)$, *unless either* $D_1 \equiv D_2$ *or* A *is coupled with* B.

Proof. Let $D_1 \equiv u | a_1 \cdots b_1 | v$, $P_1 \equiv x | a_2 \cdots b_2 | y$, and $D_2 \equiv p | c_1 \cdots d_1 | q$, $P_2 \equiv w | c_2 \cdots d_2 | z$. Without loss of generality, we assume that D_1 shares boundaries with P_2. There are two situations:

Situation 1: D_1 is adjacent to P_2, in which case we have $(w | c_2) \equiv (b_1 | v)$. Since $w | c_2$ is collapsible after removing D_2, we should have $b_2 | y$ in P_1 equal to $b_1 | v$, and thus $y = v$. However, this contradicts with the maximum alignment assumption. The exception is that either $b_1 | c_2$ or $b_2 | y$ is deleted when removing D_2. The former indicates $D_1 \equiv D_2$ by Lemma 6. For the latter, if $b_2 | y$ is completely removed by D_2, there is another copy of $b_2 | y$ in P_2, which still indicates $y = v$ and leads to a contradiction. If D_2 only removes b_2 in $b_2 | y$, then D_2 covers P_1 by Lemma 6. In this case, we have either of the following:

1. D_2 and P_1 are in the same orientation:
   In this case, $d_1 = b_2$ and $q = y$. Since $b_2|y$ is collapsible after removing D_1, and $b_2|y = d_1|q$, we must have $d_2|z$ in P_2 equal to $d_1|q$, which contradicts with the maximum alignment assumption. The only exception is that $b_2|y$ is deleted when removing D_1. In this case, (P_1, D_1) is either coupled with (P_2, D_2), or is a forward tandem repeat in the form $P_1|D_1$. The latter is impossible, otherwise after removing D_1, we should have $b_2|c_2$ collapsible, so $b_2|c_2 = p|c_1$, which contradicts with the maximum alignment assumption.

2. D_2 and P_1 are in different orientations:
   In this case, $p \equiv \overline{y}$ and $\overline{b_2} \equiv c_1 = c_2$. However, it indicates that $b_1|c_2 = b_2|\overline{b_2}$ at the boundary of $D_1|P_2$ is not collapsible after removing D_2, and thus (P_2, D_2) is not a candidate alignment. The only exception is when b_1 of $b_1|c_2$ at the boundary of $D_1|P_2$ is deleted when removing D_2, which is impossible due to Lemma 6.

Situation 2: D_1 covers P_2. After removing D_2, $(d_2|z) \equiv (b_1|v)$ in P_2 is collapsible. However, this contradicts with $v \neq y$, unless either $b_1|v$ in P_2 or $b_2|y$ in P_1 is deleted when removing D_2.

1. If $b_1|v$ in P_2 is deleted, then we either have a forward tandem repeat $P_2|D_2$, or a reverse tandem repeat $P_2|\overline{D_2}$. For the former, we must have $u = w$ and $a = c$ following similar arguments as in Lemma 6. As a result, when removing D_2, $w|c_2$ is collapsible and thus $x = w = u$, which contradicts with the maximum alignment assumption. The only exception is when (P_1, D_1) and (P_2, D_2) are coupled. For the latter, we have a reverse tandem repeat $P_2|\overline{D_2}$. Similarly, we can show that $y = \overline{p} = \overline{u}$ and $d = \overline{c}$. Therefore, $w|c$ in P_2 equals to $w|\overline{d}$, and will remain intact after removing D_2. However, after removing D_2, we should have $d|\overline{p}$ collapsible, and thus $w = p$, which contradicts with the maximum alignment assumption unless (P_1, D_1) and (P_2, D_2) are coupled.

2. if $b_2|y$ in P_1 is deleted, then first, $b_2|y$ cannot be completely deleted by D_2, otherwise there is another copy of $b_2|y$ remaining in P_2, and the same arguments that $v \neq y$ can be applied to show a contradiction; second, the y of $b_2|y$ cannot be deleted by D_2 as proved in Situation 1; third, if the b_2 of $b_2|y$ in P_1 is removed by D_2, we have $D_2 \supset P_1$, which leads to coupling because $D_1 \supset P_2$. $\qquad\square$

Lemma 8. *Given two candidate alignments (P_1, D_1) and (P_2, D_2), if at least one of them is not a forward tandem repeat, then D_1 will neither overlap with nor be adjacent to D_2. D_1 and D_2 can be coupled (i.e., $D_1 \equiv P_2$ and $D_2 \equiv P_1$), separated or representing the same region.*

Proof. Let $D_1 \equiv u|a_1 \cdots b_1|v$, $P_1 \equiv x|a_2 \cdots b_2|y$, and $D_2 \equiv p|c_1 \cdots d_1|q$, $P_2 \equiv w|c_2 \cdots d_2|z$. By Lemma 5 and Lemma 6, D_1 cannot overlap with, cover, or lie within D_2, unless both alignments are forward tandem repeats or if $D_1 \equiv D_2$. As a result, we only need to show that D_1 and D_2 are not adjacent to each other unless they are coupled. Without loss of generality, assume D_1 and D_2 are adjacent in the form $D_1|D_2 = u|a_1 \cdots b_1|c_1 \cdots d_1|q$.

Situation 1: $w|c_2$ in P_2 remains intact after removing D_1. After removing D_1, $u|v = u|c_1$ should be collapsible, and thus $u = w$. On the other hand, $w|c_2$ in P_2 is collapsible after removing D_2 and $u|a_1$ will remain intact, so we have $(u|a_1) = (w|a_1) = (w|c_2)$, which contradicts with Lemma 5. The only exception is that $w|c_2$ in P_2 is deleted when removing D_2, which indicates either (P_2, D_2) is coupled with (P_1, D_1), or (P_2, D_2) is a forward tandem repeat in the form $D_2|P_2$. The latter is impossible, because $q = c_1$, and after removing D_1, we have $u|c_1$ collapsible (because D_1 is adjacent to D_2), which means $u = d$ and thus $z = c_1 = q$, in which case (P_2, D_2) is not maximized.

Situation 2: $w|c_2$ in P_2 is completely deleted when removing D_1. In this case, we must have a copy of $w|c_2$ in P_1, and thus the same arguments for Situation 1 apply.

Situation 3: $w|c_2$ in P_2 is partially deleted when removing D_1, i.e., either w or c_2 is removed. In this case, P_2 must share boundaries with D_1, which is impossible due to Lemma 7, except for the coupling relationship or when $D_1 \equiv D_2$. $\qquad\square$

Lemma 9. *A candidate alignment (P_1, D_1) cannot be partially deleted or extended when removing another candidate alignment (P_2, D_2). Instead, either P_1 or D_1 can be completely deleted by D_2. If P_1 is deleted by D_2, then there is a third candidate alignment (P_3, D_1). If D_1 is deleted by D_2, then $D_1 \equiv D_2$.*

Proof. Let $A \equiv (P_1, D_1)$ and $B \equiv (P_2, D_2)$ denote the two candidate alignments. By Lemma 8, D_1 and D_2 may be identical, coupled, or separated. The exception is when both A and B are forward tandem repeats, in which case the statement holds true. If $D_1 \equiv D_2$, removing D_2 will completely delete D_1. If D_1 and D_2 are coupled, removing D_2 will completely delete P_1. If D_1 and D_2 are separated, deleting D_2 will only affect (P_1, D_1) if D_2 strictly covers P_1. This is because neither D_2 overlaps with P_1 nor D_2 lies within but share boundaries with P_1, according to Lemma 6, and by Lemma 7, D_2 cannot be adjacent to P_1. Assume D_1 and D_2 are separated, and let $D_1 \equiv u|a_1 \cdots b_1|v$, $P_1 \equiv x|a_2 \cdots b_2|y$, and $D_2 \equiv p|c_1 \cdots d_1|q$, $P_2 \equiv w|c_2 \cdots d_2|z$. Since P_1 is strictly within D_2, we must have a copy of P_1, denoted by $P_3 \equiv x_3|a_3 \cdots b_3|y_3$ in P_2, which will remain intact after deleting D_2. As a result, the third alignment $C = (P_3, D_1)$ must be a candidate alignment. $\qquad\square$

Using Lemma 5-9, we are now ready to prove Lemma 2-4 in Appendix A.

Proof of Lemma 2: By Lemma 9, a candidate alignment (P_1, D_1) cannot be partially removed or extended when removing other candidate alignments. We thus only need to show that, when reconstructing duplication in the reverse order, D_1 will continue to be the D segment of some candidate alignments until either D_1 is deleted or all segments matching with D_1 are deleted.

Let $D_1 \equiv u|a_1 \cdots b_1|v$ and $P_1 \equiv x|a_2 \cdots b_2|y$. Assume D_1 becomes an invalid D segment after removing a candidate alignment (P_2, D_2). If removing D_2 deletes P_1, then there is a third candidate alignment (P_3, D_1). If both P_1 and D_1 remain intact after removing D_2, then by Lemma 7 and Lemma 8, the flanking segments of P_1 and D_1 will remain intact as well. Let $D_2 \equiv p|c_1 \cdots d_1|q$ and $P_2 \equiv w|c_2 \cdots d_2|z$, removing D_2 will produce a new atomic pair $p|q$. To invalidate the D-segment role of D_1, at least one of $x|a_2$, $b_2|y$, $u|v$ pairs must become non-collapsible due to $p|q$. If $u|v$ is affected, without loss of generality, we assume $p = u$. Since $u|v$ is collapsible after removing D_1, $p|c_1$ in D_2 must equal to $u|v$ and thus $c_1 = v$. As a result, $w|c_2 = w|v$ in P_2 must equal to $u|v$, indicating $p = w = u$. This contradicts with the maximum alignment assumption. The only exception is when P_2 and D_2 are adjacent in the form $\overline{P_2}|D_2 \equiv \overline{z}|\overline{d_2} \cdots \overline{c_2}|c_1 \cdots d_1|q$, and thus $p = u = \overline{c_2}$. However, since and $v = c_1$, we have $u|v = \overline{c_2}|c_1$ non-collapsible. Similar arguments can be applied to show contradictions when either $x|a$ or $b|y$ becomes non-collapsible due to $p|q$. In conclusion, D_1 will always be the D segment of some candidate alignment until either D_1 is deleted or all segments matching with D_1 are deleted.   □

Proof of Lemma 3: We will prove this lemma by induction on the number of duplication events. First, the lemma holds trivially for the sequences with a single duplication (which must be (P, D)). Now, let us assume that the lemma holds for all duplication sequence of length less than k. We want to prove that it also holds for a sequence of duplication $O_1, \ldots, O_k$ of length k.

If $O_k = (P, D)$, then lemma holds trivially. Therefore, assume that $O_k \neq (P, D)$, and thus (P, D) is among one of $O_1, \ldots, O_{k-1}$. Let S_{k-1} be the atomic segment sequence created by $O_1, \ldots, O_{k-1}$, then according to the induction hypothesis, there exists a segment P' and a sequence of duplication $O'_1, \ldots, O'_{k-1} = (P', D)$ that also creates S_{k-1}.

Let S be the sequence created by the sequence of duplication $O'_1, \ldots, O'_{k-1}, O_k$, i.e., converted from S_{k-1} via one additional duplication O_k. Suppose that $O_k = (P_1, D_1)$, then $D_1 \neq D$ and $P_1 \neq P'$ under the no atomic boundary reuse assumption. Since D is a D-segment in S under the Lemma assumption, we can always find two alternative events $O''_{k-1} = (P'_1, D_1)$ and $O''_k = (P'', D)$ to replace $O'_{k-1} = (P', D)$ and $O_k = (P_1, D_1)$ (i.e., to switch orders of deleting D and D_1), such that S can also be created by the sequence of duplication $O'_1, \ldots, O''_{k-1}, O''_k$. This is a direct result of Lemma 9 and the fact that $D_1 \neq D$. Therefore, S can be created by k duplications with the last operation being (P'', D), even if D is generated by duplication $i(< k)$ in the real history.   □

Proof of Lemma 4: Let $P \equiv x|a \ldots b|y$ and $D \equiv p|a \ldots b|q$. If both (P, D) and (D, P) are candidate alignments in S, then by Lemma 5, no other alignments will cover either P or D unless (P, D) is a forward tandem repeat. If (P, D) is not a forward tandem repeat, $(x|a)$, $(b|y)$, $(p|a)$, $(b|q)$ must all be unique pairs in the atomic segment sequence S. In addition, we should have $x|a$ collapsible after removing D, and thus x must be unique in S. Similar arguments can show that y, p, and q are also unique in S. As a result, the two segments P and D are bounded within unique atomic segments and thus forms "two islands". So any previous duplication related with P or D segments must be completely inside of either P or D, and they do not share boundaries with P or D. The same conclusion applies even if P and D are adjacent to each other. Therefore, to change the latest duplication from $O_k = (D, P)$ to $O'_k = (P, D)$, we simply "redirect" all the duplications that are inside of D to be inside of P, and keep the rest the same. This will create a new sequence of duplication $O'_1, \ldots, O'_{k-1}, O'_k = (P, D)$ that creates S.   □

C   Duplication Complexity of Selected Gene Clusters

Name	Location	GA	OWM	NWM	LG	DOG	gaps
PRAMEF	chr1:12750851-13626366	7	23	32	48	63	3
PADI	chr1:17423413-17600526	0	0	0	0	0	0
	chr1:22775285-23112635	0	0	0	0	0	0
	chr1:25443774-25537798	0	1	1	1	1	0
CYP4	chr1:47048227-47411959	1	5	5	6	11	0
	chr1:86662627-86892926	0	0	1	1	1	0
GBP	chr1:89244904-89692274	0	5	7	9	22	0
AMY	chr1:103898363-104119006	4	10	14	14	14	0
	chr1:110861483-111018698	0	0	0	0	0	0
	chr1:119739258-119963386	0	0	3	19	20	0
HIST2H	chr1:144651745-148125604	21	41	68	101	107	0
	chr1:150451947-150599304	0	1	1	1	1	0
LCE	chr1:150776235-151067237	0	0	6	7	11	0
SPRR	chr1:151220060-151272246	0	0	0	0	1	0
SPRR	chr1:151278447-151390171	0	0	1	7	8	0
	chr1:153784948-154023311	0	5	14	24	28	0
FCRL	chr1:155406878-156042315	0	2	11	30	40	0
CD1	chr1:156417524-156593228	0	0	0	0	1	0
OR	chr1:156634961-157053841	0	0	0	0	1	0
	chr1:157512882-157835664	0	0	0	0	1	0
FC	chr1:159742726-159915333	3	3	5	6	6	0
	chr1:167848867-167968738	0	0	0	0	0	0
CFH	chr1:194914679-195244603	4	6	18	22	25	0
	chr1:205701588-205958677	1	7	12	12	13	0
ZNF	chr1:245215980-245486993	2	2	2	2	5	0
OR	chr1:245680906-246912147	1	6	23	48	55	0
	chr2:79106193-79240545	0	0	0	1	1	0
CCDC; CFC1	chr2:130461934-131153411	3	5	12	12	15	0
	chr2:166554904-167039157	0	0	0	1	3	0
	chr2:208680310-208736768	0	1	2	2	2	0
	chr2:232893923-233063157	0	3	13	21	24	0
UGT1A	chr2:234140385-234334547	0	2	13	17	23	0
	chr3:38566866-38926662	0	0	0	0	1	0
ZNF	chr3:44463068-44751808	0	1	1	2	2	0
CCR	chr3:45917359-46425558	0	0	0	0	1	0
	chr3:48977485-49396481	0	0	1	1	1	0
OR5	chr3:99254906-99898694	0	1	10	14	27	0
	chr3:134863859-134969704	0	0	0	1	1	0

Name	Location	GA	OWM	NWM	LG	DOG	gaps
	chr3:152413859-152539276	0	0	0	0	0	0
	chr3:196822567-196963470	1	1	1	1	1	1
	chr4:38451248-38507567	0	0	0	0	0	0
UGT2	chr4:68830737-70547917	2	27	51	59	82	1
CXCL	chr4:74781081-75209572	0	0	0	3	26	0
ADH	chr4:100215375-100612366	0	0	3	8	10	0
SMN	chr5:68787010-70696078	23	25	25	25	25	0
PCDH	chr5:140145736-140851366	0	0	0	1	37	0
	chr6:10322043-10743230	0	1	1	1	1	0
HIST1H; BTN	chr6:25833812-26617296	0	1	9	19	35	0
HIST1H	chr6:27561049-27970197	1	1	3	4	11	0
ZNF; OR	chr6:28161149-29664934	0	0	10	19	33	0
TRIM	chr6:29786467-30568761	0	2	29	44	58	0
BAT	chr6:31267292-31607879	0	4	12	17	28	0
HLA-D	chr6:32514542-32891079	0	1	14	21	26	0
HLA-D	chr6:33082752-33265289	0	0	0	7	14	0
GSTA	chr6:52711832-52960243	0	7	13	27	33	0
TAAR	chr6:132951558-133008844	0	0	0	0	0	0
	chr6:100791897-101275095	0	0	9	17	18	0
	chr6:169347092-169825478	0	0	0	0	0	0
LOC	chr7:71966977-72466918	1	5	8	8	8	0
CCL; CTF2; PMS2	chr7:73565093-76526339	21	31	38	40	45	1
	chr7:86869277-87034269	0	0	0	1	1	0
	chr7:98915207-99500181	0	0	10	20	26	0
	chr7:142134143-142186482	0	1	4	4	4	0
	chr7:142469761-142919050	0	0	0	0	0	0
OR	chr7:143005241-143760083	7	9	9	11	17	0
ZNF	chr7:148389924-149094267	0	4	9	15	17	0
GIMAP	chr7:149794678-150079280	0	4	4	5	5	0
DEF	chr8:6769157-6902786	1	1	1	1	13	0
DEFB10; DEFB	chr8:7069563-7953918	5	8	8	10	14	1
	chr8:22933046-23139154	0	0	6	21	30	0
	chr8:82518183-82604430	0	0	0	0	0	0
ZNF; ZNF	chr8:145901725-146244938	0	0	0	0	2	0
IFN	chr9:21048760-21471698	0	11	15	20	41	0
OR13	chr9:106305453-106535416	0	0	2	2	3	0
OR	chr9:124279100-124603579	0	0	1	1	2	0
	chr9:134962296-135122729	0	0	0	0	0	0
AKR1C	chr10:4907977-5322660	0	5	7	13	32	0
	chr10:26458036-27007198	0	4	7	17	19	0
	chr10:53701853-54315804	0	0	0	1	1	0
SFTPA	chr10:80936018-81672884	6	7	8	10	12	1
	chr10:88319645-89246594	2	2	3	3	3	0
IFIT	chr10:91051661-91168336	0	0	0	0	1	0
	chr10:96426730-96897127	1	2	18	18	20	0
	chr10:118205218-118387999	0	0	1	3	7	0
	chr10:135086124-135244057	2	2	2	2	2	0
	chr11:1065614-1239359	0	0	0	0	0	1
OR5; HB; TRIM	chr11:4124149-6177952	4	6	10	10	27	

Name	Location	GA	OWM	NWM	LG	DOG	gaps
OR	chr11:6745853-6899767	0	0	1	1	2	0
	chr11:24900251-25670383	0	0	0	0	1	0
OR4	chr11:48193633-48622537	0	0	7	17	20	0
	chr11:48865105-49870196	1	12	15	15	18	0
OR	chr11:54833085-56562513	0	1	14	46	61	0
OR	chr11:57390332-58032285	0	1	1	1	2	0
OR	chr11:58833693-59274730	0	0	2	2	6	0
	chr11:66900400-67551984	0	2	4	4	4	0
MMP	chr11:102067847-102343167	0	0	0	0	0	0
OR	chr11:123129479-123988274	0	3	5	7	15	0
	chr12:9099391-9319709	0	0	0	0	0	0
KLR	chr12:10446112-10497748	0	1	1	2	3	0
TAS2R	chr12:10845284-11475585	0	6	26	36	64	0
	chr12:20846959-21313050	0	0	0	11	24	0
KRT	chr12:50852169-51586146	0	2	4	8	15	0
OR	chr12:53795147-54317866	0	0	1	2	2	0
	chr12:55040623-55490902	0	0	0	0	2	0
	chr12:111828405-111931464	0	0	0	0	0	0
ZNF; ZNF	chr12:132011584-132289534	0	0	0	0	0	0
	chr13:19614743-19695656	0	0	0	0	0	0
	chr13:51634776-51849914	0	1	1	2	2	0
OR	chr14:19250951-19781765	0	0	0	1	3	0
RNASE	chr14:20319257-20525050	0	3	6	8	8	0
	chr14:20692977-21208956	1	1	1	2	3	0
C14orf	chr14:23177922-23591420	1	5	8	9	11	0
	chr14:24044573-24173288	0	0	0	0	0	0
C14orf	chr14:73073807-73175062	0	1	1	3	3	0
SERPINA	chr14:93850088-94034351	0	0	1	1	1	0
SERPINA	chr14:94099676-94182828	0	0	0	0	0	0
	chr14:105101878-105397048	2	17	20	20	21	0
CHRNA; KIAA	chr15:26168691-30570226	15	38	47	56	58	2
CYP1; DKFZ	chr15:71687352-74071019	2	14	23	26	28	0
	chr16:1211147-1279180	0	2	2	2	2	0
ZNF	chr16:3105811-3428601	0	0	0	0	4	0
	chr16:20234773-20711192	2	6	6	6	7	0
LOC	chr16:28560127-29404514	3	6	6	6	8	0
MT	chr16:55181257-55275655	0	0	0	4	18	0
	chr16:85101437-85170740	0	0	0	0	0	0
	chr16:88526416-88690103	0	0	0	0	1	0
OR	chr17:2912380-3289105	1	3	4	5	10	0
	chr17:6501152-6854467	0	1	1	1	1	0
MYH	chr17:10145620-10499991	1	2	7	11	25	0
	chr17:22979762-23370074	0	2	4	4	5	0
NF1; EVI2	chr17:25940349-27337990	3	9	10	10	10	0
CCL	chr17:29605831-29711075	0	0	0	0	0	0
CCL	chr17:31334805-31886998	4	7	7	8	9	1
KRT	chr17:36069761-37038364	0	9	13	20	30	0
	chr17:59292402-59355509	0	4	5	5	5	0
ABCA	chr17:64375713-64805977	0	1	1	1	3	0

Name	Location	GA	OWM	NWM	LG	DOG	gaps
CD300	chr17:70033428-70220651	0	0	0	0	2	0
DS	chr18:26828138-26991601	0	0	0	0	0	0
DS	chr18:27160523-27356213	0	0	0	0	0	0
	chr18:41459658-41573640	0	0	0	0	0	0
SERPINB	chr18:59406881-59805500	0	1	2	2	3	0
	chr19:230508-1050902	0	0	0	0	1	0
	chr19:6377406-7037708	1	4	5	6	8	0
ZNF; OR	chr19:8569586-9765797	3	5	15	24	34	1
OR	chr19:14771021-15113863	0	0	0	3	11	0
CYP4F	chr19:15508827-15669145	0	0	0	1	9	0
CYP4F; OR10H	chr19:15699700-15970865	0	1	2	7	26	0
	chr19:39695418-40633289	0	2	13	20	25	0
ZNF	chr19:40976726-43450858	0	7	13	18	27	0
CYP2	chr19:46016475-46404199	0	5	14	17	19	0
ZNF	chr19:49031476-49676451	0	1	5	9	33	0
	chr19:49840790-50069615	0	0	0	0	0	0
	chr19:55457577-55842758	0	0	0	2	4	0
KLK	chr19:56014236-56276734	0	0	1	1	3	0
KIR; LILR	chr19:59404199-60117280	0	16	30	43	65	0
CST	chr20:23560786-23885538	0	12	19	26	35	0
C20orf	chr20:31084573-31102526	0	1	1	1	1	0
WFDC	chr20:43531807-43853954	0	0	0	1	2	0
	chr20:44190604-44564928	0	0	0	0	0	0
KRTAP	chr21:30642250-30735038	0	0	0	1	2	0
KRTAP	chr21:30774233-30910843	0	0	0	1	1	0
KRTAP1	chr21:44783567-44947268	0	0	4	9	15	0
	chr22:18594272-19312230	3	4	6	6	6	1
	chr22:20705392-23410020	3	26	52	74	118	0
	chr22:30379202-31096691	0	4	5	7	8	0
APOBEC3	chr22:37674922-37828933	0	4	12	19	26	0

Ab Initio Whole Genome Shotgun Assembly with Mated Short Reads

Paul Medvedev[1] and Michael Brudno[1,2]

[1] Department of Computer Science and
[2] Donnelly Centre for Cellular and Biomolecular Research
University of Toronto, Canada
{pashadag,brudno}@cs.toronto.edu

Abstract. Next Generation Sequencing (NGS) technologies are capable of reading millions of short DNA sequences both quickly and cheaply. While these technologies are already being used for resequencing individuals once a reference genome exists, it has not been shown if it is possible to use them for *ab initio* genome assembly. In this paper, we give a novel network flow-based algorithm that, by taking advantage of the high coverage provided by NGS, accurately estimates the copy counts of repeats in a genome. We also give a second algorithm that combines the predicted copy-counts with mate-pair data in order to assemble the reads into contigs. We run our algorithms on simulated read data from *E. Coli* and predict copy-counts with extremely high accuracy, while assembling long contigs.

1 Introduction

The problem of genome assembly has perhaps been more controversial than any other topic within computational biology, leading to alternative approaches to genome sequencing and the creation of two human genomes. Initially, a BAC-by-BAC approach to genome sequencing was favored for constructing longer genomes. While this approach was much more expensive than the alternate whole genome shotgun method, it was considered unlikely that *ab initio* whole genome shotgun assembly was feasible. The development of effective shotgun assembly algorithms capable of assembling a mammalian genome, such as the Celera assembler[16] and Arachne[3], has revolutionized sequence assembly, allowing large genomes to be sequenced much cheaper than was previously thought possible. Currently, the field of genome sequencing is undergoing another major change, with the development of Next Generation Sequencing (NGS) technologies, such as Solexa, 454 and AB SOLiD. While the new technologies can currently yield reads only 25-200 basepairs long, they dramatically reduce the cost of sequencing per nucleotide and significantly speed up data acquisition, with nearly 1 billion nucleotides sequenced in one run (2-3 days) on a Solexa machine. While the novel technologies have already made great improvements

M. Vingron and L. Wong (Eds.): RECOMB 2008, LNBI 4955, pp. 50–64, 2008.

to the problem of resequencing (the determination of the genomes of various individuals once the initial, reference, genome has been built), it has not been shown whether very short reads can be used for *ab initio* genome sequencing – the determination of a completely unknown genome.

1.1 Background

One of the original approaches to genome assembly was to find the shortest common superstring of the reads, that is, to assemble a genome with minimal length. The problem of modeling genome assembly in this way is that most genomes have repeats – multiple identical, or nearly identical, stretches of DNA – while the shortest solution would include each of these repeats only once in the assembled genome. This problem is known as over-collapsing the repeats. One way of addressing this problem is to build representative strings or structures for each repeat and allow the assembly algorithm to use these multiple times. This intuition led to the development of graph-theoretic methods for sequence assembly, where the edges of the graph "spell" some string, and by walking the edges of the graph it is possible to recreate the genome.

In their EULER assembler[19], Pevzner, Tang and Waterman had the insight that by dividing the reads into shorter k-long stretches (called k mers), all of the instances of a repeat collapse into a single set of vertices. They represent each read as a walk on a de Bruijn graph, and search for a superwalk that contains all the reads. This approach was later expanded to use A-Bruijn graphs [18], where the initial subdivision into k-mers is not necessary. Myers introduced an alternative model of sequence assembly, using a string graph [15]. Instead of dividing the reads into k-mers, the algorithm starts by building an overlap graph – a graph where vertices correspond to reads and edges correspond to overlaps. Through the process of removing redundant edges, he is able to classify all edges as either unique, required or optional, and the goal of the assembly is to find the shortest walk which respects all the edge constraints.

Because walks on graphs can be elegantly defined using the concept of balance around vertices (each vertex must be entered and left an equal number of times), network flow methods have been suggested for genome assembly. Though network flow alone is not able to resolve the problem of long repeats, it is able to estimate the number of times a read appears in the genome (its copy-count). In the de Bruijn graph formulation, Pevzner and Tang [17] formulate the problem of determining copy-counts as the minimum cost circulation problem. Myers suggests a similar method to determine the copy-counts in the context of a string graph [15]. However, he augments Pevzner and Tang's approach by placing constraints on the copy-counts prior to solving the flow. As in the Celera assembler, Myers determines whether a contig (represented by an edge) is present uniquely in the genome by modeling the reads on a contig as a Poisson arrival process and calculating the probability that the arrival rate for an edge is twice as high as for the genome as a whole. If this probability is low ($p < 10^{-6}$), the edge (contig) is labeled unique, and the flow through this edge is set to be one. Another kind of constraint is placed on every edge that has an interior vertex. Since it must be traversed at least once if the read corresponding to the interior vertex is to take part in the reconstruction of the genome, the flow is constrained to be at least one on this edge.

Network flow techniques alone are insufficient to assemble a genome in the presence of long repeats which are not spanned by any single read. One of the key pieces that has allowed for whole genome shotgun assembly of mammalian genomes are matepairs – pairs of reads which come from opposite strands, at an approximately known distance in the source genome. Matepairs can be generated by taking a piece of DNA of a known size (called an insert) and generating reads from its two ends. Matepairs allow for the spanning of repeats, allowing the assembler to join together long genomic regions even in the presence of a repeat which is not spanned by any read. The typical approach is to build initial contigs (chains of edges in the overlap graph), and then attempt to join them using information from the matepairs. An alternative approach was demonstrated by Pevzner and Tang in the double-barreled version of the EULER program (EULER-DB[17]). They search for all paths in the de Bruijn graph connecting the two reads of a matepair. If there exists only one with a length approximately equal to the length of the insert, it is replaced by a direct edge. This approach has the disadvantage that it requires an algorithm to find all paths of (approximately) a given length between two nodes, which is a difficult computational problem, and scales poorly with the size of the de Bruijn graph.

Sequence assembly using NGS data is a rapidly developing area. Several methods have been recently suggested for *ab initio* sequencing using short reads; many of these appeared after this paper was submitted. We briefly describe these here. SSAKE [20] is an assembler that uses a simple algorithm for building contigs by greedily extending existing overlaps. VCAKE [12] extended SSAKE to work with error-prone, rather than perfect, data. Another approach based on elongating existing contigs is the SHARCGS assembler [6]. The Shorty assembler [11] uses a de Bruijn graph approach in combination with matepairs to assemble a small bacteria – the 600 Kb *Mycoplasma genitalium*. Chaisson and Pevzner [5] have adapted EULER to use short reads. Their approach shows high accuracy and contig sizes for the slightly longer (120 bp) reads generated by the 454 sequencers. They also use matepair information in a manner identical to the EULER-DB algorithm. Another promising, though yet unpublished, tool is the Velvet assembler [21].

All of the previous work on genome assembly shares a major assumption: the goal of the assembly problem is to minimize the length of the genome. While parsimony is usually used to justify this assumption, it is well-known that repeats are ubiquitous in eukaryotic genomes, and even bacterial genomes have sections that are present multiple times. Because of over-collapsing, any repeating region of length longer than the read length may be underrepresented in the assembled genome. For read lengths of 25 nucleotides, which is what we study in this paper, the number of such repeats is very large. We therefore propose an alternate optimization criteria, as we describe below.

1.2 Contributions

In this paper, we introduce two new methods of genome assembly that are tailored specifically to short read data. First, we believe that the overall goal of an assembler should be not to minimize the length of the genome, but to maximize the likelihood that the genome was the source of the various reads. Unlike the case of sequencing by hybridization, where the only available information is whether a certain k-mer is present in the genome, whole genome shotgun sequencing samples the genome, and hence k-mers that are present more often in the genome are more likely to be sampled. For an individual read, however, Sanger style sequencing does not have sufficient coverage to take full advantage of these frequencies. The greatest advantage that the NGS technologies give is high coverage – on a single run a bacterial genome can get as much as 250x coverage. This number makes it possible to not only determine whether a particular read is present in a genome, but also to statistically estimate its copy-count. We formulate the problem of genome assembly as maximizing the likelihood of the observed read frequencies, rather than minimizing the length of the genome. This problem can be formulated as a minimum cost bidirected flow (biflow) problem with convex costs, and we show that it can be effectively solved with a generic flow solver for the case of bacterial genomes, achieving copy counts that are accurate more than 99.99% of the time.

Second, to improve the lengths of the assembled contigs, we introduce a novel technique for taking advantage of matepair information. Our method is based on the simple Dijkstra's shortest path algorithm. In contrast to EULER-DB, we do not search for all the paths between mated reads, but rather, we search only for the existence of short paths between some pairs of reads. Because the paths we search for are bounded by a small length that is independent of the genome size (the maximum variation in the insert size), our algorithm scales extremely well for large genomes and high coverage.

2 Methods

In Sections 2.1 through 2.4, we present the steps of our copy-count prediction algorithm. In Section 2.5, we give our algorithm for repeat resolution using matepairs.

2.1 Building the Transitively Reduced Bidirected Overlap Graph

Our algorithm models the double-stranded nature of DNA during genome assembly by using the elegant bidirected graph framework. Bidirected graphs are a generalization of directed graphs that were introduced by Edmonds in [7]. A bidirected graph is different from a directed graph in that the edges have orientations on each of the ends, rather than on the whole edge. This leads to three types of edges:

- edges with one arrow pointing into its vertex and the other pointing out of it vertex.

– edges with both arrows pointing out of their respective vertices.
– edges with both arrows pointing into their respective vertices.

A walk in a bidirected graph is defined as a a sequence $x_1, e_1, \ldots, e_{k-1}, x_k$ where e_i is an edge incident to vertices x_i and x_{i+1}, and for all $2 \leq i \leq k - 1$, e_{i-1} and e_i have opposite orientations at x_i. Informally, if a walk enters a node on an in-edge, then it must exit on an out-edge, and if it enters an on out-edge, then it must exist and an in-edge (see Figure 1A for an example).

Bidirected overlap graphs were first introduced by Kececiouglu [13]. An overlap graph is a graph where each vertex corresponds to a read and each edge corresponds to an overlap between reads. In a bidirected overlap graph, each vertex corresponds to a double-stranded read (the read and its reverse complement), and each edge corresponds to one of the three ways that double-stranded reads can overlap each other. Any walk can be traversed in either of two directions, so just like a walk in a directed overlap graph spells a string that contains each of the reads, a walk in a bidirected overlap graph spells a double-stranded string that contains each of the double-stranded reads. Thus, the original double-stranded genome corresponds to a walk in the bidirected overlap graph that visits every vertex at least once (assuming error-free reads and complete coverage). For a more extended discussion of bidirected graphs in general and bidirected overlap graphs in particular, we refer the reader to [14].

The first step of our assembly algorithm is to build a bidirected overlap graph. We add an edge between two reads if they overlap by at least o_{min} characters, where o_{min} is a parameter to our algorithm. We then perform transitive edge reduction, where we remove any overlap that is spelled by two shorter overlaps. This procedure is identical to the one described in [15]. While the set of possible double-stranded strings spelled by the graph remains unchanged, the reduction drastically reduces the number of edges. The result is what we refer to as the transitively reduced bidirected overlap graph.

2.2 Convex Min-Cost Biflow

Given the (transitively reduced) bidirected overlap graph as constructed above, we now describe how to use convex min-cost biflow to estimate the copy-counts

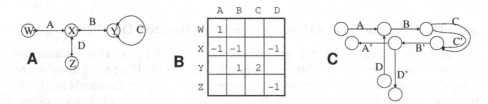

Fig. 1. A. This is an example of a bidirected graph. The sequence $W, A, X, B, Y, C, Y, B, X, D, Z$ is a walk, while W, A, X, D, Z is not. **B.** The corresponding edge incidence matrix, with the zero entries ommited. **C.** This is the associated monotonized directed graph.

of each read. The copy-count of a read is the number of times it appears in the original genome.

Let $G = (V, E)$ be a bidirected graph. Let $l : E \to \mathbb{N}$ and $u : E \to \mathbb{N}$ be lower and upper bounds associated with the edges. The function $f : E \to \mathbb{N}$ is called a flow if for every edge, $l(e) \leq f(e) \leq u(e)$, and for every vertex, the flow along the in-edges is the same as the flow along the out-edges. Given a bidirected graph with lower and upper bounds on the edges and a cost $c_e \in \mathbb{R}$ associated with each edge, the (linear) min-cost biflow problem is to find a flow that minimizes $\sum c_e f(e)$. We discuss algorithms for this problem in Section 2.4.

We take advantage of two additional variations on the min-cost biflow problem. The first allows for having lower and upper bounds for the flow going through each vertex, as well as adding a cost function on the vertices as well as the edges. Such a problem can be reduced to the min-cost biflow problem as follows. Take every vertex v and split it into two vertices v^+ and v^-. Reconnect any edge that was pointing into v to be pointing into v^-. Similarly, reconnect any edge that was pointing out of v to be pointing out of v^+. As a final step, add an edge from v^- to v^+. Now, assign any lower/upper bounds, as well as any costs, associated with v to the edge from v^- to v^+. After repeating this procedure for every vertex, a flow on the transformed graph corresponds to a flow on the original graph, and vice-versa. This transformation is based on a similar transformation on directed graphs [1].

The second equivalent variation is called the convex min-cost biflow problem. Here, the cost c_e associated with an edge e is no longer a real number but rather a convex function $c_e : \mathbb{N} \to \mathbb{R}$, and the goal is to minimize $\sum_e c_e(f(e))$. Such a minimization function is called *separable convex* because it is a sum of convex functions on each of the variables, independently. In the directed case, this problem is polynomially equivalent to the linear min-cost flow problem by modeling each convex function with piecewise-linear approximations. The same reduction holds in the bidirected case.

Before defining our flow problem, we make a modification to the overlap graph by adding a supersource and supersink to the graph. This is the standard way to convert from a flow to a circulation problem.For a thorough discussion of this method, as well as for descriptions and proofs of the above reductions, we refer the reader to a text on network flow, e.g. [1].

In our assembly algorithm, we define a convex min-cost biflow problem on the modified transitively reduced bidirected overlap graph, with bounds and costs on both the edges and the vertices. Each vertex has a lower bound of 1 since it represents a read that must be present in the genome at least once. All other lower bounds are 0 and all upper bounds are infinity. We specify convex costs for the vertices, which we describe in the next subsection, and add prohibitively large costs to the edges from/to the supersource/sink so that their usage is minimized. Next, we solve for the flow (which we describe in detail in Section 2.4). Since any flow can be decomposed into a collection of walks, our flow represents a (non-contiguous) assembly of the genome, and the flow going through each vertex represents the number of time the read is present in the assembly.

2.3   Maximizing the Global Read-Count Likelihood

Let G be a circular genome of length $N(G)$, and let g_i denote the number of times the k-mer i appears in G. Probabilistically, the dataset of n reads corresponds to a set of outcomes from n independent trials. In each trial, a position is uniformly sampled from G and the outcome of the trial is the k-mer beginning at that position. For a given i, the probability that the outcome of a single trial is i is simply $\frac{g_i}{N(G)}$. Let the random variable X_i denote the number of trials whose outcome is i. There are 4^k such variables, and when considered independently of each other, they each follow the binomial distribution. When taken together, their joint distribution is exactly the multinomial distribution, given by

$$P[X_1 = x_1, X_2 = x_2, \ldots, \text{ and } X_{4^k} = x_{4^k}] = \frac{n!}{\prod x_i} \prod_i \left(\frac{g_i}{N(G)}\right)^{x_i}$$

For the assembly problem, G is not known but the results of the n trials are known. Thus, we can consider the likelihood of the parameters of the distribution (g_i) given the outcome of the trials (x_i), which we call the **global read-count likelihood**:

$$L[g_1, \ldots, g_{4^k} | x_1, \ldots, x_{4^k}] = \frac{n!}{\prod x_i} \prod \left(\frac{g_i}{N(G)}\right)^{x_i}$$

In our approach, we attempt to assemble the genome with the maximum global read-count likelihood. Equivalently, we minimize the negative log of this likelihood. Within the biflow framework, g_i corresponds to the flow through a vertex (k-mer) in the overlap graph, and we want to find a flow that minimizes $-\log L$. In order to formulate this as a convex min-cost biflow problem, we need $-\log L$ to be a separable convex function. That is, we need to find convex functions c_i such that $-\log L = \sum c_i(g_i)$. Unfortunately, since the multinomial distribution has the constraint that $N(G) = \sum g_i$, this is not possible.

However, as the number of trials goes to infinity, the X_i random variables become independent. Because the number of trials (sampled k-mers) is typically large, we can approximate the multinomial distribution as the product of the individual binomial distributions of each X_i. Since in the binomial approximation the length of the genome $N(G)$ is a constant that is independent of each g_i, we can replace it by N, which is the length of the actual genome from which the reads were sampled. The approximate length of the actual genome can be ascertained through one of a number of biological experiments, or through an Expectation-Maximization type approach. For our experiments, we assume that the genome size is known.

The resulting approximation for L is thus

$$L[g_1, \ldots, g_{4^k} | x_1, \ldots, x_{4^k}] \approx \prod P[X_i = x_i] = \prod \binom{n}{x_i} \left(\frac{g_i}{N}\right)^{x_i} \left(1 - \frac{g_i}{N}\right)^{n - x_i}$$

Now we can write $-\log L = K \cdot \sum c_i(g_i)$, where K is some positive constant independent of all g_i, and

$$c_i(g_i) = -(x_i \log g_i) - (n - x_i)\log(N - g_i)$$

We let c_i be the convex cost functions for the vertices of our min-cost biflow problem, and reduce it to a linear min-cost biflow problem by approximating the convex function, as described in the previous section. We will now describe our approach for solving the resulting linear min-cost biflow.

2.4   Efficient Algorithm for (Linear) Min-Cost Biflow

The min-cost biflow problem was formulated by Edmonds, who showed that it is equivalent to perfect b-matchings [7]. Edmonds' work was later extended by Gabow [8], who gave the fastest to-date algorithm for sparse graphs, which runs in time $O(|V|^2 \log^2(|V|))$ in the worst case. Unfortunately, no efficient implementation of this or similar algorithms exists, and the worst-case running time is prohibitive for a large graph, such as the overlap graph of a genome. In this section, we give a much faster 2-approximation algorithm that allows us to efficiently solve min-cost biflow problems with optimal results in most cases.

One of the easiest ways that directed network flow for a graph $G(V, E)$ can be solved is through a reduction to a linear program (LP). The reduction is based on building the $|V| \times |E|$ edge incidence matrix $I_{|V||E|}$ for the graph. Every column of I corresponds to $e \in E$, and every row corresponds to $v \in V$. The cell $I_{m,n}$ is 1 if the edge n is an in-edge of vertex m, it is -1 if it is an out-edge from m, and 0 if it is not incident on m. The edge incidence matrix of a graph can be viewed as the constraint matrix for an LP where the optimal LP solution corresponds to the minimum flow in the graph.

Incidence matrices based on directed graphs are Totally Unimodular (TU), leading to LPs that always have integral solutions. Because in bidirected graphs an edge may be an in-edge or an out-edge on both of its ends, the resulting incidence matrix may have two 1s (or two -1s) in a column. It is also possible to have a 2 or a -2 if it is the only non-zero entry in its column (this corresponds to a loop). Figure 1b gives an example. The resulting matrices are known as binet matrices [2], and have the property that the optimal solution of the LP is guaranteed to be half-integral (a multiple of 0.5).

Our algorithm is based on the recent result by Hochbaum [10], who demonstrates a reduction from a binet matrix to a TU matrix by monotonization: doubling the number of columns and rows. Solving the LP defined by the new TU matrix is equivalent to solving it in the original binet matrix. However the new TU matrix corresponds to a directed graph, and one can find the min-cost flow in directed graphs using algorithms that are much faster than general LP solvers. We now formulate the monotonization procedure of Hochbaum in terms of the underlying bidirected graph.

For every vertex v of the original bidirected graph we introduce two vertices v_1 and v_2 in the new directed graph. For every in-edge of v we create two directed "twin" edges, one of which points into the v_1 vertex and the other points out of the v_2 vertex. For all out-edges of v, we again create twin edges, one of which points out of v_1 and the other into v_2. An example of the transformation is given in Figure 1c. We transfer all of the bounds and costs on the original edges to the respective twin edges, and after finding the min-cost flow in the directed

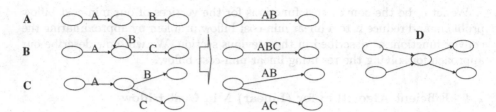

Fig. 2. A, B, and **C** demonstrate the three cases of graph simplification described in Section 2.5. Case A is a chain, case B a loop attached to a chain, and case C is a split vertex. A join vertex case is symmetrical and is not shown. The three simplifications are shown to the right. In all cases, the new graph can "spell" the exact same strings as the initial graph. **D.** This is a conflict node. By iterative application of cases A, B and C, we generate a graph where all remaining vertices are of type D.

graph we transfer the results to the original bidirected graph by adding the flows through the pairs of twin edges and dividing by two. Because the procedure above is equivalent to the monotonization procedure of Hochbaum, it has the same provable properties, e.g. that the optimal result is half integral and that the monotonized flow is at worst a 2-approximation to the optimal integral flow. This reduction allows us to convert our bidirected flow problem into a directed flow problem, for which many efficient algorithms have been developed, e.g. the network simplex algorithm. We are also able to take advantage of off-the-shelf packages for solving network flow within our implementation.

2.5   From Flow to Contigs

At this point of our algorithm, we have found a flow on the overlap graph, as described above. In general, any flow can be decomposed into a collection of walks, which, in our case, correspond to the assembled contigs. Since there is an exponential number of decompositions possible, we use a heuristic to find one where the length of the walks (contigs) is large and the accuracy of the contigs is high.

Graph simplification. In many cases, it can be inferred that certain walks will appear as a subwalk in any decomposition. First, we remove all edges with flow zero from the overlap graph. Next, by applying the following the three rules to every vertex v, we can greatly simplify the overlap graph (see Figure 2):

Case A. There is exactly one edge going into v and exactly one edge going out of v. The flow on both edges is the same. We can merge the two edges and remove v from the graph.

Case B. There are exactly two edges going out of v and two edges going into v, and exactly one of the edges going out of v is also going into v (a loop). The flow on all three edges is the same. We can merge the three edges and remove v from the graph.

Case C. There is exactly one edge going into v and $m > 1$ edges going out of v (v is a split vertex), or there is exactly one edge going out of v and $m > 1$ edges going into v (v is a join vertex). The flow on the in(out) edge is equal to the sum of the flows on the out(in) edge. We can split the in(out) edge into m copies, merge each one with one of the out(in) edges, and remove v from the graph.

We call a vertex v removable if it falls into one of the above cases, and a **conflict vertex** otherwise. For every removable vertex in the graph, we perform one of the three operations above. It can be shown that after at most $2|V|$ operations, all the remaining vertices are conflict vertices. In practice, this process reduced the number of edges in the overlap graph by over 10^5 fold.

Conflict node resolution algorithm. Once the graph contains only conflict vertices, we attempt to resolve each in turn by finding pairs of edges that are incident on the vertex with opposite orientations and are supported by matepairs. For each vertex we do a breadth first search in both the in and out directions, recording all of the reads that are within a specified distance threshold. We skip any read that was initially on an edge that had been split (during case C of the previous step), as it no longer has a unique position in the overlap graph. We now have two sets of vertices, L and R, corresponding to reads that were observed on the in side of a vertex and out side of a vertex respectively (see Figure 3). The high coverage provided for by NGS methods allows us to concentrate our analysis on reads only a short distance away from the conflict vertex. For each of the reads found, we locate their matepairs in the graph (treating the forward and reverse matepairs separately) and run an all-pairs bounded shortest path algorithm from all the mates of L to all the mates of R. Because the overlap graph is sparse, the most efficient algorithm for all-pairs shortest path is to run Dijkstra's algorithm from every vertex. Furthermore, we terminate Dijkstra's algorithm when all vertices within the bounding distance have been explored: if we expect that the true size of the insert will vary by at most E from the expected size, than the bounding distance is $2E$.

To resolve conflict vertices we implement a simple greedy matching algorithm. All of the edges incident on a particular vertex are separated into two classes depending on their direction at the node – in or out. For every pair of (in, out) edges, we compute the number of mates that are within the bounding distance from each other. If a significant fraction of one edge's matepairs are within this distance from the matepairs of another edge on the opposite side (a matching condition), the two edges are joined into a common edge. We handle any half-integral edges by allowing either of the edges to get matched to the integral edge incident to the conflict vertex. The process is repeated until no more pairs of edges that satisfy the matching condition are found at the current vertex.

After every conflict vertex has been considered, the graph simplification steps described in the previous section are run again, as new removable vertices may be created during the matching process. The matching procedure is then iterated for a set number of steps, or until convergence.

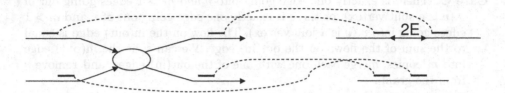

Fig. 3. Resolution of a conflict vertex: we find all reads within a pre-specified distance from the conflict vertex, and locate their mates on the graph. Because the reads are close together, if they spell the same path we expect their mates to also be close together. If the distance between the mates is less than twice the error of the insert size ($2E$) we consider the two mates to support each other. In the example the top left edge will be joined with the top right, and bottom left with the bottom right.

3 Results

We implemented a prototype assembly algorithm for short reads using the algorithms described above. We have experimented with both CPLEX and CS2 [9] for solving network flow, and found that while the running times are comparable, CS2 uses less memory; consequently we used it for all of the experiments below. To simplify the implementation, we model the convex cost function using a three-piece linear approximation (see [1] for details). The overall running time of our algorithm is approximately 1 hour on a single machine.

3.1 Description of the Dataset

Because we were unable to obtain any real data from the Solexa system, we generated synthetic read data from the *E. coli* genome, which has a total length of 4.6 megabytes. We uniformly sampled the genome to find the location of the first read of a matepair, and then sampled the second read at a distance within 10% of the expected insert size, also uniformly. The reads generated were always of length 25 and error-free (the importance of the assumptions of error-free reads and uniform coverage is elaborated upon in the Discussion). The coverage rate used was varied from 50 to 100X, though we also tried one test with 200x coverage (a single run of the Solexa system generates 1Gb of data, or greater than 200x coverage for *E. coli*). The minimum overlap length (o_{min}) was varied from 17 to 21. The exact datasets used are summarized in Table 1.

3.2 Read Count Results

To evaluate the accuracy of our maximum likelihood flow solving algorithm we compared the flow going through every vertex in the overlap graph to the number of times that the corresponding read appears in the original genome. The results are presented in Table 1. For the vast majority of the reads we correctly predicted their copy count in the genome, with the fraction of misestimated

Table 1. The first four columns describe the datasets used in the evaluation of the assembly algorithm. The insert size was simulated with a uniform error of up to 10%. The right side shows the deviations of the predicted read copy-counts from their true values. While half-integral flows were observed with some parameter settings (too low coverage, too low o_{min}), the flow was always integral over all runs with the parameters shown.

Dataset	Coverage	Insert Size	o_{min}	-2	-1	0	+1	+2	+3
50x3k	50	3000	17	4	397	3937038	170	18	6
75x3k	75	3000	19	0	9	4324061	28	3	0
75x6k	75	6000	19	0	7	4324665	22	0	0
100x3k	100	3000	21	0	2	4466328	6	0	0
100x6k	100	6000	21	0	2	4466636	23	0	0
200x6k	200	6000	19	0	0	4547426	4	0	0

Table 2. Evaluation of the assembly quality for the various datasets. Length is the N50/N90 score, number is the number of contigs longer than the N50/N90 length. The errors are computed as the total length of errors over the total size of all contigs.

Data set	length(kb)	N50 Contig number	1 error per	length(kb)	N90 Contig number	1 error per	Total contigs
50x3k	23.4	53	165k	7.1	189	148k	803
75x3k	25.8	48	177k	7.8	174	154k	731
75x6k	25.1	49	93k	7.9	176	96k	732
100x3k	27.9	48	178k	7.9	171	159k	700
100x6k	25.8	49	106k	7.9	174	109k	736
200x6k	25.8	48	154k	8.0	174	158k	727

counts varying between 10^{-4} and 10^{-6}, depending on the coverage. When our algorithm mispredicted the number of occurrences, the error was typically small compared to the true frequency of the read. We also note that the results show only slight improvement past 75x coverage.

3.3 Overall Assembly Results

In order to estimate the quality of the assembly resulting from matepair information, we take every edge of the graph after the conflict node resolution and generate the sequence which it spells. As per convention, we compute the N50 and N90 scores as the length of the shortest contig such that 50 and 90 percent of the original genome is in longer contigs, and the number of such contigs. To check for the presence of errors in the assembly, each contig was aligned to the reference *E. coli* genome. The number of errors in a contig was computed as the number of local alignments that is required to completely tile it minus one. The results are summarized in Table 2.

Overall the length of the contigs which contained 50 and 90 percent of the genome varied between 23-28k and 7-8k, respectively, while the error rate was about one error every 100-180k in the longer N50 contigs, and one error every

100-160k for the N90 contigs. These errors illustrate a weakness of a greedy matching algorithm, which may be mislead by two well-matched edges that contradict many other good matchings. While the contig sizes are short by the standard of whole genome assembly with Sanger reads, they compare favourably with the results that Chaisson et al. [4] obtained on *Neisseria meningitis* (genome length 2.2Mb) with 70 nucleotide reads, albeit without matepairs: in their experiments they required 344 contigs to achieve 95% coverage of the genome, while our algorithm required 206 contigs to cover 95% of *E. coli*, a genome which is twice as long as *N. meningitis*. These results demonstrate the power of matepair information in resolving the proper layout of the genome, even in the case of very short reads.

4   Discussion

In this paper we explore the potential for *ab initio* whole genome shotgun sequencing with very short, mated reads, similar to those that are produced by novel sequencing technologies such as Solexa or the AB SOLiD system. We demonstrate that 25 nucleotide reads, while a significant challenge to assemble, can in fact be used to construct contigs of a reasonable length by using a combination of a network flow algorithm that is able to accurately capture the frequencies at which the various reads occur in the genome and the further use of matepairs to resolve the paths in the resulting graph. While the current algorithm does not yet allow for the sequencing of an unknown bacterial genome using NGS reads, it does indicate that *ab initio* whole genome shotgun sequencing is indeed possible with read data even as short as 25 nucleotides, in the presence of matepair information. Two potential avenues towards this goal may include the use of matepair information to connect the various contigs into supercontigs, and improvements to the algorithm used to resolve the conflict nodes. At the same time, we believe that the general approach of using convex network flow to estimate frequencies of strings in a genome is a more general technique with other applications in computational biology, such as building repeat libraries for newly-sequenced genomes.

In our experiments we make two major assumptions – that the reads are error-free, and that the genome sequencing rate is uniform. We believe that the first of these assumptions is not fundamental, and a limited amount of error in the reads can be overcome using methods similar to the ones developed for the EULER assembler [19]. Moreover, the high coverage rate should improve the correction accuracy of these methods. The second assumption, however, is more essential to the accuracy of our algorithm. In the case of non-uniform coverage of certain areas in the genome (in particular it is suspected that the Solexa machines may under-sample homo-polymer runs) our algorithm may be less accurate at predicting the copy-counts, which may have significant effects in downstream analyses. We believe that these effects can be neutralized if the biases of the sequencing apparatus are known. For example, each read's observed frequency can be adjusted depending on its sequence. The exploration of the exact biases of the NGS platforms and the correction for these is an important avenue for future research.

Acknowledgments

We are grateful to Gene Myers and Pavel Pevzner for useful discussion during the development of these algorithms. This work was funded by an NSERC Discovery Grant, an equipment grant from Canada Foundation for Innovation, and NIH Grant RO1-GM81080.

References

1. Ahuja, R.K., Magnanti, T.L., Orlin, J.B.: Network flows: theory, algorithms, and applications. Prentice-Hall, Inc., Upper Saddle River, NJ, USA (1993)
2. Appa, G., Kotnyek, B.: A bidirected generalization of network matrices. Networks 47(4), 185–198 (2006)
3. Batzoglou, S., Jaffe, D.B., Stanley, K., Butler, J., Gnerre, S., Mauceli, E., Berger, B., Mesirov, J.P., Lander, E.S.: Arachne: a whole-genome shotgun assembler. Genome Res. 12(1), 177–189 (2002)
4. Chaisson, M., Pevzner, P.A., Tang, H.: Fragment assembly with short reads. Bioinformatics 20(13), 2067–2074 (2004)
5. Chaisson, M.J., Pevzner, P.A.: Short read fragment assembly of bacterial genomes. Genome. Published online before print Doi:10.1101/gr.7088808 (2007)
6. Dohm, J., Lottaz, C., Borodina, T., Himmelbauer, H.: SHARCGS, a fast and highly accurate short-read assembly algorithm for de novo genomic sequencing. Genome Res. 17, 1697–1706 (2007)
7. Edmonds, J.: An introduction to matching. In: Notes of engineering summer conference, University of Michigan, Ann Arbor (1967)
8. Gabow, H.N.: An efficient reduction technique for degree-constrained subgraph and bidirected network flow problems. In: STOC, pp. 448–456 (1983)
9. Goldberg, A.V.: An efficient implementation of a scaling minimum-cost flow algorithm. J. Algorithms 22(1), 1–29 (1997)
10. Hochbaum, D.S.: Monotonizing linear programs with up to two nonzeroes per column. Oper. Res. Lett. 32(1), 49–58 (2004)
11. Chen, J., Skiena, S.: Assembly For Double-Ended Short-Read Sequencing Technologies. In: Mardis, E., Kim, S., Tang, H. (eds.) Advances in Genome Sequencing Technology and Algorithms, pp. 123–141. Artech House Publishers (2007)
12. Jeck, W.R., Reinhardt, J.A., Baltrus, D.A., Hickenbotham, M.T., Magrini, V., Mardis, E.R., Dangl, J.L., Jones, C.D.: Extending assembly of short DNA sequences to handle error. Bioinformatics 23, 2942–2944 (2007)
13. Kececioglu, J.D.: Exact and approximation algorithms for DNA sequence reconstruction. PhD thesis, University of Arizona, Tucson, AZ, USA (1992)
14. Medvedev, P., Georgiou, K., Myers, G., Brudno, M.: Computability of models for sequence assembly. In: WABI, pp. 289–301 (2007)
15. Myers, E.W.: The fragment assembly string graph. In: ECCB/JBI, p. 85 (2005)

64   P. Medvedev and M. Brudno

16. Myers, E.W., Sutton, G.G., Delcher, A.L., Dew, I.M., Fasulo, D.P., Flanigan, M.J., Kravitz, S.A., Mobarry, C.M., Reinert, K.H.J., Remington, K.A., Anson, E.L., Bolanos, R.A., Chou, H.-H., Jordan, C.M., Halpern, A.L., Lonardi, S., Beasley, E.M., Brandon, R.C., Chen, L., Dunn, P.J., Lai, Z., Liang, Y., Nusskern, D.R., Zhan, M., Zhang, Q., Zheng, X., Rubin, G.M., Adams, M.D., Venter, J.C.: A Whole-Genome Assembly of Drosophila. Science 287, 2196–2204 (2000)
17. Pevzner, P.A., Tang, H., Waterman, M.S.: An Eulerian path approach to DNA fragment assembly. Proceedings of the National Academy of Sciences 98, 9748–9753 (2001)
18. Pevzner, P.A., Tang, H.: Fragment assembly with double-barreled data. In: ISMB (Supplement of Bioinformatics), pp. 225–233 (2001)
19. Pevzner, P.A., Tang, H., Tesler, G.: De novo repeat classification and fragment assembly. In: RECOMB, pp. 213–222 (2004)
20. Warren, R.L., Sutton, G.G., Jones, S.J., Holt, R.A.: Assembling millions of short DNA sequences using SSAKE. Bioinformatics 23, 500–501 (2007)
21. http://www.ebi.ac.uk/~zerbino/velvet/

Orchestration of DNA Methylation

Howard Cedar

Hebrew University, Jerusalem
cedar@md2.huji.ac.il

Abstract. DNA methylation plays an important role in gene regulation. In order to gain a better understanding of the rules governing this epigenetic modification, we have used microarray technology to map DNA methylation in the human genome. This analysis has helped decipher the DNA sequences involved in setting up the basic global methylation pattern in the early embryo and has revealed the full range of methylation changes that occur in a programmed manner during development. These studies also help explain how specific CpG island genes are targeted for de novo methylation in cancer.

About the keynote speaker. Dr. Howard Cedar was born in New York, received a B.Sc. degree in mathematics from M.I.T. and M.D., Ph.D. degrees from N.Y.U. After an internship he continued as a Public health fellow at the N.I.H. In 1973 he immigrated to Israel where he became a full Professor of Molecular Biology in 1981. Dr. Cedar has received a number of awards including the Israel Prize and the Wolf Prize for Medicine, and is a member of the Israel Academy of Sciences.

M. Vingron and L. Wong (Eds.): RECOMB 2008, LNBI 4955, p. 65, 2008.
© Springer-Verlag Berlin Heidelberg 2008

BayCis: A Bayesian Hierarchical HMM for Cis-Regulatory Module Decoding in Metazoan Genomes

Tien-ho Lin[1,*], Pradipta Ray[1,*], Geir K. Sandve[2], Selen Uguroglu[3], and Eric P. Xing[1,**]

[1] School of Computer Science, Carnegie Mellon University, Pittsburgh, PA, USA
[2] Dept of Computer and Information Science, Norwegian University of Science and Technology, Trondheim, Norway
[3] Dept of Computer Science and Engineering, Sabanci University, Istanbul, Turkey
epxing@cs.cmu.edu

Abstract. The transcriptional regulatory sequences in metazoan genomes often consist of multiple *cis-regulatory modules* (CRMs). Each CRM contains locally enriched occurrences of binding sites (motifs) for a certain array of regulatory proteins, capable of integrating, amplifying or attenuating multiple regulatory signals via combinatorial interaction with these proteins. The architecture of CRM organizations is reminiscent of the grammatical rules underlying a natural language, and presents a particular challenge to computational motif and CRM identification in metazoan genomes. In this paper, we present BayCis, a Bayesian hierarchical HMM that attempts to capture the stochastic syntactic rules of CRM organization. Under the BayCis model, all candidate sites are evaluated based on a posterior probability measure that takes into consideration their similarity to known BSs, their contrasts against local genomic context, their first-order dependencies on upstream sequence elements, as well as priors reflecting general knowledge of CRM structure. We compare our approach to five existing methods for the discovery of CRMs, and demonstrate competitive or superior prediction results evaluated against experimentally based annotations on a comprehensive selection of *Drosophila* regulatory regions. The software, database and Supplementary Materials will be available at http://www.sailing.cs.cmu.edu/baycis.

1 Introduction

Rules determining the spatio-temporal variations of gene expression in multi-cellular organisms are believed to be encoded as "*cis*-regulatory sequences", known to account for a large portion of a metazoan genome [15]. While recent years have seen substantial progress in *in silico* prediction of protein coding sequences from metazoan genomes, our understanding of the vocabulary and rules governing *cis*-regulatory sequences is limited, and remains a major open problem.

---

* The first two authors in the list contributed equally to the paper and should be acknowledged as co-first authors.
** Correspondence should be addressed.

M. Vingron and L. Wong (Eds.): RECOMB 2008, LNBI 4955, pp. 66–81, 2008.

Unlike prokaryotes or uni-cellular organisms like yeast, metazoan transcription factor binding sites (TFBS, also known as motifs) are usually neither located immediately upstream of the proximal promoter element, nor are they distributed uniformly and independently in the extended surrounding region. Instead, the distributions of these motifs exhibit apparent general principles referred to as *modular organizations* – being organized into a series of discrete regions of roughly 200-1000 bp in length, each of which controls a distinct aspect of a gene's expression pattern [3]. Each CRM consists of a locally enriched collection of motifs of certain combination and ordering, capable of integrating, amplifying, or attenuating multiple regulatory signals via combinatorial physical interaction with multiple transcriptional regulatory proteins (i.e., TFs) [2]. Furthermore, it is believed that there also exist dependencies among CRMs so that coordinations between regulatory signals can be orchestrated.

Motif models of TFBSs for a single transcription factor have existed for many years, currently the most common model being the position weight matrix (PWM) introduced more than twenty years ago [25]. In recent years, focus has shifted from predicting TFBSs for a single TF towards predicting CRMs comprising several TFBSs, often for several distinct TFs. Several models have been proposed, making use of certain architectural features of the CRMs. Some of these models apply comparative genomic methods for CRM prediction [12,16,22,23]. These approaches are, however, restricted to very closely related organisms, because non-coding sequences are hard to align and more subject to events like duplication and shuffling which make orthology prediction difficult. A large number of CRM and motif prediction algorithms, including the one we propose in this paper thus rely on single species data.

One line of methods for the discovery of CRMs count the number of matches (of some minimal strength) to given motif patterns within a certain window of DNA sequence [19,21,20,4]. From a modeling point of view, this family of algorithms assumes that motifs are uniformly and independently distributed within each window; an *ad hoc* window size needs to be specified, and careful statistical analysis of matching strength is required to determine a good cutoff or scoring scheme [21,10]. Rajewsky *et al.* addressed the issue of compensating the matching scores for co-occurring weak motif sites using an updatable word frequency measure, leading to higher scores for motifs co-occurring more frequently within a given window size [1] [19].

A second line of methods takes an entirely different approach by modeling the occurrences of motifs and CRMs as the output of a first-order hidden Markov process. This approach alleviates the necessity of both the window size and the score cutoff, and takes into account not only the strengths of motif matches, but also the spatial distances between matches (arguably more informative than co-occurrence within a

---

[1] Their algorithm also contains an important extension for unsupervised CRM prediction, where representations of novel motifs are estimated directly from input DNA sequences. However, under a modular formulation of the CRM prediction problem (cf. the LOGOS model [30]), prediction of motif instances from given representations, and estimation of motif representations from predicted instances, can be treated as two orthogonal sub-problems to be solved separately and coupled as components of a higher-level joint model with estimates exchanged in iterative fashion. In this paper, we only focus on CRM prediction given motif representations and defer implementing the fully autonomous *de novo* motif-finding program to a later paper.

window). The hidden Markov model (HMM) translates to a set of soft specifications of the expected CRM length and the inter-CRM distance (i.e., in terms of geometric distributions). However, since training data for fitting the HMM parameters hardly exist, these parameters typically have to be specified based on empirical guesses. HMMs and similar models that captures TFBS distributions, as well as intra-CRM and inter-CRM backgrounds, have been used in several CRM discovery methods, e.g. in Cister [7], Cluster-Buster [6], CisModule [31] and EMCModule [9]. As these methods employ a general inter-motif background, they do not infer any ordering between motifs. This model is extended to include distinct motif-to-motif transition probabilities in the methods Stubb [24] and Module Sampler [27].

In this paper, we present a new method, *BayCis*, which implements a Bayesian hierarchical HMM for CRM search. *BayCis* represents a step further along the direction of HMM-based CRM models. It uses a more sophisticated HMM model that is intended to capture, to a reasonable degree, the detailed syntactic structure of CRM and cis-regulatory regions containing CRMs. By combining general intra-CRM background, motif specific background surrounding motif instances, as well as specific motif-to-motif transitions, it allows couplings between motifs to be captured. We also introduced more advanced approaches to model the background, using separate inter-CRM, intra-CRM and motif-specific higher-order Markov backgrounds. Furthermore, inter-motif distances may be modeled with more flexible distributions (rather than only simple geometric distributions). Finally, as detailed in the following sections, we treat parameters of the HMM grammar as stochastic variables for which Bayesian priors are applied, instead of regarding the state-transition parameters of the HMM grammar as fixed parameters that solely rely on empirical default values or user specification like in previous methods. This technique in principle alleviates user specification of model parameters (although advanced users could choose to decide the "strength" of the priors, or define their own priors). On the computational front, we developed an efficient variational inference algorithm for posterior inference of sequence annotation and Bayesian parameter estimation. This algorithm enjoys a desirable convergence guarantee and is much more efficient than the classical Gibbs sampling methods without compromising much accuracy.

BayCis has several advantages over existing methods for CRM discovery. The explicit model of CRMs makes architectural assumptions clear, and supports rich interpretation of results by analyzing likelihoods at states and transitions. The sophisticated modeling, including motif-to-motif specific transitions and several distinct background states should allow more specific CRM predictions at the same level of sensitivity. Finally, by relying on soft priors instead of hard specification of model parameters, the Bayesian approach adds generality and user convenience to the method.

2   Methods

To model the complex architecture of metazoan transcriptional regulatory sequences (TRS), we propose to use a *hierarchical hidden Markov model* (hHMM) that can encode a set of stochastic syntactic rules presumably underlying the CRM organizations and motif dependencies. A first-order Markov process over a hierarchy of states allows us to describe the structure of regulatory regions at different levels of granularity, offering more modeling power than existing methods.

2.1   A Hierarchical HMM of TRS

As first proposed in [5], the hHMM is an extension of the classical HMM for modeling domains with hierarchical structures. In an hHMM, all hidden states are not equal, but follow a hierarchical organization that constrains stochastic transitions among states—transitions are only permissible for (certain pairs of) states at the same level or adjacent levels in the hierarchy; different states can emit either single observations or strings of observations, depending on their position in the state hierarchy; and the strings emitted from the non-leaf states in the hierarchy are themselves governed by a sub-hHMM (or more generally, by an arbitrary generative model, which would further extend the overall model beyond an hHMM).

An hHMM can explicitly capture nested generative structures (e.g., TRS → CRM → Motif → Single Nucleotide Site) underlying complex sequential data, and dependencies among elements at different levels of granularity (e.g., motif versus motif, site versus site, etc.), which makes it a powerful and natural approach to model genomic regions harboring transcriptional regulatory sequences. Fig. 1 shows an example of an hHMM encoding typical hierarchical structures of the metazoan TRSs we are concerned with in this study. At the top (i.e., coarsest) level, this hHMM represents a TRS as a con-

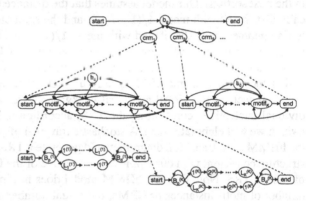

Fig. 1. The BayCis hHMM state transition diagram with 3-level hierarchy. Circular nodes represent functional states in DNA sequences, and round boxes represent start and end states in each sub-model. CRM and motif states are sub-models invoked by higher level models. Arrows between nodes represent permissible state transtions, including horizontal transitions denoted as black arrows, and verticle transitions denoted as dashed arrows.

catenation of long stretches of sequences corresponding to global backgrounds and CRMs. We can think of this top level as an HMM whose states emit whole CRMs and inter-CRM (global) background sequences. Formally, we let $Q^1 \equiv \{b_g, c_1, c_2, \ldots, c_I\}$ denote the set of these possible states. At the next level, each CRM is represented as a sequence of motifs and intra-CRM (local) background states. Accordingly we have $Q^2 \equiv \{b_c, m_1, m_2, \ldots, m_K\}$. At a finer level below, each motif is represented as a sequence of buffer states and nucleotide sites. (We will explain shortly why we include non-motif buffer states at this level.) Accordingly, we define $Q^3 \equiv \mathbb{B} \cup (\cup_i \mathbb{M}_i)$, where $\mathbb{B}$ corresponds to the non-motif buffer states padding right before and after the motif sequences and $\mathbb{M}_i$ corresponds to all possible sites within motif i. More specifically, we define: $\mathbb{M}_i \equiv \mathbb{M}_i^f \cup \mathbb{M}_i^r$, where $\mathbb{M}_i^f = \{1^{(i)} \ldots L_i^{(i)}\}$ is the set of all possible sites within motif i on the forward DNA strand, and $\mathbb{M}_i^r$ is the set of all possible sites within motif i if it is on the reverse complementary DNA strand.; $\mathbb{B} \equiv \mathbb{B}^p \cup \mathbb{B}^d$, where $\mathbb{B}^p = \{b_p^{(1)}, \ldots, b_p^{(K)}\}$ denotes the set of *proximal-buffer* states associated with each type of motif [2], and $\mathbb{B}^d = \{b_d^{(1)}, \ldots, b_d^{(K)}\}$ denotes the set of *distal-buffer* states associated with each type of motif.

The possible transitions between these states are made explicit by the arrows in the hierarchical state diagram in Fig. 1. (To make the hHMM model well-defined, we also introduce *dummy* states START and END at appropriate levels to enable instantiation of state-traversal, and proper termination of subsequences at each level.) The biological motivation for such a state hierarchy is that we expect to see occasional motif clusters in a large ocean of global background sequences (represented by state b_g); each motif instance in a cluster is like an island in a sea of intra-cluster background sequences (b_c); and adjacent motifs may be statistically coupled (we will elaborate on this point in the next section). Our model assumes that the distance between clusters is geometrically distributed with mean $1/(1 - \beta_{g,g})$, and the span of the intra-cluster background is also geometrically distributed with mean $1/(1 - \beta_{c,c})$. These modeling choices are intended to not only reflect our uncertainty about the CRM structure, but also to offer substantial flexibility to accommodate potential 1st-order syntactic characteristics within the CRMs. In this hHMM, only the bottom-level motif-site and motif-buffer states, as well as the global and local background states, are capable of emitting individual nucleotides constituting the TRS, according to a stochastic emission model (which we will elaborate later). A stochastic traversal of the hHMM states according to the hHMM state-transition diagram would generate a TRS of arbitrary length but with a structure consistent with our empirical knowledge of the functional organization of the metazoan TRS. Note that this hHMM model does not impose rigid constrains on the number of motif instances or CRMs; the actual number of instances is determined by the posterior distribution of the hHMM states given the observed sequence. Also note that we have not included functional states related to gene annotation and basic promoters, but such extensions are straightforward if co-identification of CRMs and genes is desired.

Given the observed sequences, and proper (i.e., biologically meaningful) construction of the state space and its hierarchical organization, one can infer the latent state-traversal path, which correspond to a plausible annotation or segmentation of the input sequence, using a number of exact posterior inference algorithms. The original algorithms given by [5] is a variant of the inside-outside algorithm for stochastic context free grammar, and takes $O(T^3 Q^D)$, where T is the length of the sequence, Q is the total number of states, and D is the depth of the hierarchy. A linear time algorithm was developed by [17] based on a transformation of hHMM into an equivalent dynamic Bayesian network. It is also possible to flatten the hHMM to an HMM with a block-structured sparse transition, and use a modified forward-backward algorithm for linear-time inference. In section 2.3 and

---

[2] Here, proximal-buffer refers to the background sites immediately next to the proximal-end of the motif. For consistency, orientations are defined with respect to the initial position of the input sequence. That is, the 1st position of the input sequence corresponds to the proximal end, and the last position corresponds to the distal end.

Supplementary Materials, we exploit this strategy, and develop an efficient algorithm for inference and learning under a Bayesian extension of hHMM to be described in the sequel.

Motif bigram via hHMM. An hHMM not only encodes hierarchical segmental structures in a sequence, but it can also be used to capture dependencies between sequence elements at different levels of granularity at a cost much smaller than that would be needed by a "flat" Markovian model which must resort to heavily parameterized high-order conditional probabilities. For example, we can capture the dependencies between neighboring CRMs in a TRS by modeling transitions between the CRM states. Of particular importance in this paper, we use hHMM to capture the dependencies between occurrences of motifs within a CRM. As discussed earlier, the spatial arrangement of motifs within a CRM may encode intricate combinatorial transcriptional regulatory signal. Thus modeling at least 1st-order dependencies between motifs may be beneficial to the unraveling of motifs in long TRS bearing complex regulatory function, as well-known in the case of Drosophila enhancers. Note that a direct transition between trivially defined motif states (e.g., last site of motif i and first site of motif j) would suggest that coupled motifs always occur right next to each other, which is biologically not always true. To capture possible dependencies between motifs in the vicinity of each other, we define the emission of a motif state (in $\mathbb{Q}^1$) to contain not only the motif sequence itself, but also non-motif sequences denoted as proximal and distal buffers. Such an emission can be understood as an extended instance of a motif, which we referred to as a *motif envelope*. Thus cross-background (i.e., high-order) dependencies between motifs can be captured by immediate (i.e., 1st-order) dependencies between the motif envelopes.

We write $A_2 \equiv \{a_{i,j}\}$ as the stochastic matrix for transitions among states in $\mathbb{Q}^2$, which defines a *bigram* of motifs (and their local backgrounds) within CRMs. The length of the proximal and distal buffers of a motif is geometrically distributed with mean $1/(1 - \alpha_{i,i})$ and $1/(1 - \beta_{i,i})$, and can be generated via self-transitions of the corresponding states at the third level (i.e., in $\mathbb{Q}^3$) with probability $\alpha_{i,i}$ and $\beta_{i,i}$, respectively. Then with equal probability $\alpha_{i,m}/2$, a proximal buffer state $b_p^{(i)}$ reaches the start states $1^{(i)}$ (resp. $L_i^{(i')}$) of motif i on the forward (resp. reverse) strand, deterministically passes through all internal sites of motif i, and transitions to the distal-buffer state $b_d^{(i)}$, thereby stochastically generating a non-empty motif envelope [3]. Each b_d^i has probability $\beta_{i,j}$ of transitioning to the proximal-buffer state of another motif j (or of the same motif when $j = i$) to concatenate another motif envelope, or it may choose to pad with some inter-cluster background before adding more envelopes, with probability $\beta_{i,c}$. All distal-buffer states also have probability $\beta_{i,g}$ of returning to the global background, terminating a CRM.

Spacer length distribution via GhHMM. A *spacer* is the interval seperating adjacent motif instances, modeled as b_c, b_p, and b_d states in BayCis. It has been suggested that the

---

[3] The distinction between proximal and distal buffers avoids generating empty envelops (otherwise, a single buffer state wont be able to remember if a motif has been generated beyond k positions prior to the current position under a k-th order Markov model).

range of spacer length is under selection forces according to comparative genomics data of several *Drosophila* species [13]. Empirically, we found that the distribution of spacer lengths can be approximated by a negative binomial distribution (see figure in Supplementary Materials), whereas under an hHMM, the state durations of cluster backgrounds is distributed as a goemetric distribution, which is not a good approximation of the space length distribution. In Supplementary Materials, we describe a generalized hierarchical hidden Markov model (GhHMM) which implements an approximate negative binomial distribution of spacer lengths by joining several geometrically distributed cluster background states.

The emission models: PWM and higher-order Markov background. Once the hHMM enters the motif-site states, we resort to a *motif model* to generate the nucleotides at the corresponding sites. To maintain our focus on the hHMM and relevant algorithmic issues, we only consider the scenario of searching for known motifs in this paper (although extending our model for *de novo* motif detection is straightforward based on, for example, the LOGOS framework [30]). For motif model we choose the classical product-multinomial (PM) model, which can be represented by a PWM [25].

Several previous studies have stressed the importance of using a richer background model for the non-motif sequences [26,11]. In accordance with these results, BayCis uses a standard global *k*-th order Markov model for the emission probability of the global background state. For the intra-CRM states, we used locally estimated Markov models. Since the models are defined to be *local*, the conditional probability of a nucleotide at position t is now estimated only from a window of length $2d$ centered at t. These probabilities can still be computed off-line and stored for subsequent uses, by using a careful bookkeeping scheme (i.e., using a "sliding-window" to compute the local Markov model of each successive position, each with a constant "update cost" based on the previous one).

2.2   Bayesian hHMM

One caveat of the standard HMM approach for CRM modeling is the difficulty of fitting the model parameters, such as the state-transition probabilities, due to rarity of fully annotated CRM-bearing genomic sequences. In principle, using the Baum-Welsh algorithm one can learn the maximal-likelihood (ML) estimates of the model parameters directly from the unannotated sequences while analyzing them. In practice, however, such a completely likelihood-driven approach tends to result in spurious results, such as overestimation of the motif and CRM frequencies and poor stringency of the learned models for potential motif patterns. Previous methods tried to overcome this by reducing the number of parameters needed as much as possible, and by setting them according to some good guesses of the motif/CRM frequencies or CRM sizes [7]. But as a result, such remedies compromise the expression power of the already simple HMM, and risk mis-representing the actual CRM structures. In the following, we propose a Bayesian approach that introduces the desired "soft constraints" and smoothing effect for an HMM of rich parameterization, using only a small number of *hyper-parameters*. This approach defines a posterior probability distribution of all possible value-assignments of the HMM parameters, given the observed un-annotated sequences and empirical prior distributions of the parameters that reflect general knowledge of CRM structures. The resulting model allows probabilistic queries (i.e., estimating the probability of a functional state) to be answered based on the aforementioned posterior distribution rather than on fixed given values of the HMM parameters.

We assume that the self-transition probability of the global background state $\beta_{g,g}$, and the total probability mass of transitioning into a motif-buffer state $\sum_{k \in B^p} \beta_{g,k}$ (note that $\beta_{g,g} = 1 - \sum_{k \in B^p} \beta_{g,k}$), admit a beta distribution, $Beta(\xi_{g,1}, \xi_{g,2})$. We choose a small value for $\frac{\xi_{g,2}}{\xi_{g,1} + \xi_{g,2}}$, corresponding to a prior expectation of a low CRM frequency. Similarly, we define a beta prior $Beta(\xi_{c,1}, \xi_{c,2})$ for the self- and total motif-buffer-going transition probabilities $[\beta_{c,c}, \sum_{k \in B^p} \beta_{c,k}]$ associated with the intra-cluster background state; and another beta prior $Beta(\xi_{p,1}, \xi_{p,2})$ for the self- and motif-going transition probabilities $[\alpha_{i,i}, \alpha_{i,m}]$ associated with the proximal-buffer state of a motif. Finally, we assume that for the distal-buffer state, the self-transition probability, the total mass of transition probabilities into a proximal-buffer state, the probability of transitioning into the intra-cluster background, and the probability of transitioning into the global background, $[\beta_{i,i}, \sum_{k \in B^p} \beta_{i,k}, \beta_{i,c}, \beta_{i,g}]$, admit a 4-dimensional gamma distribution, $Gamma(\xi_{d,1}, \xi_{d,2}, \xi_{d,3}, \xi_{d,4})$.

To define priors for the GhHMM parameters, the GhHMM with a single cluster background state (b_c) is considered as an HMM with several cluster background states $(\{b_c^1, \cdots, b_c^{q_{cr}}\})$ sharing the same self-transition probability $\beta_{c,c}$. Similar to other background states, we define a beta prior $Beta(\xi_{c,1}, \xi_{c,2})$ on the total probability mass of transitions into motif-buffer states $\sum_{k \in B^p} \beta_{c,k}$ (note that $\beta_{c,c} = \sum_{k \in B^p} \beta_{c,k}$).

Note that due to conjugacy between the prior distributions described above and the corresponding transition probabilities they model, the hyper-parameters of the above prior distributions can be understood as *pseudo-counts* of the corresponding transitioning events, which can be roughly specified according to empirical guesses of the motif and CRM frequencies. But unlike the standard HMM approach, of which the transition probabilities are fixed once specified, the hyper-parameters only lead to a soft enforcement of the empirical syntactic rules of CRM organization in terms of prior distributions, allowing controlled posterior update of the HMM transition probabilities while analyzing the un-annotated sequences. For the BayCis hHMM, we specify the hyper-parameters (i.e., the pseudo-counts) using estimated frequencies of the corresponding state-transition events, multiplied by a "prior strength" N, which corresponds to an imaginary "total number of events" from which the estimated frequencies are "derived". That is, for the beta priors, we let $[\xi_{[\cdot,1]}, \xi_{[\cdot,2]}] = [1 - \omega_{[\cdot]}, \omega_{[\cdot]}] \times N$, where the "·" in the subscript denotes either the g, c, or p state, and $\omega_{[\cdot]}$ is the corresponding frequency. For the gamma prior, we let $[\xi_{d,1}, \xi_{d,2}, \xi_{d,3}, \xi_{d,4}] = [\omega_{d,1}, 1 - \sum_j \omega_{d,j}, \omega_{d,2}, \omega_{d,3}] \times N$. Overall, we need to specify 7 hyper-parameters (of course one can use different "strengths" for different priors, with a few additional parameters), a modest increase compare to, say, 3 needed in Cister [7].

2.3 Inference and Learning

We have developed an efficient algorithm called *modified FB-algorithm* for inference on a "flattened" hHMM, which reduces the time complexity of the standard forward-backward algorithm from $O(K^2 \bar{L}^2 T)$ to $O(K^2 T)$. Identification of motifs/CRMs is based on posterior decoding. We also developed a *variational EM algorithm* for Bayesian inference and parameter estimation under our Bayesian hHMM and GhHMM, which is much more efficient than the traditional MCMC sampling approaches. Due to space limit, details of these algorithmic innovations are given in the Supplementary Materials.

3   Results

We evaluated BayCis on both synthetic transcriptional regulatory sequences and a rich set of carefully compiled real genomic TRSs of *Drosophila melanogaster* (available at our website). The prediction performance of BayCis was compared with 5 popular published methods for supervised discovery of motifs/CRMs based on a wide spectrum of models: Cister [7], Cluster-Buster [6], MSCAN [1], Ahab [19] and Stubb [24] (all of which were applied to the real data, and two seemingly superior ones to the semi-synthetic data), which cover a wide spectrum of different models/algorithms (e.g., HMMs, windows) for motif search. We ran other methods with default parameters, specifying 500 bp CRM window where needed.

   Overall, the prediction performance of BayCis is competitive or superior to all chosen benchmark methods on this quite comprehensive selection of data sets, according to a wide assortment of performance measures. By employing sound and flexible probabilistic modeling of regulatory regions, BayCis is also able to strike a good balance between precision and recall with its default MAP solution.

3.1   Semi-realistic Synthetic TRS

Synthetic TRSs are useful in that the ground truth for motif/CRM locations is known exactly. To generate semi-realistic synthetic TRSs, we planted selected TFBS from the Transfac [29] database in simulated background sequences according to model assumptions underlying the background distribution, the inter-TFBS and inter-CRM spacer length distributions for Baycis. 30 sequences of length 20,000 bp containing 0 - 3 CRMs were generated. The CRM length is uniformly distributed between 200 and 1600 bp, while the average motif spacer length is 50 bp. Each CRM contains 3 to 6 motif

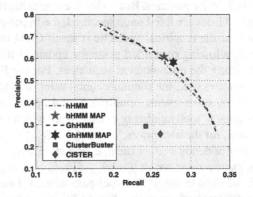

Fig. 2. The precision-recall (P/R) curves of two models of BayCis (hHMM and GhHMM) versus the P/R of default predictions by CISTER and ClusterBuster

types and about 14 motif instances. To simulate motif co-occurrence, about 25% of the motif instances in each CRM appear as predefined pairs. The background sequences inside/outside the CRM are simulated by a 3rd-order Markov model learnt from an intergenic region.

As shown in Fig. 2, the performance of BayCis using either hHMM or GhHMM is significantly better than CISTER and ClusterBuster in terms of the overall precision/recall (P/R) trade-off at the MAP prediction. The P/R curve of BayCis is also well above the default predictions from other methods. It also shows that GhHMM performs consistently better than hHMM in both precision and recall, although the difference is not very large. CISTER and ClusterBuster were chosen for the simulation study based on their good performance on real data (see next subsection).

3.2 Real Drosophila TRS

The dataset. The synthetic TRSs are generated partially based on the same model assumptions underlying BayCis, and thus the results cannot be interpreted as conclusive. A systematic investigation of the robustness of BayCis with respect to a wide spectrum of simulation conditions can be highly interesting but is beyond the scope of this short report; we will pursue this in a later full version of the paper. In this section we present an empirical evaluation based on a rich and carefully compiled Drosophila TRS dataset, although it is noteworthy that even

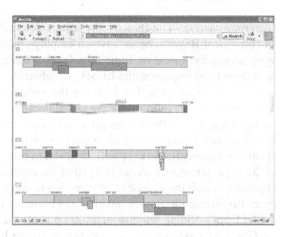

Fig. 3. Frontpage screenshot of the motif database

though we have tried our best to gather the most complete annotations for each test sequence based on footprinting results from the literature, this "gold standard" is still possibly only a subset of the ground truth.

We created a manually curated dataset containing 97 CRMs pertaining to 35 early developmental genes (see table in Supplementary Materials for details). This collection was compiled based on a filtering of all known CRMs from a number of public databases (e.g., the REDfly CRM database [8] and the Drosophila Cis-regulatory Database at the National University of Singapore [18]), through which we only chose CRMs that are at least 200 bp long, and contain at least 5 experimentally confirmed motif instances (2 CRMs with a borderline count of 4 motif instances were also included). Each test sequence consists of the CRMs pertinent to a particular gene, all intra-CRM background inbetween, with flanking regions on either side of the extremally located CRMs such that the entire sequence is at least 10 kbp long, and the boundaries of the sequence are at least 2 kbp from the extremal CRMs. We included the exonic regions of the genes only when they fell in the aforementioned selected region, and not otherwise. This database is available at http://www.sailing.cs.cmu.edu/BayCis, where the BayCis software will soon be also released. A snapshot of the interface of graphical interface of the database shown in Fig. 3, and more details are available in Supplementary Materials.

Experimental setup. BayCis is a Bayesian framework based on hHMMs and GhHMMs to model the organization and distribution of TFBS. Prior beliefs pertaining to the parameters of the model thus could be specified by the user before running on experimental data in the form of hyperparameters (i.e., pseudocounts) of the hHMM or GhHMM parameters. The PWMs of the motifs to be searched for also need to be provided because here we are interested in identifying TFBS of existing TF motifs, rather than *de novo* motif detection. As mentioned in previous sections, extending BayCis for this function is straightforward by introducing an EM step for the PWM estimation, and will be pursued in a later paper.

Hyperparameters: The choice of hyperparameters should in principle be dealt with via an "empirical Bayes scheme", which employs maximal likelihood estimates of these hyperparameters based on some fully labeled training sequences. Upon prediction on an unannotated sequence, the hHMM or GhHMM parameters themselves can be adjusted in an unsupervised fashion via the variational EM algorithm. We specify the hyperparameters as follows: for the global background, $\omega_g = 0.002$; for the inter-CRM background, $\omega_c = 0.05$; for the proximal motif buffer, $\omega_p = 0.25$; for the distal buffer hyperparameters, $\omega_{d,1} = 0.125$ (distal to global background), $\omega_{d,2} = 0.125$ (distal to clustal background), and $\omega_{d,3} = 0.25$ (distal to proximal buffer). Finally, the "strength" of the hyperparameters are set to $1/10$ of the expected counts of the transitions on a 15 kbp dataset, with the exception of ω_g which is set to $10,000$. The background probability of the nucleotide at each position was computed locally using a 2nd-order Markov model from a sliding window of 1100 bp centered at the corresponding position. For the GhHMM, based on visual inspection of spacer length distributions between motifs, we choose the parameter as $r = 2$.

Prediction scheme: BayCis provides three kinds of prediction schemes for motifs. The *maximum a posteriori* (MAP) prediction is based on the posterior probabilities of the labeling state at each site, which allows overlapping motifs. A Viterbi prediction, which gives a consistent prediction in the Bayesian setting analogous to an ML prediction under a classical setting can also be used. A third scheme is based on a simple but effective thresholding scheme where we directly predict motifs based on whether the motif states have a higher probability than the specified threshold in the posterior probabilities. For simplicity, in this paper we only present the MAP results and the P/R curve of the threshold method. Note that unlike many other scoring schemes for motif/CRM detection, such as logodds (i.e., the PSSM score) or a likelihood score regularized by word frequencies, our MAP prediction does not require a cutoff value for the scores, nor a window to measure the local concentration of motif instances, both of which are difficult to set optimally.

Evaluation measures: There is no unanimous way of evaluating the prediction performance of a motif/CRM discovery method against annotations. To avoid reliance on a single evaluation procedure and measure, we have chosen to present the performance

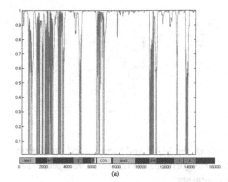
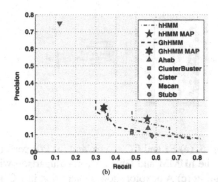

(a)   (b)

Fig. 4. Performance of BayCis (hHMM) on a representative *eve* TRS. (a) The posterior probability plot of the global background (blue), cluster background (green) and motif specific (red and other colors) states. (b) The precision versus recall performance of the MAP and thresholded predictions of the hHMM and GhHMM algorithms, as compared to those made by other methods.

of BayCis in comparison with other methods using several different evaluation proce
uures. I nis also ensures a thorough and objective presentation of results. Pur an overall
evaluation we compare the prediction performance of BayCis with other methods using both the F1-score of precision and recall, and the coefficient of correlation (CC) score at nucleotide-level [28] as single point measures (see Supplementary Materials B.3 for detailed definitions). We do this by first summing true/false positives/negatives across datasets at the nucleotide level, and then computing F1/CC from these combined counts. To present the behavior of BayCis with respect to site-level P/R, we plot the binding-site level P/R curve from different thresholds in extracting predictions, along with the P/R at MAP predictions.

Motif prediction performance. As an illustration, Fig. 4a shows a plot of the MAP prediction along the *even-skipped* gene TRS, under a particular hyperparameter setting. As revealed in the ground-truth annotation bar bellow the plot, this region contains 5 CRMs (from left to right): *stripe3+7*, *stripe2*, *stripe4+6*, *stripe1*, and *stripe5*. BayCis picks out all of them, although the CRM boundary appears to be more stringent in most cases. We believe this can be improved by adopting a more specialized cluster background model (i.e., local higher-Markov model, better GhHMM model, etc.), which we have not fully explored yet. BayCis also identifies motif-rich regions proximal and distal to the *stripe3+7* CRM, which is not reported before, and it also finds another putative motif-rich region spanning the core promoter and the CDS of *eve*, which can be a false positive or a putative CRM. The overall MAP prediction score of BayCis, and the P/R curves resulted from applying different threshold values under BayCis, are shown in Fig. 4b, along with the scores of 5 other competing algorithms in their default configurations. The BayCis MAP predictions seem significantly better than other methods, and strike a good balance between recall and precision. It is important to realize that although the threshold method can reach high precision or recall at both extremes, in practice it is very hard to pick the optimal threshold without knowing the prediction results, and typically a threshold optimal for one sequence is not necessarily good for

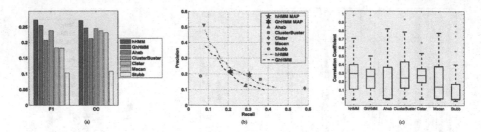

Fig. 5. (a) F1 and CC scores, and (b) P/R performances of the MAP and thresholded predictions of the hHMM and GhHMM, in comparison with other algorithms on the full *Drosophila* TRS dataset (c) A boxplot showing variation in CC across datasets

another sequence; significance-test based determination of threshold is also difficult for a complex model or large sequence. Thus, a default prediction such as MAP, which automatically finds an appropriate trade-off between precision and recall, is highly desirable.

The overall CC and F1-scores of running BayCis and five competing methods on the full set of *Drosophila melanogaster* sequences are shown in Fig. 5a. According to either measure, both the hHMM and the GhHMM version of BayCis outperforms all existing methods. The hHMM version of BayCis performs slightly better overall compared to GhHMM according to both measures. For both versions of BayCis, the MAP solution was chosen.

To look at the behavior of BayCis in the P/R landscape on our entire dataset, we plot the P/R curve resulting from different thresholds for BayCis predictions. For other methods we provide the single points in P/R landscape corresponding to their default output. As is apparent from Fig. 5b, the 5 competing methods strike different balances between precision and recall in their default output. MSCAN focuses on very high precision predictions, while Cister is geared towards high values of recall. The P/R curves of both versions of BayCis span a balanced range in the P/R landscape, with MAP estimates lying in the middle of the curves. Again, in practice the P/R values are not available for use by methods, so the balance between precision and recall has to be found based solely on the input data. Thus the ability to appropriately balance precision and recall automatically is essential.

To further investigate the prediction performance, we look at the variation of individual dataset prediction performance across all datasets. The boxplot in Fig. 5(c) shows the median CC-score for each method, as well as upper and lower quartiles and minimum/maximum values. We see that prediction scores varies much between datasets for all methods, and that the overall performance differences between methods is not very large compared to the variation of individual methods across datasets. This confirms what has long been acknowledged in the motif discovery field, that even the best performing methods will in many cases give misleading predictions (although some of the low scores may be due to lack of annotations). Among the high scoring methods (hHMM, GhHMM, Cluster-Buster and Cister), GhHMM and Cister come out as the most stable with low variance across datasets, a criterion which is useful when handling a varied set of data. The posterior expectations of the hHMM/GhHMM parameters also carry rich architectural information of each TRS we processed, and merits systematic analyses. We defer this investigation to the full paper.

4 Discussion

BayCis uses an advanced probabilistic framework to accurately model metazoan transcriptional regulatory genomic sequences — which often consist of multiple CRMs, tandemly joined by long stretches of background DNA, each containing locally enriched occurrences of binding motifs for a certain array of transcriptional regulatory proteins. Thus, we are able to detect many TFBS while avoiding too many false positives and (slightly) outperform the best of the existing methods on a comprehensive set of *Drosophila* regulatory regions. The BayCis software will soon be released on our website.

Recently, experimental results have shown that sequences immediately flanking a TFBS may contribute to the binding energy between a TF and the TFBS [14]. This suggests that sequence composition of the proximal and distal buffers of motifs may have weak type specificity, which we would like to explore in our future work. Our current TRS database for performance evaluation is still limited in size and very diverse in terms of CRM structures and complexity, which could cause BayCis to overfit certain TRS when it is applied independently to each TRS separately (as we did in this paper), using a generic set of hyperparameters that are empirically chosen. We intend to adopt a more systematic approach to fit the hyperparameters based on a small amount of labeled TRS, e.g., using a k-fold cross validation scheme. But ultimately, we believe additional TRS data will be needed to attain further performance increase. One direction of increasing input data is to combine regulatory regions of several genes that are believed to share similar CRM structure. Such gene sets should be attainable for many real scenarios where CRM discovery methods are used, could trivially be used as input to BayCis. We speculate that this could improve predictions. The limitation lies mostly in collecting such gene sets containing rich, high-quality annotations that could serve in quantitatively measuring correspondence between computational prediction and experimental determination.

Another direction is to conjoin BayCis with a phylogenetic model of motifs across species [16,22,23], and apply the integrant to orthologous TRSs. Although this limits the applicability of the approach to species where valuable orthologous sequence is available, and to the discovery of regulatory elements shared between species, we believe it could attain considerably performance gain in the cases for which it is suited.

Acknowledgements. This material is based on work supported by the Pennsylvania Dept of Health's Health Research Program under Grant No. 2001NF-Cancer Health Research Grant ME-01-739, and by an NSF CAREER Award under Grant No. DBI-054659. The authors thank Wenjie Fu for result analysis, Jostein Johansen for help with evaluating CRM predictions, and Oznur Tastan for investigating the spacer length distributions.

References

1. Alkema, W.B., Johansson, O., Lagergren, J., Wasserman, W.W.: Mscan: identification of functional clusters of transcription factor binding sites. Nucleic Acids Res. 32(Web Server issue), 195–198 (2004)
2. Berman, B.P., Nibu, Y., Pfeiffer, B.D., Tomancak, P., Celniker, S.E., Levine, M., Rubin, G.M., Eisen, M.: Exploiting transcription factor binding site clustering to identify cis-regulatory modules involved in pattern formation in the Drosophila genome. Proc. Natl. Acad. Sci. USA 99(2), 757–762 (2002)
3. Davidson, E.H.: Genomic Regulatory Systems. Academic Press, London (2001)
4. Donaldson, I.J., Chapman, M., Gottgens, B.: Tfbscluster: a resource for the characterization of transcriptional regulatory networks. Bioinformatics 21(13), 3058–3059 (2005)
5. Fine, S., Singer, Y., Tishby, N.: The hierarchical hidden Markov model: Analysis and applications. Mach Learning 32, 41–62 (1998)
6. Frith, M., Li, M., Weng, Z.: Clusterbuster:finding dense clusters of motifs in dna seqs. Nuc. Ac. Res. 31(13), 3666–3668 (2003)
7. Frith, M.C., Hansen, U., Weng, Z.: Detection of cis-element clusters in higher eukaryotic DNA. Bioinf. 17, 878–889 (2001)
8. Gallo, S., Li, L., Hu, Z., Halfon, M.: Redfly:a regulatory element database for drosophila. Bioinf. 22(3), 381–383 (2006)
9. Gupta, M., Liu, J.S.: De novo cis-regulatory module elicitation for eukaryotic genomes. Proc. Natl. Acad. Sci. USA 102(20), 7079–7084 (2005)
10. Huang, H., Kao, M., Zhou, X., Liu, J.S., Wong, W.H.: Determination of local statistical significance of patterns in Markov sequences with application to promoter element identification. Journal of Computational Biology 11(1) (2004)
11. Liu, X., Brutlag, D.L., Liu, J.: Bioprospector: Discovering conserved DNA motifs in upstream regulatory regions of co-expressed genes. Proc. of Pac. Symp. Biocomput., 127–138 (2001)
12. Loots, G.G., Ovcharenko, I., Pachter, L., Dubchak, I., Rubin, E.M.: rVista for comparative sequence-based discovery of functional transcription factor binding sites. Genome Res. 12(5), 832–839 (2002)
13. Ludwig, M.Z., Patel, N.H., Kreitman, M.: Functional analysis of eve stripe 2 enhancer evolution in Drosophila: rules governing conservation and change. Development 125(5), 949–958 (1998)
14. Maerkl, S.J., Quake, S.R.: A systems approach to measuring the binding energy landscapes of transcription factors. Science 315, 233–237 (2007)
15. Michelson, A.: Deciphering genetic regulatory codes:a challenge for fnal genomics. Pr. Nat. Acad. Sc. USA 99, 546–548 (2002)
16. Moses, A.M., Chiang, D.Y., Eisen, M.B.: Phylogenetic motif detection by expectation-maximization on evolutionary mixtures. Pac. Symp. Biocomput., 324–335 (2004)
17. Murphy, K., Paskin, M.: Linear time inference in hierarchical hmms. Adv. in Neural Inf. Proc. Sys. 14 (2002)
18. Narang, V., Sung, W.K., Mittal, A.: Computational annotation of transcription factor binding sites in D melanogaster developmental genes. In: Proceedings of The 17th International Conference on Genome Informatics (2006)
19. Rajewsky, N., Vergassola, M., Gaul, U., Siggia, E.D.: Computational detection of genomic cis-regulatory modules, applied to body patterning in the early Drosophila embryo. BMC Bioinformatics 3(30), 1–13 (2002)
20. Rebeiz, M., Reeves, N.L., Posakony, J.W.: Score: a computational approach to the identification of cis-regulatory modules and target genes in whole-genome sequence data site clustering over random expectation. Proc. Natl. Acad. Sci. USA 99(15), 9888–9893 (2002)

21. Sharan, R., Ovcharenko, I., Ben-Hur, A., Karp, R.M.: Creme: a framework for identifying cis-regulatory modules in human-mouse conserved segments. Bioinformatics 19(Suppl 1), i283–291 (2003)
22. Siddharthan, R., Siggia, E.D., van Nimwegen, E.: Phylogibbs: A gibbs sampling motif finder that incorporates phylogeny. PLoS Computational Biology 1(7), e67 (2005)
23. Sinha, S., Blanchette, B., Tompa, M.: Phyme: A probabilistic algorithm for finding motifs in sets of orthologous sequences. BMC Bioinformatics 5(170) (2004)
24. Sinha, S., Liang, Y., Siggia, E.: Stubb: a program for discovery and analysis of cis-regulatory modules. Nucleic Acids Res. 34(Web Server issue), W555–W559 (2006)
25. Staden, R.: Computer methods to locate signals in nucleic acid sequences. Nucleic Acids Res. 12(1 Pt 2), 505–519 (1984)
26. Thijs, G., Lescot, M., Marchal, K., Rombauts, S., DeMoor, B., Rouze, P., Moreau, Y.: A higher-order background model improves the detection of promoter regulatory elements by gibbs sampling. Bioinformatics 17(12), 1113–1122 (2001)
27. Thompson, W., Palumbo, M.J., Wasserman, W.W., Liu, J.S., Lawrence, T.E.: Decoding human regulatory circuits. Genome Res. 14(10A), 1967–1974 (2004)
28. Tompa, M., Li, N., Bailey, T., Church, G., DeMoor, B., Eskin, E., Favorov, A., Frith, M., Fu, Y., Kent, W., Makeev, V., Mironov, A., Noble, A., Pavesi, G., Pesole, G., Regnier, M., Simonis, N., Sinha, S., Thijs, G., van Helden, J., Vandenbogaert, M., Weng, Z., Workman, C., Ye, C., Zhu, Z.: Assessing computational tools for the discovery of transcription factor binding sites. Nat. Biotech. 23(1), 137–144 (2005)
29. Wingender, E., Dietze, P., Karas, H., Knuppel, R.: TRANSFAC: a database on transcription factors and their DNA binding sites. Nucleic. Acids. Res. 24(1), 238–241 (1996)
30. Xing, E.P., Wu, W., Jordan, M.I., Karp, R.M.: Logos: A modular Bayesian model for de novo motif detection. Journal of Bioinformatics and Computational Biology 2(1), 127–154 (2004)
31. Zhou, Q., Wong, W.H.: Cismodule: de novo discovery of cis-regulatory modules by hierarchical mixture modeling. Proc. Natl. Acad. Sci. USA 101(33), 12114–12119 (2004)

A Combined Expression-Interaction Model for Inferring the Temporal Activity of Transcription Factors

Yanxin Shi[1], Itamar Simon[2], Tom Mitchell[1], and Ziv Bar-Joseph[1,*]

[1] School of Computer Science, Carnegie Mellon University, Pittsburgh, PA, 15213
zivbj@cs.cmu.edu
[2] Dept of Molecular Biology, Hebrew University Medical School, Jerusalem, Israel

Abstract. Methods suggested for reconstructing regulatory networks can be divided into two sets based on how the activity level of transcription factors (TFs) is inferred. The first group of methods relies on the expression levels of TFs assuming that the activity of a TF is highly correlated with its mRNA abundance. The second treats the activity level as unobserved and infers it from the expression of the genes the TF regulates. While both types of methods were successfully applied, each suffers from drawbacks that limit their accuracy. For the first set, the assumption that mRNA levels are correlated with activity is violated for many TFs due to post-transcriptional modifications. For the second, the expression level of a TF which might be informative is completely ignored. Here we present the Post-Transcriptional Modification Model (PTMM) that unlike previous methods utilizes both sources of data concurrently. Our method uses a switching model to determine whether a TF is transcriptionally or post-transcriptionally regulated. This model is combined with a factorial HMM to fully reconstruct the interactions in a dynamic regulatory network. Using simulated and real data we show that PTMM outperforms the other two approaches discussed above. Using real data we also show that PTMM can recover meaningful TF activity levels and identify post-transcriptionally modified TFs, many of which are supported by other sources.

1 Introduction

Transcriptional gene regulation is a dynamic process which utilizes a network of interactions. This process is primarily controlled by transcription factors (TFs) that bind DNA and activate or repress sets of genes. Regulatory networks activate hundreds of genes as part of a biological system such as the cell cycle [1] and circadian rhythm [2], in response to internal and external stimuli [3] and during development [4]. Proper functioning of these networks is essential for all living organisms. For example, several diseases are associated with partial or complete loss of appropriate transcriptional regulation [5]. Determining accurate models for these regulatory networks is thus an important challenge.

---

* Corresponding author.

M. Vingron and L. Wong (Eds.): RECOMB 2008, LNBI 4955, pp. 82–97, 2008.
© Springer-Verlag Berlin Heidelberg 2008

A major source of information regarding these networks is gene expression data which measures the effects TFs have on their regulated targets. Many methods have been suggested for using this and other data sources for reconstructing regulatory networks. One of the key challenges faced by methods aimed at reconstructing such networks is to infer the activity levels of the factors regulating the network. While the activity levels for some TFs can be determined by looking at their expression levels, many of the master TFs are post-transcriptionally regulated and can be active even if their expression levels do not change [6].

So far, methods suggested for reconstructing regulatory networks can be divided into two major groups based on how they infer the activity levels of the TFs. The first set of methods (e.g., [7,8,9,10]) relies on the mRNA levels measured for TFs and uses these to represent the activity levels of TFs. The second group of methods (e.g., [6,11,12]) treats the activity levels of TFs as completely unobserved values and infers them from the mRNA levels of their regulated genes.

While both methods have proven useful for many different reconstruction efforts, each suffers from drawbacks that can limit their ability to accurately reconstruct the networks. The first set of methods is less appropriate in cases where TFs are post-transcriptionally modified, which may lead to activity levels that are not reflected in the mRNA levels measured for these TFs [13]. The second group of methods overcomes this problem but does not take advantage of the information from the mRNA levels of the TFs. There are many cases in which TFs' activities are reflected in their mRNA levels [10] and ignoring these levels may reduce the ability to correctly model the activity levels of these TFs.

A possible way to combine the two approaches is to measure protein levels in addition to gene expression levels [14]. However, this data cannot account for other post-transcriptional events including phosphorylation and nuclear exclusion [15]. In addition, it requires proteomics measurements that are not always available and have a limited ability to identify low abundance proteins [16].

Another approach was proposed by Nachman et al. [17] infers regulatory networks from expression data using a dynamic Bayesian network. They model the unobserved TF activity levels by hidden variables and then use a post-processing step to link these levels to known TFs based on their expression levels. Thus, this model is a variant of the second set of methods discussed above since the TF expression levels are not used when reconstructing the network.

Here we present an algorithm combining the two types of methods mentioned above during the reconstruction phase to get the 'best of both worlds': For transcriptionally regulated TFs, we infer their activity levels from their mRNA levels, and for post-transcriptionally regulated TFs, we rely on the mRNA levels of their target genes. The key insight we make is that when using time series expression data we can compare the mRNA levels of a TF with the expression levels of genes regulated by the TF in consecutive time points in order to determine whether the mRNA levels correlate with the activity levels for that TF. This allows us to determine which TFs are post-transcriptionally modified and which are not. For this we develop the Post-Transcriptional Modification Model (PTMM), a variant of factorial hidden Markov model [18] that accounts for the factor-specific correlation between

TF's mRNA levels and activity levels. For each TF we maintain a binary indicator representing whether or not this TF is post-transcriptionally modified. If a TF is post-transcriptionally modified, we treat its activity levels as unobserved variables whose values can be inferred from the observed expression levels of genes regulated by this TF using a Kalman filter based model. If a TF is not post-transcriptionally modified, we use its expression levels as prior and combine them with the expression levels of its target genes to infer its posterior activity. This posterior accounts for the noisy measurement of the TF's expression levels which might lead to slightly different protein activity levels.

We tested our method and compared it to methods that rely only on target genes or on TF's mRNA levels. Using simulated and real expression data we show that our method has higher accuracy in detecting the post-transcriptional modification events, in inferring the hidden activity levels of TFs and in predicting the regulatory relationships between TFs and genes. We also discovered some candidate post-transcriptionally modified TFs, which are validated by other sources.

2 Methods

Here we introduce the Post-Transcriptional Modification Model (PTMM) which combines time series expression data from multiple experimental conditions and static TF-gene interaction data to reconstruct temporal regulatory networks and to infer whether a TF is post-transcriptionally modified. We also introduce an associated EM algorithm to, i) learn the parameters of a PTMM, ii) infer whether or not each TF is post-transcriptionally modified and iii) infer the hidden activity levels of each TF. In addition to inferring TF activity levels PTMM can also determine new TF-gene interactions. For each known or inferred interaction the learned model assigns a condition independent weight between a TF and the genes it regulates representing the strength of this interaction.

2.1 Post-Transcriptional Modification Model (PTMM)

Let m be the number of a set of genes whose expression level is measured at a series of time points under a variety of experimental conditions (datasets). Let n represent the number of a subset of these m genes that are TFs. A PTMM defines a joint probability distribution over an observed time series of gene expression levels, unobserved time series of TF activity levels and the unobserved post-transcriptional status for each TF (modified or unmodified). We use PTMM to estimate which TFs are post-transcriptionally modified, to infer the hidden activity levels of TFs over time, to determine which genes are regulated by each TFs and to assign a weight to these regulatory interactions.

Let $G_{i,d,t}$ represent the expression level of gene i ($1 \leq i \leq m$) in dataset d at time t. Without loss of generality, we assume that the first n ($n < m$) genes encode for TFs. Let $T_{j,d,t}$ denote the (hidden) activity level of TF j (the protein product of gene j) in dataset d at time t. Each gene may be regulated by zero or several TFs. Let $w_{i,j}$ denote the weight with which gene i is regulated by TF j. A positive weight means that TF j is an activator of gene i, a negative weight implies that TF j represses gene i. A weight of zero indicates that gene i is not regulated by TF j. Similar to other methods (e.g., [19]), PTMM models the observed expression level for gene i at each time point t as the linear superposition of contributions from each of the TFs that regulates this gene. More precisely:

$$G_{i,d,t}|T_{:,d,t} \sim \begin{cases} \mathcal{N}(0 , \alpha_d^2) & \text{if gene i is not regulated by any TF in hand} \\ \mathcal{N}(\sum_{j=1}^n w_{i,j} T_{j,d,t} , \beta_d^2) & \text{otherwise} \end{cases} \tag{1}$$

where $\mathcal{N}(\mu, \sigma^2)$ represents a Gaussian distribution with mean μ and variance σ^2. In Equation 1, the expression profile of a gene over time is a noisy realization of the weighted sum of activity profiles of the TFs which regulate this gene. At each time point the expression level is modelled using a Gaussian distribution whose variance is either α_d^2 or β_d^2, depending on whether the gene is believed to be regulated by at least one TF. If it is regulated by at least one TF, then the variance β_d^2 is used to represent experimental measurement noise. If PTMM cannot assign a regulator to a gene it may be the case that the gene is regulated by TF(s) that is/are not included in the model. These genes are assumed to have a higher variance since some of their variance can be attributed to deficiencies in the model. Thus we use a different variance, α_d^2, for these genes.

For each TF j, we maintain a global binary indicator Z_j independent of experimental conditions and constant over time indicating whether this TF is post-transcriptionally modified. Z_j is a random variable following a Bernoulli distribution with parameter ρ. We treat ρ as a pre-specified constant representing the proportion of TFs that are post-transcriptionally modified. Based on this indicator, we assume that each TF follows one of these two models: i) If TF j is not post-transcriptionally modified, i.e., $Z_j = 0$, we model the activity profile of TF j as a noisy realization of its gene's expression profile with one time point lag (Figure 1(a)), i.e., $T_{j,d,t}|G_{j,d,t-1} \sim \mathcal{N}(G_{j,d,t-1}, \tau_d^2)$. τ_d^2 represents the possible experimental noise which may lead to slight differences between TF activity levels and mRNA levels. The one time point lag accounts for the time of translation from mRNA to protein. It also makes the model computationally sound preventing possible loops in the time slice model (allowing, for example, self-regulation by TFs). The first time point in each dataset is modelled by a Gaussian distribution with zero mean and variance σ_d^2. ii) The second option is that the TF j is post-transcriptionally modified, i.e., $Z_j = 1$. For these TFs the change in activity levels over time is modelled as a hidden Markov chain (Figure 1(b)). The activity level of the TF at time point t (i.e. $T_{j,d,t}$) is dependent on the activity level of this same TF at time point $t-1$ (i.e. $T_{j,d,t-1}$). This dependency is modelled as a Gaussian random walk, i.e., $T_{j,d,t}|T_{j,d,t-1} \sim \mathcal{N}(T_{j,d,t-1}, \gamma_d^2)$. The variance γ_d^2 determines the likely amount of change in the TF's activity level between consecutive time points. The activity level of each TF at the very first time point in dataset d is modelled by a Gaussian distribution with mean 0 and variance σ_d^2. This dataset-specific variance allows integrating multiple datasets in which the activity levels at the first time point for some TFs may differ from 0, e.g., cell cycle experiments. Figure 1(c) presents the full graphical model of a PTMM, using indicator variables Z_j to select between the two cases.

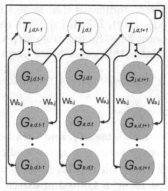

(a) no PTM case, TF j

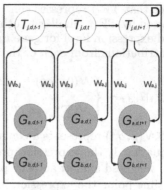

(b) PTM case, TF j

Fig. 1. Graphical model representations for (a) TFs with no post-transcriptional modification ($Z_j = 0$), (b) Post-transcriptionally modified TFs ($Z_j = 1$), and (c) The general case (complete PTMM model). Observed variables are shaded. $T_{j,d,t}$ is the (hidden) activity level of TF j at time point t in dataset d. $G_{i,d,t}$ is the observed expression level for gene i at time point t in dataset d. The edge from TF j to gene i exists if and only if gene i is regulated by TF j, i.e. $w_{i,j} \neq 0$, where $w_{i,j}$ represents the weight of each edge. The edge from gene j to its protein product, TF j, exists when there is no post-transcriptional modification for TF j. Each TF j has a global binary indicator variable Z_j indicating whether the TF is post-transcriptionally modified. D plates correspond to the D datasets.

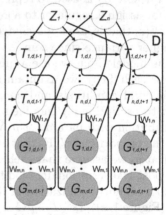

(c) overall case, all TFs

Note that within a dataset, the expression noise parameters α_d^2, β_d^2 and τ_d^2 are shared across genes/TFs, and the TF activity level smoothness term γ_d^2 is shared across TFs. We estimate different noise parameters for each dataset d, to allow for the possibility that noise levels may differ across datasets from different labs using different array platforms. On the other hand, we assume the regulation between gene i and TF j is independent of experimental conditions. That is, the weight parameters $w_{i,j}$ are shared across all datasets.

2.2 Penalized Likelihood Score

Given a set of TFs, a set of genes, and a collection of gene expression datasets, we train the PTMM by inferring which TFs are post-transcriptionally modified, the TFs activity levels, which genes are regulated by each TF, and by estimating the various PTMM parameters $w_{i,j}$, α_d, β_d, τ_d, γ_d and σ_d. These estimates are chosen to maximize a penalized complete log-likelihood score subject to the constraint that any gene be regulated by at most C TFs (i.e., it has at most C incoming edges). This constraint is motivated by the fact that recent high throughput studies find that most genes are regulated by only a few TFs [20]. The constrained penalized log-likelihood score is $Score(\mathbf{o}, \mathbf{h}, \mathbf{z} : W, \theta)$ where $\mathbf{o}, \mathbf{h}, \mathbf{z}$ represent observed gene expression levels, hidden TF activity levels and unobserved post-transcriptional modification (PTM) indicators, respectively, and θ includes all model parameters other than the regulation weights W.

$$Score(\mathbf{o}, \mathbf{h}, \mathbf{z} : W, \theta) = \log(P(\mathbf{z})) + \sum_{d=1}^{D} \log(P(\mathbf{o_d}, \mathbf{h_d} | \mathbf{z}, W, \theta)) - \lambda_1 \sum_{i=1}^{m} \sum_{j=1}^{n} |w_{i,j}|$$

$$- \lambda_2 \Big\{ \sum_{i=1}^{m} \sum_{j=1}^{n} \delta(w_{i,j} \neq 0) \big[E_{i,j} \pi_1 + (1 - E_{i,j}) \pi_0 \big]$$

$$+ \sum_{i=1}^{m} \sum_{j=1}^{n} \delta(w_{i,j} = 0) \big[E_{i,j} \pi_0 + (1 - E_{i,j}) \pi_1 \big] \Big\}$$

$$subject\ to: \quad \big(\big| \{ w_{i,j} | w_{i,j} \neq 0, 1 \leq j \leq n \} \big| \leq C \big) \quad for\ all\ i \qquad (2)$$

Here $\mathbf{o_d}$ and $\mathbf{h_d}$ are the observed expression levels for genes and the unobserved activity levels for TFs in dataset d, respectively. The score contains two regularization terms. The first imposes an L_1 penalty on the weights, encouraging most TF-gene regulation weights to be zero [21]. The second term incorporates prior knowledge from binding experiments. $E_{i,j}$ is a binary indicator which is 1 if gene i is thought a priori to be bound by TF j and 0 otherwise. $\delta(\cdot)$ is 1 if $\cdot$ is true, and 0 otherwise. π_0 is a penalty term paid when the model selects a TF-gene edge weight that is inconsistent with prior assumptions (including an edge that is not assumed a priori, or excluding one that is). π_1 is a smaller penalty term for using edges that are supported by prior assumptions from binding experiments, and for dropping edges that are not supported by binding experiments. Since we use highly trusted binding data to form our prior assumptions, we set $\pi_0 >> \pi_1$. Thus, the learned model is encouraged to assign $w_{i,j}$ weights consistent with prior knowledge, though it may depart from these priors if the incurred penalty is offset by improved data likelihood. Such departures might result from incompleteness or noise in prior binding datasets, or from the fact that only a subset of bound TFs may affect a target gene's expression [22]. π_1 and π_0 are user defined and indicate confidence in the prior assumptions regarding the binding data. λ_1 and λ_2 are constants representing the tradeoff between likelihood and regularization terms, which can be used to control the tradeoff between precision and recall in predicting TF-gene regulatory relationships.

2.3   Inference and Learning for PTMM

To learn the PTMM we use an approximate EM algorithm to attempt to maximize $Score(\mathbf{o}, \mathbf{h}, \mathbf{z} : W, \theta)$. The algorithm iteratively performs an E step in which the current model parameters W and θ are used to calculate the expected values of the hidden TF activity levels $\mathbf{h}$ and the most likely values of the unobserved

PTM indicators **z**, followed by an M step in which these activity levels **h** and indicators **z** are used to re-estimate the model parameters W and θ. These two steps are iterated until convergence.

E step: Given all model parameters, we employ a generalized mean field algorithm [23] to iteratively infer the PTM indicators **z** and the hidden activity levels **h** of TFs. This variational inference method is based on non-overlapping clusters of random variables. Specifically in the PTMM, the hidden chain of activity levels for each TF forms a cluster and each PTM indicator forms an additional cluster. The E step iterates until convergence by inferring values for one cluster assuming the current values for all other clusters.

To infer the hidden activity levels for TF j given the most likely assignment of the PTM indicators **z** and expected activity levels for all other TFs, we first examine the current assignment of Z_j: i) If $Z_j = 0$, i.e., TF j is not post-transcriptionally modified, we can compute the posterior distribution of $T_{j,d,t}$ independently for each time point t in each dataset d. In this case the prior of $T_{j,d,t}$ is a Gaussian distribution whose mean is the expression level of its corresponding gene at time point $t - 1$ in dataset d, (i.e., $G_{j,d,t-1}$) and whose variance is τ_d^2. The posterior of $T_{j,d,t}$ depends on the observed expression levels of genes regulated by TF j as well as the hidden activity levels of other TFs which have overlapping target genes due to the v-structure in directed graphical model [24]. Since we have fixed the activity levels for other TFs while inferring the level for TF j, we subtract out their assumed contributions to the observed expression levels of the target genes for TF j, in order to estimate the contribution due solely to TF j:

$$\widetilde{G}_{i,d,t} = G_{i,d,t} - \sum_{k \neq j} w_{i,k} T_{k,d,t} \tag{3}$$

where $\widetilde{G}_{i,d,t}$ is the *adjusted expression level* which represents the inferred contribution of TF j to the expression level for gene i at time point t in dataset d. The posterior of $T_{j,d,t}$ depends only on these *adjusted expression levels*. This posterior is a Gaussian distribution due to the conjugacy of Gaussian distribution to itself and it is straightforward to obtain the posterior mean and variance. ii) If $Z_j = 1$, i.e., TF j is post-transcriptionally modified, we first calculate the *adjusted expression levels* for the genes regulated by TF j. Given these *adjusted expression levels* the posterior for the activity level of TF j is no longer dependent on the activity levels of other TFs and the resulting model is equivalent to a single hidden Markov chain for TF j regulating multiple genes with their *adjusted expression levels*. Let $\overrightarrow{\widetilde{G}}_{d,t}$ denote the m-dimensional column vector for *adjusted expression levels* of all genes in dataset d at time point t. We can then write for TF j:

$$T_{j,d,t} = T_{j,d,t-1} + Q_{j,d,t}, \quad where \quad Q_{j,d,t} \sim \mathcal{N}(0, \gamma_d^2); \tag{4}$$

$$\overrightarrow{\widetilde{G}}_{d,t} = W_{:,j} \times T_{j,d,t} + R_{d,t}, \quad where \quad R_{d,t} \sim \mathcal{N}(\mathbf{0}, \Sigma_{R_d}); \tag{5}$$

where W is the m-by-n regulation weight matrix (0 indicates no edge) and $W_{:,j}$ represents the jth column of this matrix corresponding to the regulation weights associated with TF j. Here γ_d^2 determines the probable rate of change of the TF activities over time (i.e., $Q_{j,d,t}$). Σ_{R_d} is a m-by-m diagonal matrix where the ith

diagonal element is α_d^2 if $w_{i,j} = 0$ for all j and β_d^2 otherwise. It determines the variance in the noise in the observed expression levels (i.e., $R_{d,t}$).

As Equation 4 and 5 show, for post-transcriptionally modified TFs when the parameters are known the model reduces to a special case of Kalman filter [25] model with one-dimensional hidden chain. Inference on it can be done efficiently by computing the posterior of hidden variables $T_{j,d,t}$. The probabilities are all Gaussian distributed and the computation is tractable because of the conjugacy of the Gaussian distribution.

Inferring the unobserved assignment of the PTM indicator Z_j of TF j is in fact a model selection process. Given the inferred expected values of activity levels of TF j, we examine which model explains this TF better. That is, if the no PTM model (Figure 1(a)) has a higher likelihood than the PTM model (Figure 1(b)) we assign Z_j as 0. Otherwise, we assign Z_j as 1. This likelihood can be easily computed as the product of the local conditional probabilities of all nodes associated with TF j. Along with the assignment of each indicator, we can also output a confidence score for this assignment which we define as the log ratio between the likelihood scores of two different models.

Before the iterations in the E step, we need to initialize the hidden variables in the PTMM. The activity levels of TFs are initialized by a standard Kalman filter model assuming all TFs are post transcriptionally modified, and the PTM indicators for TFs are initialized by computing the correlation between the initialized TFs' activity profiles and their corresponding genes' expression profiles and setting the less correlated TFs to be post-transcriptionally modified.

M step: Given the expected activity levels of TFs and the most likely assignment of PTM indicators inferred in the E step, we learn new parameters by attempting to maximize the *Score* function subject to the constraint discussed above (number of TFs for each gene). We can calculate exact solutions for the variance terms γ, σ and τ by zeroing the partial derivatives of the penalized complete log-likelihood of data defined in Equation 2. We also calculate maximum likelihood estimates for α and β by fixing the regulation weights W.

Unlike the noise parameters the weight parameters W cannot be computed in closed form because of the limit on the number of incoming edges for each gene. Instead, we first conduct a greedy search to associate TFs with each gene, and then solve an optimization problem to obtain estimates for W. See supporting website[27] for more details.

3   Experiments and Results

We tested PTMM's performance on both simulated and real gene expression time series data. Using simulated data we show that our algorithm can indeed recover the hidden activity levels of TFs and determine whether a TF is post-transcriptionally modified. Using real data we show that by PTMM we can reconstruct meaningful TF activity profiles, detect known post-transcriptionally

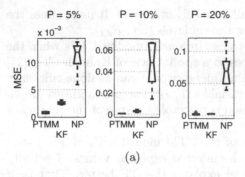

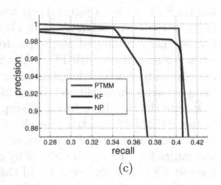

Fig. 2. Simulated data results. (a) Mean squared error (MSE) between actual and inferred hidden activity levels of TFs. The three plots correspond to different percentages of post-transcriptionally modified TFs (5%, 10% and 20%). Red line is the median. Blue box indicates upper and lower quantiles. The black bars are the range of the MSE. Outliers are plotted by "+". (b) Comparison of precision-recall curves for predicting the PTM indicators. (c) Comparison of precision-recall curves for predicting TF-gene regulatory relationships.

regulated TFs and improve the ability to determine TF-gene regulatory relationships. We also compared PTMM on both simulated and real expression data with two methods representing the two approaches mentioned in the introduction:

- Kalman Filter model (KF): This model corresponds to methods that assume that TF activity levels can only be inferred from the expression levels of its regulated genes. For this we set all PTM indicators in PTMM to 1. Thus, PTMM reduces to a Kalman filter model. We can infer the hidden activity levels of TFs efficiently by standard inference method for Kalman filter [25].
- No Post-transcriptional modification model (NP): This model corresponds to methods that use a TF's expression levels to infer its activity levels, i.e., all PTM indicators fixed to 0 in PTMM.

For comparison of predicting PTM indictors we use a post-processing step for both methods in which we compute the Pearson correlation between the inferred activity profile for each TF and the expression profile of its corresponding gene. A cutoff is applied to turn this correlation score to binary PTM indicators.

The maximum number C of regulating TFs for one gene was set to 3 in all experiments below.

3.1 Results on Simulated Data

We first synthesized n TF activation profiles using a random walk model and used a noisy version of these profiles as the mRNA levels for the TFs. Next we randomly selected P percent of TFs and set them to be post-transcriptionally modified. For these TFs we replaced their mRNA level with Gaussian noise with mean 0 (though we kept their activity levels and used it to generate the profiles for the regulated genes, see below). Finally we randomly generated a TF-gene regulation weight matrix. We used the TF activity levels and the weight matrix to generate the observed expression values for all genes and added i.i.d. random noise to each time point for each gene.

We varied the percentage of TFs that are post-transcriptionally modified. For all cases we sampled $n = 100$ TFs and $m = 1000$ genes. For noise parameters we used the values learned from real data to make the simulation realistic. The prior constants on evidences were set to $\pi_0 = 0.7$ and $\pi_1 = 0.3$ by cross validation.

Figure 2(a) presents the mean squared error between the true and inferred TF activity profiles for each of the methods. As can be seen, PTMM consistently outperformed all other methods. In all three cases, NP cannot capture the underlying activity profiles as accurately as the two other models because it tends to predict activity levels of post-transcriptionally modified TFs as their mRNA expression levels which significantly compromises its performance. KF ignores all information from a TF's expression levels and infers its activity levels solely from its regulated genes. It has better performance than NP because of the way we constructed the expression levels (linear combination of the activity levels of TFs). However, since it ignores useful data for many TFs (mRNA levels) the reconstructed profiles are still not as good as PTMM.

Figure 2(b) shows precision-recall curves of the prediction of PTM indicators. The precision-recall curves were drawn by increasing the cutoff for the prediction confidence scores (for PTMM) or correlation coefficients (for KF and NP). Again, PTMM outperformed other two methods.

We also tested the ability of all methods to predict the regulatory relationships between TFs and genes by 8-fold cross validation. Since the regulation weights outputted by all methods are continuous values, we applied a cutoff to turn the regulation weights into binary predictions of regulation. Figure 2(c) shows the precision-recall curves of regulatory relationship prediction by all three methods. Interestingly, the curve for NP starts higher indicating that when mRNA levels correspond to activity levels this model is very powerful at identifying TF-gene interactions. However, the KF model is more general and applies to both transcriptionally and post-transcriptionally regulated TF. Thus the KF curve crosses the NP curve at a recall of 35%. Our method that utilize both sources of information acts as the best of both methods. It starts out very strong (similar to NP method) but unlike NP method it remains strong for higher recall rates as well.

3.2 Yeast Expression Data

We applied PTMM to *saccharomyces cerevisiae* microarray time series data collected under 17 experimental conditions including various stresses, cell cycle and DNA damage (see supporting website [27] for complete list). The number of time points in these datasets ranges from 8 to 24. To construct the prior binding

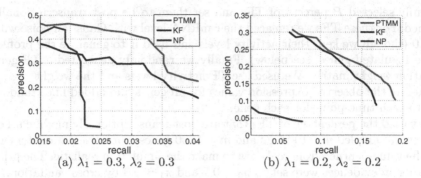

(a) $\lambda_1 = 0.3$, $\lambda_2 = 0.3$ (b) $\lambda_1 = 0.2$, $\lambda_2 = 0.2$

Fig. 3. Results for yeast expression data. Precision and recall curves for recovering TF-gene interactions in cross validation studies. The tradeoff constants used are (a) $\lambda_1 = 0.3$, $\lambda_2 = 0.3$ and (b) $\lambda_1 = 0.2$, $\lambda_2 = 0.2$.

evidence matrix E we used protein-DNA binding data from MacIsaac *et al.* [28]. Their binding data offers a list of ORFs with binding sites for each transcription factor at various binding and conservation thresholds. We only used the most confident data (binding p-value < 0.001 and motifs are conserved in at least 2 additional species). We removed TFs that had less than 5 known targets and TFs that had more than 50% missing expression values in at least one dataset. The remaining 72 TFs were used for the analysis. We set the prior knowledge $\pi_0 = 0.7, \pi_1 = 0.3$ indicating our belief in the high quality binding data. We tested our approach using cross-validation. Thus we also removed all genes that are not known to be bound by any of the 72 TFs we modelled leaving 1069 genes.

Predicting TF-gene regulatory relationships. We first tested the ability of PTMM to predict TF-gene regulatory relationships by performing 8-fold cross validation. In each fold we hid the associations of 1/8 of the 1069 genes (i.e., set the corresponding entries in the evidence matrix E to zero) and used all three methods to predict the regulatory relationships for these genes. Again, by varying a cutoff w_0 for all edges between genes and TFs we can generate precision-recall curves for all methods.

Figure 3(a) presents these curves of all three methods. The results are qualitatively similar to the simulated data results. The NP method starts strong but drops rapidly. The KF method that does not rely on TF expression level is more robust and holds for longer recall rates. Nicely, our method dominates both other methods indicating that it is indeed possible to combine both approaches for modelling regulatory networks. Note that since each gene can only be assigned to up to 3 out of 72 TFs, a precision rate of close to 50% is quite impressive. Also, it is important to remember that most ChIP-chip experiments were carried out in YPD whereas the expression data we used is primarily from stress conditions. Thus, some of our false positive might be actually correct prediction and the reason they were not identified before is due to the condition under which the experiments were carried out. This also effects the recall rate which is low for all methods. Another possible reason for the low recall rate

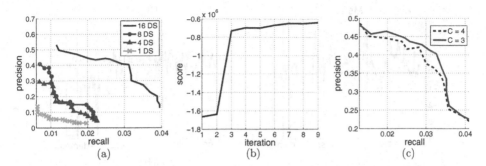

Fig. 4. Results for yeast expression data. (a) The precision-recall curves as a function of the number of expression datasets used for PTMM, ranging from 1 to 16 datasets (DS). (b) The penalized likelihood score curve for PTMM versus the number of iterations. (c) Comparison of the precision-recall curves when setting maximum number C of associated TFs for each gene to be 3 and 4.

is the protein-DNA binding data combined with strict conservation standard which makes our prior knowledge arguably incomplete. In fact, we can use the constants λ_1 and λ_2 to control the level of tradeoff between the precision and recall. Figure 3(b) shows the precision recall curves in predicting the TF-gene regulatory relationships when setting both λ_1 and λ_2 to 0.2. As can be seen, using these values the recall rate substantially improved though the precision drops. Again, the curve of PTMM outperforms the other two methods. In both plots (Figure 3(a)(b)), the fact that higher weight correlates well with correct TF-gene associations indicates that the recovered TF activity profiles are a good representation of the underlying profiles.

To test whether more data can improve the performance of our algorithm we measured precision-recall curves using different numbers of datasets. Figure 4(a) shows four curves corresponding to the performance of PTMM with 1, 4, 8 and 16 datasets. Indeed, more datasets improved both precision and recall. Figure 4(b) shows the penalized likelihood scores versus the number of iterations. As can be seen, the score converges quickly, reaching a (local) maximum after only a few iterations. Note that while this convergence may seem fast, it is a direct result of the fact that we are initializing our model with known TF-gene binding data rather than random initializations that are common in many EM applications. Figure 4(c) presents the precision-recall curves for setting the maximum number C of associated TFs for each gene to be 3 and 4. As can be seen, setting C bigger does not help improving the coverage of PTMM.

Insights into post-transcriptional modifications. Of the 72 TFs, PTMM determined that 7 are post-transcriptionally modified in at least some of the conditions we looked at. We found strong indications for differences between the transcript level and the activity level of five factors (Gcn4, Msn4, Swi5, Fkh2, Rap1). Most of these factors are known to be regulated post-transcriptionally. For example, the master regulator of amino acid starvation, Gcn4, is regulated at the translational level [12]. The stress response TF, Msn4, is regulated at the

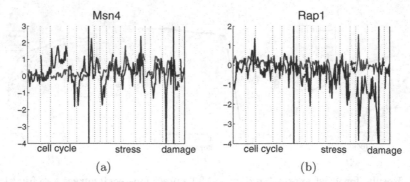

Fig. 5. The observed mRNA expression levels (blue) v.s. TF activity levels (red) inferred by PTMM for all 17 experimental conditions (separated by dashed lines) for (a) Msn4 and (b) Rap1

post-transcriptional level by phosphorylation [30] and by nucleus localization [31]. Interestingly, we found that the Msn4 activity follows its mRNA level in stress conditions but differs from them during the cell cycle (Figure 5(a)). This observation suggests that the main role of the PTMs is to prevent undesired activation of the stress response pathway. During stress when Msn4 activity is desired, it is induced and activated, whereas in other conditions such as the cell cycle, when its activity may be harmful to cell it is kept in a silent form and fluctuations in its mRNA level seem to be not important. The cell cycle regulators, Swi5 and Fkh2 are regulated by phosphorylation which is important for the nucleus localization of Swi5 [32] and for the activity of Fkh2 [33]. Finally it has been shown that the phosphorylation of Rap1 (Figure 5(b)) affects its binding to DNA [34]. As for the other two factors, we could not confirm, nor reject the prediction regarding Met32. The last TF predicted to be post-transcriptionally regulated, Mbp1 is known to be transcriptionally regulated in at least some conditions [1] and may be a false positive result. See supporting website [27] for more figures.

Validating predictions regarding TF activity levels. To further analyze the predictions regarding TF activity levels we have looked at a specific condition, yeast response to methyl-methanesulfonate stress (MMS). For this condition we have both time series expression data (by which we use to make predictions) as well as new ChIP-chip data [29] which we did not use to learn the model. In their paper Workman *et al.* classified each TF they tested as expanding in MMS (regulating more genes when compared to general growth conditions), contracting or not changing. Thus, we can use this new interaction data to determine whether the predictions made by our method agree with the activity observed by the ChIP-chip experiment, and by the mRNA levels of the TF.

Due to space limitation we only present results for 5 TFs in Table 1. The complete table can be found on the supporting website [27]. In general, we see a good agreement between the predicted activity levels and the observed binding profiles. For example, both Pdr1 and Uga3 are expanded in MMS suggesting that

Table 1. Summary of agreement between predicted and observed TF activity level in MMS

TF	conf[1]	ChIp[2]	post[3]	activity[4]	expression[5]
Pdr1	213	e	0	∧	∧
Uga3	7065	e	0	∧	∧
Gcn4	488	n	1	∧	∨
Sko1	993	n	0	∨	∨
Yap5	1396	c	0	∨	∧

[1] Confidence score given by PTMM. [2] (e)xpending, (c)ontracting or (n)either, according to ChIp-chip data in [29]. [3] PTMM result. 0 for factors determined to be transcriptionally regulated, 1 for post-transcriptionally modified. [4] Activity level inferred by PTMM. [5] Observed gene expression level. For [4,5], up (∧) , down (∨) or flat (-).

they are active. This is accurately predicted by PTMM. Interestingly, there are two TFs for which PTMM's prediction differs from their mRNA level. The first is Yap5. Even though Yap5 is not identified as post-transcriptionally modified its activity level for this condition is accurately predicted to be lower than its normal level. In contrast, its expression level is actually higher than baseline. PTMM's prediction for Yap5 is validated by the MMS ChIP-chip data which shows that Yap5 is contracting. The second is Gcn4. While the expression of Gcn4 is slightly lower than baseline, PTMM predicts that this factor is post-transcriptionally modified and its activity increases in MMS. While Workman *et al.* did not find Gcn4 to be expanding, a more recent study [12] experimentally tested Gcn4's binding in a more appropriate time point (15 minutes following MMS treatment) and found that Gcn4 greatly expands in MMS as predicted by PTMM.

4 Discussion

In this paper we developed, for the first time, a method that utilizes both expression and interaction datasets for inferring the activity of TFs. Our method uses a switching model to determine whether a TF is transcriptionally or post-transcriptionally regulated. This model is combined with a factorial HMM to fully model interactions in a dynamic regulatory network.

Factorial HMMs and variants of dynamic Bayesian networks have been suggested in the past for modelling regulatory networks [17], for modelling the activity of neurons in the brain [35] and for determining functional GO annotations in time series expression experiments [36]. However, the ability to use both interaction and expression data to model the activity of the hidden layer in these models is a novel aspect of PTMM. As we show using simulated and real expression data this allows our method to combine the best of both worlds. PTMM outperforms other methods when comparing their ability to predict new TF-gene interactions. Many of the factors predicted to be post-transcriptionally regulated are validated by prior knowledge. Our method is also successful in predicting TF activity in a new condition.

Acknowledgments. This work was supported in part by NIH grant NO1 AI-5001 and NSF CAREER award 0448453 to Z.B.J.

References

1. Spellman, P.T., et al.: Comprehensive identification of cell cycle-regulated genes of the yeast Saccharomyces cerevisiae by microarray hybridization. Mol. Biol. Cell. 9, 3273–3297 (2004)
2. Panda, S., et al.: Coordinated transcription of key pathways in the mouse by the circadian clock. Cell 109(3), 307–320 (2002)
3. Nau, G., et al.: Human macrophage activation programs induced by bacterial pathogens. PNAS 99, 1503–1508 (2002)
4. Arbeitman, M., et al.: Gene expression during the life cycle of drosophila melanogaster. Science 298, 2270–2275 (2002)
5. Theuns, J., et al.: Transcriptional regulation of Alzheimer's disease genes: implications for susceptibility. Hum. Mol. Genet. 9, 2383–2394 (2000)
6. Beer, M., et al.: Predicting gene expression from sequence. Cell 117(2), 185–198 (2004)
7. Zou, M., et al.: A new dynamic Bayesian network (DBN) approach for identifying gene regulatory networks from time course microarray data. Bioinformatics 21, 71–79 (2005)
8. Tanay, A., et al.: Computational expansion of genetic networks. Bioinformatics 17, S270–S278 (2001)
9. D'haeseleer, P., et al.: Genetic network inference: from co-expression clustering to reverse engineering. Bioinformatics 16, 707–726 (2000)
10. Segal, E., et al.: Module networks: identifying regulatory modules and their condition-specific regulators from gene expression data. Nat. Genet. 34(2), 166–176 (2003)
11. Sabatti, et al.: Bayesian sparse hidden components analysis for transcription regulation networks. Bioinformatics 22, 739–746 (2006)
12. Ernst, J., et al.: Reconstructing dynamic regulatory maps. Nature-EMBO Molecular Systems Biology 3, 74 (2007)
13. Ideker, T., et al.: Integrated genomic and proteomic analyses of a systematically perturbed metabolic network. Science 292, 929–934 (2001)
14. Kannan, et al.: A Bayesian Model That Links Microarray mRNA Measurements to Mass Spectrometry Protein Measurements. In: Speed, T., Huang, H. (eds.) RECOMB 2007. LNCS (LNBI), vol. 4453, pp. 325–338. Springer, Heidelberg (2007)
15. Bose, S., et al.: Genetic factors that regulate the attenuation of the general stress response of yeast. Genet. 169, 1215–1226 (2005)
16. Washburn, M.P., et al.: Protein pathway and complex clustering of correlated mRNA and protein expression analyses in Saccharomycs cerevisiae. PNAS 100, 3107–3112 (2003)
17. Nachman, I., et al.: Inferring quantitative models of regulatory networks from expression data. Bioinformatics 20(Suppl 1), I248–I256 (2004)
18. Ghahramani, Z., et al.: Factorial hidden Markov models. Machine Learning 29, 245–273 (1997)
19. Wang, L., et al.: Group SCAD regression analysis for microarray time course gene expression data. Bioinformatics 23, 1486–1494 (2007)

20. Harbison, C.T., et al.: Transcriptional regulatory code of a eukaryotic genome. Nature 431, 99–104 (2004)
21. Tibshirani, R.: Regression shrinkage and selection via the lasso. J. Royal Statist. Soc. B. 58, 267–288 (1996)
22. Hu, Z., et al.: Genetic reconstruction of a functional transcriptional regulatory network. Nat. Genet. 39, 683–687 (2007)
23. Xing, E., et al.: A generalized mean field algorithm for variational inference in exponential families. In: Proceedings of UAI (2003)
24. Pearl, J.: Probabilistic Reasoning in Intelligent Systems: Networks of Plausible Inference. Morgan Kaufmann, San Francisco (1988)
25. Murphy, K.: Dynamic bayesian networks: Representation, inference and learning. Ph.D. Thesis, University of California, Berkeley (2002)
26. Coleman, T.F., et al.: An interior trust region approach for nonlinear minimization subject to bounds. SIAM Journal on Optimization 6, 418–445 (1996)
27. Supporting website, http://www.cs.cmu.edu/~yanxins/ptmm
28. MacIsaac, K., et al.: An improved map of conserved regulatory sites for Saccharomyces cerevisiae. BMC Bioinformatics 7, 113 (2006)
29. Workman, C.T., et al.: A systems approach to mapping DNA damage response pathways. Science 312, 1054–1059 (2006)
30. Garreau, H., et al.: Hyperphosphorylation of Msn2p and Msn4p in response to heat shock and the diauxic shift is inhibited by cAMP in Saccharomyces cerevisiae. Microbiology 146, 2113–2120 (2000)
31. Gorner, W., et al.: Nuclear localization of the C2H2 zinc finger protein Msn2p is regulated by stress and protein kinase A activity. Genes. Dev. 12, 586–597 (1998)
32. Moll, T., et al.: The role of phosphorylation and the CDC28 protein kinase in cell cycle-regulated nuclear import of the S. cerevisiae transcription factor SWI5. Cell 66, 743–758 (1991)
33. Pic-Taylor, A., et al.: Regulation of cell cycle-specific gene expression through cyclin-dependent kinase-mediated phosphorylation of the forkhead transcription factor Fkh2p. Mol. Cell. Biol. 24, 10036–10046 (2004)
34. Tsang, J.S., et al.: Phosphorylation influences the binding of the yeast RAP1 protein to the upstream activating sequence of the PGK gene. Nucl. Acids Res. 18, 7331–7337 (1990)
35. Mitchell, T., et al.: Hidden process models. In: Proceedings of ICML (2006)
36. Shi, Y., et al.: Continuous hidden process model for time series expression experiments. Bioinformatics 23, 1459–1467 (2007)

A Fast, Alignment-Free, Conservation-Based Method for Transcription Factor Binding Site Discovery

Raluca Gordân*, Leelavati Narlikar*, and Alexander J. Hartemink

Department of Computer Science, Duke University, Durham, NC 27708-0129
{raluca,lee,amink}@cs.duke.edu

Abstract. As an increasing number of eukaryotic genomes are being sequenced, comparative studies aimed at detecting regulatory elements in intergenic sequences are becoming more prevalent. Most comparative methods for transcription factor (TF) binding site discovery make use of global or local alignments of orthologous regulatory regions to assess whether a particular DNA site is conserved across related organisms, and thus more likely to be functional. Since binding sites are usually short, sometimes degenerate, and often independent of orientation, alignment algorithms may not align them correctly. Here, we present a novel, alignment-free approach for incorporating conservation information into TF motif discovery. We relax the definition of conserved sites: we consider a DNA site within a regulatory region to be conserved in an orthologous sequence if it occurs anywhere in that sequence, irrespective of orientation. We use this definition to derive informative priors over DNA sequence positions, and incorporate these priors into a Gibbs sampling algorithm for motif discovery. Our approach is simple and fast. It does not require sequence alignments, nor the phylogenetic relationships between the orthologous sequences, and yet it is more effective on real biological data than methods that do.

1 Introduction

With recent advances in DNA sequencing technologies, the number of closely related genomes being sequenced [1, 2, 3] has increased tremendously. Consequently, this has led to an increased emphasis on comparative studies focused on detecting functional elements in intergenic DNA sequences. Functional elements, including TF binding sites, are known to evolve at a slower rate than non-functional elements, and therefore DNA sites that are well conserved in orthologous regulatory regions are considered good candidates for TF binding sites.

A plethora of algorithms use evolutionary conservation information for *de novo* TF motif discovery, either by filtering the putative regions according to their conservation levels and then applying conventional motif finders, or by incorporating the conservation information into the motif finder itself. The former

---

* These authors contributed equally to this work.

M. Vingron and L. Wong (Eds.): RECOMB 2008, LNBI 4955, pp. 98–111, 2008.

approach has a major limitation: motifs that are not well conserved are likely to be missed. Most conservation-based motif finders therefore take the latter approach. These methods can be further divided into two main categories: 1) 'single gene, multiple species', and 2) 'multiple genes, multiple species'. Methods in the first category (e.g., FootPrinter [4], the phylogenetic Gibbs sampler of Newberg et al. [5]) take as input the regulatory region of a single gene, together with its orthologs from related organisms. Methods in the second category (e.g., the method of Kellis et al. [1], Converge [6, 7], PhyloCon [8], PhyME [9], PhyloGibbs [10], OrthoMEME [11], EMnEM [12], CompareProspector [13]) are designed to search for motifs that are both over-represented in a set of given sequences (from a reference species) and conserved across related organisms. Our method falls into this category, so for the rest of the paper we will focus only on 'multiple genes, multiple species' approaches.

Most conservation-based approaches to TF binding site discovery rely on multiple or pair-wise alignments of orthologous regulatory regions to assess whether a particular DNA site is conserved across related organisms [1, 6, 7, 9, 10, 12, 13]. However, since binding sites are usually short, sometimes degenerate, and often in reverse orientation or even relocated, alignment algorithms may not correctly align the binding sites within orthologous regulatory sequences. Especially when the sequences are very divergent, the background 'noise' of diverged nonfunctional regions may be stronger than the 'signal' of conserved motifs, preventing a correct alignment. In Fig. 1 we illustrate four scenarios where motifs in orthologous sequences are not correctly aligned, and thus would most likely be missed by alignment-based motif finders. When a motif changes position or orientation, as in Fig. 1(c,d), correct alignment of motifs may even be impossible.

In consequence, motif finding algorithms based on alignments of orthologous promoter regions will only work when the promoters in the reference species align well with the promoters in the related species (e.g., this is not true for many promoters in S. cerevisiae and their orthologs in the non-sensu stricto Saccharomyces species used in our analysis). Even when the orthologous promoters align well, depending on the exact algorithm used to construct the alignments, different sites may appear to be conserved. For example, while some studies report a significant number of S. cerevisiae TF binding sites to be conserved in related Saccharomyces species [14, 15], a study by Siggia [16] found that among 407 experimentally verified binding sites in S. cerevisiae, only about half appear to be conserved in an alignment of sensu stricto promoter sequences (in his study, the sequences were aligned using a method by Morgenstern [17]).

Here, we describe a novel, alignment-free method for conservation-based motif discovery. We relax the definition of conserved DNA sites and consider a site within a reference regulatory region to be conserved in an orthologous sequence if it occurs anywhere in that sequence, irrespective of orientation. We start with a set of sequences believed to be bound by a common TF in the reference organism. Using orthologous sequences from related organisms, we compute a conservation score for each word and use it to bias our search towards conserved DNA sites. Our method outperforms current conservation-based motif discovery methods

(a) Sequence iYLR213C, bound by Mac1

```
Scer:  ...CGCCGATATTTTTGCTCACCTTTTTTTTTTGCTCATCG-AAAATTGTTATAGCG...
Spar:  ...CACCGATATTTTTGCTCACCTTTTTTTT--GCTCATCG-AAAATTGTTA--GCG...
Skud:  ...AGTCGATATTTTTGCTCATCTTTTTTTTTTGCTCATTGAAAAATTGCAATGGCG...
Sbay:  ...CAGTGAAATTTTTGCTCATCGAATTTTT--GCTCATCG---AAGTGTAAT-GCG...
```

(b) Sequence iYAR014C, bound by Tec1

```
Scer:  ...ATATATATATATATACATTCTATATATTCTTACCCAGATTCTTT-GAGGTAAGA...
Spar:  ...ATATATATATATATA-----TGTACATTCTCACCTGGATTCTTTGGGGGTAAAA...
```

(c) Sequence iYKL054C, bound by Rpn4

```
Scer:  ...TGGGGTAATTGGTAAGAGTTT-TT...GCCACTACTTTTTGCCACCATTT-CCC...
Spar:  ...TGGGGTAATTGGTAAGAGTTTCTT...GCCACTATTTTTTGCCACCATTT-CCC...
Smik:  ...-GGGGTAATTGGTAAGAGTTTCTT...GCCACTGTTTTTTGCCACCATTTTCCC...
Skud:  ...TGGGGTAATTGGTAAGAGTTCCTT...GCCACT-TTTTTTGCCACCATTT--CC...
Sbay:  ...TGTGGTAATTGGTAAGTTTTTCTT...GCCACT-TTTTTTGCCACCATTTTTCC...
Sklu:  ...GTGGGAGGGTGGCAAATTTTTCTC...GACACAGT------CCATAAGCT-GCC...
```

(d) Sequence iYMR107W, bound by **Leu3** and Ume6

```
Scer:  ...CGCCTAGCCGCCGGAGCCTGCCGGTACCGGCTTGGCTTCAGTTGCTGATCTCGG...
Smik:  ...TACCTAACAGCCGG----------TACCGGCTTGAATGCCGCCGTTGGCTTCCG...
```

Fig. 1. Examples of conserved TF binding sites in aligned [14] orthologous yeast sequences that can be missed by alignment-based motif discovery programs. The sites matching the motifs of the respective TFs are marked in color. (a) Alignment algorithms may incorrectly insert gaps in orthologous motif occurrences. (b) Non-functional regions that are conserved in closely related organisms may prevent a correct alignment of the binding sites. (c) Binding sites are sometimes free to change orientation, which is probably the case for the Rpn4 binding site in *S. kluyveri*. (d) Motifs may change their position relative to each other, as shown by the Leu3 and Ume6 sites. (The sequences in the figure correspond to *S. cerevisiae*, *S. paradoxus*, *S. kudriavzevii*, *S. mikatae*, *S. bayanus*, and *S. kluyveri*. Due to lack of experimental data, we can only assume the depicted binding sites are functional in organisms other than *S. cerevisiae*.)

in both speed and accuracy. We further show that if negative examples (*i.e.*, sequences believed *not* to be bound by the TF) are also available, we can further improve the performance of our algorithm by considering conservation across those regions as well.

2   Methods

In this section, we describe the generative formulation of motif discovery widely used to find significant motifs in sets of promoters of co-regulated genes. In earlier work [18, 19, 20, 21], we have introduced PRIORITY, a framework for incorporating additional information into motif discovery using informative positional priors. Here, we develop a method for incorporating conservation information across multiple species into our framework. It is important to note that the present paper is not about the PRIORITY framework *per se*, but rather about a simple, but clever method for exploiting conservation information for more accurate motif discovery that is orders of magnitude more efficient than methods proposed to date. Consequently, the methods introduced here can also be adapted to other motif finders beyond PRIORITY.

2.1   Sequence Model and Objective Function

Assume we have n DNA sequences X_1 to X_n believed to be commonly bound by some TF. For simplicity, we model at most one binding site in each sequence. This is analogous to the zero or one occurrence per sequence (ZOOPS) model in MEME [22]. Let Z be a vector of length n denoting the starting location of the binding site in each sequence: $Z_i = j$ if a binding site starts at location j in X_i and we adopt the convention that $Z_i = 0$ if X_i contains no binding site. We assume that the TF motif can be modeled as a position specific scoring matrix (PSSM) of length W while the rest of the sequence follows some background model parameterized by ϕ_0. The PSSM can be described by a matrix ϕ where $\phi_{a,b}$ is the probability of finding base b at location a within the binding site for $1 \le b \le 4$ and $1 \le a \le W$.

Thus if the sequence X_i is of length l_i, and X_i contains a binding site at location Z_i, we can compute the probability of the sequence given the model parameters as:

$$P(X_i \mid \phi, Z_i > 0, \phi_0) = P(X_{i,1}, \ldots X_{i,Z_i-1} \mid \phi_0) \times \left(\prod_{a=1}^{W} \phi_{a,X_{i,Z_i+a-1}} \right)$$
$$\times P(X_{i,Z_i+W}, \ldots X_{i,l_i} \mid \phi_0)$$

and if it instead does not contain a binding site as:

$$P(X_i \mid \phi, Z_i = 0, \phi_0) = P(X_{i,1}, X_{i,2} \ldots X_{i,l_i} \mid \phi_0)$$

We wish to find ϕ and Z that maximize the joint posterior distribution of all the unknowns given the data. Assuming priors $P(\phi)$ and $P(Z)$ over ϕ and Z respectively, our objective function is:

$$\underset{\phi, Z}{\arg\max}\, P(\phi, Z \mid X, \phi_0) = \underset{\phi, Z}{\arg\max}\, P(X \mid \phi, Z, \phi_0) P(\phi) P(Z) \qquad (1)$$

2.2   Optimization Strategy and Scoring Scheme

We use Gibbs sampling to sample repeatedly from the posterior over ϕ and Z with the hope that we are likely to visit those values of ϕ and Z with the highest posterior probability. Proceeding analogously to the derivation of Liu [23], collapsing ϕ, we get the final distribution for sampling Z_i:

$$P(Z_i = j \mid Z_{[-i]}, X, \phi_0) = \frac{P(Z_i = j) \times \left(\prod_{a=1}^{W} \phi_{a,X_{i,j+a-1}} \right)}{P(Z_i = 0) \times P(X_{i,j}, \ldots, X_{i,j+W-1} \mid \phi_0)}$$

for $1 \le j \le l_i - W + 1$, and $P(Z_i = j \mid X, \phi_0) = 1$ for $j = 0$, where ϕ is calculated from the counts of the sites contributing to the current alignment $Z_{[-i]}$, which is the vector Z without Z_i. In practice, we run the Gibbs sampler, which we call PRIORITY [18], for a predetermined number of iterations after apparent convergence to the joint posterior and output the highest scoring PSSM at the end. We use the single best motif to evaluate the algorithm and compare it with other popular methods.

2.3 Incorporation of Conservation Information

The Gibbs sampling technique described above has been used in several motif finders, often with additional parameters and heuristics. Usually, these motif finders assume a uniform prior over the locations Z. We will now show how conservation information across related organisms can be incorporated as an informative prior over Z.

Assume that we have sequence information from k related organisms. Thus for each sequence X_i in the original species, we have an orthologous sequence $X_i^{(s)}$ where $1 \leq s \leq k$. These sequences may be obtained via a genome alignment or by searching for regions near orthologous genes. A sequence may even be empty if no such region is found in the genome of the corresponding organism.

In this paper, we apply our method to ChIP-chip data [6] from *S. cerevisiae*. We obtain orthologous sequences from six related organisms (*S. paradoxus*, *S. mikatae*, *S. kudriavzevii*, *S. bayanus*, *S. castelli*, and *S. kluyveri*) based on the MULTIZ and BLASTZ alignments from Siepel *et al.* [14]. We describe two different ways in which this information can be used; the first uses the alignments, while the second does not.

Alignment-based conservation prior

Using multiple alignments of the seven yeast species mentioned earlier, Siepel *et al.* [14] have published a conservation track that is freely available at the UCSC genome browser. This track reports the probability of every position in the *S. cerevisiae* genome being conserved based on a program called PhastCons that fits a two-state phylogenetic HMM to aligned orthologous sequences by maximum likelihood. We use these conservation track probabilities to define a score $\mathcal{S}_T(X_i, j)$ for the W-mer at position j in the bound sequence X_i as:

$$\mathcal{S}_T(X_i, j) = \frac{1}{W} \sum_{t=0}^{W-1} Ph(X_i, j+t) \qquad (2)$$

where $Ph(X_i, j)$ is the probability of conservation reported by PhastCons at position j in sequence X_i. In practice, while computing $\mathcal{S}_T$, we scale the output of the PhastCons program linearly to lie between 0.1 and 0.9 to avoid singularities in the model. We assume that $\mathcal{S}_T(X_i, j)$ reflects the probability of the W-mer starting at position j in sequence X_i being a binding site. Note that the values $\mathcal{S}_T(X_i, j)$ themselves do not define a probability distribution over j. As mentioned earlier, we model each sequence X_i as containing at most one binding site. If X_i has no binding site, then none of the positions in X_i can be the starting location of a binding site. On the other hand, if X_i has one binding site at position j, not only must a binding site start at location j, but also no such binding site should start at any other location in X_i. Using a little algebra, we can write:

$$P(Z_i = 0) \propto 1 \quad \text{and} \quad P(Z_i = j) \propto \frac{\mathcal{S}_T(X_i, j)}{1 - \mathcal{S}_T(X_i, j)} \quad \text{for} \quad 1 \leq j \leq l_i - W + 1 \quad (3)$$

We then normalize $P(Z_i)$ so that under the assumptions of our model we have $\sum_{j=0}^{l_i-W+1} P(Z_i = j) = 1$ for $1 \leq i \leq n$. We call this prior $\mathcal{T}$.

Alignment-free conservation prior

In Section 1, we outlined some of the shortcomings of using alignments to detect conserved binding sites. Due to the short length of most binding sites, multiple alignment algorithms are likely to misalign functional sites that are actually conserved across species (Fig. 1). We therefore describe an alignment-free prior that searches orthologous sequences $X_i^{(s)}$ for occurrences of all W-mers present in X_i. We assume that a W-mer has a high probability of being conserved if it occurs in most of the orthologous sequences regardless of its orientation or specific position. We define a conservation score S_C for the W-mer at position j in the bound sequence X_i as:

$$S_C(X_i, j) = \frac{1}{k} \sum_{s=1}^{k} I[X_{ij}^W \in X_i^{(s)}] \tag{4}$$

where $I[\cdot]$ is an indicator function and X_{ij}^W denotes the W-mer at position j in sequence X_i. In other words, the score $S_C(X_i, j)$ is directly proportional to the number of orthologous sequences in which the W-mer X_{ij}^W appears. The values of S_C range from 0 to 1. To avoid singularities, as before, we scale S_C linearly so that the values lie between 0.1 and 0.9.

We have also explored refinements of this simple approach that weigh sequences based on evolutionary distance, or account for imperfect matches while searching for occurrences of W-mers in orthologous sequences. These extensions did not perform better so we stick here to the simplest version (but see Section 4 for further discussion).

As in the case of $S_T(X_i, j)$, $S_C(X_i, j)$ is only the probability of the W-mer at position j in sequence X_i being a binding site. To convert these values into a positional prior, we substitute S_C for S_T in (3). After normalizing the resulting $P(Z_i)$ as shown earlier, we get a valid prior over Z, which we call C.

Priors with a discriminative perspective

The scores S_T and S_C used to compute the priors T and C, respectively, reflect the probability that a W-mer at a certain position is conserved. While it is true that regions bound by the TF are more likely to be conserved, it does not follow that every conserved region is more likely to be bound by the profiled TF. Some conserved regions could be binding sites of other TFs or other functional DNA elements. We now describe a method for computing a prior that addresses the issue of conserved regions not specific to the profiled TF.

A ChIP-chip experiment gives rise to sequences X that are bound by the profiled TF as well as sequences Y that are not bound. Assume we are given m such unbound sequences. As in the case of X, we have orthologous sequences $Y_1^{(s)}$ to $Y_m^{(s)}$ where $1 \leq s \leq k$. We compute a discriminative score $S_{DT}(X_i, j)$ by taking into account the conservation score S_T over both sets X and Y as follows. For each W-mer in X, we ask the following question: "Of all the conserved occurrences of this W-mer, what fraction occur in the bound set?". The motivation behind this is to ensure a high score for W-mers that are conserved only in the bound set but not W-mers that are conserved in general through-out the genome. Since we only know the probability that a certain location is conserved, we count the number of conserved W-mers in expectation, weighing each occurrence of the W-mer according to how conserved it is. Using the score S_T derived over both sets X and Y, we calculate S_{DT} as:

$$S_{DT}(X_i, j) = \frac{\sum_{(q,r):X_{qr}^W = X_{ij}^W} S_T(X_q, r)}{\sum_{(q,r):X_{qr}^W = X_{ij}^W} S_T(X_q, r) + \sum_{(q,r):Y_{qr}^W = X_{ij}^W} S_T(Y_q, r)} \tag{5}$$

As in the case of $S_T(X_i, j)$ and $S_C(X_i, j)$, we convert S_{DT} into a positional prior which we call DT. Similarly, we compute the discriminative score S_{DC} using the conservation-based score S_C across X and Y, by substituting S_C for S_T in equation (5). We convert S_{DC} into a positional prior which we call DC.

Fig. 2 shows the scores S_C and S_{DC} over an intergenic sequence belonging to the sequence-set of Ste12. As can be seen, the prior computed with a discrimi-native perspective is effective in filtering out false peaks. Note that if we assume a constant level of conservation across all W-mers, then priors C and T simplify to the widely used uniform prior over Z, which we call U. Priors DC and DT, however, simplify to a special prior D that reflects the relative frequency of each W-mer in X versus both X and Y; we have shown previously [19] the bene-fits of using such a discriminative prior. We incorporate these six priors U, T, C, D, DT, and DC in PRIORITY and call the resulting programs, respectively,

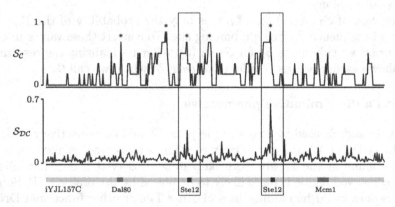

Fig. 2. Scores S_C and S_{DC} computed over intergenic region iYJL157C. Binding sites of Dal80, Ste12, and Mcm1 are shown as annotated by MacIsaac *et al.* [7]. iYJL157C belongs to the sequence-set bound by Ste12 during a ChIP-chip experiment [6]. The score S_{DC} is therefore computed from this sequence-set and a sequence-set that is not bound (see text). S_C has five big peaks, marked with asterisks. Two of them correspond to the start of Ste12 binding sites, one to the start of the Dal80 binding site. The two remaining peaks correspond to conserved A-T rich regions. However, the score S_{DC} has only two large peaks and both correspond to the start of Ste12 binding sites. This shows that prior DC is more specific to the profiled TF and effectively filters non-specific peaks corresponding to A-T rich regions or other conserved sites.

PRIORITY-$\mathcal{U}$, PRIORITY-$\mathcal{T}$, PRIORITY-$\mathcal{C}$, PRIORITY-$\mathcal{D}$, PRIORITY-$\mathcal{DT}$, and PRIORITY-$\mathcal{DC}$.

3   Results

We compiled ChIP-chip data published by Harbison *et al.* [6], who profiled the intergenic binding locations of 203 yeast TFs under various environmental conditions over 6140 intergenic regions. For each TF, we define its sequence-set X for a particular condition to be those intergenic sequences reported to be bound with p-value ≤ 0.001 in that condition. Similarly, for each TF we define Y to be all intergenic sequences bound with p-value ≥ 0.5. We consider all sequence-sets X of size at least 10 that are bound by TFs with a consensus binding motif in the literature (as used by Harbison *et al.* [6], or as reported in [24, 25]). This leaves us with 156 sequence-sets corresponding to 80 TFs profiled under various conditions. The analysis that follows is performed on those 156 sequence-sets.

It is common practice for methods to be evaluated on synthetically generated promoter data. However, in our framework, the informative priors capture information of biological relevance from true genomic sequences. Therefore, evaluating our method on simulated data is not appropriate.

3.1   Comparison of Priors

Table 1 shows the performance of the six priors when incorporated into PRIORITY[1], on the 156 ChIP-chip sequence-sets with known motifs. Three main conclusions can be drawn from the results in Table 1:

1. Overall, it appears that alignment-based conservation information (at least when used in the form of $\mathcal{T}$) is only slightly more useful than using no information. However, PRIORITY-$\mathcal{T}$ finds 10 motifs that PRIORITY-$\mathcal{U}$ does not, and PRIORITY-$\mathcal{U}$ finds 8 motifs that PRIORITY-$\mathcal{T}$ does not (data available in Supplementary Material). In examining the former 10 cases, it seems the information in the alignment helps. In most of the latter 8 cases, however, we notice that PRIORITY-$\mathcal{T}$ reports motifs with low information content. A closer examination reveals that some of them are weak matches to the literature consensus but do not satisfy our stringent success criterion. It is possible that the alignments produce misleading peaks in the prior at regions other than (or in addition to) the binding sites of the TF, thereby diluting the true motif signal. In the rest of the cases, we believe the alignment is faulty, *i.e.*, the binding sites do not get aligned correctly. Interestingly, one of these 8 sequence-sets corresponds to TF Mac1 and contains the sequence iYLR213C (see Fig. 1).

---

[1] All the results reported here were obtained with PRIORITY 2.0.0, which implements an improved sampling strategy compared to PRIORITY 1.0.0. This improves the results of baseline priors $\mathcal{U}$ and $\mathcal{D}$ over the results reported earlier [19, 20, 21].

Table 1. Number of motifs correctly identified by PRIORITY when using the six priors described in Section 2. Each version of PRIORITY is run with the default settings (motif width set to 8, and using a third order Markov model to describe the background). Then, for each of the 156 sequence-sets, the top scoring motif is compared with the literature consensus. We call an algorithm 'successful' on a particular sequence-set if this motif is less than a distance of 0.25 from the literature consensus according to the widely used inter-motif distance [6].

Priors	$\mathcal{U}$	$\mathcal{T}$	$\mathcal{C}$	$\mathcal{D}$	$\mathcal{DT}$	$\mathcal{DC}$
Number of successes	58	60	**69**	68	71	**76**

2. Our alignment-free approach, PRIORITY-$\mathcal{C}$, does significantly better than PRIORITY-$\mathcal{T}$ and PRIORITY-$\mathcal{U}$. Since the computation of S_C depends only on the presence of W-mers across orthologous sequences, this approach is impervious to the alignment artifacts described in Fig. 1, and hence seems to better pick up the true motif signal.
3. In each of the three priors $\mathcal{U}$, $\mathcal{T}$, and $\mathcal{C}$, adopting a discriminative perspective helps find the true motif in many more instances. PRIORITY-$\mathcal{DC}$ does the best: it finds the true motif in 76 sequence-sets across 50 TFs. In fact, there is no sequence-set on which PRIORITY-$\mathcal{DC}$ fails to find the true motif but PRIORITY-$\mathcal{D}$ or PRIORITY-$\mathcal{DT}$ is successful. This shows that, at least on these sequence-sets, conservation information used in this manner does not harm motif discovery.

Since PRIORITY-$\mathcal{T}$ is not much better than PRIORITY-$\mathcal{U}$ (nor is PRIORITY-$\mathcal{DT}$ much better than PRIORITY-$\mathcal{D}$), we will henceforth focus on the performance of our alignment-free motif finders PRIORITY-$\mathcal{C}$ and PRIORITY-$\mathcal{DC}$.

3.2 PRIORITY-$\mathcal{C}$ and -$\mathcal{DC}$ Are More Accurate than Current Conservation-Based Methods

In this section we compare the results of PRIORITY-$\mathcal{C}$ and PRIORITY-$\mathcal{DC}$ with the results of six conservation-based motif finders: MEME_c [6], a method of Kellis *et al.* [1], Converge [7], PhyloCon [8], PhyME [9], and PhyloGibbs [10]. All methods fall into the 'multiple genes, multiple species' category, and thus search for motifs that both over-represented in a set of bound sequences from a species of reference, and conserved across related species. We did not compare with other methods from this category [11, 12, 13, 26] due to one or more of the following reasons: some are so computationally expensive that running them on all 156 sequence-sets was practically impossible; some are designed for only two related organisms; some have been reported to perform worse than methods we include in our analysis; and some were simply not available. We provide more detailed descriptions of all algorithms in the Supplementary Material, along with specific reasons why an algorithm was not selected for comparison in cases where that applies.

Table 2. Number of successfully identified motifs for different conservation-based methods. For each of the 156 sequence-sets, we use the same criterion of success as in Section 3.1.

Program	Description	Number of successes
MEME_c	alignment-based; masks non-conserved bases and then applies MEME	49
Kellis *et al.*	alignment-based; searches for significantly conserved 3-gap-3 motifs, then extends them	56
Converge	alignment-based; uses EM; incorporates conservation and evolutionary distances into the model	66
PhyloCon	locally aligns conserved regions into profiles, compares profiles and merges them using a greedy approach	19
PhyME	alignment-based; uses EM; evolutionary model accounts for binding site specificities	21
PhyloGibbs	alignment-based; similar to PhyME, but uses Gibbs sampling; searches for multiple motifs simultaneously	54
PRIORITY-$\mathcal{C}$	alignment free; incorporates a prior based on conserved W-mers into a Gibbs sampler	60
PRIORITY-$\mathcal{DC}$	alignment-free; incorporates a prior based on conserved W-mers in both bound *and* unbound sequences	76

Table 2 shows the results of PRIORITY-$\mathcal{C}$ and PRIORITY-$\mathcal{DC}$ compared to the six conservation-based methods described above. For MEME_c and the method of Kellis *et al.* we use the results reported by Harbison *et al.* [6]; for Converge we use the results reported by MacIsaac *et al.* [7]. We ran PhyloCon version 3b with the default parameter setting and the parameter s set to 0.5, as in [7]. However, unlike [7], we did not preprocess the data or postprocess the results reported by PhyloCon. Both PhyME (version 1.2) and PhyloGibbs (version 1.0) were run with their respective default settings, a motif width of 8, and a third order Markov model to describe the background. As recommended by the authors of these programs, we used LAGAN [27] and Sigma [28] to compute alignments for PhyME and PhyloGibbs, respectively.

These results show that our algorithm PRIORITY-$\mathcal{DC}$ is more effective at finding the true motif than the other methods. Even when negative examples (*i.e.*, sequences believed *not* to be bound by the TF) are not available, PRIORITY with the simple conservation prior $\mathcal{C}$ still performs better than all six methods; when negative examples are available, the performance is higher yet.

3.3 PRIORITY-$\mathcal{C}$ and -$\mathcal{DC}$ Are Orders of Magnitude Faster than Current Conservation-Based Methods

PRIORITY with the conservation priors outperforms other methods not only in terms of accuracy, but also speed. In Fig. 3 we show a log-scale plot of the running

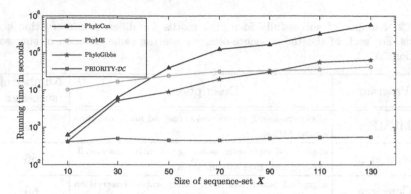

Fig. 3. Log-scale plot of running times of conservation-based algorithms on sequence-sets of increasing size. Running times for each algorithm include preprocessing steps (*i.e.*, alignment computation for PhyME and PhyloGibbs, and prior computation for PRIORITY-$\mathcal{DC}$). All programs were run on a 3.06GHz Intel Pentium 4 processor.

time of PhyloCon, PhyME, PhyloGibbs, and PRIORITY-$\mathcal{DC}$ for sequence-sets of varying size. Since the running times of PRIORITY-$\mathcal{C}$ and PRIORITY-$\mathcal{DC}$ are comparable (with minor differences in the prior computation making PRIORITY-$\mathcal{C}$ slightly faster), we only show the times for PRIORITY-$\mathcal{DC}$.

The running time of PRIORITY-$\mathcal{DC}$ varies only slightly with increasing number of sequences, and PRIORITY-$\mathcal{DC}$ is faster than PhyloCon, PhyME, and PhyloGibbs on all sequence-sets. On sets of 50 or more sequences, our algorithm becomes 2-3 orders of magnitude faster than the other three methods.

4   Discussion

We have presented a fast motif discovery algorithm that uses sequence conservation across related organisms without relying on alignments. Our method outperforms currently used conservation-based programs in both speed and accuracy.

We are not the first to use alignment-free conservation across species to find motifs. Elemento and Tavazoie [29] look for conserved regulatory elements by scanning a pair of related genomes for highly enriched W-mers, on the order of 400. Then they use a hypergeometric distribution to evaluate the significance of each of these W-mers in bound ChIP-chip sets. Using this method they are able to assign a W-mer that matches to the true motif to only 15 TFs. Since they limit their analysis to reporting W-mers, it is possible that they are not able to find TF motifs that have greater sequence variation. In contrast, though our scores $\mathcal{S}_{\mathcal{DC}}$ are also computed over W-mers, we use them only to construct positional priors; our Gibbs sampler returns a PSSM. In addition, the approach of Elemento and Tavazoie is limited to pairs of related organisms, and thus the choice of organisms becomes crucial for the success of the algorithm.

In this paper, we show how multiple unaligned genomes can be successfully used for motif discovery. Our method can be applied to any number of genomes.

For instance, we independently computed six variant $\mathcal{DC}$ priors using: only the single closest species (*S. paradoxus*); the two closest species (*S. paradoxus* and *S. mikatae*); the three closest species (*S. paradoxus*, *S. mikatae*, and *S. kudriavzevii*); and so on. PRIORITY-$\mathcal{DC}$ consistently found 69 or more motifs with each of these variant priors. The general trend indicated that more organisms improve performance.

The *sensu stricto* species (*S. paradoxus*, *S. mikatae*, *S. kudriavzevii*, and *S. bayanus*) provide most of the conservation information in the priors. However, since these species are closely related to *S. cerevisiae*, their intergenic regions may contain many non-functional conserved sites, simply because not enough evolutionary time has passed since the species diverged from their common ancestor. This does not pose a problem for our conservation-based algorithm because the information in the cobound sequences helps reduce the space of putative TF binding sites to those conserved DNA sites that also appear in most of the cobound sequences. Furthermore, the more distantly related species *S. castelli* and *S. kluyveri* provide some of the sequence divergence necessary for filtering out the conserved non-functional sites. According to a study by Cliften *et al.* [30], only a small number of the intergenic regions in the *S. castelli* and *S. kluyveri* genomes can be aligned to *S. cerevisiae* regions, and only after the corresponding orthologous genes have been identified. Even then, the conserved regulatory sites may be hard, if not impossible, to align correctly. Hence, alignment-based motif finders may not be able to fully exploit the information provided by the two distantly related species, while our alignment-free algorithm can.

Our conservation-based approach is much faster than current methods. It only needed a few minutes to compute a motif, even on the largest sequence-set, while other methods required days or in some cases months. Interestingly, other methods become slower precisely because they use conservation information, but our method actually speeds up: the informative prior computed from conservation information facilitates rapid convergence to the posterior, as evidenced by the fact that PRIORITY-$\mathcal{DC}$ reaches convergence faster than PRIORITY-$\mathcal{U}$ (data not shown).

In Fig. 3 we showed that PRIORITY-$\mathcal{DC}$ scales well with the size of the sequence-set. A similar analysis can be done by keeping the size of the sequence-set fixed but varying the number of orthologs for each sequence. The running time for PRIORITY-$\mathcal{DC}$ varies only slightly when we increase the number of orthologous sequences, while the running time of other conservation-based methods increases substantially (data available in the Supplementary Material).

Currently, the derivation of our conservation-based priors does not take phylogenetic information into account, mainly because high-quality phylogenetic trees are usually hard to compute. However, when such a tree is available, our algorithm can easily incorporate the phylogenetic information into the priors, by weighting the sequences in each organism (and thus the occurrences of W-mers in these sequences) according to the evolutionary distance between that organism and the reference organism. We have derived such a weighting scheme for the *Saccharomyces* species using the phylogenetic tree reported by Siepel *et al.*

[14]. However, conservation priors computed using the weighted sequences did not show any improvement over the initial conservation priors, C and DC.

One potential limitation of our approach is that the conservation priors are computed by counting only exact matches between the W-mers in the reference genome and W-mers in the related genomes. We have also tried computing priors similar to C and DC that allow for one mismatch when searching for conserved words. Since we do not know *a priori* the position in which a mismatch may occur, we allowed it to be anywhere in the W-mer. For example, an 8-mer was defined as "conserved" in an orthologous sequence if the sequence contained either an exact match to that 8-mer or any of the 24 8-mers that differed at exactly one position. The effect of allowing one mismatch was that the signal of truly conserved sites was mixed with random noise due to the 24 8-mers, and overall these priors were not as effective as C and DC. Allowing for more than one mismatch may further dilute the signal of conserved sites. However, prior knowledge about the structure of the binding site (for example, when we know we should be searching for a gapped motif) may be used to restrict the mismatches to certain positions.

Here, we have successfully applied our algorithm on seven *Saccharomyces* species. We believe our approach is even more useful on higher organisms, where motif finding has proven difficult due to longer promoters and smaller fraction of functional elements. We are planning to apply our method on data from higher organisms, including worm, fly, and human.

Supplementary Material can be found at http://www.cs.duke.edu/~amink/.

References

[1] Kellis, M., et al.: Sequencing and comparison of yeast species to identify genes and regulatory elements. Nature 432, 241–254 (2003)
[2] Cliften, P., et al.: Finding functional features in *Saccharomyces* genomes by phylogenetic footprinting. Science 301, 71–76 (2003)
[3] Clark, A., et al.: Proposal for *Drosophila* as a model system for comparative genomics (2003),
    http://flybase.net/.data/docs/CommunityWhitePapers/GenomesWP2003.html
[4] Blanchette, M., Tompa, M.: FootPrinter: a program designed for phylogenetic footprinting. Nucleic Acids Research 31, 3840–3842 (2003)
[5] Newberg, L.A., et al.: A phylogenetic Gibbs sampler that yields centroid solutions for *cis*-regulatory site prediction. Bioinformatics 23, 1718–1727 (2007)
[6] Harbison, C., et al.: Transcriptional regulatory code of a eukaryotic genome. Nature 431, 99–104 (2004)
[7] MacIsaac, K.D., et al.: An improved map of conserved regulatory sites for *Saccharomyces* cerevisiae. BMC Bioinformatics 7, 113 (2006)
[8] Wang, T., Stormo, G.D.: Combining phylogenetic data with co-regulated genes to identify regulatory motifs. Bioinformatics 19, 2369–2380 (2003)
[9] Sinha, S., Blanchette, M., Tompa, M.: PhyME: A probabilistic algorithm for Finding Motifs in Sets of Orthologous Sequences. BMC Bioinformatics 5, 170 (2004)

[10] Siddharthan, R., Siggia, E.D., van Nimwegen, E.: PhyloGibbs: A Gibbs sampling motif finder that incorporates phylogeny. PLoS Comp. Biol. 1, e67 (2005)

[11] Prakash, A., Blanchette, M., Sinha, S., Tompa, M.: Motif discovery in heterogeneous sequence data. In: PSB 2004, pp. 348–359 (2004)

[12] Moses, A., Chiang, D., Eisen, M.: Phylogenetic motif detection by expectation-maximization on evolutionary mixtures. In: PSB 2004, pp. 324–335 (2004)

[13] Liu, Y., et al.: Eukaryotic regulatory element conservation analysis and identification using comparative genomics. Genome Research 14, 451–458 (2004)

[14] Siepel, A., et al.: Evolutionarily conserved elements in vertebrate, insect, worm, and yeast genomes. Genome Res. 15, 1034–1050 (2005)

[15] Chin, C., Chuang, J.H., Li, H.: Genome-wide regulatory complexity in yeast promoters: Separation of functionally conserved and neutral sequence. Genome Res. 15, 205–213 (2005)

[16] Siggia, E.: Computational methods for transcriptional regulation. Current Opinion in Genetics & Development 15, 214–221 (2005)

[17] Morgenstern, B.: A space-efficient algorithm for aligning large genomic sequences. Bioinformatics 16, 1531–1539 (2000)

[18] Narlikar, L., Gordân, R., Ohler, U., Hartemink, A.: Informative priors based on transcription factor structural class improve de novo motif discovery. Bioinformatics 392, e384–e392 (2006)

[19] Narlikar, L., Gordân, R., Hartemink, A.: Nucleosome Occupancy Information Improves de novo Motif Discovery. In: Speed, T., Huang, H. (eds.) RECOMB 2007. LNCS (LNBI), vol. 4453, pp. 107–121. Springer, Heidelberg (2007)

[20] Narlikar, L., Gordân, R., Hartemink, A.: A Nucleosome-Guided Map of Transcription Factor Binding Sites in Yeast. PLoS Computational Biology 3, e215 (2007)

[21] Gordân, R., Hartemink, A.: Using DNA duplex stability information to discover transcription factor binding sites. In: PSB 2008, vol. 13, pp. 453–464 (2008)

[22] Bailey, T., Elkan, C.: Fitting a mixture model by expectation maximization to discover motifs in biopolymers. In: ISMB 1994, pp. 28–36 (1994)

[23] Liu, J.: The collapsed Gibbs sampler with applications to a gene regulation problem. Journal of the American Statistical Association 89, 958–966 (1994)

[24] Dorrington, R.A., Cooper, T.G.: The DAL82 protein of Saccharomyces cerevisiae binds to the DAL upstream induction sequence (UIS). Nucleic Acids Research 21, 3777–3784 (1993)

[25] Jia, Y., Rothermel, B., Thornton, J., Butow, R.A.: A basic helix-loop-helix-leucine zipper transcription complex in yeast functions in a signaling pathway from mitochondria to the nucleus. Molecular and Cellular Biology 17, 1110–1117 (1993)

[26] Li, X., Wong, W.H.: Sampling motifs on phylogenetic trees. PNAS 102, 9481–9486 (2005)

[27] Brudno, M., et al.: LAGAN and Multi-LAGAN: efficient tools for large-scale multiple alignment of genomic DNA. Genome Res. 13, 721–731 (2003)

[28] Siddharthan, R.: Sigma: multiple alignment of weakly-conserved non-coding DNA sequence. BMC Bioinformatics 7, 143 (2006)

[29] Elemento, O., Tavazoie, S.: Fast and systematic genome-wide discovery of conserved regulatory elements using a non-alignment based approach. Genome Biology 6, R18 (2005)

[30] Cliften, P.F., et al.: Surveying Saccharomyces genomes to identify functional elements by comparative DNA sequence analysis. Genome Res. 11, 1175–1186 (2001)

The Statistical Power of Phylogenetic Motif Models

John Hawkins and Timothy L. Bailey

Institute for Molecular Bioscience, QLD 4072,
The University of Queensland, Australia
j.hawkins@imb.uq.edu.au,t.bailey@imb.uq.edu.au,
http://www.imb.uq.edu.au

Abstract. One component of the genomic program controlling the transcriptional regulation of genes are the locations and arrangement of transcription factors bound to the promoter and enhancer regions of a gene. Because the genomic locations of the functional binding sites of most transcription factors is not yet known, predicting them is of great importance. Unfortunately, it is well known that the low specificity of the binding of transcription factors to DNA makes such prediction, using position-specific probability matrices (motifs) alone, subject to huge numbers of false positives. One approach to alleviating this problem has been to use phylogenetic "shadowing" or "footprinting" to remove unconserved regions of the genome from consideration. Another approach has been to combine a phylogenetic model and the site-specificity model into a single, predictive model of conserved binding sites. Both of these approaches are based on alignments of orthologous genomic regions from two or more species. In this work, we use a simplified, theoretical model to study the statistical power of the later approach to the prediction of features such as transcription factor binding sites. We investigate the question of the number of genomes required at varying evolutionary distances to achieve specified levels of accuracy (false positive and false negative prediction rates). We show that this depends strongly on the information content of the position-specific probability matrix and on the evolutionary model. We explore the effects of modifying the structure of the phylogenetic model, and conclude that placing the target genome at the root of the tree has a negligible effect on the power predicted by the model. Hence, as it is much easier to calculate, we can use this as an approximation to phylogenetic motif scanning using real trees. Finally we perform an empirical study and demonstrate that the performance of current phylogenetic motif scanning programs is far from the theoretical limit of their power, leaving ample room for improvement.

1 Introduction

Phylogenetic motif models are probabilistic models of sequence features. They are a natural extension of the probabilistic motif models used in computational biology to represent and identify sequence features such as transcription factor

M. Vingron and L. Wong (Eds.): RECOMB 2008, LNBI 4955, pp. 112–126, 2008.

binding sites (TFBSs), splice junctions and binding domains in DNA, RNA and protein molecules, respectively (1; 2). Phylogenetic motif models extend the usefulness of standard motifs by leveraging the knowledge that important features in biological sequences tend to evolve more slowly than the neutral rate, a standard assumption of comparative genomics. Phylogenetic motif models are a refinement of the idea of phylogenetic footprinting (3) and shadowing (4), key tools in the arsenal of comparative genomics. This study examines the statistical power provided by phylogenetic motif models for identifying sequence features as a function of the number of comparative genomes, their average evolutionary distance and the information content of the motif.

Standard probabilistic motif models assume that sequence features have a fixed length, and that the frequencies of the letters (e.g., base or residue) that occur at each position in an occurrence of the feature are independent. This allows the motif model to be completely described by a single position-specific probability matrix (PSPM), M, where M_a^j gives the probability of observing letter a at position j in the motif. Thus, the motif model defines the probability of any sequence, x, of the correct length, as the product of the corresponding terms in M, written here as $Pr(x|M)$.

Phylogenetic motif models extend standard motif models to allow them to define the probability of a *multiple alignment*, rather than of a sequence. In addition to the motif model, M, they incorporate a model of evolution (substitution model, e.g., Jukes-Cantor or Hasegawa-Kishino-Yano (HKY) (5)), E, and a phylogenetic tree, T. Each sequence in the alignment is associated with one leaf in the tree, as they are assumed to be orthologous (descended from a common, ancestral sequence.) In essence, the model treats each column of the multiple alignment as though it were a "letter", and defines the probability of the alignment (with the same length as the motif) as the product of the probabilities of the individual columns. Under the model the probability of an alignment *column* is the probability of observing the letters in the column assuming the evolutionary substitution model and assuming that the sequences (and their ancestors) have been under purifying selection to maintain the frequencies given in the corresponding column of the motif, M^j. Thus, the model is a direct generalization of the standard probabilistic motif model and it defines the probability of the multiple alignment column, σ, corresponding to the jth position in the motif, here written $Pr(\sigma|M^j, E, T)$. (Since our model assumes that alignment columns are independent, this is easily generalized to the probability of the complete alignment of length L by taking the product of the column probabilities.)

The focus of this paper is on the theoretical limits on the utility of phylogenetic motif models for *identifying* genomic features when the motif is known, here referred to as "motif search". In the past few years, algorithms have been developed that use phylogenetic motifs for motif search, notably the Monkey algorithm (6; 7) and Motiph (unpublished, available as part of the Meta-MEME software http://metameme.sdsc.edu). These tools make more sophisticated use of the information implicit in an alignment of orthologous sequences than tools such as the UCSC Genome Browser (8), the ECR Browser (9), and ConSite (10), because they explicitly use a model of substitution and the evolutionary relationships and distances specified by a phylogenetic tree.

Despite the existence of such tools and their intuitive usefulness, little is known about the limits of their ability to detect genomic features. This is mainly due to the difficulty and expense of obtaining "gold standard" sets of all known, functional instances of a feature in a genome. Lacking such a gold standard, it is difficult to validate the "false positive" (FP) rate of a model since one doesn't know how many of the supposed false positive predictions may be real. Similarly, if true instances of a feature are missing from the validation set, one cannot accurately estimate the "false negative" (FN) rate of a model.

An important biological application where this problem is particularly acute is in the identification of transcription factor binding sites (TFBSs), where it is well known that standard probabilistic motif search suffers from overwhelming numbers of false positive predictions (the so-called "Futility Theorem" (11).) The evolutionary motif search algorithms already mentioned were developed in large part specifically to overcome this problem, but little or no data is available as to the extent to which they succeed.

In this paper we develop a theoretical framework for analyzing the statistical power of an evolutionary motif model used in motif search in a "target" genome. We assume that the search uses the standard approach for scoring putative sites in the multiple alignment–the log-odds score–the logarithm of the ratio of the probability of the site given the evolutionary motif model or given the neutral ("background") model, respectively. Our framework allows us to compute, for any specified motif and evolutionary model, the number of comparative genomes required in order to achieve given FP and FN rates. Conversely, we can compute a theoretical ROC-like curve for a motif, plotting FN rate as a function of FP rate for a given number of genomes at a given evolutionary distance from the target genome.

To compute the theoretically achievable FN and FP rates of a motif, we must estimate the distributions of the log-odds scores under the motif and background models, respectively. To make this computation feasible, we make the following simplifying assumption–that each of the comparative genomes is at the same evolutionary distance from the target genome. This assumption allows the phylogeny to be represented by a star topology, and makes the probabilities of the letters in each of the genomes independent, given the letter at the root of the tree. It also makes the contributions to the log-odds score from each genome additive, and allows us to parameterize a problem with a single distance, D, the length of each of the branches in the star tree.

This is the same approach as taken by Eddy (12), who studied the simpler problem of determining if a column or set of columns in a multiple alignment was *conserved*, as opposed to our goal of identifying if a set of columns is a *conserved instance of the particular feature type* defined by the motif model. We show the validity of the simplifying assumption of a phylogenetic star by computing FP and FN rates for an actual species tree for four yeast species. This allows our results to be directly applicable to existing phylogenetic motif search algorithms such as Monkey. In what follows, we will refer to Eddy's goal as estimating the statistical power of *phylogenetic footprinting*.

Our theoretical analysis quantifies the maximum sensitivity and specificity of phylogenetic motifs during motif search under ideal conditions. We assume that we have a correct alignment of orthologous sequences. We presume that we know the substitution rate of the motif, R_M, in reference to the neutral substitution rate, which is our metric of evolutionary distance. In some cases, we add an additional assumption that the background substitution rate R_B, varies from the neutral rate. This variation allows us to investigate searching for motifs within sequence regions that are more conserved than neutral sequence. We assume that the evolutionary substitution model is correct. In these assumptions, we mirror the analysis of Eddy (12). We further assume that the feature of interest is accurately represented by the probabilistic motif, M, and the non-site positions are accurately modeled by a 0-order Markov model with parameters B. Of course, we also assume that the underlying premise of phylogenetic motif search is correct–that motif sites (features) are under identical purifying selection in each of the comparative organisms and their common ancestor. We summarize our model parameters as $\theta_M = \{M, E, T, R_M\}$ for the motif model, and θ_{B} = $\{B, E, T, R_B\}$ for the background model.

Our framework allows us to explore a number of factors affecting the statistical power of phylogenetic motif models. We demonstrate that identifying conserved motifs across phylogenies requires fewer genomes than the number predicted for phylogenetic footprinting. We show that a small part of the improvement is due to the use of the Halpern-Bruno modification in the model of the substitution probabilities, and that the majority of the improvement comes from having *site-specific probability distributions* in the model of evolution. We demonstrate that the information content of a motif has an inverse relationship with the number of required genomes, up until about 17 bits. We provide estimates of the number of genomes needed when the motif is less conserved, and we explore the difficulty encountered when searching for motifs in genomes with large regions evolving more slowly than the neutral rate. Finally, we also explore the affects of the topology of the phylogenetic model. Our results suggest that placing the target genome at the center of the star has little effect on the statistical power estimates, thus validating the model as a method for estimating the statistical power of phylogenetic motif scanning with real phylogenetic trees.

2 Methods

2.1 Phylogenetic Motif Model

Our phylogenetic motif model involves computing a log-odds ratio of an alignment column σ of N sequences given evolutionary models of the motif and background θ_M and θ_B, respectively. (Since log-odds scores are additive, this generalizes easily to the score for an alignment of length L by summing the

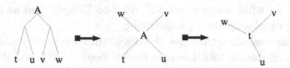

Fig. 1. Transformation from a real species tree to the 'target-centric' phylogenetic model

scores of the individual columns.) When aligned with the jth position in the motif, the log-odds score for this column is written as

$$S(\sigma) = \log \frac{Pr(\sigma|\theta_{M^j})}{Pr(\sigma|\theta_B)}.$$

The two models θ_M and θ_B incorporate the frequencies in the position-specific probability matrix (PSPM) of the motif, M, the background frequencies of the residues, B, different substitution rates for the two models R_M and R_B respectively, and an evolutionary model for calculating the substitution probabilities. (M^j represents the frequencies in position j of the motif.) We use a phylogenetic star tree with equal branch lengths, D, to describe the relationship between the genomes. By placing the first genome in the center of the star (as Eddy does (12)), we are able to produce a dynamic programming solution that computes the probability distribution in time linear to the motif parameters. We verify the accuracy of this in the later part of the study by placing an unknown ancestor in the center (as has been done in other work (13; 14; 6)) and calculating the probability distribution by brute force, which is practical only for trees containing fewer than nine or ten genomes.

When we place the target genome in the center of the star, the score function is

$$S(\sigma) = \log \frac{Pr(\sigma_1|M^j) \prod_{i=1}^{N} Pr(\sigma_i|\theta_{M^j}, \sigma_1)}{Pr(\sigma_1|B) \prod_{i=1}^{N} Pr(\sigma_i|\theta_B, \sigma_1)},$$

which can be rewritten as

$$S(\sigma) = \log \frac{Pr(\sigma_1|M^j)}{Pr(\sigma_1|B)} + \sum_{i=2}^{N} \log \frac{Pr(\sigma_i|\theta_{M^j}, \sigma_1)}{Pr(\sigma_i|\theta_B, \sigma_1)},$$

where σ_1 is the target genome's letter (Fig. 1), and the σ_i are the letters in the other genomes. Note that, when the target genome is in the center of the star, the probability of the site in the target (first) genome is defined completely by the motif model, M, and background model, B, and does not involve the evolutionary substitution model. We compute both of these scores using the "pruning algorithm" of Felsenstein (5).

In this study, we use the HKY (15) substitution model to calculate the substitution probabilities for both the background and the motif evolutionary models.

(Our analysis allows any of the standard substitution models, and our implementation incorporates the Jukes-Cantor, Kimura 2-parameter, F81, F84, HKY and Tamura-Nei models). For the motif evolutionary model, we apply the Halpern-Bruno modification (16), using the appropriate column of the motif PSPM as the equilibrium frequencies.

We use the parameter settings of the HKY model employed in MONKEY, so that the transition-transversion ratio is set to 3.8, and the background distribution, B, is set to $B_A = B_T = 0.3$ and $B_C = B_G = 0.2$. These values are very similar to the ones employed by Eddy in his numerical verification of his phylogenetic footprinting study using an HKY-generated sample (12). The scoring function we use in this study is identical to that of MONKEY (6).

Two assumptions of independence that simplify the process of calculating the probability distributions required in computing the distribution of the log-odds score, S. Firstly, the assumption of a phylogenetic star with the target genome in the center means that each genome evolves from the target independently, hence the probability of N genomes, is the probability of the first $N - 1$ genomes times the probability of seeing the Nth genome. Secondly, the fact that we assume independence between the positions within the motif, means that the probability distribution for the score considering only the first m columns in the multiple alignment is the probability of seeing the first $m - 1$ columns times the probability of the mth column.

These assumptions allow us to apply dynamic programming to calculate a discretized approximation to the score probability distributions (17). We calculate the distribution under both the assumption that we are dealing with a conserved motif, and under the assumption that we are dealing with a neutral sequence. We are then able to generate the cumulative distributions under each model and determine if, for the given number of genomes, there is an S score threshold that satisfies the false positive and false negative criteria.

Our algorithm is linear in the length of the motif, L, the maximum number of genomes to be tested, G, however it is quadratic in the size, s, of the discretized distribution, so it has computational complexity $O(LGs^2)$. We found that to obtain reliable cumulative distributions we needed to use a discretization size, s, of $2 \cdot 10^4$. Hence, the computation time for long motifs or large numbers of genomes can be extensive. For example, to produce the two cumulative probability distributions for four genomes at evolutionary distances of 0.19 and 0.31 the algorithm takes on the order of 5 minutes on a 2 gigahertz workstation. However, to calculate the number of required genomes over 100 different evolutionary distances with a low information content motif requires 80 hours of computing time on the same workstation.

We validated the cumulative distributions by generating Q-Q plots, in which we sample 10,000 alignments generated under the two models, and plotted the p-values of each log-odds score predicted by our model against those suggested by the random sampling. The Q-Q plots show that the estimated p-values are very accurate (data not shown).

2.2 Motifs and Information Content

In this paper, we use motifs from the JASPAR (18) and SCPD (19) databases. These databases contain "count" matrices, computed by aligning known TFBSs and counting the number of occurrences of each nucleotide in each position in the known sites. We convert these counts to a probability matrix, M, by normalizing each column to sum to one. To account for small-sample errors, we add a "pseudocount", equal to 0.375, to each count before normalizing. (This value was determined to be optimal for normalizing TFBS motifs by Frith *et al.* (20).)

To calculate the information content of the motif, we use the same derivation of the Shannon entropy employed in the calculation of sequence LOGOs (21). The information content (in bits) of a DNA PSPM, M, of length L, is given as

$$IC(M) = \sum_{j=1}^{L} \left(2 + \sum_{a \in \mathcal{A}} M_a^j \log_2(M_a^j) \right).$$

This is equal to the average log-odds score of motif instances when the background base distribution is uniform.

3   Results

3.1   Required Number of Comparative Genomes for Typical TFBS Motifs

In our first study, we take two TFBS motifs from the JASPAR database (18), both of length eight, but with different information contents. We compare the theoretical number of genomes required for accurately detecting sites of each of these motifs (computed by our approach) with the number of genomes required to simply predict the conservation of a length-eight region by phylogenetic footprinting. Our chosen motifs are the length-eight motifs with the highest and lowest information content in the JASPAR database: MA0033 (FOXL1) and MA0024 (E2F1), respectively. The LOGOs for these two motifs are shown in the inset in Fig. 2. We use a single setting for the statistical power, which is the most stringent used in Eddy's study: an FP rate of 10^{-4} and an FN rate of 10^{-2}, and we place the target genome in the center of the phylogenetic star.

The results are shown in Fig. 2. For both high and low information content motifs, fewer genomes are required to predict, at the given level of accuracy, conserved *motif sites* than to predict which regions are conserved. As one might expect, the higher information content motif has more statistical power, requiring fewer genomes.

We conducted simulations in order to investigate the effects of the main differences between our approach and the numerical simulations performed by Eddy. Firstly, that when we use the HKY evolutionary model, we have site specific probability distributions that affect the substitution probabilities. Secondly, we use the the Halpern-Bruno (HB) modification when generating the substitution

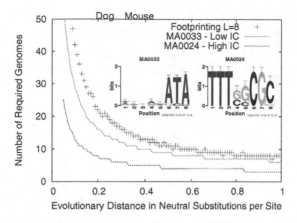

Fig. 2. Results for the length-eight motifs: FP $= 10^{-4}$ and FN $= 10^{-2}$. The plot of points is a reproduction of Eddy's results for a length-eight conserved region. The middle line shows the results for our model using the low information content JASPAR motif MA0033. The lower line shows the results of our model using the high information JASPAR motif MA0024. We place the target genome at the center of the star and use the parameters outlined in Section 2.1. The x position of first letters of 'Dog' and 'Mouse' correspond to their approximate evolutionary distances to Human.

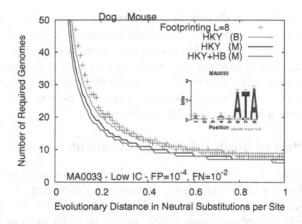

Fig. 3. Results for the length-eight motif MA0033 with different evolutionary models: FP $= 10^{-4}$ and FN $= 10^{-2}$. The plot of points is a reproduction of Eddy's results for a length-eight conserved region. The three line plots show the results for our model using the HKY model or the HKY with the Halpern-Bruno modification. The equilibrium frequencies for the model(s) are shown in parentheses—"B" (background) or "M" (motif). The x position of first letters of 'Dog' and 'Mouse' correspond to their approximate evolutionary distances to Human.

probabilities. We use the low information content motif MA0033 to test the cumulative effect of adding each of the these features to the evolutionary model. We first plot results generated using our model with the HKY using the background

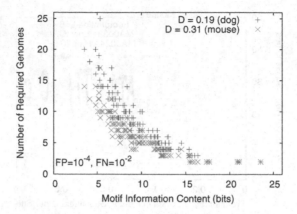

Fig. 4. The effect of motif information content on the minimum number of genomes required. Each point, (X, Y), represents an experiment using a single JASPAR motif. X is the information content of the motif, and Y is the minimum number of genomes required at a given evolutionary distance, D, in order to achieve a prediction accuracy of FP $= 10^{-4}$ and FN $= 10^{-2}$. We place the target genome at the center of the star and use the HKY model with the Halpern-Bruno modification for the motif substitution rates.

model as the equilibrium distribution (rather than using the motif). As can be seen in Fig. 3, these results are barely distinguishable from Eddy's footprinting results. This shows, as one would expect, that using a phylogenetic motif model with no information content is essentially the same as phylogenetic footprinting. When we modify HKY to use the site specific frequencies from the columns of the model, we see an improvement in statistical power, and an incremental improvement when we apply the HB modification.

3.2 The Effect of Motif Information Content

In order to obtain a greater indication of the influence of information content over the number of required genomes, we conduct a second study using all motifs from the JASPAR database (18). We calculate the number of genomes required to achieve a FP rate of 10^{-4} and a FN rate of 10^{-2} for each of the 123 motifs and plot it against the information content of the motif.

We place the first genome in the center of the star and we use two different evolutionary distances (D)–0.19 and 0.31–the values of the independent branch lengths chosen by Eddy as representative distances corresponding to human-dog and human-mouse inter-genomic distances. The results are both shown as a scatter plot in Fig. 4. The most notable result is that there is a strong, general trend for the number of required genomes to decrease with information content. However, as the information content reaches and exceeds 17 bits, the plots reach the limiting value of two genomes for both evolutionary distances.

3.3   Empirical Validation of the Model

The decision to place the target sequence in the center of a phylogenetic star is biologically implausible. However, given time-reversible models of evolution, we can rearrange a phylogenetic tree into a star with the target species in the center, such that the total independent branch length of the tree is conserved (12). The rearrangement simplifies the mathematics, permitting the kind of analysis presented in this paper. However, we wish to verify that the transformation of the tree does not drastically change the results of phylogenetic motif scanning. Furthermore, we wish to see how close to the theoretical limits of statistical power the current phylogenetic motif scanning programs are currently performing.

As a case study for exploring these issues, we chose the problem of identifying TFBSs within a multiple alignment of four yeast genomes, which has been analysed in a number of previous studies (22; 23; 12). We use the Kellis *et al.* (22) four Yeast genome data set, including phylogenetic tree for our analysis. We take the total independent branch length of 0.963 for this tree and divide it by three, giving a branch length of $D = 0.321$ in the target-centric tree.

In order to evaluate the effect of our target-in-the-center simplification of the tree, we developed an alternative program that calculates the theoretical power of phylogenetic motif scanning on *any* tree for a (small) specified number of genomes. We chose a number of TFBSs from the SCPD (19) database (ABF1, MCM1, RAP1, REB1 and URS1H) and calculated their theoretical statistical power, using both the real tree and the target-in-center transformed tree. We also computed the theoretical power of using each of these motifs in a simple motif scan (without comparative genomes). We use two phylogenetic motif scanning programs, MONKEY (6; 7) and Motiph, to scan for these same TFBSs in the

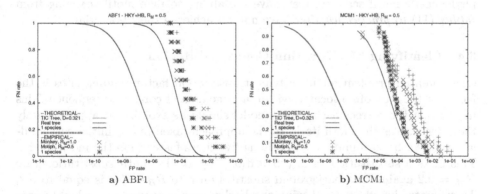

a) ABF1   b) MCM1

Fig. 5. Comparison of theoretical and empirical power of phylogenetic motifs. The plot shows ROC-like curves for two of the experiments performed with SCPD motifs. Panel **a** shows results for ABF1 and panel **b** shows results for MCM. The lines show the theoretical limit of the statistical power using the real tree, the target-in-center tree ("TIC"), or in a simple motif scan ("1-species"). The points show the empirical performance of the phylogenetic motif scanning programs Motiph and Monkey, and a simple motif scan ("1-species").

aligned yeast sequences, and we plot the statistical power of the results on the same graphs. We plot the results as FP versus the FN rate, allowing us to visualise the tradeoff between these two measures of accuracy.

The results for motifs ABF1 and MCM1 are shown in Fig. 5, and are representative of what we see for all five motifs (data not shown). We can draw several conclusions from these plots. Firstly, the two curves for theoretical power of phylogenetic motif scanning are almost identical, and actually overlay each other in the plots. This demonstrates that using the target-in-the-center transformation yeilds an excellent estimate of the statistical power of models using the true tree. Secondly, the theoretical increase in power of phylogenetic motif scanning over simple motif scanning is close to three orders of magnitude (compare pink and blue plots in each panel). Thirdly, although simple motif scanning is performing very close to its theoretical limits, the same is not true for phylogenetic motif scanning. In fact, in the examples we study here, the phylogenetic motif scanners perform worse than simple motif scanning at low FN (high sensitivity) thresholds.

The poor emprical performance of the phylogenetic motif model scanning algorithms in Fig. 5 could be due to any of a number of factors. Firstly, our "gold standard" of known TFBSs, SCPD, is probably incomplete, which will lead to erroneous "false positives". Secondly our alignments are no doubt imperfect, causing real binding sites to be missed by phylogenetic motif model scanning (but not by simple motif scanning). Thirdly, it is well known that functional binding sites are often *not conserved* (24; 7), once again causing phylogenetic motif model scanning to miss binding sites. Finally, false positives may be caused by highly conserved regions that contain sequences that match the motif, but are not *functional* binding sites. Whatever, the reason we can surmise that, although phylogenetic motif scanners may have the ability to take motif scanning from *futility* (11) to *practicality*, they have not yet achieved their potential.

3.4   Identifying Motifs within Conserved Regions

One potential problem for identifying transcription factor binding sites is the fact that they are often located within a large area of conserved sequence. This means that the correct background model should be evolving at approximately the same rate as the motif itself. In the approach taken in our initial case studies, we use a background model evolving five times faster than the motif model. This gives a substitution ratio ("S Ratio") $\frac{R_M}{R_B}$ of the motif substitution rate $R_M = 0.2$ against the background substitution rate $R_B = 1.0$ is equal to 0.2. To evaluate the effect of allowing the background sequence to be highly conserved, we generate two alternative versions of the plots shown in Fig. 2 for the high information content motif MA0024. We keep the motif substitution rate constant at 0.2 and modify the background substitution rate so that we obtain substitution ratios of 0.5 and 1.0. The results are shown in Fig. 6.

As would be expected, the number of required genomes increases as we attempt to identify TFBSs within highly conserved sequence. In the worst case scenario, the background sequence is as conserved as the motif itself (ignoring

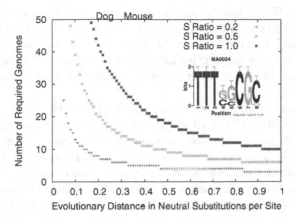

Fig. 6. Effect of changing the background substitution rate. Using the high information content motif MA0024 we searched for the number of genomes required to satisfy $FP = 10^{-4}$ and $FN = 10^{-2}$. The three curves correspond to three different setting of the motif to background substitution rates. The value 0.2 is identical to our first study (Fig. 2), the value 0.5 corresponds to a background rate of 0.4, i.e. a region that is evolving at twice the rate of the motif and at 40% of the neutral rate. The results for a value of 1.0 correspond to a background substitution rate of 0.2, i.e. a background region that is as conserved as the motif being sought, and they are both evolving at 20% of the neutral rate. First letters of 'Dog' and 'Mouse' correspond to approximate distance to Human.

the possibility that it might be more conserved) and we require on the order of 3 to 6 times the number of genomes to achieve the same statistical power.

4   Discussion

Phylogenetic motif models are a specialization of profile phylogenetic hidden Markov models (25). The introduction of phylogenetic relationships has been responsible for considerable improvement in the performance of *de novo* motif discovery algorithms (26; 14; 13). However, the advantages of phylogenetic motif models for motif search are less clear. Even with the use of phylogenetic motif models and/or phylogenetic footprinting, transcription factor binding sites have remained difficult to identify due to their short lengths, low-specificity motifs and their presence inside highly conserved promoter regions. We have sought to analyse each of these limiting factors and present theoretical results on the number of genomes required at a variety of evolutionary distances to achieve reasonable statistical power.

In the first set of simulations we saw that fewer genomes are required to achieve given target levels of statistical significance in phylogenetic motif search than for phylogenetic footprinting. The amount of reduction depends on the information content of the motif. Compared to phylogenetic footprinting, phylogenetic motif search typically requires only about 50% as many genomes with high-information TFBS motifs, and 90% as many with low-information motifs. We showed that the size of the improvement is not merely a matter of the specificity of the motif, but depends on the particulars of the evolutionary model. The use of site specific frequencies from the motif provide the majority of the improvement, and the use of the Halpern-Bruno modification to the substitution model adds additional power.

In the second set of simulations we explicitly explored the relationship between the information content of a motif and the number of genomes required to achieve a given statistical power. We saw the number of genomes required is inversely proportional to the information content of the motif. However, at an information content of 17 bits or more, we reach the limiting case of comparing 2 genomes. This means that for high information content motifs, the evolutionary distance between comparative genomes is much less important than for low information content motifs.

In the third study we performed an empirical investigation to validate the model and test the validity of the target-in-the-center simplification. We observed that placing the target genome in the center of a phylogenetic star produced results almost indistinguishable from the theoretical results using the phylogenetic tree for four yeast species. When we compared these results with the theoretical power of performing simple motif scanning, we saw that phylogenetic motif scanners have the potential to improve performance by up to three orders of magnitude, which would take the performance of the model from *futility* (11) to *practicality*. However, when we compared these theoretical curves with empirical results of the accuracy of two phylogenetic motif scanning models MONKEY and Motiph, we observed that the empirical results are far worse than theory predicts. In fact, the empirical results are, in general, no better than simple motif scanning. This disappointing result may be due to a number of factors: bad alignments, incomplete knowledge of the real TFBSs (thus false assignment of false positives), divergence or drift (7; 24) of binding sites, or the prevalence of highly conserved regions that contain sequences that are indistinguishable from functional binding sites. It remains to be seen whether these problems can be identified and overcome, or whether they are simply inherent problems with the data.

In the final set of simulations we saw that, under the worst case scenario where a TFBS is evolving at the same rate as the surrounding promoter region, the number of genomes required increases significantly to between three and six times the number required when the motif is evolving five times slower. This result may explain to some extent the great difficulty that has been encountered in identifying TFBSs accurately, while at the same time providing an upper limit on how many genomes we need at a given distance to identify these elusive features.

We intend to use the tool developed here to create a database of ROC-like curves for a wide variety of motifs from the JASPAR and transfac SCPD databases. These will be provided via the WWW, and will provide estimates of the theoretical false positive versus false negative rates for each motif, on a given set of genomes at different, fixed evolutionary distances.

Acknowledgement

JH is funded by Australian Research Council grant DP0770471. TLB is funded by NIH grant RO-1 RR021692-01.

Refernces

[1] GuhaThakurta, D.: Computational identification of transcriptional regulatory elements in DNA sequence. Nucleic Acids Res. 34(12), 3585–3598 (2006)

[2] Stormo, G.D.: DNA binding sites: representation and discovery. Bioinformatics 16(1), 16–23 (2000)

[3] Gumucio, D.L., Heilstedt-Williamson, H., Gray, T.A., Tarlé, S.A., Shelton, D.A., Tagle, D.A., Slightom, J.L., Goodman, M., Collins, F.S.: Phylogenetic footprinting reveals a nuclear protein which binds to silencer sequences in the human gamma and epsilon globin genes. Mol. Cell Biol. 12(11), 4919–4929 (1992)

[4] Boffelli, D., McAuliffe, J., Ovcharenko, D., Lewis, K.D., Ovcharenko, I., Pachter, L., Rubin, E.M.: Phylogenetic shadowing of primate sequences to find functional regions of the human genome. Science 299(5611), 1391–1394 (2003)

[5] Felsenstein, J.: Evolutionary trees from DNA sequences: a maximum likelihood approach. J Mol. Evol. 17(6), 368–376 (1981)

[6] Moses, A.M., Chiang, D.Y., Pollard, D.A., Iyer, V.N., Eisen, M.B.: MONKEY: identifying conserved transcription-factor binding sites in multiple alignments using a binding site-specific evolutionary model. Genome Biol. 5(12), R98 (2004)

[7] Moses, A.M., Pollard, D.A., Nix, D.A., Iyer, V.N., Li, X.Y., Biggin, M.D., Eisen, M.B.: Large-scale turnover of functional transcription factor binding sites in drosophila. PLoS Comput. Biol. 2(10), e130 (2006)

[8] Kent, W.J., Sugnet, C.W., Furey, T.S., Roskin, K.M., Pringle, T.H., Zahler, A.M., Haussler, D.: The human genome browser at UCSC. Genome Res. 12(6), 996–1006 (2002)

[9] Loots, G.G., Ovcharenko, I.: rVISTA 2.0: evolutionary analysis of transcription factor binding sites. Nucleic Acids Res. 32(Web Server issue), W217–W221 (2004)

[10] Sandelin, A., Wasserman, W.W., Lenhard, B.: ConSite: web-based prediction of regulatory elements using cross-species comparison. Nucleic Acids Res 32(Web Server issue), W249–W252 (2004)

[11] Wasserman, W.W., Sandelin, A.: Applied bioinformatics for the identification of regulatory elements. Nat. Rev. Genet. 5(4), 276–287 (2004)

[12] Eddy, S.R.: A model of the statistical power of comparative genome sequence analysis. PLoS Biol. 3(1), e10 (2005)

[13] Siddharthan, R., Siggia, E.D., van Nimwegen, E.: PhyloGibbs: a gibbs sampling motif finder that incorporates phylogeny. PLoS Comput. Biol. 1(7), e67 (2005)

[14] Sinha, S., Blanchette, M., Tompa, M.: PhyME: a probabilistic algorithm for finding motifs in sets of orthologous sequences. BMC Bioinformatics 5, 170 (2004)

[15] Hasegawa, M., Kishino, H., Yano, T.: Dating of the human-ape splitting by a molecular clock of mitochondrial dna. J Mol Evol 22(2), 160–174 (1985)

[16] Halpern, A.L., Bruno, W.J.: Evolutionary distances for protein-coding sequences: modeling site-specific residue frequencies. Mol. Biol. Evol. 15(7), 910–917 (1998)

[17] Staden, R.: Searching for patterns in protein and nucleic acid sequences. Methods Enzymol. 183, 193–211 (1990)

[18] Sandelin, A., Alkema, W., Engström, P., Wasserman, W.W., Lenhard, B.: JAS-PAR: an open-access database for eukaryotic transcription factor binding profiles. Nucleic Acids Res. 32(Database issue), D91–D94 (2004)

[19] Zhu, J., Zhang, M.Q.: SCPD: a promoter database of the yeast Saccharomyces cerevisiae. Bioinformatics 15(7-8), 607–611 (1999)

[20] Frith, M.C., Hansen, U., Spouge, J.L., Weng, Z.: Finding functional sequence elements by multiple local alignment. Nucleic Acids Res. 32(1), 189–200 (2004)

[21] Schneider, T.D., Stephens, R.M.: Sequence logos: a new way to display consensus sequences. Nucleic Acids Res. 18(20), 6097–6100 (1990)

[22] Kellis, M., Patterson, N., Endrizzi, M., Birren, B., Lander, E.S.: Sequencing and comparison of yeast species to identify genes and regulatory elements. Nature 423(6937), 241–254 (2003)

[23] Cliften, P., Sudarsanam, P., Desikan, A., Fulton, L., Fulton, B., Majors, J., Waterston, R., Cohen, B.A., Johnston, M.: Finding functional features in saccharomyces genomes by phylogenetic footprinting. Science 301(5629), 71–76 (2003)

[24] Borneman, A.R., Gianoulis, T.A., Zhang, Z.D., Yu, H., Rozowsky, J., Seringhaus, M.R., Wang, L.Y., Gerstein, M., Snyder, M.: Divergence of transcription factor binding sites across related yeast species. Science 317(5839), 815–819 (2007)

[25] Siepel, A., Haussler, D.: Combining phylogenetic and hidden markov models in biosequence analysis. J Comput Biol. 11(2-3), 413–428 (2004)

[26] Moses, A.M., Chiang, D.Y., Eisen, M.B.: Phylogenetic motif detection by expectation-maximization on evolutionary mixtures. In: Pac Symp. Biocomput., pp. 324–335 (2004)

Transcriptional Regulation and Cancer Genomics

Edison Liu

Genome Institute of Singapore, Singapore
liue@gis.a-star.edu.sg

About the Keynote Speaker. Professor Edison Liu graduated from Stanford University and its medical school. From 1987–96, he was professor of medicine, biochemistry and epidemiology at the University of North Carolina, and Director of its Specialized Program of Research Excellence in Breast Cancer. From 1996–2001, Prof Liu was the Division Director of Clinical Sciences (Intramural program) at the US National Cancer Institute. In 2001, he was appointed the executive director for the Genome Institute of Singapore He has received the Rosenthal Award from the American Association of Cancer Research and the Brinker International Award from Susan Komen Foundation for his breast cancer research. He is currently the President of the Human Genome Organization (HUGO).

M. Vingron and L. Wong (Eds.): RECOMB 2008, LNBI 4955, p. 127, 2008.
© Springer-Verlag Berlin Heidelberg 2008

Automatic Recognition of Cells (ARC) for 3D Images of *C. elegans*

Fuhui Long[1,*], Hanchuan Peng[1], Xiao Liu[2], Stuart Kim[2], and Gene Myers[1]

[1] Janelia Farm Research Campus, Howard Hughes Medical Institute, Ashburn, Virginia, USA
[2] Department of Developmental Biology, Stanford University, Stanford, California, USA
* longf@janelia.hhmi.org

Abstract. The development of high-resolution microscopy makes possible the high-throughput screening of cellular information, such as gene expression at single cell resolution. One of the critical enabling techniques yet to be developed is the automatic recognition or annotation of specific cells in a 3D image stack. In this paper, we present a novel graph-based algorithm, ARC, that determines cell identities in a 3D confocal image of *C. elegans* based on their highly stereotyped arrangement. This is an essential step in our work on gene expression analysis of *C. elegans* at the resolution of single cells. Our ARC method integrates both the absolute and relative spatial locations of cells in a *C. elegans* body. It uses a marker-guided, spatially-constrained, two-stage bipartite matching to find the optimal match between cells in a subject image and cells in 15 template images that have been manually annotated and vetted. We applied ARC to the recognition of cells in 3D confocal images of the first larval stage (L1) of *C. elegans* hermaphrodites, and achieved an average accuracy of 94.91%.

1 Introduction

Automatic recognition of the identities of individual cells in 3D microscopy images is indispensable for the high-throughput analysis of cellular information, such as gene expression levels and cell morphology, at the single cell level. One example is our recent work on high-throughput whole-animal single-cell gene expression analysis for *C. elegans* [1] based on a 3D digital atlas of the nuclei of this animal [2]. Currently cell recognition is accomplished by expert manual annotation, which is extremely labor intensive and basically untenable for a large number of images. Using a small set of, say a dozen or so, manually annotated images of the same organism as *templates*, we demonstrate that it is possible to extract cellular information such as location and relative spatial relationship of individual cells from these templates, and automatically assign names to cells in any new image of the same organism provided it is sufficiently stereotyped, which *C. elegans* most certainly is. But this is not the only application, for example, the embryonic and larval neurons of *D. melanogaster* are highly stereotyped, as are many other early developmental patterns. Figure 1 illustrates this problem schematically.

M. Vingron and L. Wong (Eds.): RECOMB 2008, LNBI 4955, pp. 128–139, 2008.
© Springer-Verlag Berlin Heidelberg 2008

It is challenging to develop such automatic cell recognition technique for several reasons. First, individual cells in an image need to be segmented to high accuracy as a precursor, and this is difficult when the image quality is limited and cells are tightly clustered. Second, it is common that an image of an entire organism (e.g. *C. elegans*), or a particular tissue of an organism (e.g. the mushroom body of a fly brain) may contain hundreds or thousands of cells. Thus the scale of the problem presents a challenge to traditional graph matching techniques [4~14], which have been successfully applied in applications such as face recognition [11], object tracking [12], image retrieval [13], and image registration [14] to find correspondences between two sets of spatial points, each usually containing less than a hundred objects. Finally, due to the imperfection of staining and the resolution-limit of the imaging, an expert annotator can only annotate the subset of cells in a template image that are large enough and strongly stereotyped in location. Thus the problem becomes a subset-matching problem, which is more difficult than the case where both the subject image and the template images have the same number of cells.

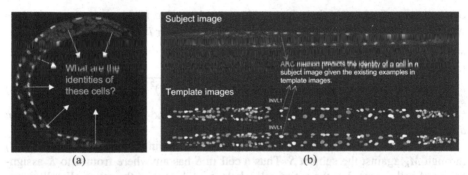

Fig. 1. (a) is a raw image of *C. elegans*, and in (b) we illustrate the cell recognition problem on image stacks where the worm has first been straightened [3], and then size and orientation normalized and segmented as described in our earlier work [2]. Cells in a template are colored so that locally it is clear which cells have the same identity between instances.

For this problem, our new method, called Automatic Recognition of Cells (ARC), is developed below in three layers of increasing refinement or power as follows. In Section 2 we introduce a basic framework of a two-stage bipartite matching that first matches cells in a subject image against the annotated cells in *each* template image, and then matches cells in the subject image to a unique cell by considering assignment scores based on the first-level matching results. In Section 3 we introduce and constrain the possible matchings to observe relative spatial invariant relationships discovered in the training stacks, specifically, the anterior-posterior (AP), left-right (LR), dorsal-ventral (DV) invariant relationships. In Section 4 we introduce a marker-based strategy in which a fiducial framework of alternately-labeled marker cells is automatically annotated with very high confidence, and then these are used to triangulate and constrain the annotation of the remaining cells.

We applied ARC to the 3D confocal images of the first larval stage (L1) of *C. elegans* hermaphrodite that has ~558 cells [1]. For this problem we manually annotated 351 cells in 15 templates. Most of the un-annotated cells are small neurons in the head of the organism. Our results show that using our marker-guided, AP/LR/DV-constrained, two-stage bipartite matching, we achieved 94.91% accuracy in searching for these 351 cells in an initially unsegmented image stack.

2 Two-Stage Bipartite Graph Matching

Given a subject image S in which cells are to be recognized by a computer program and a template image T in which cell identities have been annotated by biologists, we can formulate our problem as bipartite matching. Consider the directed bigraph $G = (V^S \cup V^T, E)$ consisting of two disjoint vertex sets, V^S for the subject S and V^T for the template T, and all edges $E = V^S \times V^T$ from V^S to V^T. Later we will restrict E to be a subset of all the possible pairings. In the first stage we find a minimal cost, maximal matching M between V^S and V^T. That is, we minimize a cost function

$$E^1 = \sum_{s \to t \in M} D^1(s \to t) \tag{1}$$

over all sets of edges M for which adding another edge to M gives a set of edges which is no longer a matching, i.e. $\forall s \to t \; (out(s) = 1 \text{ or } in(t) = 1)$ where out and in are the out- and in-degree of a vertex. $D^1(s \to t)$ is the distance between cells s and t, i.e.,

$$D^1(s \to t) = \|p_s - p_t\| = \sqrt{(x_s - x_t)^2 + (y_s - y_t)^2 + (z_s - z_t)^2} \tag{2}$$

where $p_c = (x_c, y_c, z_c)$ is the coordinate of the cell c. We find M by using the Hungarian algorithm [15].

If we have K template images T_1 through T_K, we obtain K maximal matchings M_1 through M_K against the subject S. Thus a cell in S has anywhere from 0 to K assignments of cell names. Let the set of cell labels $L = \bigcup_k V^{T_k}$ be the set of all cell names used in some template. Note carefully, that not every cell annotation in L is necessarily labeled in a template. We then use the second stage bipartite matching to determine the unique identity of each cell in S, by finding the minimum cost E^2, maximal match $M^* \subseteq V^S \times L$ with respect to the cost function D^2 defined as follows:

$$D^2(s \to t) = \left(\sum_{s \in V^S} N(s \Leftrightarrow t) \right) - N(s \Leftrightarrow t) \tag{3}$$

where $N(s \Leftrightarrow t) = |\{k : s \to t \in M_k\}|$ is the number of times that s is assigned to t. In summary, the first stage finds the best matching of subject cells to the cells of each template based on minimizing Euclidean distances, and the second stage finds the best matching of subject cells to a *label* by, in effect maximizing the number of template cells that support the assignment. Because the bipartite matching minimizes a global cost and guarantees a one-to-one mapping, it is superior to a simple majority vote scheme. Note that the result does not depend on the processing order of the templates.

3 Imposing AP/LR/DV Constraints

The bipartite matching scheme in §2 does not consider the relative spatial relationship among vertices within V^S or within V^T. For example suppose a pair of cells (a,b) in the subject S should be mapped to a pair of cells (u,v) in the template T, with a to u, b to v, where it is always the case that u is anterior to v in *all* the templates. The unconstrained bipartite matching is free to match a to v and b to u and this is likely wrong. To solve this problem, we propose using invariant anterior-posterior (AP), left-right (LR), and dorsal-ventral (DV) relationships between cells to prune the possible match edges, i.e. the set of edges E between V^S and V^T in the bipartite graph model.

3.1 Deriving the Intrinsic AP/LR/DV Relationships between Cells

The intrinsic AP/LR/DV relationships among cells are derived from the template images. Let us take the AP relationship as an example. We compute the $|L|\times|L|$ adjacency matrix $\mathbf{AP}_k$ for each template T_k, where $\mathbf{AP}_k(u,v) = 1$ if cell u is anterior to cell v, or either of u or v is in $L - V^{T_k}$, and 0 otherwise. Then the consensus AP adjacency matrix, denoted $\mathbf{AP}$, can be obtained by applying the simple element-wise AND operation, $\wedge$, on the $\mathbf{AP}_k$, i.e., $\mathbf{AP} = \mathbf{AP}_1 \wedge \mathbf{AP}_2 \wedge ... \wedge \mathbf{AP}_K$. In this matrix, $\mathbf{AP}(u,v) = 1$ if and only if cell u is always anterior to cell v in all K templates, and 0 otherwise (we are assuming that every label is used in at least one template). In the same way, we also compute the LR/DV adjacency matrix $\mathbf{LR}$ and $\mathbf{DV}$ to describe the intrinsic LR and DV relationships among cells across different images.

3.2 Constructing AP/LR/DV Adjacency Matrices for a Subject Image

Given a matching M that maps cells in the subject S to cells in a template T_k, we may construct AP/LR/DV adjacency matrices for S, denoted $\mathbf{ap}$, $\mathbf{lr}$, and $\mathbf{dv}$, as follows. If a pair of cells a and b in S are recognized as cells u and v in T_k, respectively, under the bipartite matching M, i.e., $a \rightarrow u \in M$ and $b \rightarrow v \in M$, and cell a is anterior to b in the subject image then we set $\mathbf{ap}(u,v) = 1$. We also set $\mathbf{ap}(u,v) = 1$ if u or v is in $L - V^{T_k}$. Otherwise $\mathbf{ap}(u,v) = 0$. Similarly, we compute the LR/DV adjacency matrices $\mathbf{lr}$ and $\mathbf{dv}$. In brief, the spatial relationships of the subject are mapped to the template via the matching M.

3.3 Selecting Wrongly Recognized Cells and Pruning Impossible Edges of the Bipartite Graph

Given $a \rightarrow u \in M$ and $b \rightarrow v \in M$, if $\mathbf{ap}(u,v)=1$ and $\mathbf{ap}(v,u)=0$, but $\mathbf{AP}(u,v)=0$ and $\mathbf{AP}(v,u)=1$, then it is the case that cells a and b in the subject, where a is anterior to b, are labeled as cells u and v, with u always posterior to v in the templates they occur in. Thus at least one of the cells a and b in the subject is matched incorrectly. More generally, we may compute a contradiction matrix $\mathbf{C}$ using the 6 adjacency-matrices:

$$\mathbf{C} = \mathbf{C}_{ap} \vee \mathbf{C}_{lr} \vee \mathbf{C}_{dv} \tag{4}$$

$$\mathbf{C}_r = [(\mathbf{R}) \wedge (\neg \mathbf{R}^T) \wedge (\neg \mathbf{r}) \wedge (\mathbf{r}^T)] \vee [(\neg \mathbf{R}) \wedge (\mathbf{R}^T) \wedge (\mathbf{r}) \wedge (\neg \mathbf{r}^T)] \tag{5}$$

where $\vee, \wedge, \neg$ are the element-wise OR, AND, and NOT operations, respectively, and ^{T} is matrix transposition. $\mathbf{R}$ represents adjacency matrices AP, LR, DV, and $\mathbf{r}$ represents adjacency matrices ap, lr, dv respectively. Moreover, when $\mathbf{r}$ equals say $\mathbf{ap}$ in

Eq. (5) then **R** is **AP**. Observe that $C(u,v) = 1$ if and only if one or more of the AP, LR, or DV relationships of cells a and b in the subject image are contradictory to those of cells u and v in the template. Thus at least one of a and b is wrongly recognized.

Based on the contradiction matrix **C**, we select, with high confidence, the cells in the subject that are wrongly labeled by M. We then cut the edges between these cells in the subject image and their mappings in the template and rerun the bipartite matching. To select the cells that are most likely to be wrongly recognized, we count, for each cell a in the subject S, the number of cells in S that have a contradictory AP/LR/DV relationships with cell a, i.e.,

$$conflict(a) = |\{b \mid C(u,v) = 1, a{\rightarrow}u \in M, b{\rightarrow}v \in M\}| \tag{6}$$

We then take the most conflicted cell and remove the edge between it and its assigned vertex in M. We then compute using the Hungarian algorithm [15] a new M with respect to the reduced bipartite graph. This process is repeated until $\Sigma_a \, conflict(a)$ does not decrease for t_{max} sequential steps ($t_{max}=3$ for the results reported, but other values yielded similar results.). Once terminated, one takes as the answer the matching M that gives the minimum $\Sigma_a \, conflict(a)$.

We have thus far not identified M as M^* or one of the M_k. We actually find conflicts for each stage 1 matching M_k and should technically speak of C_k. That is, we produce the best subject to template matching for each template using the matrices **AP**, **DV**, and **LR** that represent the invariant relationships over all the templates. Thereafter, we proceed with compute M^* in stage 2 as before. Algorithm 1 in Appendix shows the pseudo-code of the AP/LR/DV constrained two-stage bipartite matching.

4 Marker-Based Recognition

The recognition approach above treats each cell equally and matches them all together at once. However, biologists usually use markers to aid cell identification. For instance, in the manual annotation of cells in *C. elegans*, our biologists first assigned names to the body wall muscle cells that were stained separately with GFP. With these marker cells labeled, the biologists then annotated both the ventral motor neurons and intestinal cells by examining their positions relative to the marker cells. After that the biologists used relative triangulation to annotate most other cells in trunk. Therefore, in addition to AP/LR/DV-constrained bipartite matching, in the following we present a hierarchical strategy similar to that of the biologists by first identifying marker cells and then using these marker cells to aid the identification of other cells.

4.1 Recognition of Muscle Cells

In the L1 larval stage of *C. elegans*, there are 81 body wall muscle cells and 1 depressor cell distributed along the entire worm body from the head to the tail. In our data (see §5 for details), most muscle cells, lit up by GFP in a separate frequency channel, are well separated from each other and thus are easier to segment and recognize, compared to other cells. We thus first use the AP/LR/DV-constrained bipartite matching to recognize just these 82 cells in the GFP channel. In this case, the adjacency matrices, **AP, LR, DV, ap, lr, dv** and an assignment are computed only for these 82 cells.

4.2 Identifications of Additional Markers

Once the identities of 81 body wall muscle cells and the 1 depressor cell have been determined, we use them as markers to identify cells that can be uniquely determined according to their relative positions with respect to these muscle cells. For this purpose, we again make use of adjacency matrices of template images and of the subject image. However, at this stage we only care about the relative relationship of cells to be recognized with respect to the markers. Thus we use sub-matrices of the six adjacency matrices. Using **AP** as an example, we extract 2 sub-matrices, denoted as $\mathbf{AP}_{(pxq)}$ and $\mathbf{AP}_{(qxp)}$. The sub-matrix $\mathbf{AP}_{(pxq)}$ contains p rows and q columns. The p rows correspond to the p cells in the template to be matched by the cells in the subject image. The q columns correspond to the q marker cells (i.e., 82 in this example). Note that $p+q = N^T$ is the total number of cells annotated in the template images. The sub-matrix $\mathbf{AP}_{(qxp)}$ contains q rows and p columns, corresponding to q marker cells and p cells to be matched by cells in the subject image. The combination of $\mathbf{AP}_{(pxq)}$ and $\mathbf{AP}_{(qxp)}$ reflects the relative AP relationship between a cell and a marker. More specifically, if we denote $\mathbf{B}_1 = \mathbf{AP}_{(pxq)} \wedge (\neg\mathbf{AP}_{(qxp)}^{\mathrm{T}})$, and $\mathbf{B}_2 = (\neg\mathbf{AP}_{(pxq)}) \wedge (\mathbf{AP}_{(qxp)}^{\mathrm{T}})$, then cell u is anterior to marker v if $\mathbf{B}_1(u,v) = 1$ and $\mathbf{B}_2(v,u) = 0$, posterior to marker v if $\mathbf{B}_1(u,v) = 0$ and $\mathbf{B}_2(v,u) = 1$, and can be either posterior or anterior to marker v if $\mathbf{B}_1(u,v) = 0$ and $\mathbf{B}_2(v,u) = 0$. Similarly, we compute $\mathbf{LR}_{(pxq)}$, $\mathbf{LR}_{(qxp)}$, $\mathbf{DV}_{(pxq)}$ and $\mathbf{DV}_{(qxp)}$.

We also extract the sub-matrices of **ap, lr,** and **dv**, denoted as $\mathbf{ap}_{(rxq)}$, $\mathbf{ap}_{(qxr)}$, $\mathbf{lr}_{(rxq)}$, $\mathbf{lr}_{(qxr)}$, $\mathbf{dv}_{(rxq)}$ and $\mathbf{dv}_{(qxr)}$. The r rows (in $\mathbf{ap}_{(rxq)}$, $\mathbf{lr}_{(rxq)}$, $\mathbf{dv}_{(rxq)}$) or r columns (in $\mathbf{ap}_{(qxr)}$, $\mathbf{lr}_{(qxr)}$, and $\mathbf{dv}_{(qxr)}$) correspond to the r cells in the subject image to be recognized (note that $r \geq p$). The q columns (in $\mathbf{ap}_{(rxq)}$, $\mathbf{lr}_{(rxq)}$, $\mathbf{dv}_{(rxq)}$) or q rows (in $\mathbf{ap}_{(qxr)}$, $\mathbf{lr}_{(qxr)}$, and $\mathbf{dv}_{(qxr)}$) correspond to the q cells in the subject image that have been recognized as markers (i.e, 82 muscle cells in this example). Note that $r+q = N^S$ is the total number of segmented cells in the subject image S. With these adjacency sub-matrices available, we further derive three matrices:

$$\mathbf{H}_{(rxp)}^{(\mathbf{r})} = [h^{(\mathbf{r})}]_{(rxp)} = [(\mathbf{r}_{(rxq)}) \wedge (\neg\mathbf{r}_{(qxr)})^{\mathrm{T}}] \times [(\neg\mathbf{R}_{(pxq)})^{\mathrm{T}} \wedge (\mathbf{R}_{(qxp)})]$$
$$+ [(\neg\mathbf{r}_{(rxq)}) \wedge (\mathbf{r}_{(qxr)})^{\mathrm{T}}] \times [(\mathbf{R}_{(pxq)})^{\mathrm{T}} \wedge (\neg\mathbf{R}_{(qxp)})] \tag{7}$$

where $\times$ is matrix multiplication operation. **R** represents adjacency matrices **AP, LR, DV,** and **r** represents adjacency matrices **ap, lr, dv** respectively, similar to Eq. (5).

We then binarize $\mathbf{H}_{(rxp)}^{(\mathbf{r})}$, obtaining $\mathbf{C}_{(rxp)}^{(\mathbf{r})}$:

$$\mathbf{C}_{(rxp)}^{(\mathbf{r})} = [c^{(\mathbf{r})}(a,u)]_{(rxp)} = \begin{cases} 1 & \text{if } h^{(\mathbf{r})}(a,u) > 0 \\ 0 & \text{if } h^{(\mathbf{r})}(a,u) = 0 \end{cases} \tag{8}$$

If an element $c^{(\mathbf{ap})}(a,u)$ in $\mathbf{C}_{(rxp)}^{(\mathbf{ap})}$ is 1, then the AP relationships between cell a and the marker cells in the subject image are different from those between cell u and the marker cells in the template. Thus cell a should not be recognized as cell u. The edge between a and u in the bipartite graph should be cut. On the contrary, if $c^{(\mathbf{ap})}(a,u) = 0$, then the AP relationships between cell a and the marker cells in the subject image are consistent with those between cell u and the markers in the template. Thus cell a can be recognized as cell u. The edge between a and u in the bipartite graph should be kept. Similar explanation applies to $c^{(dv)}(a,u)$ and $c^{(lr)}(a,u)$.

Considering AP, LR, DV relationships all together, the contradictory matrix is computed as

$$\mathbf{C}_{(r \times p)} = \mathbf{C}_{(r \times p)}^{(ap)} \vee \mathbf{C}_{(r \times p)}^{(lr)} \vee \mathbf{C}_{(r \times p)}^{(dv)} \tag{9}$$

We then search for pair-wise cells (a,u) in matrix $\mathbf{C}_{(r \times p)}$, such that $\mathbf{C}(a,u) = 0$, and $\forall x \neq u$, $\mathbf{C}(a,x)=1$, $\forall x \neq a$, $\mathbf{C}(x,u)=1$. This condition means cell a in the subject image can only be recognized as cell u in the template and at the same time cell u can only be assigned to cell a. In another word, cell a can be uniquely identified based on its relative position with respect to the markers. Cells thus identified are added to the set of markers. For those pair-wise cells (a,u) such that $\mathbf{C}(a,u) = 1$, we cut the edge between them in the bipartite graph by setting the distance between a and u to infinity.

After expanding the marker set, we repeat the above process until no new marker cell can be found. The remaining cells that cannot be uniquely determined according to their relative relationship with respect to markers are then recognized using AP/LR/DV constrained bipartite matching as described in §3. Algorithm 2 in Appendix shows the pseudocode of the marker-guided, AP/LR/DV constrained, two-stage bipartite matching.

5   Experiments

We applied our ARC method to the 3D images of newly hatched first larval stage hermaphrodites of *C. elegans* that were acquired using a Leica confocal microscope with 63x/1.4 oil lens. We used DAPI to stain the nuclei of all 558 cells, and GFP to stain the nuclei of the 81 body wall muscle cells and 1 depressor muscle cell (see Figure 1 for example data). As a worm body usually curves in 3D, we developed an automated approach to straighten a curved worm body into a canonical rod shape to facilitate later image comparison across different individuals [3] (see example in Figure 1 (b)). We then segmented cells in 3D using adaptive thresholding, the watershed algorithm, and a region grouping method (the details of the method [2] are beyond the scope of this paper, thus they are omitted). After that, we normalized each worm image, making the sizes and the orientations of different worms the same. This step maps the coordinates of the cells into a standard space defined by AP, LR, and DV axes. Finally, we annotated cells in a set of images with the aid of a 3D annotation tool called WANO developed by us (see Figure 1 (b) for schematic example of 3D annotated templates). Cells in the nerve ring of the head are small and tightly clustered and so very difficult to annotate solely based on our current images without developmental or cell-specific staining information. We annotated the subset of 351 cells out of the 558 cells, that exclude most neurons in the pharynx. These annotated cells include all the body wall muscle cells distributed along the entire worm body, 99 cells in the trunk where cell densities are relatively low, and 170 additional cells of different types in the head and tail. Thus our purpose was to match a subset of ~558 segmented regions in a subject image against the 351 annotated cells in templates.

One of the key ideas in this paper is to use the relative location relationships among cells to constrain the possible matching. We computed and analyzed the AP, DV, LR adjacency matrices from template images. Figure 2 illustrates the invariant AP relationship. For clarity of displaying, we only show the AP relationships of 181 cells by plotting the AP adjacency matrix as a graph after transitive reduction. It can be seen that many cells have strong AP relationships, apparently due to the stereotypy of *C. elegans* cells. These relationships, as well as the DV and LR relationships were used to constrain the possible mappings between cells in a subject image and the templates.

We used 15 image stacks and leave-one-out cross validation scheme to test our recognition method. In other words, we repeated the experiment 15 times. Each time we took one image as the subject image and the remaining 14 as the template images. Our purpose was to identify from all the segmented cells in each subject image the 351 cells that had been annotated in the templates. We compared our three approaches: two-stage bipartite matching (BM), AP/LR/DV constrained two-stage bipartite matching, and marker guided AP/LR/DV constrained two-stage bipartite matching.

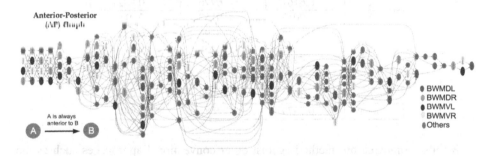

Fig. 2. Illustration of the invariant AP relationship of cells. For clarity of visualization, only the transitive reduction of the AP adjacency matrix is shown here for a set of 181 cells, including all 82 muscle cell markers and all cells in the trunk of L1-stage *C. elegans*. In this figure, left is anterior and right is posterior. An arrow always points from left to right (i.e. anterior to posterior).

The results in Table 1 show that using only the spatial coordinates of cells without considering the relative relationships between cells, the bipartite matching can only achieve an average of 73.79% accuracy in recognizing the 351 cells from the ~558 segmented regions (the second column in Table 1). When adding AP/LR/DV constraints to tailor edges in the bipartite graph, the accuracy improved ~5% (the third column in Table 1). With the combined use of marker-guides and AP/LR/DV-constrained bipartite matching, the average recognition accuracy improved significantly to 94.91% (the fourth column in Table 1). In this case, the average recognition rate of muscle cells (markers) is 99.81% (all 100% except that for stack S_{13}, which is 97.56%) (not shown in Table 1). Thus the average recognition rate of the remaining 269 cells is 93.42%. This indicates that the accuracy improvement is not merely due to the increased number of muscle cells that are correctly recognized in a separate channel but due to the marker guided scheme.

Table 1. Comparison of the recognition rates of the two-stage bipartite matching (BM), AP/LR/DV constrained two-stage BM, and marker-guided-AP/LR/DV-constrained-two-stage BM. The rates are produced by leave-one-out cross validation on 15 image stacks.

Image stack	Two-stage BM	AP/LR/DV con-strained BM	Marker guided AP/LR/DV constrained BM
S_1	0.7114	0.7771	0.9486
S_2	0.7593	0.8166	0.9341
S_3	0.7607	0.8319	0.9829
S_4	0.7721	0.8205	0.9288
S_5	0.7892	0.8689	0.9259
S_6	0.5244	0.6074	0.9799
S_7	0.8054	0.8084	0.9581
S_8	0.7216	0.7994	0.9731
S_9	0.6161	0.6726	0.9821
S_{10}	0.9017	0.9153	0.9186
S_{11}	0.8328	0.8396	0.9147
S_{12}	0.6944	0.6458	0.9271
S_{13}	0.7550	0.8177	0.9459
S_{14}	0.7229	0.7971	0.9571
S_{15}	0.7009	0.8034	0.9601
mean	0.7379	0.7881	0.9491

We also compared our method against other conventional approaches such as the K-Nearest-Neighbor (KNN) classifier and soft assignment approach [7]. The KNN approach finds for each cell in the subject image the K closest cells in the templates and then use majority vote to determine cell identities. The method did not yield a leave-one-out accuracy higher than 60%, much lower than our results showed in Table 1. The soft assignment method is computationally very expensive for big graphs. Thus we tested the recognition of 99 trunk cells. Despite the low number and low density of these cells which makes the task easier than our original matching problem, our results show that the average recognition rate using soft assignment is no higher than 68%.

We further analyzed for each cell to be recognized, how many times in the 15 images it is wrongly recognized. We then computed distribution of the cells and plotted the percentage of cells as a function of the number of images in which a cell was wrongly recognized. The result is shown in Figure 3. Among the 351 cells, 71% (the left most bar) of them are correctly recognized in all the 15 images, 90% (the sum of the three left most bars) are correctly recognized in 13 to 15 of the15 images. In the worst cases, there are two cells wrongly recognized in 7 images and another two cells wrongly recognized in 9 images (the two right-most bars). Those cells do not have a fixed local spatial relationship with respect to their neighboring cells and are in the head where cells are more densely clustered in the animal.

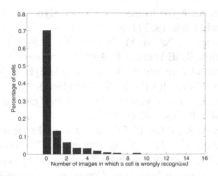

Fig. 3. The percentage of cells $P(k)$ incorrectly recognized in k of the 15 images

Overall, the experimental results show that our method can achieve high recognition accuracy despite the difficulty of the problem. To further improve the recognition accuracy, we will use additional cell information, such as size, shape, and gene expression levels. In fact, although our method currently only uses spatial coordinates and the relative spatial relationships between cells, our scheme is general enough to incorporate this additional cell information for further improvement.

Acknowledgement. The authors thank Andrew Fire for providing reagents and advice. Three-dimensional image stacks were generated in the Cell Sciences Imaging Facility of Stanford University. The authors acknowledge the financial support of the Larry L. Hillblom Foundation for XL. The work of XL and SK was funded by the Ellison Medical Foundation and the NIH.

References

1. Riddle, D., Blumenhal, T., Meyer, B., Priess, J.R.: C. Elegans II. Cold Spring Harbor Laboratory Press, New York (1997)
2. Long, F., Peng, H., Liu, X., Kim, S., Myers, E.W.: A 3D digital cell atlas for the first larval stage of C. elegans hermaphrodite, HHMI JFRC Technical Report (2007), (Also Appear on 2007 Int. Conference of C. elegans)
3. Peng, H., Long, F., Liu, X., Kim, S., Myers, E.: Straightening C.elegans Images. Bioinformatics 24(2), 234-242 (2008)
4. Conte, D., Foggia, P., Sansone, C., Vento, M.: Thirty Years of Graph Matching in Pattern Recognition. International Journal of Pattern Recognition and Artificial Intelligence 18(3), 265–298 (2004)
5. Ullman, J.R.: An Algorithm for Subgraph Isomorphism. J. Assoc. Comput. Mach. 23, 31–42 (1976)
6. Llados, J., Marti, E., Villanueva, J.J.: Symbol Recognition by Error-Tolerant Sub-Graph Matching Between Region Adjacency Graphs. IEEE Trans. Patt. Anal. Mach. Intell. 23, 1137–1143 (2001)

7. Gold, S., Rangarajan, A.: A Graduated Assignment Algorithm for Graph Matching. IEEE Trans Patt. Anal. Mach. Intell. 18, 377–388 (1996)
8. Christmas, W.J., Kittler, J., Petrou, M.: Structural Matching in Computer Vision Using Probabilistic Relaxation. IEEE Trans. Patt. Anal. Mach. Intell. 17(8), 749–764 (1995)
9. Umeyama, S.: An Eigendecomposition Approach to Weighted Graph Matching Problems. IEEE Trans. Patt. Anal. Mach. Intell. 10, 695–703 (1988)
10. Wilson, R.C., Hancock, E.R.: Structural Matching by Discrete Relaxation. IEEE Trans. Patt. Anal. Mach. Intell. 19, 634–648 (1997)
11. Kotropoulos, C., Tefas, A., Pitas, I.: Frontal Face Authentication Using Morphological Elastic Graph Matching. IEEE Trans. Imag. Process. 9, 555–560 (2000)
12. Chen, H.T., Lin, H., Liu, T.L.: Multi-object Tracking Using Dynamical Graph Matching. In: Proc. 2001 IEEE Computer Soc. Conf. Computer Vision and Pattern Recognition, pp. 210–217 (2001)
13. Berretti, S., Del Bimbo, A., Vicario, E.: Efficient Matching and Indexing of Graph Models in Content-based Retrieval. IEEE Trans. Patt. Anal. Mach. Intell. 23, 1089–1105 (2001)
14. Ton, J., Jain, A.K.: Registering Landsat Images by Point Matching. IEEE Trans. Geoscience and Remote Sensing 27(5), 642–651 (1989)
15. Munkres, J.: Algorithms for the Assignment and Transportation Problems. Journal of the Society of Industrial and Applied Mathematics 5(1), 32–38 (1957)

Appendix

Algorithm 1: Cell recognition using AP/LR/DV constrained two-stage bipartite matching
Input: A subject image S with N^S segmented cell regions and template images T_k, $k \in [1,K]$, with N^{T_k} annotated cells each, and a threshold t_{max}.
  Output: Matching matrix M^* and cost value E^2
  1.   Compute the intrinsic adjacency matrices **AP, LR, DV**.
  2. $\forall a, \forall u$, Set $N(a \Leftrightarrow u) = 0$
  3.   FOR EACH T_k
  4.   { Compute distance $D^1(a \to u)$ using Eq. (2)
  5.   Set $t = 0$, $minerr = \infty$
  6.   WHILE ($t < t_{max}$)
  7.   { Compute matchings M_k using the first stage bipartite matching
  8.   Compute adjacency matrices **ap, lr, dv** from the subject image S using M_k
  9.   Compute contradiction matrix **C** using Eqs. (4)~(5)
  10.   Select wrongly matched a and set $D^1(a \to u) = \infty$ for $a \to u \in M_k$
  11.   IF $\sum_a conflict(a) \leq minerr$
  12. $minerr = \sum_a conflict(a)$, $MB_k = M_k$
  13.   ELSE
  14. $t = t+1$ }
  15. $M_k = MB_k$
  16. $N(a \Leftrightarrow u) = N(a \Leftrightarrow u) + 1$, if $a \to u \in M_k$ }
  17.   Compute $D^2(a \to u)$ using Eq. (3)
  18. Compute the matching M^* and cost value E^2 using $D^2(a \to u)$ and the second stage bipartite matching

Algorithm 2: Cell recognition using marker guided, AP/LR/DV constrained, two-stage bipartite matching

Input: A subject image S with N^S segmented cell regions and K template images T_k, $k \in [1, K]$, each with N^{T_k} annotated cells, and a threshold t_{max}.

Output: Matching matrix M^* and cost value E^2

1. Compute adjacency matrices **AP, LR, DV** from template images T_k, $k \in [1, K]$.
2. Recognize muscle cells in the GFP channel by calling *Algorithm 1*.
3. Let $U = \{$all segmented regions in $S\}$, $U_m = \{$recognized muscle cells in $S\}$, $V = \{$annotated cells in templates$\}$, $V_m = \{$annotated muscle cells in templates$\}$
4. WHILE (new markers detected)
5. $\quad \{ \ U = U \setminus U_m, r = |U|, V = V \setminus V_m, p = |V|$
6. $\qquad$ Compute contradiction matrix $\mathbf{C}_{(r \times p)}$ using Eqs. (7)~(9)
7. $\qquad$ Prune edges in the bipartite graph using $\mathbf{C}_{(r \times p)}$
8. $\qquad$ Detect new markers, $U_m = U_m \cup \{$new markers in $S\}$,
   $\qquad\qquad V_m = V_m \cup \{$new markers in $T \}$
   $\qquad \}$
9. $\forall a \in U \setminus U_m, \forall u \in V \setminus V_m$, set $N(a \Leftrightarrow u) = 0$
10. FOREACH T_k
11. $\quad \{$Compute distance $D^1(a \to u)$ using Eq. (2)
12. $\quad$ Set $t = 0$, $minerr = \infty$
13. $\quad$ WHILE $(t < t_{max})$
14. $\quad\quad \{$ Compute matching matrix M_k using the first stage bipartite matching
15. $\qquad$ Compute adjacency matrices **ap, lr, dv** from subject image S using M_k
16. $\qquad$ Compute contradiction matrix $\mathbf{C}$ using Eqs. (4)~(5)
17. $\qquad$ Select wrongly matched a's and set $D^1(a \to u) = \infty$ for $a \to u \in M_k$
18. $\qquad$ IF $\sum_a conflict(a) \leq minerr$
19. $\qquad\quad minerr = \sum_a conflict(a)$, $MB_k = M_k$
20. $\qquad$ ELSE
21. $\qquad\quad t = t+1 \}$
22. $\qquad M_k = MB_k$
23. $\qquad N(a \Leftrightarrow u) = N(a \Leftrightarrow u) + 1$, if $a \to u \in M_k\}$
24. Compute $D^2(a \to u)$ using Eq. (3)
25. Compute the matching M^* and cost value E^2 using $D^2(a \to u)$ and the second stage bipartite matching

Spectrum Fusion: Using Multiple Mass Spectra for De Novo Peptide Sequencing

Ritendra Datta[1] and Marshall Bern[2]

[1] Penn State University, University Park, PA 16801, USA
datta@cse.psu.edu
[2] Palo Alto Research Center, Palo Alto, CA 94304, USA
bern@parc.com

Abstract. We report on a new algorithm for combining the information from several mass spectra of the same peptide. The algorithm automatically learns peptide fragmentation patterns, so that it can handle spectra from any instrument and fragmentation technique. We demonstrate the utility of the algorithm, and the power of multiple spectra, by showing that combining pairs of spectra (one CID and one ETD) greatly improves *de novo* sequencing success rates.

1 Introduction

There are two basic approaches to peptide sequencing by tandem mass spectrometry (MS/MS): *database search* [11], which identifies the sequence by finding the closest match in a protein database, and *de novo sequencing* [3], which attempts to compute the sequence from the spectrum alone. *De novo* sequencing actually started first, but with the explosion of genomically derived protein sequences, database search quickly became the dominant approach, because it can make identifications from lower-quality spectra with less complete fragmentation. There are, however, good reasons to continue the pursuit of *de novo* sequencing. First, if databases are of low quality, as in the case of unsequenced organisms [19], then *de novo* analysis can outperform database search. Second, sometimes the unknown peptide—rather than a parent protein—is itself the object of interest, e.g., toxins [2], neurotransmitters, and hormones. Third, protein design techniques, such as directed mutation and recombination, often produce proteins without knowing exactly which gene produced them, and the protein sequence may have to be obtained directly through chemical and mass spectrometric sequencing.

Algorithm designers factor *de novo* sequencing into two subproblems: *candidate generation* and *scoring*. Candidate generation typically uses a graph algorithm, such as a longest- or best-path algorithm [6,8,17,22], to compute 1000s of possible sequences. The scoring phase then scores each of these candidates, using more detailed information such as the fragmentation propensities of residues [10,16], mass measurement recalibration [5], and so forth, that would be difficult to incorporate into the candidate generation phase. Because *de novo* sequencing requires essentially complete fragmentation, new fragmentation techniques (microwave assisted acid hydrolysis, IRMPD, ECD, and ETD) offer the best hope of performance improvement. The most interesting of the new techniques is electron-transfer dissociation (ETD) [20], because it is commercially

M. Vingron and L. Wong (Eds.): RECOMB 2008, LNBI 4955, pp. 140–153, 2008.
© Springer-Verlag Berlin Heidelberg 2008

available, fast enough to be used with ion-trap instruments, and gives quite different, and hence complementary, information to the standard technique of collision-induced dissociation (CID). ETD reduces charge as it induces fragmentation, and hence it gives good fragmentation for highly charged (+3 and +4) parent ions, not-so-good fragmentation for +2 parents, and neutralizes and loses +1 parents.

In this paper, we give a generic algorithm for *spectrum fusion*, combining information from more than one spectrum of the same peptide. We apply the algorithm to *de novo* sequencing of peptides from pairs of spectra collected on a Thermo Electron LTQ instrument, run in a mode that alternates between CID and ETD fragmentation. We show significant improvements in *de novo* sequencing, approximately doubling the number of sequences that could be identified exactly.

Spectrum fusion has previously been done with special-case algorithms. Zhang and McElvain [24] used CID MS/MS and MS^3 pairs, and Bandeira et al. used overlapping [2] or differentially modified [1] MS/MS spectra, for *de novo* sequencing. Finally, and most relevant to the present work, Zubarev and collaborators pioneered the use of CID/ECD pairs for *de novo* sequencing [18]. ECD (electron-capture dissociation) gives similar spectra to ETD, but is less efficient, so it cannot generally be used with ion-trap instruments, only with expensive FTICR instruments. These instruments have about 100 fold better mass accuracy and resolution than ion-trap instruments, but take about 10 times longer to acquire each spectrum. Due to the high mass accuracy, *de novo* sequencing on FTICR is much easier than on ion-trap instruments. Indeed, Savitski et al. [18] achieve reasonable results from CID/ECD pairs using a simple greedy algorithm to compute a single approximate longest path.

2   Algorithms

We developed a generic algorithm for combining the information from multiple fragmentation spectra of the same peptide. The output is a *synthetic spectrum* with peaks at integer masses, representing the likelihood that the mass is equal to the sum of the (integer parts of) amino acid residue masses of a prefix of the peptide sequence. The synthetic spectrum is used as the input for candidate generation. By building a synthetic spectrum containing only prefixes, rather than prefixes and suffixes, we use spectrum fusion both to improve fragmentation completeness and to separate prefixes from suffixes. We previously applied graph partitioning [5] to peak separation; with complementary spectra (CID/ETD pairs) a global approach like graph partitioning is unnecessary. We have not yet applied spectrum fusion to scoring; for this phase we used ByOnic [4], our database-search tool, and scored the multiple spectra independently.

We demonstrate spectrum fusion on CID/ETD pairs, but the algorithm could be applied to other combinations of spectra and or other types of biomolecules (for example, glycans). The fusion algorithm is fully automated, so that the algorithm determines the information in the various spectra and peaks, with minimal dependence on prior knowledge. CID fragmentation patterns and peak intensities have been mapped [10,16,21], but no such statistical studies have been published for ETD.

At the core of our spectrum fusion algorithm lies a supervised learning phase that relieves dependence on prior knowledge. A sample consists of C MS/MS spectra and the corresponding peptide (reliably identified by database search and knowledge of the biological material), where C is the number of spectra per peptide. Let us denote by $\mathcal{S}_{i,c}$, $c \in \{1, \cdots, C\}$, the spectrum of type c for sample i, consisting of mass over charge (m/z) values and corresponding peak intensities. Let the measured parent mass be M_i and the "ground truth" peptide string be $\mathcal{P}_i$. (Following convention, M_i denotes M+H mass, the sum of residue masses + 19 for water + 1 for proton.) Thus our labeled data, which can be split into training and test sets, consists of tuples of the form $(\mathcal{S}_{i,1}, \cdots, \mathcal{S}_{i,C}, M_i, \mathcal{P}_i)$; the task is to predict prefix masses of $\mathcal{P}_i$ from $\mathcal{S}_{i,C}$ and M_i.

We make the simplifying assumption that the parent mass M_i is correct to ± 0.5 Daltons (Da). In reality, the estimated parent mass M_i' from ion-trap instruments may be off by 1 or 2 Da for peptides of charge +2, and by as much as ± 7 Da for peptides with greater charge. Accurate parent masses, however, can be obtained in various ways: by a high-resolution single-MS scan with another mass analyzer (e.g, Orbitrap), by a "zoom scan" with the same ion-trap mass analyzer, or by software parent mass correction. CID/ETD pairs offer improved software correction; see the Appendix. The training phase of spectrum fusion consists of two steps (Sections 2.1 and 2.2):

Training: (**Input:** Training data $(\mathcal{S}_{i,1}, \cdots, \mathcal{S}_{i,C}, M_i', \mathcal{P}_i)$, **Output:** Features, Model)
1. *Feature Selection:* Pick informative features across the different types of spectra.
2. *Statistical Learning:* Learn a probabilistic model on the selected features.

Then, an unknown sample $(\mathcal{S}_{i,1}, \cdots, \mathcal{S}_{i,C}, M_i')$ is processed through a pipeline which includes spectrum fusion, candidate generation, and scoring (Sections 2.3 and 2.4):

De Novo Sequencing: (**Input:** Test data $(\mathcal{S}_{i,1}, \cdots, \mathcal{S}_{i,C}, M_i')$, **Output:** Guess $\mathcal{P}'_i$)
1. *Parent Mass Correction:* Correct the reported parent mass M_i' to get M_i.
2. *Spectrum Fusion:* Generate one synthetic spectrum $\widehat{\mathcal{S}}_i$ from spectra $(\mathcal{S}_{i,1}, \cdots, \mathcal{S}_{i,C})$; the synthetic spectrum should ideally contain most integer prefix masses, but few suffix or noise masses.
3. *Candidate Generation:* Compute the K best paths in $\widehat{\mathcal{S}}_i$ to generate candidates.
4. *Scoring:* Score candidate peptides by summing ByOnic scores for each spectrum.

2.1 Feature Selection Using Offset Frequency Functions

The dominant peaks in CID spectra are well-understood and have known, fixed, relationships to prefix and suffix masses. For example, if a prefix mass (sum of residue masses of an initial subsequence of the peptide) is w, then there will usually be a spectral peak (called the "b-ion") at $w + 1$ and another peak at $w + 2$ (the "isotope peak" containing one ^{13}C). Our goal was to introduce an algorithmic framework that would automatically "learn" these features. For this problem, machine learning offers many advantages: it can learn weights and dependencies among features; it can learn features for new techniques such as ETD; and it can make use of more, and more subtle, features.

Our features were selected from the *offset frequency function* (OFF), proposed by Dančik et al. [8]. The OFF is a histogram, giving the number of times a spectral peak is observed at a given integer mass offset from a known prefix or suffix mass. Suppose an ETD training-set spectrum $\mathcal{S}$ consists of observed peaks at $\{s_1, \cdots, s_n\}$, and the ground-truth peptide contains real-valued prefix masses at $w_1, w_2, \cdots, w_k$. We compute the OFF as follows. If a prefix mass w_j is separated from s_i by an integer mass δ

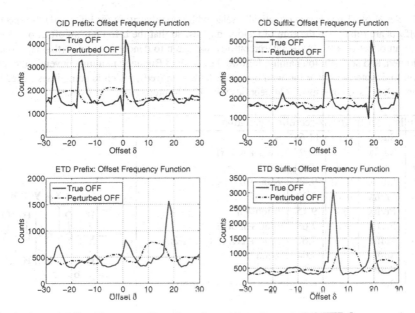

Fig. 1. Plots of Offset Frequency Functions for prefix/suffix and CID/ETD fragmentation types

(within a tolerance ϵ, set to 0.2 Da for our ion-trap data), then we increment the count for δ for the combination (ETD, prefix). Similarly, if $M - 19 - w_j$, where M is the parent mass of the peptide, is separated from s_i by δ, then we increment the count for δ for (ETD, suffix). With an estimate of M accurate to the closest integer, the suffix OFF provides valuable corroborating evidence to the presence of a prefix at w. (Without an accurate M, the suffix OFF would still be useful but somewhat blurry.)

We determine which δ offsets are most informative in an automated manner. For each combination of prefix/suffix and type of spectrum (CID, ETD, etc.), we generate a *perturbed OFF* in which the theoretical prefix and suffix masses are shifted by a uniform random sample in the range $1 - 10$ Da; this OFF is used to estimate the statistics of insignificant OFF counts. We limit the computation of both OFFs, true and perturbed, to ± 30 Da offsets, since δ values beyond this range are unlikely to be of any interest (Figure 1). Let the true OFF be $\{v_{-30}, \cdots, v_0, \cdots, v_{+30}\}$, and the deliberately perturbed OFF be $\{v'_{-30}, \cdots, v'_0, \cdots, v'_{+30}\}$, where each v or v' represents a count. We estimate the mean μ and variance σ^2 of insignificant OFF counts from the perturbed OFF, and compute a *z-score* for each count in the true OFF to determine informative peaks. More specifically, we say that δ is an informative offset if its z-score $(v_\delta - \mu) / \sigma$ exceeds 3, where $\mu = \frac{1}{61} \sum_{j=-30}^{30} v'_j$, and $\sigma^2 = \frac{1}{60} \sum_{j=-30}^{30} (v'_j - \mu)^2$. We denote the final set of informative offsets across prefixes/suffixes and all fragmentation methods by $\widehat{\Delta}$, which is a set of D offsets, $\{T_1, T_2, \cdots, T_D\}$. Each offset T_d is specified by a triple of fragmentation type (CID, ETD, etc.), orientation (prefix or suffix), and integer offset δ. These triples comprise our set of selected *features*, for use in the model-based classification step that follows. As shown in Table 1, automatic feature selection discovered

Table 1. Set $\widehat{\Delta}$ of 21 features automatically selected by our algorithm for ETD and CID spectra. Mass offsets are relative to the sum of residue masses, so that the b-ion has an offset of +1 and the y-ion an offset of +19. In the interpretations, we use n to denote a neutron (that is, ^{13}C), H for hydrogen, and so forth. Ion naming follows the standard Biemann naming convention, but we use z-ion to denote the stable ion with one extra hydrogen (red), which is sometimes called the "z+1 ion". Data used consists of 724 peptides with 12,574 prefixes and suffixes.

Frag'n	Pref/Suff	Offset	z-score	Interpretation
CID	Suffix	+19	11.68	y-ion
ETD	Suffix	+4	10.97	z-ion + n or H
CID	Prefix	+1	10.44	b-ion
CID	Suffix	+20	9.45	y-ion + n
ETD	Prefix	+18	9.39	c-ion
CID	Prefix	+2	9.15	b-ion + n
ETD	Prefix	+19	7.68	c-ion + n or H
ETD	Suffix	+5	7.62	z-ion + 2H, H + n, or 2n
ETD	Suffix	+3	7.04	z-ion
ETD	Suffix	+19	6.54	y-ion
CID	Prefix	−16	6.77	b-ion − H_2O + n or NH_3
CID	Prefix	−17	6.36	b-ion − H_2O
CID	Suffix	+2	5.72	y-ion − H_2O + n or NH_3
ETD	Prefix	+17	5.71	c-ion − H ??
CID	Suffix	+1	5.68	y-ion − H_2O
CID	Prefix	−27	4.75	a-ion
CID	Suffix	+21	4.43	y-ion + 2n
CID	Prefix	−15	4.24	b-ion − H_2O + 2n or NH_3 + n
CID	Prefix	+3	3.93	b-ion + 2n
ETD	Suffix	+20	3.54	y-ion + n or H
ETD	Prefix	+20	3.45	c-ion + 2H, H + n, or 2n

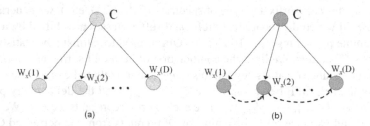

Fig. 2. (a) In the naïve Bayes structure, each binary attribute variable $W_x(d)$ (presence/absence of a peak at offset d) depends only on the class variable C. (b) The TAN structure allows each attribute variable an in-degree of up to 2, thus generalizing (a) to allow more dependencies.

and ranked all the well-known ions such as b- and y-ions in CID spectra, and also discovered some less-known phenomena, such as the high intensity of isotope (or "neutral-gain") peaks and the rarity of neutral losses in ETD spectra.

2.2   Combining Features with a Tree-Augmented Naïve Bayes Network

With the set $\widehat{\Delta} = \{T_1, \cdots, T_D\}$ of informative offsets automatically determined, the next step is to use them to learn a statistical model to perform a binary classification: For each integer x within the range of 1 to parent mass M, decide if x is a prefix mass or not. The feature vector for this binary classifier consists of a length-D binary vector W_x, with each element (attribute variable) indicating the absence or presence (0 or 1) of a peak (within a mass tolerance ϵ) at the corresponding offset. More specifically, a CID prefix entry of W_x is set to 1 if there is a peak in the CID spectrum at position $x + \delta$, and a suffix entry is set to 1 if there is a peak at $M - 19 - x + \delta$, where δ are the informative integer offsets. ETD entries of W_x are treated analogously.

There is one subtlety in converting real-valued-masses to integer-masses in the spectra. A real-valued mass M should be rounded to the closest integer of $0.9995M$ in order to remove the characteristic mass defect (fractional part) of a peptide. Thus 1814.1 Da rounds to 1813. We used this correction wherever required, so that integer amino acid residue masses sum to M+H peptide masses (minus 19).

One straightforward approach to the classification problem would be to use a naïve Bayes classifier [9], meaning that, given the class variable, each feature is assumed conditionally independent of every other feature (Figure 2). Denoting by indicator variable prf_x whether x is a prefix mass or not, the class conditional is written as follows:

$$\text{Prob}(prf_x = 1 \mid W_x) \propto \text{Prob}(prf_x = 1) \cdot \prod_{d=1}^{D} Pr(W_x(d) \mid prf_x = 1).$$

To do away with the proportionality constant, classification can be done based on some thresholding of the odds $\text{Prob}(prf_x = 1 \mid W_x)/\text{Prob}(prf_x = 0 \mid W_x)$. This model is computationally efficient and easy to estimate, but the assumption of conditional independence of the attributes is too strong. There are obvious dependencies among the attribute variables, as isotope peaks are almost sure to co-occur with peptide peaks. This motivated us to explore a generalization of the naïve Bayes classifier to a richer network structure that allows dependencies. One such generalization is the *tree-augmented* naïve Bayes classifier (TAN) [14], shown in Figure 2(b), which allows each attribute variable to depend on the class variable along with at most one other variable. The TAN model has many attractive properties: (1) An in-degree of two should capture the most important interactions among spectral peaks; an isotope or neutral-loss peak depends on the monoisotopic peak, but has little dependence upon peaks at other cleavages. (2) Unlike general Bayes nets, TAN is efficiently estimated in polynomial time, and inferencing is fast once the structure is known. (3) TAN is simple enough to be visually interpretable.

We estimate the TAN structure based on the polynomial-time *Construct-TAN* algorithm from [14], which in turn is based on earlier work on second-order product approximation of discrete joint distributions [7]. Learning the TAN structure for prefix/non-prefix classification is presented in Algorithm 1. The learning time is $O(n^2D)$, where n is the total number of positive and negative examples (binary vectors) used for training. Note that the TAN so constructed is optimal, in the sense that of all network structures possible given the TAN restrictions, the one obtained maximizes the likelihood given

---

Algorithm 1. Learning a TAN Structure for Prefix/Non-Prefix Classification

---

Require: Selected features $\widehat{\Delta} = \{T_1, \cdots, T_D\}$, spectra of multiple types $(\mathcal{S}_{i,1}, \cdots, \mathcal{S}_{i,C}, M_i')$ for known peptides.

1: To create positive examples, for each peptide P_i, and each prefix position x for it, compute length D binary vector W_x based on presence/absence of the selected features $\widehat{\Delta} \rightarrow$ *True vectors.*

2: To create negative examples, perturb each such true prefix position by a uniformly sampled integer between (1-10) to simulate non-prefix positions, and compute binary vector W_x' as before $\rightarrow$ *Perturbed vectors.*

3: Empirically compute *conditional mutual information* between attribute pairs, $j, k \in \{1, \cdots, D\}, j \neq k$:

$$I_P(T_j; T_k \mid prf) = \sum_{\substack{w(j) \in \{0,1\} \\ w(k) \in \{0,1\} \\ p \in \{0,1\}}} \mathrm{Prb}\big(w(j), w(k), prf{=}p\big) \cdot \log \frac{\mathrm{Prb}\big(w(j), w(k) \mid prf{=}p\big)}{\mathrm{Prb}\big(w(j) \mid prf{=}p\big)} \mathrm{Prb}\big(w(k) \mid prf{=}p\big)$$

where $\mathrm{Prb}(- \mid prf = 1)$ are estimated from *true vectors*, and $\mathrm{Prb}(- \mid prf = 0)$ from *perturbed vectors*.

4: Build full undirected graph $\mathcal{G}$ with nodes $\{T_1, \cdots, T_D\}$, edge weight $-I_P(T_j; T_k \mid prf)$ between T_j and T_k.

5: Apply Prim's Algorithm to find the minimum spanning tree in $\mathcal{G}$ (with the -ve edge weights $\Rightarrow$ a *max.* spanning tree).

6: Select a node in $\mathcal{G}$ arbitrarily and set all edges outward from it, to get directed graph $\mathcal{G}'$.

7: Add class variable prf_x to $\mathcal{G}'$ as a node, and direct edges from it to each $T_j \rightarrow$ desired TAN structure (e.g., Fig. 2 (b))

---

the training data. The structure can be described by the set of parents of each attribute T_d, which includes the class variable prf, and at most one other attribute, which we refer to as $T_{d'}$. An example TAN structure estimated over CID/ ETD pairs is shown in Figure 3. It is worth noting that PepNovo [13] also uses a tree of dependencies for scoring CID spectra; however, PepNovo's dependencies were determined manually and only the weights were learned automatically.

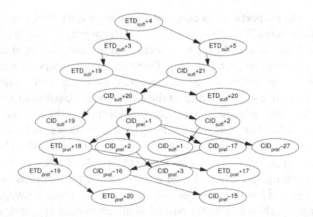

Fig. 3. The TAN structure for the CID/ETD spectra pairs, estimated over 724 peptides using Algorithm 1. Dependencies are interpretable, e.g. , the top of the TAN shows that the monoisotopic z-ion and the +2 isotope both depend upon the +1 isotope. The connection from $CID_{pref}+1$ to $CID_{pref} - 27$ implies that a-ion depends upon b-ion. Class variable prf is omitted to avoid clutter.

The training process is completed by empirical estimation of $\mathrm{Prob}\big(W(d) \mid prf{=}1\big)$ and $\mathrm{Prob}\big(W(d) \mid prf{=}0\big)$ if T_d has only one parent, or $\mathrm{Prob}\big(W(d) \mid prf{=}1, W(d')\big)$

and $\mathrm{Prob}\big(W(d) \mid prf{=}0, W(d')\big)$ if T_d has two parents. As suggested in [14], the empirical estimates are smoothed to avoid poor estimates from a limited sample size N:

$$\mathrm{Prob}\big(W(d) \mid prf_x = 1\big) \ \leftarrow\ \frac{N}{N+5}\mathrm{Prob}\big(W(d) \mid prf_x = 1\big) + \frac{5}{N+5}\mathrm{Prob}\big(W(d)\big).$$

Probability estimates for attributes with two parents are smoothed similarly. Here we have made a reasonable choice of prior for the conditional, the marginal distribution. The classification of a mass x as prefix or not is ultimately based on the *odds* ϕ_x,

$$\phi_x = \frac{\mathrm{Prob}\big(prf_x{=}1 \mid W_x\big)}{\mathrm{Prob}\big(prf_x{=}0 \mid W_x\big)} = \frac{\mathrm{Prob}\big(prf_x{=}1\big)}{\mathrm{Prob}\big(prf_x{=}0\big)} \cdot \prod_{d=1}^{D} \frac{\mathrm{Prob}\Big(W_x(d) \mid prf_x{=}1[\,,W_x(d')]\Big)}{\mathrm{Prob}\Big(W_x(d) \mid prf_x{=}0[\,,W_x(d')]\Big)}.$$

2.3 Producing a Fusion Spectrum

The trained classifier can be used to combine multiple spectra, $S_{i,c}$, $c \in \{1, \cdots, C\}$, with integer parent mass M_i, into a single synthetic spectrum $\widehat{S}_i$ (Figure 4), which contains most prefix masses but as few other types of peaks as possible. The steps:

1. Initialize spectrum $\widehat{S}_i$ to be an empty (no-peak) spectrum with mass range 1 to M_i.
2. In each $S_{i,c}$, $c \in \{1, \cdots, C\}$, the intensities at x are replaced by rank-based intensities, namely $\max\{0,\ 200 - rk(x)\}$, where $rk(x)$ is descending order rank.
3. For each integer x in the range 1 to M_i, a binary vector W_x is created based on the feature set $\widehat{\Delta}$ using spectra $S_{i,c}$, $c \in \{1, \cdots, C\}$, indicating presence/absence of peaks at the D offsets. The vector W_x is used to compute the odds ϕ_x as above.
4. The $M_i/10$ positions with greatest $\phi_x > 1$ (odds in favor of position being a prefix mass) are picked to be synthetic peaks. Peak intensities are set to the sum of intensities corresponding to the D offsets in the various spectra (with nothing added if no peak is present at a given offset).
5. Some peaks are removed since they cannot be prefix masses: (a) those from 1 to 56 Da, (b) those from $M_i - 19 - 56$ to M_i, and (c) masses within 250 Da of either 0 or $M_i - 19$ that cannot be completed with a sum of amino acid residue masses.
6. Each peak x is then compared with its complementary peak $M_i - 19 - x$, and the peak with lower intensity is removed. We found this step to be very effective in eliminating suffixes while retaining prefixes.

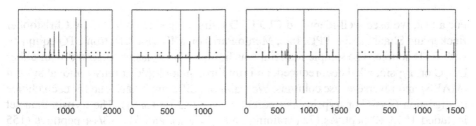

Fig. 4. Sample synthetic spectra generated by Naïve Bayes (below, inverted) vs. TAN (above, upright). *Note:* TAN spectra contain many more true prefix masses (red dots) at high intensity.

2.4 De Novo Sequencing Steps

The final steps in our *de novo* sequencing algorithm are standard steps with some minor modifications. We first construct a *spectrum graph* [8] on the synthetic integer-mass spectrum $\widehat{S}_i$. Denoting the peak masses in $\widehat{S}_i$ by V_i, we make a node for each integer in $\mathcal{V}_i = V_i \cup \{0\} \cup \{M_i - 19\}$. For each $a, b \in \mathcal{V}_i$, we add a directed edge $a \to b$ if $b - a$ equals the mass of one of the amino acids, and label it accordingly.

In order to use a standard K-shortest-path algorithm [12] (with implementation [15]) to generate candidates, rather than a special-purpose algorithm [6,17], we devised a special edge-weighting scheme. Denoting by $Int(x)$ the intensity of a peak in $\widehat{S}_i$ at position x (trivially 0, if a peak is absent), we have

$$Wt.(a \to b) = \underbrace{\frac{50}{(b-a)}}_{\text{Path hop control}} - \underbrace{(Int(a) + Int(b))}_{\text{Prefix intensity sum}} - \underbrace{\frac{1}{2}\left(Int(M_i - 19 - a) + Int(M_i - 19 - b)\right)}_{\text{Suffix intensity sum}}$$

The intensity terms in the equation above are negated to convert a longest path problem into one of shortest path. The suffix intensities are added (with lower weight 1/2), because we observed that some suffix peaks remained in the artificial spectrum $\widehat{S}_i$, and we wanted to take advantage of their presence by treating them as corroborative evidence. Suppose a peptide is AGPTRK, and let a and b correspond to the mass of prefixes AGP and AGPT respectively. While the high intensities at a and b do suggest amino acid T, peaks at their complementary positions (suffixes TRK and PTRK) also support the occurrence of T. The first term "path hop control" introduces a small bias toward paths with fewer hops. This bias helps avoid generating peptide candidates containing long sequences of low mass amino acids, so that, e.g., N is generally favored over GG (having same mass) and K over AG and GA. The numerator arbitrarily controls this bias, and a choice of 50 worked well. The bias toward fewer hops can be explained by the fact that $\frac{1}{x} + \frac{1}{y} > \frac{1}{x+y}$, when $x, y > 0$. This means that, everything else remaining same, an edge of length $(x + y)$ is favored over two edges of lengths x and y.

Each of the K candidate peptides is then scored using ByOnic [4]. ByOnic scores each candidate against each of the C different original spectra, and we simply sum these scores and pick the candidate $\mathcal{P}'_i$ with the highest total. While more complex functions of the C scores can be used, we have not experimented with any variants.

3 Experiments

For a test, we used well-identified CID/ETD pairs of spectra collected by Christopher Becker and Shanhua Lin (PPD, Inc., Menlo Park) on a Thermo Electron LTQ equipped with ETD source. The sample material was human blood plasma, digested with either Lys-C or trypsin, alkylated, reduced, and run through multiple affinity removal system (MARS) and reverse-phase columns. We trained 4 different TANs, one for each choice of digestion (Lys-C or trypsin) and parent charge (+2 or +3/+4). The trypsin data set included 1543 +2 peptides (724 training and 719 test), and 317 +3/+4 peptides (155 training and 162 test). The Lys-C data set included 1025 +2 peptides (520 training and 505 test), 539 +3 peptides (274 training and 264 test), and 178 +4 peptides (50 training

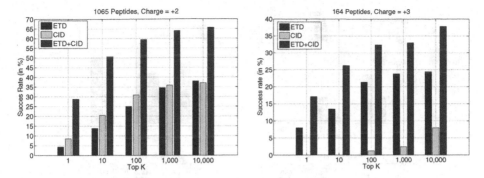

Fig. 5. Comparison of our candidate generation algorithm using only one type of spectrum versus the same algorithm using CID/ETD pairs. A successful trial was one in which the correct peptide was in the list of the top K generated candidates. The average length of the $+3$ charged peptides is 17, making this a challenging data set.

and 128 test). In all our tests, we considered L and I interchangeable, since they have exactly the same mass and similar chemistry, but we considered K and Q different since their masses differ slightly and chemistries differ considerably.

Figure 5 compares the performance of our candidate generation algorithm run on single spectra, either CID or ETD, with the performance of the same algorithm using CID/ETD pairs. Even with a single input spectrum, the TAN output is enriched in prefix peaks and depleted in suffix and noise peaks, yet candidate generation is much worse than with CID/ETD pairs. We found that for $+2$ peptides, CID spectra have better fragmentation than ETD spectra, with a median of 75% of the possible b- and y-ions present among the top 200 peaks compared to 44% of the c- and z-ions for ETD. A median of 90% of the cleavages were represented by either a b- or y-ion for CID compared to 81% represented by either a c- or z-ion for ETD. Separating prefix from suffix peaks is somewhat easier for ETD spectra than for CID spectra, due to fewer noise peaks and the frequent co-occurrence of z- and y-ions for ETD suffixes, so that the ETD candidate generation catches up with CID when the number of paths K is large (Figure 5, left). For $+3$ parents, CID spectra had worse fragmentation, with medians of 44% of possible b- and y-ions and 71% of cleavages, compared to 64% and 86% for ETD. In a CID/ETD pair, a median of 93% (respectively, 88%) of cleavages are represented by at least one of the four possibilities (b-, y-, c-, z-ion) for $+2$ (respectively $+3$) parents. Better fragmentation, along with easier prefix and suffix separation, gave quite dramatic improvement in candidate generation performance for both $+2$ and $+3$ peptides.

Figure 6 compares the results of our complete *de novo* sequencing pipeline versus PepNovo [13], which we believe to be the best available *de novo* sequencing program for ion-trap spectra. This experiment is not meant to be a fair comparison of algorithms or software, but rather an assessment of the "bottom-line" advantage of CID/ETD pairs over single CID spectra. Savitski et al. [18] reported that CID/ECD pairs were advantageous, but did not attempt to quantify the performance improvement offered by multiple spectra. In Figure 6, we see that the number of exactly correct peptides increases by about a factor of 2, yet remains at a modest level, and that partially correct peptides

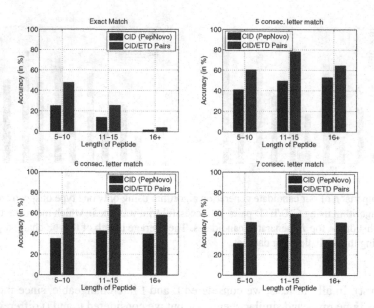

Fig. 6. Comparison of the results of our algorithm, which makes use of ETD/CID spectra together, with the results of PepNovo [13] run on the CID spectra only. We counted exact matches (every letter correct except for L and I swaps), and longest consecutive sequences of correct letters.

(with 5, 6, or 7 correct consecutive letters) increases by more than 50%. About 70% of the peptides of lengths 11–15 have 6 or more consecutive letters correct.

4 Discussion

Although *de novo* sequencing has been used for various experiments, it is unlikely to become the technique of choice for well-studied complex samples, such as human plasma or tissue. For these samples, it will continue to play an important niche role by identifying polymorphisms and alternate splices.

For complex samples from highly variable organisms such as pathogens, *de novo* sequencing will likely play a more central role. Savitski et al. [18] argue that ECD/CID pairs provide the first "proteomics-grade" *de novo* sequencing, meaning that their processing pipeline makes approximately as many peptide identifications as would a CID-only, database-search strategy employing the same FTICR instrument. The slow duty cycle of CID/ECD-FTICR, however, limits the number of MS/MS spectra that can be acquired on one run. We expect *de novo* sequencing using CID/ETD on LTQ ion-trap instrument to actually outperform CID/ECD-FTICR, if the measure of success is the number of distinct peptides with useful sequence tags (say 6+ letters) per unit time.

Finally, *de novo* sequencing probably is indeed the technique of choice for studies of relatively simple mixtures of peptides or proteins with high biological activity, such as toxins and neurotransmitters. Such peptides are often from unsequenced organisms, have been heavily processed post-translationally, and may vary from strain to strain or even from individual to individual. Such samples are also important enough to warrant extra work, such as the acquisition of multiple fragmentation spectra per peptide, perhaps many more than two spectra per peptide. The proposed spectrum fusion algorithm would then be useful in automatically extracting the information in the suite of spectra.

References

1. Bandeira, N., Tsur, D., Frank, A., Pevzner, P.A.: Protein identification by spectral networks analysis. Proc. Natl. Acad. Sci. USA 104, 6140–6145 (2007)
2. Bandeira, N., Clauser, K.R., Pevzner, P.A.: Assembly of peptide tandem mass spectra from mixtures of modified proteins. Molecular Cell. Proteomics 6, 1123–1134 (2007)
3. Bartels, C.: Fast algorithm for peptide sequencing by mass spectrometry. Biomedical and Environmental Mass Spectrometry 19, 363–368 (1990)
4. Bern, M., Cai, Y., Goldberg, D.: Lookup peaks: a hybrid of de novo sequencing and database search for protein identification by tandem mass spectrometry. Anal. Chem. 79, 1393–1400 (2007)
5. Bern, M., Goldberg, D.: De novo analysis of peptide tandem mass spectra by spectral graph partitioning. J. Computational Biology 13, 364–378 (2006)
6. Chen, T., Kao, M.-Y., Tepel, M., Rush, J., Church, G.M.: A dynamic programming approach to de novo peptide sequencing by mass spectrometry. J. Computational Biology 8, 325–337 (2001)
7. Chow, C.K., Liu, C.N.: Approximating discrete probability distributions with dependence trees. IEEE Trans. on Information Theory 14, 462–467 (1968)
8. Dančik, V., Addona, T.A., Clauser, K.R., Vath, J.E., Pevzner, P.A.: De novo peptide sequencing via tandem mass spectrometry. J. Computational Biology 6, 327–342 (1999)
9. Duda, R.O., Hart, P.E., Stork, D.G.: Pattern Classification. Wiley-Interscience, Chichester (2000)
10. Elias, J.E., Gibbons, F.D., King, O.D., Roth, F.P., Gygi, S.P.: Intensity-based protein identification by machine learning from a library of tandem mass spectra. Nature Biotechnology 22, 214–219 (2004)
11. Eng, J.K., McCormack, A.L., Yates III., J.R.: An approach to correlate tandem mass spectral data of peptides with amino acid sequences in a protein database. J. Am. Soc. Mass Spectrom. 5, 976–989 (1994)
12. Eppstein, D.: Finding the k shortest paths. SIAM J. Computing 28, 652–673 (1998)
13. Frank, A., Pevzner, P.: PepNovo: De Novo Peptide Sequencing via Probabilistic Network Modeling. Anal. Chem. 77, 964–973 (2005)
14. Friedman, N., Geiger, D., Goldszmidt, M.: Bayesian Network Classifiers. Machine Learning 29, 131–163 (1997)
15. Graehl, J.: Implementation of David Eppstein's k Shortest Paths Algorithm., http://www.ics.uci.edu/~eppstein/
16. Havilio, M., Haddad, Y., Smilansky, Z.: Intensity-based statistical scorer for tandem mass spectrometry. Anal. Chem. 75, 435–444 (2003)
17. Ma, B., Zhang, K., Hendrie, C., Liang, C., Li, M., Doherty-Kirby, A., Lajoie, G.: PEAKS: powerful software for peptide de novo sequencing by tandem mass spectrometry. Rapid Comm. in Mass Spectrometry 17, 2337–2342 (2003), http://www.bioinformaticssolutions.com
18. Savitski, M.M., Nielsen, M.L., Kjeldsen, F., Zubarev, R.A.: Proteomics-Grade de Novo Sequencing Approach. J. Proteome Research, 2348–2354 (2005)
19. Shevchenko, A., et al.: Charting the proteomes of organisms with unsequenced genomes by MALDI-quadrupole time-of-flight mass spectrometry and BLAST homology searching. Anal. Chem. 73, 1917–1926 (2001)

20. Syka, J.E., Coon, J.J., Schroeder, M.J., Shabanowitz, J., Hunt, D.F.: Peptide and protein sequence analysis by electron transfer dissociation mass spectrometry. Proc. Natl. Acad. Sci. USA 101, 9528–9533 (2004)
21. Tabb, D.L., Smith, L.L., Breci, L.A., Wysocki, V.H., Lin, D., Yates III., J.R.: Statistical characterization of ion trap tandem mass spectra from doubly charged tryptic digests. Anal. Chem. 75, 1155–1163 (2003)
22. Taylor, J.A., Johnson, R.S.: Implementation and uses of automated de novo peptide sequencing by tandem mass spectrometry. Anal. Chem. 73, 2594–2604 (2001)
23. Venable, J.D., Xu, T., Cociorva, D., Yates III., J.R.: Cross-correlation algorithm for calculation of peptide molecular weight from tandem mass spectra. Anal. Chem. 78, 1921–1929 (2006)
24. Zhang, Z., McElvain, J.S.: De novo peptide sequencing by two-dimensional fragment correlation mass spectrometry. Anal. Chem. 72, 2337–2350 (2000)

Appendix: Parent Mass Correction

In the context of spectrum fusion, the problem of parent mass correction is as follows. The MS instrument outputs a real-valued, nominal parent mass M_i' for peptide $\mathcal{P}_i$, and our goal is to correct it to ± 0.5 Da. (A precise estimate of the parent mass is not required by our algorithm.) Once again, we have multiple spectra per peptide available for this classification problem, and once again, we use the M+H convention, meaning that we would like the sum of the residue masses plus 19 Daltons, to account for water and one proton. Empirically, we find that the true parent mass M_i is never beyond ± 7 Da of M_i'. Therefore, in each case, there are only 15 candidate integers in the neighborhood of M_i' that could be the correct answer.

1. **Complementary Peak Sums:** We take pairwise sums of the 30 tallest peaks in the original spectra. (We make adjustments for the M+H convention, and the fact that the sum of an ETD complementary pair is one Dalton greater than the sum of a CID pair.) For those sums that fall within $M_i' \pm 7$, we add the sum of intensities for that pair to the "intensity" of the parent mass candidate.

2. **Known Peak Positions:** We observed that spectra often contain peaks at positions that directly reflect the true mass M_i. For example, CID spectra of peptides with parent charge +2 often contain a peak at $(M_i + 1 - 18)/2$, representing the entire peptide minus water, doubly charged. ETD spectra of peptides with parent charge +3 often contain peaks at $M_i + 2$ and $M_i - 15$, for the entire peptide with two protons reduced to hydrogens, and the entire peptide with two protons reduced, minus ammonia. We developed a simple method to automatically deduce such mass cues from different types of spectra, given some training data. Here, we hypothesize in turn each of the 15 candidates as the actual mass, and seek these known cues in the spectra. If a cue is found, a score equal to the intensity of that peak is added to the parent mass candidate.

3. **Suffix/Prefix Alignment:** As in [23], we complement all the peaks in a spectrum relative to the candidate parent mass, and thereby create a mirror spectrum. We expect this spectrum to align with the original spectrum best when the correct mass candidate was used for complementing. Therefore, we take dot products of the

intensities of the original and mirror spectra (separately, for each type of spectrum) and add the dot products as scores for the 15 mass candidates.

Finally, we normalize each of the three scores by the respective maximum values among the candidates, and sum the scores to determine the winning candidate.

Parent mass correction performed in this manner is fairly effective. For the $+2$ charged peptides, the closest integer to the nominal M+H parent mass (after correction for mass defect) matched the correct mass only 28% of the time. In contrast, our correction leads to about 90% accuracy. For the $+3$ charged peptides, the nominal mass is correct only about 4% of the time. (Due to isotopes shifting the center of the single-MS peak, the nominal mass is most commonly one or two Daltons too high.) In contrast, our mass correction leads to about 75% accurate parent mass estimation. We also found out that a majority of the times that our mass correction made wrong estimates, even a correct mass estimate would not lead to successful *de novo* sequencing (5 or more consecutive letters correct), suggesting poor quality of such spectra.

A Fragmentation Event Model for Peptide Identification by Mass Spectrometry

Yu Lin[1], Yantao Qiao[1], Shiwei Sun[1], Chungong Yu[1], Gongjin Dong[1], and Dongbo Bu[1,2,*]

[1] Bioinformatics Group, Key Lab of Intelligent Software Systems,
Institute of Computing Technology, Chinese Academy of Sciences, China[**]
[2] Bioinformatics Lab, University of Waterloo, Ontario, N2L 3G1, Canada

Abstract. We present in this paper a novel fragmentation event model for peptide identification by tandem mass spectrometry. Most current peptide identification techniques suffer from the inaccuracies in the predicted theoretical spectrum, which is due to insufficient understanding of the ion generation process, especially the b/y ratio puzzle.

To overcome this difficulty, we propose a novel fragmentation event model, which is based on the abundance of fragmentation events rather than ion intensities. Experimental results demonstrate that this model helps improve database searching methods. On LTQ data set, when we control the false-positive rate to be 5%, our fragmentation event model has a significantly higher true positive rate (0.83) than SEQUEST (0.73). Comparison with Mascot exhibits similar results, which means that our model can effectively identify the false positive peptide-spectrum pairs reported by SEQUEST and Mascot.

This fragmentation event model can also be used to solve the problem of missing peak encountered by *De Novo* methods. To our knowledge, this is the first time the fragmentation preference for peptide bonds is used to overcome the missing-peak difficulty.

Availability: http://www.bioinfo.org.cn/MSMS/.

1 Introduction

Computational proteomics has become an emerging field arising from the demand of high throughput analysis to identify all proteins for creating a catalogue of information. One of the major technical advances in this newborn field is the use of an analytical instrument known as a tandem mass spectrometer (MS/MS) [26, 31]. In an MS/MS experiment, proteins of interest are first selected and digested into peptides by an enzyme, such as trypsin. Next these resultant peptides are separated in a mass analyzer by their mass to charge ratio (*m/z value*). During the subsequent collision-induced dissociation (CID) step, the peptides are

---

[**] To whom correspondence should be addressed. Correspondence may also be addressed to Shiwei Sun. email: bdb@ict.ac.cn, dwsun@ict.ac.cn. The first two authors contributed equally to this paper.

M. Vingron and L. Wong (Eds.): RECOMB 2008, LNBI 4955, pp. 154–166, 2008.

further fragmented into charged ions. The m/z value and intensities of these ions are measured and recorded as an experimental MS/MS spectrum [1].

To explain the complicated peptide fragmentation process, many hypotheses have been proposed, amongst which the *mobile proton* tenet is the most widely accepted. Under this hypothesis, the most common pathway is charge-directed backbone fragmentation, where at least one proton is not localized to the arginine side chains. These mobile protons can migrate along the peptide backbone to an amide carbonyl oxygen, leading to cleavage of the associated peptide bond and the production of a b ion or a y ion. Both the b ion and y ion can form a set of derivatives: a ion, b-NH_3 ion, b-H_2O ion, y-NH_3 ion, y-H_2O ion, etc. Other theoretically possible backbone ions, such as c, x, z ions, are not typically generated under low energy CID conditions [19, 23, 25, 30]. As in previous works [1, 5, 22], this paper also adopts the *mobile proton* hypothesis.

1.1 Related Work

There are two types of approaches for peptide identification through tandem mass spectrum: database searching [1, 4, 8, 21, 24, 28] and *De Novo* method [2, 3, 5, 10, 12, 14, 16] (See [17] for a recent review). Generally, database searching methods are considered more accurate than *De Novo* methods. Recently, sequence tagging approaches have been proposed to reduce the computation requirement of *De Novo* methods. This technique is particularly useful for the peptides with post transcription modifications [11, 18].

Theoretical spectrum prediction has received intensive studies because of its importance to both database searching and *De Novo* methods. Besides the promising chemical kinetic model [30], several statistical prediction models have been proposed. For example, V. Dancik *et al* employed an offset frequency function to learn the ion types tendency and ion intensity threshold from experimental spectra [1, 5]. J. R. Yates III *et al* attempted to identify statistical trends in spectrum peak intensities and to put them into a chemical context [26]. J. E. Elias *et al* applied a probability decision tree approach to identify the important factors influencing peptide fragmentation from a total of 63 attributes [6]. F. Schutz fitted a linear model to spectra, in which the influence of amino acid types and cleavage site are reflected [22].

These efforts are helpful to both understanding and simulating the complicated fragmentation process during mass spectrometry. However, most of these prediction methods suffer from the following two limitations:

- Ion intensity is difficult to predict. The reason is that when a peptide bond breaks, it is unclear whether a b ion or an y ion will be generated, needless to say the other derivative ions. Take, for example, the doubly charged peptides in the A8_IP dataset [9], the ratio of the intensity of a b ion to that of the corresponding y ion, called b/y *ratio* in this paper, varies from spectrum to spectrum. Therefore, the intensity of b ion or y ion cannot be accurately estimated even if the sum of these two intensities is known.

– An appropriate framework that considers the factors influencing peptide fragmentation remains a challenge to theoretical spectrum prediction. In essence, peptide fragmentation is a stochastic process governed by complicated physical and chemical rules, and is affected by many factors, such as the position of the fragmentation site, and the cleavage preference for specific peptide bonds. It is still unclear to what extent each factor affects the fragmentation process during the mass spectrometry.

In our previous works [27, 29], we designed an iterative algorithm to quantify the factors influencing peptide fragmentation, and employed relative entropy to measure the similarity between theoretical spectrum and experimental spectrum. These earlier works focused on predicting theoretical spectra accurately; the focus of the current paper is on how to overcome the difficulties of the theoretical spectrum prediction framework.

1.2   Our Contributions

The above-mentioned limitations have inspired us to directly employ the fragmentation pattern rather than the theoretical spectrum prediction framework. Briefly, for a peptide, we attempt to explore its fragmentation pattern instead of its theoretical spectrum, and thus our model is based on abundance of fragmentation events rather than the ion intensities.

Our contributions within this paper are as follows:

1. We proposed a novel fragmentation event model for peptide identification. In order to compare one experimental spectrum and one candidate peptide of length L, we first represented the experimental spectrum as one L-dimension vector. Also we predict the L-dimension theoretical fragmentation vector by using a statistical model for the candidate peptide. Finally, we employed Jensen-Shannon divergence, a variant of relative entropy, to measure the similarity between the theoretical and experimental fragmentation event vectors.

2. We used this fragmentation event model to improve both database searching and De Novo methods. For database searching methods, we attempt to validate the peptide-spectrum pairs reported by SEQUEST and Mascot. In this validation step, the reported peptide-spectrum pairs are re-ranked with respect to Jensen-Shannon divergence between the theoretical and experimental fragmentation patterns. Experimental results on both LTQ and QSTAR data sets suggest that this re-ranking strategy can effectively identify the false-positive pairs. In addition, we applied the fragmentation event model to solve the problem of missing peak that encountered by De Novo methods. To our knowledge, this is the first time the fragmentation preference for peptide bonds is used to overcome the missing-peak problem.

We implemented our model and related score scheme into an open source package PI (Peptide Identifier), which can be downloaded freely from http://www.bioinfo.org.cn/MSMS/.

2   Methods

The procedure of our fragmentation event model has three main steps:

Step 1. Deriving the experimental fragmentation event vector from an experimental spectrum;

Step 2. Predicting the theoretical fragmentation event vector for a candidate peptide;

Step 3. Applying Jensen-Shannon divergence to measure the similarity of these two vectors.

The details of these three steps are described below.

2.1   Deriving the Experimental Fragmentation Event Vector

Before describing our fragmentation event model, we briefly introduce the notations used in this paper as follows:

Let $A = \{a_1, a_2, ..., a_{20}\}$ be the amino acids set; each amino acid $a \in A$ has a molecular mass $m(a)$. A peptide is a string of amino acids, denoted as $P = p_1 p_2 ... p_L$, where $p_j \in A$. The breakage of the i-th peptide bond p_i-p_{i+1}, $i = 1, 2, ..., L - 1$, is referred to as the i-th fragmentation event e_i. This fragmentation event will typically generate two ions: one is b_i ion from the N-terminal peptide $P_i = p_1 p_2 ... p_i$, and the other is y_{L-i} ion from the C-terminal peptide $\bar{P}_i = p_{i+1} p_{i+2} ... p_L$. Both of these ions may form a set of derivative ions: the loss of a carbon monoxide from the C-terminus of b_i forms an a_i ion; the loss of a water from b_i forms ion b_i-H_2O; and the loss of an ammonia from side chains of b_i forms ion b_i-NH_3. The situation is similar for the C-terminal ions.

A spectrum, which consists of a series of peaks, is denoted as $S = \{(x_i, h_i) | i = 1, 2, ..., |S|\}$, where x_i is the m/z value of a peak, and h_i is the abundance of this peak. We say a peak (x_i, h_i) is explained by an ion with m/z value m if $|x_i - m| \leq \delta$, where δ is the precision of the mass spectrum S. Note that a peak may be explained by multiple ions. In this case, the contribution of each ion to this peak is calculated according to Gaussian mixed model [27].

Given a candidate peptide with length L and the ion precision δ, an experimental spectrum can be converted into an experimental fragmentation event vector $V^E =< v_1^E, v_2^E, ..., v_{L-1}^E, v_L^E >$, where v_i^E is the abundance of fragmentation event $E_i (i = 1, ..., L - 1)$, i.e., the total contributions of the ions generated by this fragmentation. We use v_L^e to represent the noise of this spectrum, i.e., the total intensity of the unexplained ions divided by L. This vector is normalized to make $\sum_{i=1}^{L} v_i^E = 1$.

2.2   Predicting the Theoretical Fragmentation Event Vector

In this subsection, we will describe the method to predict the theoretical fragmentation pattern based on the *mobile proton* hypothesis. Compared with the uniform models that SEQUEST [8] and Mascot [21] adopt, we take into account more factors with significant influence on peptide fragmentation, including the fragmentation position and the cleavage preference for specific peptide bonds, i.e., the possibility that a bond breaks.

In this method, we perform the following learning procedure for different peptides of lengths $L = 7, 8, ..., 18$. The other cases were not considered due to the insufficient number of spectra. For each peptide of length L, the learning procedure can be described as follows: for a peptide $P = p_1 p_2, ..., p_L$, let f_i, $i = 1, ..., L - 1$, denote the preference for fragmentation at position i, and $E(Xaa, Yaa)$ denote the possibility that the peptide bond Xaa-Yaa breaks. Under a reasonable assumption that these two factors are mutually independent, the theoretical abundance of the fragmentation event e_i, denoted as v_i^T, can be estimated to be proportional to $f_i \times E(p_i, p_{i+1})$, i.e., $v_i^T = \alpha \times f_i \times E(p_i, p_{i+1})$, for $i = 1, ..., L - 1$ and $v_L^T = 0$, where α is a scale factor.

The remaining difficulty is how to derive the parameters f_i and $E(Xaa, Yaa)$. Here, a learning formulation is adopted to derive these parameters from a training set containing peptide-spectrum pairs with high confidence. Furthermore, we propose an optimization strategy to solve this learning problem. The basic idea of this strategy is to minimize the difference between the experimental fragmentation event vector V^E and the theoretical fragmentation event vector V^T. By minimizing the difference, we can assign reasonable values for the parameters f_j and $E(Xaa, Yaa)$.

More specifically, for the k-th peptide-spectrum pair $(P^{(k)}, S^{(k)})$ in the training set, we first calculate the experimental event vector $V_k^E = < v_{k,1}^E, v_{k,2}^E, ..., v_{k,L}^E >$, then estimate the theoretical event vector $V_k^T = < v_{k,1}^T, v_{k,2}^T, ..., v_{k,L}^T >$ from $P^{(k)}$ by $v_{k,i}^T = \alpha_k \times f_i \times E(p_i^{(k)}, p_{i+1}^{(k)})$, where α_k is a scale parameter for spectrum $S^{(k)}$. Our optimization objective is the sum of square of the difference between the theoretical and experimental event vectors. Now, we can wrap up everything into the following non-linear programming problem:

$$min \sum_{k=1}^{K} \sum_{i=1}^{L-1} (v_{k,i}^T - v_{k,i}^E)^2$$

$$s.t. \ v_{k,i}^T = \alpha_k \times f_i \times E(p_i^{(k)}, p_{i+1}^{(k)}) \tag{1}$$

$$\sum_{i=1}^{L-1} f_i = 1 \tag{2}$$

$$\sum_{i=1}^{L-1} \alpha_k \times f_i \times E(p_i^{(k)}, p_{i+1}^{(k)}) = 1. \quad k = 1, 2, ..., K. \tag{3}$$

$$f_i \geq 0, \alpha_k \geq 0, E(Xaa, Yaa) \geq 0$$

Here, restriction (1) describes the process to estimate the theoretical fragmentation event vector; restriction (2) and (3) are normalization requirements.

We designed an effective iteration method to solve this high-rank non-linear optimization problem. The convergence of this iteration strategy is also guaranteed. Generally, the iteration method terminates in no more than 10 iteration loops. Experimental results suggest that this method is both robust and effective. Please refer to [27] for more details.

2.3 Applying Jensen-Shannon Divergence to Measure the Similarity of Two Vectors

In our previous work [29], we utilized the relative entropy to measure the similarity between theoretical spectrum and experimental spectrum. Relative entropy can also be employed to measure the similarity of theoretical fragmentation event vector V^T and experimental fragmentation event vector V^E, i.e., $H(V^E, V^T) = \sum_{i=1}^{L} v_i^E \times ln\frac{v_i^E}{v_i^T}$. In fact, relative entropy measures the likelihood that peptide P exhibits the experimental fragmentation event pattern V^E if we treat the spectrum generating process as a repeat trial.

Relative entropy is an ideal measure; however, we found from practical experience that this measure suffers from some limitations. Specifically, $H(V^E, V^T)$ is undefined if $v_i^E \neq 0$ and $v_i^T = 0$ for any $i \in \{1, 2, \ldots, L\}$. However, this case is very common due to the peak-missing in both theoretical spectrum and experimental spectrum. To overcome this restriction, we adopt Jensen-Shannon divergence $JS(V^E, V^T) = \frac{1}{2}(H(V^E, M) + H(V^T, M))$, where $M = \frac{1}{2}(V^E + V^T)$ [13] instead of relative entropy in this paper. $JS(V^E, V^T)$ is nonnegative and equal to zero when $V^E = V^T$. The smaller the Jensen-Shannon divergence between V^E and V^T is, the more likely that the spectrum was generated by the corresponding peptide.

We also performed comparison of Jensen-Shannon divergence with another similarity measure, the *Cosine Coefficient* (CC). CC is defined as:

$$CC(V^E, V^T) = \frac{V^E \cdot V^T}{\|V^E\|\|V^T\|},$$

where $V^E \cdot V^T$ is the inner product of V^E and V^T, and $\|V^E\|$ and $\|V^T\|$ are the norm of vector V^E and V^T, respectively.

3 Results

3.1 Experiments on Improving Database Searching Methods

As one of the applications, our model can be used to validate the peptide identification results of SEQUEST and Mascot. During the protein identification process, SEQUEST compares a given spectrum against peptides in a database, and reports a set of peptide-spectrum pairs ordered by their confidence scores. However, since SEQUEST employs a simple theoretical spectrum prediction model, there are always false positive pairs in the identification results. Here, we adopt the reverse-database criteria of false-positive [20]; that is, a peptide-spectrum pair is labeled as a false positive if the peptide appears in the reverse database. Our goal is to improve peptide predictions by distinguishing these false positive pairs.

Based on the fragmentation event model, we apply a re-ranking strategy to achieve this goal. More specifically, for each spectrum-peptide pair reported by SEQUEST, we first calculate the experimental and theoretical fragmentation patterns. Next we sort the peptide-spectrum pairs according to Jensen-Shannon divergence between these patterns. Ideally, a false positive pair will be given relatively high score.

In this experiment, LTQ and QSTAR data sets [7] were used to test the re-ranking strategy. The data sets have precision $\delta = 0.5 Dalton$ and $\delta = 0.2 Dalton$, respectively. For the LTQ data set, SEQUEST reported 13,599 peptide-spectrum pairs. The first $2,000$ pairs, which were reported by SEQUEST with high confidence ($Xcorr > 4.194$, $DeltaCn > 0.29$), were used to train our fragmentation event model. The other 11,599 pairs were used as the testing set, which contains 186 false positive pairs. Similarly, for the QSTAR data set, we chose the first 2,000 peptide-spectrum pairs as the training set, and the other 6,414 reported pairs as the testing set, which contains 450 false positive pairs (See http://www.bioinfo.org.cn/MSMS/ for supplementary material).

We evaluated the performance of the following three combinations of prediction models and scoring functions: the first one being the fragmentation event model with Jensen-Shannon divergence, the second one being the fragmentation event model with cosine coefficient scoring function, and the last one being the theoretical spectrum model [27] (with b/y ratio assumed to be 1) with Jensen-Shannon divergence. These three combinations are denoted as $PI^{Event+JS}$, $PI^{Event+CC}$, and $PI^{Spectrum+JS}$, respectively. For these combinations and SEQUEST, the relationship between the false positive rate and true positive rate in the testing set is graphically shown in Figure 1 as receiver-operating characteristic (ROC) plots. From this Figure, we can see that if we control the false-positive rate to be 5%, $PI^{Event+JS}$ has a significantly higher true positive rate (0.8355) than SEQUEST (0.7301), $PI^{Event+CC}$ (0.7011), and $PI^{Spectrum+JS}$ (0.5906). This demonstrates that Jensen-Shannon divergence is better than cosine coefficient, and our fragmentation event model is better than the theoretical spectrum model that assumes b/y ratio to be 1. Hence, this re-ranking technique can help detect the false-positive pairs reported by SEQUEST.

As a concrete example, a false-positive pair is shown in Figure 2 to demonstrate the advantage of $PI^{Event+JS}$ over SEQUEST. SEQUEST reports this pair with a high confidence (Rank=3384-th, $Xcorr = 3.41$, $DeltaCn = 0.295$). In contrast, among the entire $11,599$ peptide-spectrum pairs, this pair is ranked $9,906$-th by $PI^{Event+JS}$. Intuitively, Figure 2 reveals a significant dissimilarity between the theoretical and the experimental fragmentation patterns. Precisely, the Jensen-Shannon divergence between these two fragmentation event vectors is 0.206, which is significantly large with respect to the peptide length $L = 13$.

To illustrate the advantage of $PI^{Event+JS}$ over $PI^{Spectrum+JS}$, a true-positive pair is shown in Figure 3. This pair achieves a rank of $2,783$-th from $PI^{Event+JS}$ and a rank of $10,778$-th from $PI^{Spectrum+JS}$. The similarity between the theoretical and experimental fragmentation patterns is clearly demonstrated in Figure 3 (left-hand side, Jensen-Shannon divergence: 0.070). However, if we adopt the assumption that b/y ratio is 1, we will miss this true positive pair due to the insignificant similarity between the theoretical and experimental spectrum (See Figure 3, right-hand side. Jensen-Shannon divergence: 0.441).

We compared the above three combinations with Mascot, and obtained similar observations (See Figure 4). These observations suggest that our model can help improve the accuracy of SEQUEST and Mascot on both LTQ and QSTAR data sets.

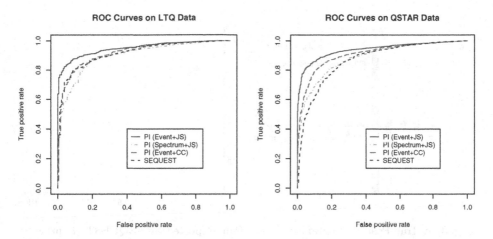

Fig. 1. Performance of Validating SEQUEST's Results on LTQ Data Set (left-hand side) and QSTAR Data Set (right-hand side). The relationship between the false positive rate and true positive rate is graphically shown as receiver-operating characteristic (ROC) plots for SEQUEST, $PI^{Event+JS}$, $PI^{Event+CC}$ and $PI^{Spectrum+JS}$.

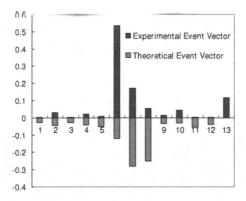

Fig. 2. A False Positive Peptide-spectrum Pair Reported by SEQUEST (Peptide Sequence: MSKDNEISPSLLR, Spectrum ID: Band7c.1890.1890.3.dta, $Xcorr = 3.41$, $DeltaCn = 0.295$). This pair achieves a rank of 3384-th from SEQUEST; however, this pair is ranked $9,906$-th by $PI^{EVENT+JS}$ due to the insignificant similarity between experimental and theoretical fragmentation patterns (Jensen-Shannon divergence: 0.206).

3.2 Experiments on Improving De Novo Sequencing Methods

In this subsection, we describe the application of fragmentation event model to improve De Novo methods. The dataset used to test our algorithm is A8_IP [9]. Each spectrum in A8_IP has been annotated with a matched peptide. In this proof-of-concept experiment, we restricted our analysis to the doubly charged peptides (2744 peptide-spectrum pairs in total). These pairs are further randomly divided into two parts: one is a training set with 1803 peptide-spectrum

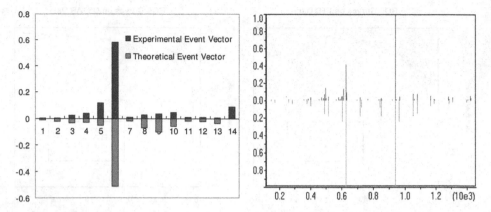

Fig. 3. A True Positive Peptide-spectrum Pair Reported by SEQUEST (Peptide Sequence: NGLDDIPQLLDDIK, Spectrum ID: Band5b.3108.3108.2.dta, $Xcorr = 2.57$, $DeltaCn = 0.315$). The fragmentation pattern (left-hand side, Jensen-Shannon divergence: 0.070) exhibits more significant similarity than the spectrum model (right-hand side, Jensen-Shannon divergence: 0.441). On both sides, the experimental patterns are shown above the axis, while the theoretical ones are shown below the axis.

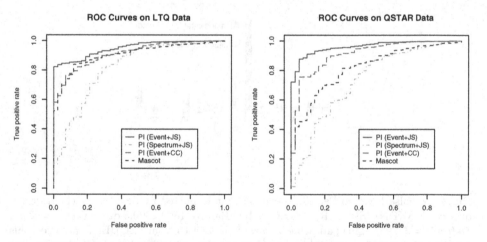

Fig. 4. Performance of Validating Mascot Results on LTQ Data Set (left-hand side) and QSTAR Data Set (right-hand side). The relationship between the false positive rate and true positive rate is graphically shown as receiver-operating characteristic (ROC) plots for Mascot, $PI^{Event+JS}$, $PI^{Event+CC}$ and $PI^{Spectrum+JS}$.

pairs, and the other is a testing set with 941 peptide-spectrum pairs (See http://www.bioinfo.org.cn/MSMS/ for supplementary material).

On the training set, we applied the optimization approach described in Section 2.2 to learn the fragmentation preference for bonds. In the testing set, we focused on a total of 181 bonds suffering from the missing-peak difficulty; that is, the fragmentation abundance of this bond is zero while the fragmentation

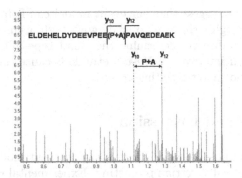

Fig. 5. A Missing-peak Example in Spectrum of Peptide *ELDEHELDYDEEVPEEPA-PAVQEDEAEK*. The gap between y_{10} and y_{12} is equal to the sum of the mass of amino acid P and A. However, the relative order of these two amino acids is undetermined since no cleavage event at this site is observed.

abundance of the neighboring bonds are not zero. Intuitively, this means there is a gap of W Daltons in the spectrum. In such cases, most *De Novo* methods can easily determine these two amino acids Xaa and Yaa based on the restriction $m(Xaa) + m(Yaa) = W$. However, a *De Novo* method itself generally cannot determine the relative order of these two amino acids, i.e., whether the bond is Xaa-Yaa or Yaa-Xaa [15]. An example is shown in Figure 5, where the abundance of the fragmentation event between P and A, represented as (P+A), is zero.

Here, we apply our event model along with Jensen-Shannon divergence to determine the order of these two neighboring amino acids. The basic idea can be described as follows: suppose the unknown bond is the i-th one of peptide $P = p_1p_2\ldots p_ip_{i+1}\ldots p_L$, and we attempt to determine whether p_i-$p_{i+1} = Xaa$-Yaa or p_i-$p_{i+1}=Yaa$-Xaa. These possible peptide choices are denoted as P_{XaaYaa} and P_{YaaXaa}, respectively. Notice that these two peptide choices differ in 3 bonds, i.e., peptide P_{XaaYaa} has bonds p_{i-1}-Xaa, Xaa-Yaa, and Yaa-p_{i+2}, while peptide P_{YaaXaa} has bonds p_{i-1}-Yaa, Yaa-Xaa, and Xaa-p_{i+2}. Since peptide bonds show a significant heterogeneity in cleavage preference (See Table 1 in [27]), we can determine which one of the possible peptides is more likely based on the fragmentation patterns. In detail, for these two possible peptides, we predict the theoretical fragmentation event vector, and calculate the Jensen-Shannon divergence between the theoretical fragmentation event vector and the experimental counterpart. Finally, the relative order can be determined by using the following rule:

$$p_ip_{i+1} = \begin{cases} XaaYaa, & if\ JS(V^E, V^T_{XaaYaa}) < JS(V^E, V^T_{YaaXaa}) - \epsilon \\ YaaXaa, & if\ JS(V^E, V^T_{YaaXaa}) < JS(V^E, V^T_{XaaYaa}) - \epsilon \end{cases}$$

Here, ϵ is a threshold to control the significance of the difference between the two Jensen-Shannon divergence scores. We achieve a prediction accuracy of 69.10% when ϵ is 0, and a prediction accuracy of 82.86% when ϵ is 0.008.

These experiments suggest that our event model can help determine the relative order of two neighboring amino acids with high accuracy, even if there is no peak available to directly determine the bond type. Thus, our fragmentation event model can improve *De Novo* methods because the existing *De Novo* methods alone cannot determine this order [15].

4   Conclusion and Discussion

In this paper, we proposed a novel fragmentation event model to overcome the difficulties in theoretical spectrum predicting. Experimental results demonstrate that this model can help to identify the false-positive results of both SEQUEST and Mascot. Moreover, our event model can also help to solve the problem of missing peak encountered by *De Novo* methods.

An interesting question arises about the reason why fragmentation event model performs better than the traditional theoretical spectrum model. The advantage of fragmentation event model seems counterintuitive since a theoretical spectrum contains more items than a fragmentation event vector. We attempt to explain as follows:

First, fragmentation pattern is more robust than theoretical spectrum. Though there are more information in a theoretical spectrum than that in a fragmentation event vector, the information in a theoretical spectrum is difficult to restore because of the long-standing b/y ratio puzzle (from personal communication with Vicki H. Wysocki). Our survey of A8_IP [9] data set suggests that the b/y ratio varies a lot from ion to ion. In particular, 50% of ions have a b/y ratio below $\frac{1}{2}$ or over 2, and 20% of ions have a b/y ratio of 0 or infinity. In contrast, fragmentation pattern is more robust since b/y ratio is not necessary in this framework.

Second, a potential advantage of fragmentation event model is that setting threshold is much easier relative to the theoretical spectrum framework. Jensen-Shannon divergence under our fragmentation event model is related to the peptide length only, and independent of the peptide mass and the number of ions. Therefore, the similarities of different peptide-spectrum pairs are comparable so long as these peptides share the same length. Thus we need only to set a threshold for each possible peptide length, which is much easier compared with setting the threshold under the theoretical spectrum framework.

In summary, although at the current stage the ion generating process is still unclear, the coarse-grained fragmentation pattern framework appears to be a robust strategy.

Acknowledgment

This work was supported by National Sciences Foundation of China under grants 60496320, 30500104 and 30570393, National Key Basic Research and Development Program under grants 2003CB715900, and an opening task of Shanghai

Key Laboratory of Intelligent Information Processing Fudan University with No.IIPL-04-001.
We appreciate Richard Jang and Annie Lee for their helps to polish this paper.

References

1. Bafna, V., Edwards, N.: Scope: a probabilistic model for scoring tandem mass spectra against a peptide database. Bioinformatics 17(1), S13–S21 (2001)
2. Bartels, C.: Fast algorithm for peptide sequencing by mass spectroscopy. Biomed Environ Mass Spectrom 19(6), 363–368 (1990)
3. Chen, T., Kao, M.Y., Rush, J., Church, G.M.: A dynamic programming approach to de novo peptide sequencing via tandem mass spectrometry. J. Comput. Bio. 8(3), 325–337 (2001)
4. Craig, R., Beavis, R.C.: Tandem: matching proteins with tandem mass spectra. Bioinformatics 20, 1466–1467 (2004)
5. Dancik, V., Addona, T.A., Clauser, K.R., Vath, J.E., Pevzner, P.A.: De novo peptide sequencing via tandem mass spectrometry. J. Comput. Bio. 6(3–4), 327–342 (1999)
6. Elias, J.E., Gibbon, F.D., King, O.D., Roth, F.P., Gygi, S.P.: Intensity-based protein identification by machine learning from a library of tandem mass spectra. Nat. Biotechnol. 23(2), 214–214 (2004)
7. Elias, J.E., Hass, W., Faherty, B.K., Gygi, S.P.: Comparative evaluation of mass spectrometry platforms used in large-scale proteomic investigations. Nature Methods 2(9), 667–675 (2005)
8. Eng, J.K., McCormack, A.L., Yates, J.R.: An approach to correlate tandem massspectral data of peptides with amino acid sequences in a protein database. J. Am. Soc. Mass. Spect. 5, 976–989 (1994)
9. Resing, K.A., et al.: Improving reproducibility and sensitivity in identifying human proteins by shotgun proteomics. Anal. Chem. 76(13), 3556–3568 (2004)
10. Frank, A., Pevzner, P.A.: Pepnovo: de novo peptide sequencing via probabilistic network modeling. Anal. Chem. 77(4), 964–973 (2005)
11. Frank, A., Tanner, S., Bafna, V., Pevzner, P.A.: Peptide sequence tags for fast database search in mass-spectrometry. J. Proteome. Res. 4(4), 1287–1295 (2005)
12. Hines, W.M., Falick, A.M., Burlingame, A.L., Gibson, B.W.: Patternbased algorithm for peptide sequencing from tandem high energy collision-induced dissociation mass spectra. J. Am. Soc. Mass. Spect. 3, 326–336 (1992)
13. Lin, J.: Divergence measures based on the shannon entropy. IEEE Trans. on Information Theory 37(1), 145–151 (1991)
14. Lu, B., Chen, T.: A suboptimal algorithm for de novo peptide sequencing via tandem mass spectrometry. J. Comput. Bio. 10(1), 1–12 (2003)
15. Lu, B., Chen, T.: Algorithms for de novo peptide sequencing via tandem mass spectrometry. Drug Discovery Today: BioSilico 2, 85–90 (2004)
16. Ma, B., Zhang, K., Hendrie, C., Li, M., Doherty-Kirby, A., Lajoie, G.: Peaks: powerful software for peptide de novo sequencing by tandem mass spectrometry. Rapid Commun. Mass Spectrom. 17(20), 2337–2342 (2003)
17. Matthiesen, R.: Methods, algorithms and tools in computational proteomics: a practical point of view. proteomics 7(16), 2815–2832 (2007)
18. Matthiesen, R., Bunkenborg, J., Stensballe, A., Jensen, O.N.: Database-independent, database-dependent, and extended interpretation of peptide mass spectra in vems v2.0. Proteomics 4(9), 2583–2593 (2004)

19. Paizs, B., Suhai, S.: Towards understanding the tandem mass spectra of protonated oligopeptides. 1: mechanism of amide bond cleavage. J. Am. Soc. Mass. Spect. 15(1), 103–113 (2004)
20. Peng, J., Elias, J.E., Thoreen, J.E., Licklider, L.J., Gygi, S.P.: Evaluation of multidimensional chromotography coupled with tandem mass spectrometry (lc/lc-ms/ms) for large-scale protein anaysis: the yeast proteome. J. Proteome. Res. 2(1), 43–50 (2003)
21. Perkins, D.N., Pappin, D.J., Creasy, D.M., Cottrell, J.S.: Probability-based protein identification by searching sequence databases using mass spectrometry data. Electrophoresis 20(18), 3551–3567 (1999)
22. Schutz, F., Kapp, E.A., Simpson, R.J., Speed, T.P.: Deriving statistical models for predicting peptide tandem ms product ion intensities. Proteomics 31, 1479–1483 (2003)
23. Tabb, D.L., Smith, L.L., Breci, L.A., Wysocki, W.H., Yates, J.R.: Statistical characterization of ion trap tandem mass spectra from doubly charged tryptic peptides. Anal. Chem. 75(5), 1155–1163 (2003)
24. Wan, Y., Chen, T.: A Hidden Markov Model Based Scoring Function for Mass Spectrometry Database Search. In: Miyano, S., Mesirov, J., Kasif, S., Istrail, S., Pevzner, P.A., Waterman, M. (eds.) RECOMB 2005. LNCS (LNBI), vol. 3500, pp. 163–173. Springer, Heidelberg (2005)
25. Wysocki, V.H., Tsaprailis, G., Smith, L.L., Breci, L.A.: Mobile and localized protons: a framework for understanding peptide dissociation. J. Mass Spectrom 35(12), 1399–1406 (2000)
26. Yates, J.R.: Mass spectrometry and the age of the proteome. J. Mass Spectrom 33(1), 1–19 (1998)
27. Yu, C., Lin, Y., Sun, S., Cai, J., Zhang, J., Bu, D., Zhang, Z., Chen, R.: An iterative algorithm to quantify factors influencing peptide fragmentation during tandem mass spectrometry. J. Bioinform. Comput. Biol. 5(2), 297–311 (2007)
28. Zhang, N., Aebersold, R., Schwikowski, B.: Probid: A probabilistic algorithm to identify peptides through sequence database searching using tandem mass spectral data. Proteomics 2(10), 1406–1412 (2002)
29. Zhang, Z., Sun, S., Zhu, X., Chang, S., liu, X., Yu, C., Bu, D., Chen, R.: A novel scoring schema for peptide identification by searching protein sequence databases using tandem mass spectrometry data. BMC Bioinformatics 7(222) (2006)
30. Zhang, Z.Q.: Prediction of low-energy collision-induced dissociation spectra of peptides. Anal. Chem. 76(14), 3908–3922 (2004)
31. Zhu, H., Bilgin, M., Snyder, M.: Proteomics. Annu. Rev. Biochem. 72, 783–812 (2003)

A Bayesian Approach to Protein Inference Problem in Shotgun Proteomics

Yong Fuga Li[1], Randy J. Arnold[2], Yixue Li[3], Predrag Radivojac[1], Quanhu Sheng[1,3], and Haixu Tang[1]

[1] School of Informatics, Indiana University, Bloomington, IN 47408, USA
[2] Department of Chemistry, Indiana University, Bloomington, IN 47405, USA
[3] Key Lab of Systems Biology, Shanghai Institutes for Biological Sciences, Chinese Academy of Sciences, Shanghai, China

Abstract. The protein inference problem represents a major challenge in shotgun proteomics. Here we describe a novel Bayesian approach to address this challenge that incorporates the predicted peptide detectabilities as the prior probabilities of peptide identification. Our model removes some unrealistic assumptions used in previous approaches and provides a rigorous probabilistic solution to this problem. We used a complex synthetic protein mixture to test our method, and obtained promising results.

1 Introduction

In shotgun proteomics, a complex protein mixture derived from a biological sample is directly analyzed via a sequence of experimental and computational procedures [1,2,3,4]. After protease digestion, liquid chromatography (LC) coupled with tandem mass spectrometry (MS/MS) is typically used to separate and fragment peptides from the sample, resulting in a number of MS/MS spectra. These spectra are subsequently searched against a protein database to identify peptides present in the sample [5,6]. Many peptide search engines have been developed, among which Sequest [7], Mascot [8] and X!Tandem [9] are commonly used. However, after a reliable set of *peptides* is identified, it is often not straightforward to assemble a reliable list of *proteins* from these peptides. This occurs because some identified peptides, referred to as the *degenerate peptides*, are shared by two or more proteins in the database. As a result, the problem of determining which of the proteins are indeed present in the sample, known as the *protein inference problem* [10], often has multiple solutions and can be computationally intractable. Nesvizhskii and colleagues first addressed this challenge using a probabilistic model [11], but different problem formulations and new solutions have recently been proposed as well [10,12,13].

Previously, we introduced a combinatorial approach to the protein inference problem that incorporates the concept of *peptide detectability*, i.e. the probability of a peptide to be detected (identified) in a standard proteomics experiment, with the goal of finding the set of proteins with the minimal number of *missed peptides* [12]. As in the other combinatorial formulations [13], the *parsimony*

M. Vingron and L. Wong (Eds.): RECOMB 2008, LNBI 4955, pp. 167–180, 2008.
© Springer-Verlag Berlin Heidelberg 2008

condition was chosen only for convenience reasons, without theoretical justification. Furthermore, parsimonious formulations often lead to the *minimum cover set problem*, which is NP-hard. Thus, heuristic algorithms following greedy [12] or graph-pruning strategies [13] are used to solve the protein inference problem without performance guarantee.

In this paper, we address protein inference by proposing two novel Bayesian models that take as input a set of identified peptides from any peptide search engine, and attempt to find a most likely set of proteins from which those identified peptides originated. The basic model assumes that all identified peptides are correct, whereas the advanced model also accepts the probability of each peptide to be present in the sample. Compared with the previous probabilistic models, such as ProteinProphet [11], both of our models differ in two key aspects. First, our approach incorporates peptide detectability [14] since it has been recently shown that even among the peptides that belong to the same protein, some peptides are commonly observed, while some others are not [14,15]. This results in the fact that the peptides not identified by peptide search engines may have significant impact on the final solution. Second, previous models assume that the posterior probability of each peptide is independent of other peptides and can be computed separately. Although this assumption significantly simplifies the computation of the protein posterior probabilities, it is inconsistent with the Bayesian model of a shotgun proteomics experiment (see Materials and Methods). We relax this assumption and adopt Gibbs sampling approach to estimate protein posterior probabilities. The results of this study provide evidence that our models achieve satisfactory accuracy and can be readily used in protein identification.

2   Materials and Methods

To illustrate the challenge of protein inference, we define the *protein configuration graph* (Fig. 1(a)), i.e. a bipartite graph in which two disjoint sets of vertices represent the proteins in the database and the peptides from these proteins, respectively, and where each edge indicates that the peptide belongs to the protein. We emphasize that the protein configuration graph is independent of the proteomics experiment, and thus can be built solely from a set (database) of protein sequences. Therefore, in contrast to the bipartite graph used previously [13], where only the identified peptides and the proteins that contain those peptides were represented, our model also considers the non-identified peptides. Protein configuration graph is partitioned into *connected components*, each representing a group of proteins (e.g. homologous protein families) sharing one or more (degenerate) peptides. If there are no degenerate peptides in the database, each connected component will contain exactly one protein and its peptides. In practice, however, the protein configuration graph may contain large connected components, especially for protein databases of higher animals or those containing closely related species.

Given that the protein configuration graph can be interpreted as a Bayesian network with edges pointing from proteins into peptides, it is straightforward

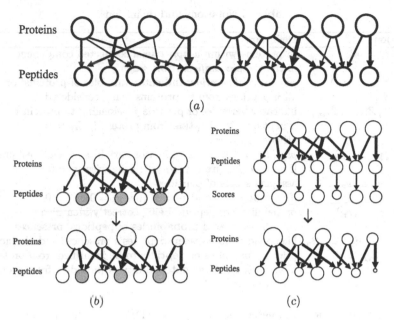

Fig. 1. (a) A protein configuration graph consisting of two connected components; (b) basic Bayesian model for protein inference, in which peptides are represented as a vector of indicator variables: 1 (grey) for identified peptides, and 0 (white) for non-identified peptides; (c) advanced Bayesian model for protein inference, in which each peptide is associated to an identification score (0 for non-identified peptides). Sizes of circles reflect prior/posterior probabilities.

to show that protein inference can be addressed separately for each individual connected component. In this approach, the peptide identification results are first mapped to the protein configuration graph. We use a vector of indicator variables $(y_1, ..., y_j, ..., y_n)$, referred to as the *peptide configuration* to denote a set of identified peptides. Given the peptide configuration, a connected component of the protein configuration graph is called *trivial* if it contains no identified peptides. Clearly, in this case protein inference is simple – none of the proteins should be present in the sample. Therefore, the protein inference problem can be reduced to finding the most likely *protein configuration* $(x_1, ..., x_i, ..., x_m)$ by analyzing *non-trivial* components only. In the basic model, all identified peptides are assigned equal probabilities ($= 1$) (Fig. 1(b)), whereas in the advanced model different probabilities are considered for different identified peptides depending on the associated identification scores $(s_1, ..., s_j, ..., s_n)$ (Fig. 1(c)). Notation and definitions used in this study are summarized in Table 1.

2.1 Basic Bayesian Model

In the basic probabilistic model, we assume that each identified peptide has an equally high prior probability to be present in the sample and low false discovery rate (FDR) in the results of peptide identification. In practice, even though this

Table 1. Notations and definitions

Notation	Definition
$(1, ..., i, ..., m)$	m proteins within a *non-trivial* connected component of the *protein configuration graph*
$(x_1, ..., x_i, ..., x_m)$	*protein configuration*: indicator variables of proteins' presences
$(1, ..., j, ..., n)$	all n peptides from m proteins being considered
$(Z_{11}, ..., Z_{ij}, ..., Z_{mn})$	indicator variables of peptide j belonging to protein i if peptide j is a peptide from protein i, $Z_{ij} = 1$; otherwise $Z_{ij} = 0$
$(y_1, ..., y_j, ..., y_n)$	*peptide configuration*: indicator variables of peptides' presences if peptide j is present, $y_j = 1$; otherwise $y_j = 0$
$(s_1, ..., s_j, ..., s_n)$	assigned scores of peptides if peptide j is not identified (i.e. $y_j = 0$), $s_j = 0$
$(r_1, ..., r_j, ..., r_n)$	probabilities of peptide being correctly identified also the estimated probabilities of peptides' presences
$(LR_1, ..., LR_j, ..., LR_n)$	likelihood ratio between peptides's presences and absences
$(d_{11}, ..., d_{ij}, ..., d_{mn})$	prior probabilities of peptides to be identified from proteins if $Z_{ij} = 1$, $d_{ij} =$ the detectability of peptide j from protein i; otherwise, $d_{ij} = 0$

assumption does not completely hold, peptide FDRs are usually controlled at a low level (e.g. 0.01) by either a heuristic target-decoy search strategy [16,13,17] or by probabilistic modeling of random peptide identification scores [18,19,20]. In the next section, we extend this basic model to a more realistic model in which we incorporate different probabilities for different identified peptides that are estimated based on the peptide identification scores.

Consider m proteins and n peptides from these proteins within a non-trivial connected component of the protein configuration graph. Each protein i is either present in the sample or absent from it, which can be represented by an indicator variable x_i. Therefore, any solution of the protein inference problem corresponds to a vector of indicator variables, $(x_1, ..., x_m)$, referred to as a protein configuration. Given the set of identified peptides from peptide search engines (peptide configuration $(y_1, ..., y_n)$), our goal is to find the *maximum a posteriori* (MAP) protein configuration, that is the configuration that maximizes the posterior probability $P(x_1, ..., x_m | y_1, ..., y_n)$. Using the Bayes' rule, this posterior probability can be expressed as

$$P(x_1, ..., x_m | y_1, ..., y_n) = \frac{P(x_1,...,x_m)P(y_1,...,y_n | x_1,...,x_m)}{\sum_{(x_1,...,x_m)}[P(x_1,...,x_m)P(y_1,...,y_n | x_1,...,x_m)]}$$
$$= \frac{P(x_1,...,x_m) \prod_j [1 - Pr(y_j=1 | x_1,...,x_m)]^{1-y_j} Pr(y_j=1 | x_1,...,x_m)^{y_j}}{\sum_{(x_1,...,x_m)} P(x_1,...,x_m) \prod_j [1 - Pr(y_j=1 | x_1,...,x_m)]^{1-y_j} Pr(y_j=1 | x_1,...,x_m)^{y_j}}$$

$$(1)$$

where $P(x_1, ..., x_m)$ is the prior probability of the protein configuration. Assuming the presence of each protein i is independent of other proteins, this prior probability can be computed as

$$P(x_1, ..., x_m) = \prod_i P(x_i) \qquad (2)$$

$Pr(y_j = 1|x_1, ..., x_m)$ is the probability of peptide j to be identified by shotgun proteomics given the protein configuration $(x_1, ..., x_m)$. Assuming that different proteins are present in the sample independently of one another and ignoring the competition of peptides for ionization and MS/MS fragmentation, we can compute it as

$$Pr(y_j = 1|x_1, ..., x_m) = 1 - \prod_i [1 - x_i Pr(y_j = 1|x_i = 1, x_1 = ... = x_{i-1} = x_{i+1} = ... = x_m = 0)]$$

(3)

where $Pr(y_j = 1|x_i = 1, x_1 = ... = x_{i-1} = x_{i+1} = ... = x_m = 0)$ is the probability of peptide j to be identified if only protein i is present in the sample. As we previously showed, for a particular proteomics platform (e.g. LC-MS/MS considered here), this probability, referred to as the *standard peptide detectability* d_{ij}, is an intrinsic property of the peptide (within its parent protein), and can be predicted from the peptide and protein sequence prior to a proteomics experiment [14]. Combining equations above, we can compute the posterior probabilities for protein configurations as

$$P(\pi_1, ..., \pi_m|y_1, ..., y_n) = \frac{\prod_i P(x_i) \prod_j \{[\prod_i (1 - \tau_i d_{ij})]^{1-y_j} [1 - \prod_i (1 - x_i d_{ij})]^{y_j}\}}{\sum_{(x'_1, ..., x'_m)} \prod_i P(x'_i) \prod_j \{[\prod_i (1 - x'_i d_{ij})]^{1-y_j} [1 - \prod_i (1 - x'_i d_{ij})]^{y_j}\}}$$

(4)

Hence, protein inference is equivalent to finding the MAP protein configuration by maximizing the above function

$$(x_1, ..., x_m)_{MAP} = argmax_{(x_1, ..., x_m)} P(x_1, ..., x_m|y_1, ..., y_n)$$

(5)

Sometimes, we are also interested in the marginal posterior probability of a specific protein i to be present in the sample, which can be expressed as,

$$P^o(x_i) = P(x_i|y_1, ..., y_n) = \sum_{x_1, ..., x_{i-1}, x_{i+1}, ..., x_m} P(x_1, ..., x_m|y_1, ..., y_n)$$

(6)

2.2 Advanced Bayesian Model

The basic model described above assumes all identified peptides have equal probability ($= 1$) of being correctly identified. Here we relax this assumption by introducing a peptide identification score s_j for each peptide j, which is output by peptide search engines. We assume the peptide identification score is highly correlated with the probability of a peptide being correctly identified and their relationship (denoted by $r_j = Pr(y_j = 1|s_j)$) can be approximately modeled using probabilistic methods adopted by some search engines such as Mascot [8] or post-processing tools such as PeptideProphet [18]. Our goal is to compute $P(x_1, ..., x_m|s_1, ..., s_n)$ by enumerating all potential peptide configurations

$$P(x_1, ..., x_m | s_1, ..., s_n) = \sum_{(y_1, ..., y_n)} [P(x_1, ..., x_m | y_1, ..., y_n) P(y_1, ..., y_n | s_1, ..., s_n)]$$
$$= \sum_{(y_1, ..., y_n)} [\frac{P(x_1, ..., x_m)}{P(s_1, ..., s_n)} P(y_1, ..., y_n | x_1, ..., x_m) P(s_1, ..., s_n | y_1, ..., y_n)]$$

$$(7)$$

Assuming that s_j is independent of the presences of the other peptides (i.e. $(y_1, ..., y_{j-1}, y_{j+1}, ..., y_n)$) and applying Bayes' rule, we have

$$P(s_1, ..., s_n | y_1, ..., y_n) = \prod_j \frac{P(y_j | s_j) P(s_j)}{P(y_j)} = \prod_j \frac{(1 - r_j)^{(1 - y_j)} r_j^{y_j} P(s_j)}{P(y_j)} \quad (8)$$

Combining these equations, we can compute the posterior probability of protein configurations as

$$P(x_1, ..., x_m | s_1, ..., s_n) =$$

$$\frac{\sum_{(y_1, ..., y_n)} \{\prod_i P(x_i) \prod_j \{[\prod_i (1 - x_i d_{ij})]^{1 - y_j} [1 - \prod_i (1 - x_i d_{ij})]^{y_j} \frac{(1 - r_j)^{(1 - y_j)} r_j^{y_j}}{P(y_j)}\}\}}{\sum_{(x_1', ..., x_m')(y_1, ..., y_n)} \{\prod_i P(x_i') \prod_j \{[\prod_i (1 - x_i' d_{ij})]^{1 - y_j} [1 - \prod_i (1 - x_i' d_{ij})]^{y_j} \frac{(1 - r_j)^{(1 - y_j)} r_j^{y_j}}{P(y_j)}\}\}}$$

$$(9)$$

For most of the work presented here, we do not assume any prior knowledge about the protein presence in the sample. Therefore, in equations 4 and 9, $P(x_i)$ is regarded as constant (i.e. 0.5) for all proteins. In practice, prior knowledge, such as the species which the sample is from, the number of candidate proteins, and known protein relative quantities or protein families that are likely present in the sample, can be directly integrated into our Bayesian models (see results section for a simple demonstration).

Similar to the basic model, we can also compute the posterior probability of a specific protein i present in the sample as

$$P(x_i | s_1, ..., s_n) = \sum_{x_1, ..., x_{i-1}, x_{i+1}, ..., x_m} P(x_1, ..., x_m | s_1, ..., s_n) \quad (10)$$

and the marginal probability of a peptide j in the sample (see Appendix).

2.3 Adjustment of Peptide Detectabilities

An adjustment of the predicted peptide detectabilities is necessary when applying them here, since the predicted standard peptide detectabilities (denoted as d_{ij}^0) reflect the detectability of a peptide under a standard proteomics experimental setting, in particular, under fixed and equal abundances (i.e. q^0) for all proteins [14]. Assuming that the abundance of protein i in the sample mixture is q_i instead of q^0, the *effective* detectability of peptide j from this protein should be adjusted to

$$d_{ij} = 1 - (1 - d_{ij}^0)^{q_i/q^0} \quad (11)$$

Although we do not know q_i explicitly, since the total probability of observing a peptide j is given by r_j (or y_j for basic model), we can estimate q_i by solving the equation $\sum_j d_{ij} = \sum_j Z_{ij} r_j$ for a specific protein i. We note that this adjustment method may immediately lead to a new approach to absolute protein quantification [15]. However, we will address the evaluation of its performance in our future work. Here, our goal is to utilize it to adjust the predicted standard peptide detectabilities based on the estimated protein abundances.

2.4 Gibbs Sampling

Given a protein configuration graph, the peptide detectabilities (d_{ij}) and the probabilities of peptide presence in the sample (r_j), the posterior distribution of protein configurations can be computed directly from equations 4 or 9, depending on which Bayesian model is used. This brute force method, which has computational complexity of $O(2^m)$, is very expensive and only works for small connected components in the protein configuration graph.

Gibbs sampling is a commonly used strategy to rapidly approximate a high dimensional joint distribution that is not explicitly known [21,22]. We adopted this algorithm to achieve the optimal protein configuration with the MAP probability. The original Gibbs sampling algorithm considers one individual variable at a time in the multi-dimentional distribution. It, however, often converges slowly and is easily trapped by local maxima for long time. Several techniques have been proposed to improve the search efficiency of Gibbs sampling algorithm, such as *random sweeping*, *blocking* and *collapsing* [22]. Because in our case each variable x_i to be sampled has small search space (i.e. $\{0, 1\}$), we applied the blocking sampling technique in our Gibbs sampler algorithm.

Without increasing the computational complexity, we adopt a novel *memorizing* strategy that keeps a record of all (as well as the maximum) posterior probabilities (and the corresponding protein configurations) among all configurations we evaluated during the sampling procedure, and report the maximum solution in the end. The memorized posterior probabilities are also used to calculate the marginal posterior protein probabilities in equation 6 and 10. Due to the page limits, the sketch of the block Gibbs sampling algorithms and the memorizing approach for the basic and advanced Bayesian models are described in the Appendix.

2.5 Datasets

We used two datasets from different sources that are both generated using mixtures of model proteins. Therefore, we know the proteins in these samples. The first dataset is used only for the training of the detectability predictor, while the other dataset was used for testing the protein inference methods. The first dataset from a mixture (Sample A) of 13 standard proteins was prepared at 1 μM final digestion concentration for each protein except human hemoglobin which is at 2 μM, combined with buffer, reduced, alkylated, and digested overnight with trypsin [14]. Peptides were separated by nano-flow reversed-phase liquid chromatography gradient and analyzed by mass spectrometry and tandem mass spectrometry in a Thermo Electron (San Jose, CA) LTQ linear ion trap mass spectrometer. The second mixture (Sample Sigma49) was cleaned up by gel electrophoresis, reduced, alkylated, and digested in-gel with trypsin. Tandem mass spectra for doubly-charged precursor ions were obtained from the website at Vanderbilt University website [13] and searched against human sequences in Swiss-Prot using Sequest [7].

3   Results

We implemented two Bayesian approaches described in the Methods section and tested them on the Sigma49 sample. The peptide detectability predictors were trained using Sample A following the method described previously [14]. Similarly as in [13], prior to the protein inference, 13388 MS/MS spectra acquired from Sigma49 sample in one LC/MS experiment were searched against the human proteome in Swiss-Prot database (version 54.2). PeptideProphet [18] was then used to assign a probability score for each identified peptide. For the basic Bayesian model, 152 unique peptides with minimum PeptideProphet probability score 0.95 were retained as identified peptides, while for the advanced model, we retained 443 peptides with minimum probability score 0.05. We tried two methods to set the probability r_j for each peptide identification. In the first method, we directly use the probability for each identified peptide reported by PeptideProphet. Since PeptideProphet does not consider peptide detectability, we implemented the second method which converts the PeptideProphet probability into a likelihood ratio LR_j and then apply our models.

The conversion is done by $LR_j = Pr_{PP}(y_j = 1)/[c \times (1 - Pr_{PP}(y_j = 1))]$, where $Pr_{PP}(y_j = 1)$ is the PeptideProphet probability, and c is the ratio between the prior probabilities of the peptide's presence and absence. For both models, we used block size 3 in the Gibbs sampler.

Table 2 compares the results for the Sigma49 sample from the basic and advanced Bayesian models with that from ProteinProphet [11] and the minimum missed peptide (MMP) approach we proposed previously [12] on the Sigma49 sample. Sigma49 sample was prepared by mixing 49 human proteins, among which 44 proteins contain at least one peptide that can be identified by shotgun proteomics. In addition, 9 keratin proteins and 4 other proteins are categorized as the "keratin contamination" and "bonus" proteins, respectively, and are believed to be present in the sample due to contamination.

From the results, we observed that using detectabilities to adjust Peptide-Prophet probability improves the performance of the probabilitic models. For example, the advanced model (ABLA) achieved 0.83 and 1.0 for the precision and recall, respectively, whereas directly using ProteinProphet probability (ABPA) achieved 0.66 and 0.98. This indicates that peptide detectability is a useful concept in protein inference. The adjustment of detectability also improves accuracy (see F measures) of the protein inference (e.g. BBA vs. BB or ABLA vs. ABL), implying that the predicted peptide detectabilities need to be adjusted by peptide quantities in real proteomics experiments. We also tried to incorporate a simple method for estimating protein prior probabilisties. In ABLA, we set the

Table 2. Protein inference results on the Sigma49 dataset using minimum missed peptide approach (MMP), ProteinProphet (PP), basic Bayesian model (BB), basic Bayesian model with detectability adjustment (BBA), advanced Bayesian model using raw PeptideProphet probabilities (ABP), ABP after detectability adjustment (ABPA), advanced Bayesian model using converted probability scores (ABL), ABL after detectability adjustment (ABLA), and ABLA with estimated protein prior probabilities (ABLAP). All results are evaluated based on the true positive (TP), false positive (FP) and false negative (FN) numbers of proteins, and the precision (PR), recall (RC) and F-measure (F) in two categories of true proteins in the sample: model proteins, and model proteins plus all contaminations. MAP solutions were used as positive proteins for our probabilistic models; and 0.5 cutoff was used for ProteinProphet.

	MMP	PP	BB	BBA	ABP	ABPA	ABL	ABLA	ABLAP
TP	39/45	41.5/47.5	39/47	37/43	35/39	43/49	37/41	44/50	43/49
FP	6/0	7.5/1.5	16/8	6/0	4/0	22/16	4/0	9/3	6/0
FN	5/12	2.5/9.5	5/10	7/14	9/18	1/8	7/16	0/7	1/8
PR	0.87/1	0.85/0.97	0.71/0.85	0.86/1	0.9/1	0.66/0.75	0.9/1	0.83/0.94	0.88/1
RC	0.89/0.79	0.94/0.83	0.89/0.82	0.84/0.75	0.8/0.68	0.98/0.86	0.84/0.72	1/0.88	0.98/0.86
F	0.88/0.88	0.89/0.90	0.79/0.84	0.85/0.86	0.84/0.81	0.79/0.81	0.87/0.84	0.91/0.91	0.92/0.92

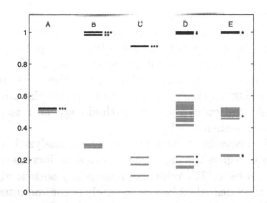

Fig. 2. Protein inference using ABLA method from 5 (A-E) connected components in the protein configuration graph built from Sigma49 dataset. Proteins in the same components are ordered in the same column based on their marginal posterior probabilities. Proteins in the MAP configuration are highlighted in black. The true proteins in Sigma49 sample are labelled by "***" for model proteins, "**" for non-karetin contaminations and "*" for karetin contaminations.

protein prior probabilities as 0.5 for all proteins, whereas in ABLAP, we set it as 0.2, i.e. the ratio of the expected number of proteins in the sample (i.e. 48) and the total number of candidate proteins. Comparing their results, we observed that incorporating protein prior probabilities further improves the performance of protein inference, in which the F measure improves from 0.91 to 0.92. Overall, ABLAP and ABLA models outperform the other methods. However, it is hard to draw a firm conclusion from the experiments on a relatively simple protein mixture. Further comparative analysis of these models using more complex (e.g. with hundreds of proteins), but well elucidated samples like Sigma49, is needed.

Fig. 2 illustrates the results of ABLA on 5 connected components in the protein configuration graph built from the Sigma49 dataset. The model proteins and likely contaminant proteins in the sample received higher marginal posterior probabilities than the other proteins, and the MAP configuration contains mostly true proteins. PeptideProphet cannot resolve the correct protein assignment in component A and C. We note that component A consists of three proteins (P51965, Q96LR5 and Q969T4) which share only one identified peptide (often referred to as "single-hit wonders"). ABLA algorithm correctly assigns the true model protein (P51965) as the MAP configuration over the other two proteins.

4   Discussion

In this study we proposed and evaluated a new methodology for protein inference in shotgun proteomics. The two Bayesian models proposed herein attempt to find the set of proteins that is most likely to be present in the sample. The new approach has three advantages over the existing methods: (1) it calculates or, if global optimum is not reached, approximates a MAP solution for the set of proteins present in the sample and can also output the probability of each protein to be present in the sample; (2) it can output the marginal probabilities of the identified peptides to be present in the sample, given the entire experiment; (3) the Gibbs sampling approach used to approximate the posterior probabilities of protein configuration is a proven methodology, and its performance and convergence has been well-studied.

It is common in proteomics for a sample to be analyzed multiple times in order to increase coverage of the proteome as well as to increase confidence in low sequence coverage proteins [23]. While not specifically addressed, the application of the Bayesian models described here adequately accommodates such data since peptide detectability, used to calculate prior probabilities, should assign lower values to those peptides not identified in all the replicate analyses. In addition, higher mammals often contain multiple very similar homologous proteins due to recent gene duplications. These proteins are almost impossible to differentiate using shotgun proteomics, if some but not all of these proteins are present in the sample. As a result, although the MAP protein configuration will contain at least one of these proteins, they each can receive a low marginal probability (e.g. < 0.5). While we have not explicitly addressed this problem here, we note that the proposed models can be easily modified to consider a given set of proteins as a group and then compute the probability of their presence as a whole. We will test this functionality in future implementation of the models.

While we show that the new methodology is accurate and useful, we note that the current detectability predictor, which was trained on a small number of doubly-charged fully tryptic peptide ions, poses a limitation of this approach. Therefore, it does not fully accommodate the results of all shotgun proteomics experiments, which are known to produce singly- and triply-charged ions as well as peptide ions with missed cleavages that are readily identified. Furthermore, we are currently using an inaccurate method to adjust detectability with different peptide quantities. Future improvements in detectability prediction, peptide confidence estimation, as well as the detectability adjustment may lead to further improvement of the Bayesian models described here.

Acknowledgements

We acknowledge the support of the NCI grant U24 CA126480-01 to F. Regnier, RJA, PR, HT et al. HT and RJA acknowledge the support from NIH/NCRR grant 5P41RR018942. HT is partially supported by NSF award DBI-0642897. PR is partially supported by NSF award DBI-0644017.

References

1. Aebersold, R., Mann, M.: Mass spectrometry-based proteomics. Nature 422, 198–207 (2003)
2. McDonald, W.H., Yates, J.R.: Shotgun proteomics: integrating technologies to answer biological questions. Curr. Opin. Mol. Ther. 5(3), 302–309 (2003)
3. Kislinger, T., Emili, A.: Multidimensional protein identification technology: current status and future prospects. Expert Rev. Proteomics 2(1), 27–39 (2005)
4. Swanson, S.K., Washburn, M.P.: The continuing evolution of shotgun proteomics. Drug Discov. Today 10(10), 719–725 (2005)
5. Marcotte, E.M.: How do shotgun proteomics algorithms identify proteins?. Nat. Biotechnol. 25(7), 755–757 (2007)
6. Nesvizhskii, A.I.: Protein identification by tandem mass spectrometry and sequence database searching. Methods Mol Biol 367, 87–119 (2007)
7. Yates, J.R., Eng, J.K., McCormack, A.L., Schieltz, D.: Method to correlate tandem mass spectra of modified peptides to amino acid sequences in the protein database. Anal Chem 67, 1426–1436 (1995)
8. Perkins, D.N., Pappin, D.J., Creasy, D.M., Cottrell, J.S.: Probability-based protein identification by searching sequence databases using mass spectrometry data. Electrophoresis 20(18), 3551–3567 (1999)
9. Craig, R., Beavis, R.C.: TANDEM: matching proteins with tandem mass spectra. Bioinformatics 20(9), 1466–1467 (2004)
10. Nesvizhskii, A.I., Aebersold, R.: Interpretation of shotgun proteomic data: the protein inference problem. Mol Cell Proteomics 4(10), 1419–1440 (2005)
11. Nesvizhskii, A.I., Keller, A., Kolker, E., Aebersold, R.: A statistical model for identifying proteins by tandem mass spectrometry. Anal Chem 75(17), 4646–4658 (2003)
12. Alves, P., Arnold, R.J., Novotny, M.V., Radivojac, P., Reilly, J.P., Tang, H.: Advancement in protein inference from shotgun proteomics using peptide detectability. In: PSB 2007: Pacific Symposium on Biocomputing, pp. 409–420. World Scientific, Singapore (2007)
13. Zhang, B., Chambers, M.C., Tabb, D.L.: Proteomic Parsimony through Bipartite Graph Analysis Improves Accuracy and Transparency. J Proteome Res. 6(9), 3549–3557 (2007)

14. Tang, H., Arnold, R.J., Alves, P., Xun, Z., Clemmer, D.E., Novotny, M.V., Reilly, J.P., Radivojac, P.: A computational approach toward label-free protein quantification using predicted peptide detectability. Bioinformatics 22(14), 481–488 (2006)
15. Lu, P., Vogel, C., Wang, R., Yao, X., Marcotte, E.M.: Absolute protein expression profiling estimates the relative contributions of transcriptional and translational regulation. Nat. Biotechnol. 25(1), 117–124 (2007)
16. Elias, J.E., Haas, W., Faherty, B.K., Gygi, S.P.: Comparative evaluation of mass spectrometry platforms used in large-scale proteomics investigations. Nat. Methods 2(9), 667–675 (2005)(Comparative Study)
17. Elias, J.E., Gygi, S.P.: Target-decoy search strategy for increased confidence in large-scale protein identifications by mass spectrometry. Nat. Methods 4(3), 207–214 (2007) (Evaluation Studies)
18. Keller, A., Nesvizhskii, A.I., Kolker, E., Aebersold, R.: Empirical statistical model to estimate the accuracy of peptide identifications made by MS/MS and database search. Anal Chem 74(20), 5383–5392 (2002)
19. Wu, F.-X., Gagne, P., Droit, A., Poirier, G.G.: RT-PSM, a real-time program for peptide-spectrum matching with statistical significance. Rapid Commun Mass Spectrom 20(8), 1199–1208 (2006)
20. Bern, M., Goldberg, D.: Improved ranking functions for protein and modification-site identifications. In: Speed, T., Huang, H. (eds.) RECOMB 2007. LNCS (LNBI), vol. 4453, pp. 444–458. Springer, Heidelberg (2007)
21. Geman, S., Geman, D.: Stochastic relaxation, Gibbs distributions, and the Bayesian restoration of images. IEEE Trans. on Pattern Analysis and Machine Intelligence 6, 721–741 (1984)
22. Liu, J.S.: Monte Carlo strategies in scientific computing. Springer, Heidelberg (2002)
23. Brunner, E., Ahrens, C.H., Mohanty, S., Baetschmann, H., Loevenich, S., Potthast, F., Deutsch, E.W., Panse, C., de Lichtenberg, U., Rinner, O., Lee, H., Pedrioli, P.G.A., Malmstrom, J., Koehler, K., Schrimpf, S., Krijgsveld, J., Kregenow, F., Heck, A.J.R., Hafen, E., Schlapbach, R., Aebersold, R.: A high-quality catalog of the Drosophila melanogaster proteome. Nat Biotechnol. 25(5), 576–583 (2007)

Appendix

Note: In the two algorithms presented below, we have

$$F(x_{v_1}, ..., x_{v_c}, y_{w_1}, ..., y_{w_d}) = \prod_{i \in v} P(x_i) \prod_{j \in N^+(v) \cup w} P(y_j|x_{N^-(j)}) \prod_{j \in w} \frac{P(y_j|s_{N^+(j)})}{P(y_j)} \tag{12}$$

where $N^+(.)$ and $N^-(.)$ refer to the nodes that the current node(s) are linked to and linked from, respectively; $v = (v_1, ..., v_c)$ and $w = (w_1, ..., w_d)$ are the block indices for x and y, repectively; w is empty for the basic model; $P(y_j|x_{N^-(j)})$ can be computed by 6; and $Pr(y_j = 1|s_{N^+(j)}) = r_j$.

$$\begin{aligned} P(y_j) &= \sum_{(x_1,...,x_m)}[P(y_j|x_1, ..., x_m)P(x_1, ..., x_m)] \\ &= [1 - \prod_i(1 - Pr(x_i = 1)d_{ij})]^{y_j}[\prod_i(1 - Pr(x_i = 1)d_{ij})]^{(1-y_j)} \end{aligned} \tag{13}$$

Algorithm 1. Gibbs sampler for protein inference using the basic model

Input : Peptide configuration $(y_1, ..., y_n)$ and peptide detectabilities $[d_{ij}]$
Output: MAP protein configuration $(x_1, ..., x_m)$

Initialize $(x_1, ..., x_m)$ randomly ;
$MaxPr \leftarrow 0$;
Normalizing factor $T \leftarrow 1$;
while *not converge* **do**
    $v = (v_1, ..., v_t) \leftarrow$ a random t-block from $(1, ..., m)$;
    $T \leftarrow \frac{Value(x_1,...,x_m)}{F(x_{v_1},...,x_{v_t})}$;
    for all $(v_1, ..., v_t)$ **do**
        Compute $F(x_{v_1}, ..., x_{v_t})$;
        $Value(x_1, ..., x_m) \leftarrow F(x_{v_1}, ..., x_{v_t}) \times T$;
        if $Value(x_1, ..., x_m) > MaxPr$ **then**
            $MaxPr \leftarrow Value(x_1, ..., x_m)$;
            $(x_1^{Max}, ..., x_m^{Max}) \leftarrow (x_1, ..., x_m)$;
            $(x_{v_1}^{Max}, ..., x_{v_t}^{Max}) \leftarrow (x_{v_1}, ..., x_{v_t})$;
        end
    end
    Sample $(x'_{v_1}, ..., x'_{v_t})$ from normalized $F(x_{v_1}, ..., x_{v_t})$;
    $(x_{v_1}, ..., x_{v_t}) \leftarrow (x'_{v_1}, ..., x'_{v_t})$;
end
Report $MaxPr$, $(x_1^{Max}, ..., x_m^{Max})$, and compute marginal probabilities ;

Algorithm 2. Gibbs sampler for protein inferencing using the advanced model

Input   : Peptide prior probabilities $(r_1, ..., r_n)$ and peptide detectabilities $\{d_{ij}\}$
Output: MAP protein configuration $(x_1, ..., x_m)$

Initialize $(x_1, ..., x_m)$ and $(y_1, ..., y_n)$ randomly ;
$MaxPr \leftarrow 0$;
Normalizing factor $T \leftarrow 1$;
while *not converge* **do**
    $c \leftarrow$ a random number between 0 and t ;
    $(v_1, ..., v_c) \leftarrow$ a random c-block from $(1, ..., m)$;
    $d \leftarrow t - c$;
    $(w_1, ..., w_d) \leftarrow$ a random d-block from $(1, ..., n)$;
    Compute normalizing factor $T \leftarrow \frac{Value(x_1,...,x_m;y_1,...,y_n)}{F(x_{v_1},...,x_{v_c},y_{w_1},...,y_{w_d})}$;

    for *all* $(x_{v_1}, ..., x_{v_c})$ *and* $(y_{w_1}, ..., y_{w_d})$ **do**
        Compute $F(x_{v_1}, ..., x_{v_c}; y_{w_1}, ..., y_{w_d})$;
        memorizing: $Value(x_1, ..., x_m, y_1, ..., y_n) \leftarrow F \times T$;
        if $Value(x_1, ..., x_m, y_1, ..., y_n) > MaxPr$ **then**
            $MaxPr \leftarrow Value(x_1, ..., x_m, y_1, ..., y_n)$;
            $(x_1^{Max}, ..., x_m^{Max}) \leftarrow (x_1, ..., x_m)$;
            $(x_{v_1}^{Max}, ..., x_{v_c}^{Max}) \leftarrow (x_{v_1}, ..., x_{v_c})$;
            $(y_1^{Max}, ..., y_n^{Max}) \leftarrow (y_1, ..., y_n)$;
            $(y_{w_1}^{Max}, ..., y_{w_d}^{Max}) \leftarrow (y_{w_1}, ..., y_{w_d})$;
        end
    end
    Sample $(x'_{v_1}, ..., x'_{v_c}; y'_{w_1}, ..., y'_{w_d})$ from normalized $F(x_{v_1}, ..., x_{v_c}; y_{w_1}, ..., y_{w_d})$;
    $(x_{v_1}, ..., x_{v_c}) \leftarrow (x'_{v_1}, ..., x'_{v_c})$;
    $(y_{w_1}, ..., y_{w_d}) \leftarrow (y'_{w_1}, ..., y'_{w_d})$;
end
Report $MaxPr$, $(x_1^{Max}, ..., x_m^{Max})$, and compute marginal probabilities ;

De Novo Sequencing of Nonribosomal Peptides

Nuno Bandeira[1], Julio Ng[1], Dario Meluzzi[1], Roger G. Linington[2],
Pieter Dorrestein[1], and Pavel A. Pevzner[1]

[1] University of California, San Diego, USA
[2] University of California, Santa Cruz, USA
ppevzner@cs.ucsd.edu

Abstract. While nonribosomal peptides (NRPs) are of tremendous
pharmacological importance, there is currently no technology capable
of high-throughput sequencing of NRPs. Difficulties in sequencing NRPs
slow down the progress in elucidating the non-ribosomal genetic code
and negatively affect various screening programs aimed at the discovery of natural compounds of medical importance. We propose to employ
multi-stage mass-spectrometry (MS^n) for the data acquisition, followed
by alignment-based heuristic algorithms for data analysis. Since mass
spectrometry based analysis of NRPs is fast and inexpensive, this approach opens the possibility of high-throughput sequencing of many unknown NRPs accumulated in large screening programs.

Keywords: Cyclic Peptides Sequencing De novo Algorithm.

1 Introduction

The classical protein synthesis pathway (translation of template mRNA into
proteins/peptides) is not the only mechanism for cells to assemble amino acids
into peptides. The alternative *Non Ribosomal Peptide Synthesis* is performed
by a large multi-enzyme complex (called Non Ribosomal Peptide Synthetase or
NRPS) that represents both the biosynthetic machinery and the mRNA-free
template for the biosynthesis of secondary metabolites (see [1,2,3] for recent reviews). NRPS gene clusters produce relatively short (up to 50 aa) nonribosomal
peptides (NRP) that are not directly inscribed in the genomic DNA and thus
cannot be inferred with traditional DNA-based sequencing techniques. NRPs
are of tremendous pharmacological importance since they were optimized during millions of years of evolution to play important roles in chemical defense
and communication for producing organisms. Starting from penicillin, NRPs
and other natural products have an unparallel track record in pharmacology:
9 out of the top 20 best-selling drugs were either inspired by or derived from
natural products. NRPs have some naturally evolved features that are applicable to the modulation of protein function in human systems, making them
excellent lead compounds for the development of novel pharmaceutical agents.
In particular, NRPs include antibiotics (penicillin, cephalosporin, vancomycin,

M. Vingron and L. Wong (Eds.): RECOMB 2008, LNBI 4955, pp. 181–195, 2008.
© Springer-Verlag Berlin Heidelberg 2008

etc.), immunosuppressors (cyclosporin), antiviral agents (luzopeptin A), antitumor agents (bleomycin), toxins (thaxtomin), and many peptides with yet unknown functions.

When DNA sequencing is not available, biologists use either Edman degradation or tandem mass spectrometry (MS^2) to sequence ribosomal peptides. However, neither of these approaches works for nonribosomal peptides since they differ from ribosomal peptides in many respects: (i) they often represent non-linear structures of amino acids, e.g., cyclic, tree-like, and branch-cyclic peptides, (ii) they often contain non-standard amino acids increasing the number of possible building blocks from 20 to several hundreds, (iii) they often have a non-standard backbone, and (iv) they are often modified. Each of these complications renders traditional Edman degradation and MS^2 peptide sequencing approaches useless, leaving NMR as the only technology capable of analyzing NRPs [4,5,6,7]. The use of NMR for NRP sequencing is time-consuming, difficult to automate (there are currently no software tools for automatic interpretation of NRPs from NMR data), and error-prone (see [7,8] for examples of errors in NMR sequencing). As a result, the extremely difficult total chemical synthesis remains the only reliable way to sequence and validate NRPs [9]. For example, Patrick Harran won the 2007 Hackerman Award in Chemical Research for his pioneering work on diazonamide A, a rare marine NRP. In the process of synthesizing diazonamide A, he discovered that the initial structure reported for the molecule was flawed [10].

An efficient and automated way to sequence NRPs will immediately benefit all searches for natural compounds of medical importance as well as studies of the still poorly understood mechanisms of the nonribosomal peptide synthesis. Currently, the prediction of the chemical structure of the unknown NRPs is not possible even if all genes involved in non-ribosomal synthesis are identified [11]. Furthermore, efficient NRP sequencing will aid biosynthetic engineering efforts to reprogram the NRP assembly lines in *E. coli* (most microbes producing NRPs are not amenable to cultivation). For example, the recent success in production of antitumor NRPs in *E. coli* required analysis of genetically engineered NRPs [12].

This paper introduces a combination of experimental and computational protocols that enable a mass-spectrometry based approach to sequencing NRPs. To the best of our knowledge, it is the first attempt to de novo sequence NRPs using mass spectrometry. Previous studies were limited to detection or resequencing of NRP, i.e., sequencing of new NRP variants when the major NRP variant was known. In an early attempt to use mass spectrometry for NRP analysis, Barber et al., 1992 [13] analyzed variants of tyrothricin, an antimicrobial agent produced by *Bacillus brevis*. Tyrothricin is a mixture of different NRPs and three major components of this mixture were previously identified. These three components were used as reference points to derive six other variant NRPs using mass spectrometry. In a recent study, Hitzeroth et al., 2005 [14] resequenced new variations of streptocidins on a MALDI-TOF-MS using information about previously sequenced streptocidins [15,16]. However, as the authors of [14] remarked this resequencing strategy is limited to peptides with pure amino acid sequence but grows more difficult when modifications are present. In another application of mass spectrometry to NRP analysis, Redman et al., 2003 [17] developed an algorithm for identification of cyclic peptides from combinatorial libraries. This approach amounts to accurate scoring of all candidates from the predefined library and cannot be applied to de novo sequencing since the set of all possible peptides grows exponentially with the length of peptide.

In the following sections, we employ *multi-stage* mass-spectrometry (MS^n) for de novo sequencing of cyclic peptides. We first describe the NRP-Sequencing algorithm for reconstructing cyclic peptides from a single MS^3 spectrum. We then extend the experimental protocol by incorporating MS^4 and even MS^5 spectra to score putative MS^3 reconstructions against all MS^4/MS^5 spectra. Finally, we describe the NRP-Assembly approach that assembles MS^4/MS^5 spectra and further integrates the resulting contig with all non-assembled spectra. The choice of a particular approach for analyzing NRPs depends on the specifics of the peptide, its fragmentation properties, accuracy of the mass spectrometer, etc. In the remainder of the paper, we do not distinguish between MS^4 and MS^5 spectra and refer to them as MS^n spectra. We note that although multi-stage mass spectrometry recently emerged as a valuable tool for peptide identification [18,19], this is the first report on using multi-stage mass-spectrometry for de novo peptide sequencing and sampling as many as 5 stages of mass-spectrometry (previous studies were mainly limited to 2 stages).

2   Sequencing Cyclic Peptides

We start by analyzing the simplest version of the NRP sequencing problem when the NRP is a cyclic peptide. Below we use NRP Seglitide, a somatostatin receptor antagonist, as an illustration. Seglitide is more potent than somatostatin for inhibition of insulin, glucagon and growth hormone release, and it is used experimentally in the treatment of Alzheimer's disease. The structure of Seglitide is *Cyclic*(N-methyl-Ala-Tyr-D-Trp-Lys-Val-Phe).[1]

For a cyclic peptide $P = p_1 \ldots p_n$ it results in n possible linear peptides $P_i = p_i \ldots p_n p_1 \ldots p_{i-1}$ with the same parent mass (Fig.1). The mixture of these peptides is further subjected to the next stage of mass spectrometry (MS^3) resulting in the difficult problem of interpreting an MS^3 spectrum of n different (but related) peptides. The theoretical MS^3 spectrum $Spectrum(P)$ of the cyclic peptide P is thus the superposition of the theoretical spectra $Spectrum(P_i)$ of linear peptides P_i as shown in Figure 1. Therefore, reconstructing the circular peptide P from its theoretical spectrum $Spectrum(P)$ amounts to the circular version of the classical *Partial Digest Problem (PDP)* [20].

While the complexity status of linear PDP remains unknown (a pseudo-polynomial algorithm for PDP is described in [21]), a simple branch-and-bound algorithm works well in practice [20]. However, it appears that the circular version of PDP may be harder than its linear version, in particular, the Rosenblatt-Seymour pseudo-polynomial algorithm for linear PDP [21] does not

---

[1] We remark that tandem mass-spectrometry (MS^2) amounts to simply breaking (linearizing) the cyclic peptide and does not generate any useful information.

generalize for circular PDP. While reconstructing the cyclic peptide P from its theoretical $Spectrum(P)$ is already a hard problem, reconstructing P from its experimental MS^3 spectrum S is much more difficult. In practice, the contributions of different linear versions of P to the experimental spectrum are highly non-uniform. For example, if a certain bond (e.g., before p_i) has a low propensity for breakage in the mass spectrometer, the spectrum P_i may not contribute any peaks to the MS^3 spectrum S. Such missing peaks combined with many noise peaks make the reconstruction very hard (the PDP problem is known to be NP-hard for noisy inputs [22]) and lead to the following *Cyclic Peptide Sequencing Problem* that is similar to the NP-hard problem of peptide sequencing in the presence of internal ions [23].

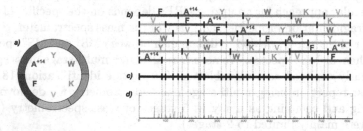

Fig. 1. Analysis of the cyclic peptide Seglitide. **a)** The circular structure of Seglitide is schematically illustrated with each residue represented by a different color (slice sizes not scaled to corresponding masses of the residues). A^{+14} denotes a non-standard residue with integer mass 71+14=85 Da. **b)** MS^2 fragmentation of Seglitide generates up to 6 linear peptides representing different rotated variants of the same cyclic peptide. **c)** Theoretical spectrum for Seglitide by superposition of the fragment masses of the linearized peptides. For simplicity, only prefix masses (*b*-ions) are shown here. **d)** Experimental spectrum of Seglitide resulting from a mixture of 6 linear peptides (the peaks corresponding to prefix ions are shown in red).

Cyclic Peptide Sequencing Problem (CPSP). Given an experimental MS^3 spectrum S, find a cyclic peptide P maximizing the number of shared masses between S and the theoretical spectrum of P.

Since the branch-and-bound approach to solving CPSP is prohibitively time-consuming, we describe some alignment-based heuristics that take advantage of the specifics of the particular CPSP instances arising in NRP studies.

Sequencing cyclic peptides using MS^3 spectra. Pevzner et al., 2000 [24] introduced *spectral convolution* and *spectral alignment* for revealing similarities between related but different spectra. We argue that since an experimental MS^3 spectrum of a cyclic peptide is a superposition of multiple spectra of linearized peptides, spectral *auto-convolution* and *auto-alignment* should reveal key features (e.g. amino acid composition and true peaks) for the identification of the peptide.

The spectral convolution between spectra S and S' is defined as the number of masses s in S such that $s - x$ is also a mass in S' (for every parameter x).[2] Also, the *cyclic convolution* $Conv(S, S', x)$ of spectra S and S' is defined as the number of masses s in S such that either $(s - x)$ or $(s - x) + ParentMass(S)$ is also a mass in S'. The auto-convolution $Conv(S, x)$ of a spectrum S is simply the cyclic convolution of S with itself. Figure 2c presents the auto-convolution of the MS^3 spectrum for Seglitide, a 6 amino acid long cyclic peptide A^{+14}YWKVF (integer residue masses are 85, 163, 186, 128, 99 and 147, respectively). As expected, peaks of the auto-convolution reveal neutral losses (e.g., the peak at 18 corresponds to H_2O losses). However, in the case of cyclic peptides, there are many other high-scoring peaks. For example, the largest peak $Conv(S, 85) = 14$ corresponds to the mass of amino acid A^{+14} (auto-alignment of the spectrum S with offset A^{+14} reveals many aligned peaks). Other amino acids in Seglitide also correspond to high peaks: $Conv(S, 163) = 10$, $Conv(S, 186) = 8$, $Conv(S, 128) = 8$, $Conv(S, 99) = 8$, and $Conv(S, 147) = 8$. We remark that in the interval between 50 and 200Da there are only 4 other peaks with $Conv(S, x) \geq 8$ (at offsets 78, 81, 103 and 191) indicating that spectral convolution can be used to derive the set of amino acid masses present in the circular peptide.

The auto-alignment of the spectrum S with offset x is defined as the set of peaks $\{s \cdot (s - x) \subset S\}$. We view auto-alignment (denoted S_x) as a virtual spectrum with parent mass equal to $ParentMass(S) - x$. The auto-alignment of Seglitide's MS^3 spectrum of with offset 85Da (maximum peak revealed by spectral convolution) corresponds to the alignment between A^{+14}YWKVF and YWKVFA^{+14}. Similarly to the spectral alignment of spectra from different peptides [25], one would expect auto-alignment to mostly reflect *either* prefix *or* suffix ion fragments from the linearized peptides A^{+14}YWKVF and YWKVFA^{+14} (with the number of noisy peaks greatly reduced). The separation of prefix (e.g., b-ions) and suffix (e.g., y-ions) ladders by spectral alignment is important since it significantly simplifies spectral interpretation and enables accurate de novo peptide sequencing [26,27]. However, it turns out that interpretation of auto-alignments (of cyclic peptides) is more complex than interpretation of spectral alignments of different (linear) peptides. While auto-alignment reduces the noise, it does not separate prefix and suffix ladders, i.e., auto-alignment contains both prefix *and* suffix ladders. This is caused by the fact that the MS^3 spectrum contains peaks from *both* A^{+14}YWKVF and YWKVFA^{+14}. Thus, the b-ions from A^{+14}YWKVF match the b-ions of YWKVFA^{+14} and, moreover, the y-ions from YWKVFA^{+14} match the y-ions of A^{+14}YWKVF (with the same offset 85 for both b- and y-ions).

When the set of possible amino acid masses is known in advance (like in traditional peptide sequencing), one can interpret the auto-alignment S_x of the MS^3 spectrum S using either the *anti-symmetric path* approach [28] or the *spectral*

---

[2] While the standard spectral convolution simply counts the number of peaks separated by mass x, in the case of *scored spectra* (represented as vectors $S = (s_1 \ldots s_n)$ and $S' = (s'_1 \ldots s'_n)$) reflecting peak intensities or other characteristics) the spectral convolution is defined as $\sum_{i=1,n} s_i \cdot s'_{i-x}$.

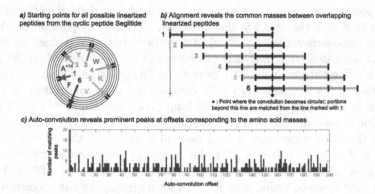

Fig. 2. Auto-convolution of MS³ spectra of Seglitide. **a)** A single fragmentation of Seglitide generates up to six linear peptides overlapping in a circular fashion. **b)** The auto-convolution between a base spectrum (peptide 1) and another spectrum S represents the matching peaks in the overlapping part as well as matching peaks in the non-overlapping part between suffix of S (after the line marked with ●) and the prefix of S (after the line marked with †). **c)** Since the MS³ spectrum of Seglitide is a superposition of multiple linearized peptides (see Figure 1), the auto-convolution of this spectrum has prominent peaks for all offsets corresponding to masses of amino acids (shown in red). Some additional offsets with comparable numbers of matched peaks are caused by the presence of neutral losses from peptide fragments (H_2O water at 18Da) or spurious matches. The peak at 0 is truncated.

partitioning approach [27]. Indeed, de novo peptide sequencing of S_x (with offset 85) promptly returns YWKVF as the top scoring de novo reconstruction (assuming that the overlapping portion is formed by standard amino acids as in the case of Seglitide). However, in practice the amino acid composition of cyclic peptides is hardly ever known since almost all NRPs contain non-standard amino acids and unknown modifications. While Marfey's analysis [29] for deriving amino acid composition can alleviate this problem, it only recovers the masses of the residues that react with Marfey's reagent. Spectral auto-convolution (Fig. 2c) represents a computational (rather than experimental) approach to deriving amino acid composition (residue masses) of cyclic peptides. In contrast to Marfey's approach, which typically misses some amino acids, it tends to over-predict the set of amino acids in the cyclic peptide. However, combined with de novo peptide sequencing, spectral auto-convolution leads to the successful reconstruction of cyclic peptides. For Seglitide, considering all x with $Conv(S,x) \geq 8$ in the interval from 50 to 200 Da³ results in the alphabet of only 10 amino acid masses {78, 81, 85, 99, 103, 128, 147, 163, 186, 191} including the masses of all amino acids in Seglitide (shown underlined). Considering all x with $Conv(S,x) \geq 7$ increases the size of the alphabet by another 3 masses but still brings the correct reconstruction to the top of the list of de novo reconstructions.

---

³ In most cases, the masses of amino acids found in NRPs fall into this interval.

The text box below outlines the NRP-Sequencing algorithm, which operates in different modes depending on the knowledge of the possible amino acids composition of the cyclic peptide. In particular NRP-Sequencing can accept as input the standard amino acid masses (Mode 1), derived amino acid masses from the auto-convolution functions (Mode 2), and the set of all integer amino acid masses from 50 to 200Da (Mode 3).

Input: MS^3 spectrum S of an (unknown) cyclic peptide, set of MS^n spectra, parameter k (maximum number of candidate amino acid masses) and p (minimum percentage of top de novo score to report a suboptimal peptide)
Output: Ranked list of candidate peptide reconstructions

$PeptideList = \emptyset$
Select top k peaks $x_1 \ldots x_k$ in the auto-convolution $Conv(S, x)$ in the $[50, 200]$ interval
For $i = 1$ to k
    Set $x = x_i$ and construct the auto-alignment S_x
    De novo sequence S_x using masses $x_1 \ldots x_k$ and find highest scoring peptide P
    For every suboptimal peptide P' such that $Score(P') \geq p \cdot Score(P)$
        Append x to P' and add the resulting circular peptide to $PeptideList$
Re-score each peptide P' in $PeptideList$ by matching P' against all MS^n spectra
Output peptides from $PeptideList$ in the decreasing order of their scores

---

Algorithm for de novo NRP sequencing.

Assembling MS^n spectra. In this section we describe the reconstruction of cyclic peptides by *assembling* MS^n spectra (rather than just matching MS^n spectra against putative interpretations of the MS^3 spectrum). MS^n spectra are less likely to contain a mixture of peptides, as opposed to MS^3. The only cases of mixture peptides in MS^n is when two fragments of the cyclic peptide have identical parent masses. In particular, mixed MS^n spectra are common if there are repeated amino acid masses in the cyclic peptide as illustrated in Figure A-1a. However, MS^n spectra are typically of lower quality than MS^3 spectra (lower proportion of b- and y-ions and increased noise) and the assembly result might not cover the entire cyclic peptide. Therefore, it still remains a challenge to fill in the gaps where there is low coverage by MS^n spectra.

In the absence of noise and in the presence of perfect b- and y- ladders, pairwise alignment separates b-ions from y-ions [30,26]. However, missing peaks, noise peaks and mixed MS^n spectra result in spurious pairwise alignment that may confuse the assembly process. Our analysis of Seglitide spectra revealed that, although we do obtain many correct pairwise alignments, it is not easy to separate them from spurious alignments. Given that we only generated a limited amount of mass spectrometry data, we cannot train a rigorous statistical model to separate true alignments from spurious ones.

NRP-Assembly attempts to combine pairwise alignments into multiple alignment by finding pairwise alignments with *compatible* offsets (see [30]). The triple alignments (3-contigs) can be constructed from pairwise alignments by adding a third spectrum that aligns with compatible offsets to both spectra in pairwise alignments (Figure A-1b). This process can be iteratively applied to form larger and larger contigs. In practice, we found that this method works better than any score cutoff approach to filter out spurious alignments. For example, a spurious pairwise alignment ends up in a spurious 3-contig only if there are two other spurious pairwise alignments with compatible offsets (an unlikely event). NRP-Assembly constructs contigs of size 5, in which all pairwise alignments have compatible shifts. A consensus spectrum is constructed from these contigs and candidate sequences are derived by de novo sequencing. These candidate sequences are re-scored using the remaining MS^n spectra that were not part of the initial contig.

Results. Figure 3 shows the results of the NRP-Sequencing and NRP-Assembly algorithm applied to the Seglitide data. The tests were performed in 3 modes ("known", "discovered", and "unknown") for selecting the alphabet of amino acid masses for the de novo sequencing stage of the algorithms. Not surprisingly, the best sequencing results are achieved in the computational experiment when the set of amino acid masses is known - constraining the set of amino acid masses greatly reduces the number of possible peptides. The number of generated peptides grows significantly in the "discovered" and "unknown" modes. Although it was always possible to generate the correct peptide, it turns out that the correct peptide tends to score lower than some incorrect peptides. However, the high-scoring incorrect peptides are similar to the correct peptide. The main reason behind the high-ranking incorrect peptides returned by NRP-Assembly is the presence of high intensity unexplained peaks in the consensus spectrum (see Figure 3a), a consequence of the low signal-to-noise ratio in MS^n spectra for $n > 3$. In addition, both NRP-Sequencing and NRP-Assembly generate large numbers of putative peptides in the "unknown" mode since in this case the unrestrained use of amino acid masses, noise peaks, and missing b/y-ions relegate the correct peptide to low ranks.

While these problems may later be mitigated by a more realistic scoring scheme, using additional MS^n spectra to re-rank all putative de novo peptides readily singles out the correct de novo peptide. Most surprisingly, the "strength in numbers" of this MS^n-matching approach is discriminating enough to even select the correct peptide from the large sets of of possible peptides generated in the "unknown" mode.

The results of NRP-Sequencing and NRP-Assembly applied to a novel peptide (Compound X^4) are shown in Figure 4. While the overall shape of the auto-convolution is similar to Seglitide, the larger set of possible reconstructions and the slightly larger set of amino acid masses used by the top de novo sequences are consistent with the larger parent mass of Compound X. In contrast with Seglitide, the cyclic peptides recovered by NRP-Sequencing and NRP-Assembly were mostly but not completely identical: the LNG subpeptide returned by NRP-Sequencing was replaced by MF^{+6} subpeptide (of the same mass) in the peptide returned by NRP-Assembly. Nevertheless, the consensus spectrum

---

[4] We name this Compound X for the lack of a better name. Initial NMR experiments revealed that this cyclic peptide contains the following masses $Cyclic$(161.08-141.08-71.04-131.04-163.07-127.10-57.02-113.08)

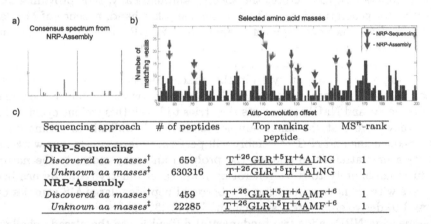

a) Seglitide MS4 spectra aligned by NRP-Assembly

b) Resulting consensus spectrum

Sequencing approach	# peptides	De novo rank	MS^n rank
NRP-Sequencing			
known aa masses[†]	20	1	1
discovered aa masses[‡]	290	2	1
unknown aa masses[°]	173894	23388	1
NRP-Assembly			
known aa masses[†]	60	45	1
discovered aa masses[‡]	2793	2203	1
unknown aa masses[°]	104430	31589	1

Fig. 3. De novo sequencing of Seglitide performed either by self-aligning the MS^3 spectrum (NRP-Sequencing) or by assembling MS^n spectra (NRP-Assembly) as illustrated on the left. The statistics are given for "known"([†]), "discovered"([‡]), and "unknown" ([°]) modes of generating the alphabet of amino acid masses. The first column shows the total number of generated peptides; the second column shows the rank of the correct peptide; the third column shows the rank of a correct de novo sequence after re-ranking the generated sequences by their match scores to the MS^n spectra.

a) Consensus spectrum from NRP-Assembly

Number of matching peaks

b) Selected amino acid masses

- NRP-Sequencing
- NRP-Assembly

Auto-convolution offset

Sequencing approach	# of peptides	Top ranking peptide	MS^n-rank
NRP-Sequencing			
Discovered aa masses[†]	659	T^{+26}GLR^{+5}H^{+4}ALNG	1
Unknown aa masses[‡]	630316	T^{+26}GLR^{+5}H^{+4}ALNG	1
NRP-Assembly			
Discovered aa masses[†]	459	T^{+26}GLR^{+5}H^{+4}AMF^{+6}	1
Unknown aa masses[‡]	22285	T^{+26}GLR^{+5}H^{+4}AMF^{+6}	1

Fig. 4. De novo sequencing of the Compound X. The sequenced spectrum was obtained by aligning the MS^3 spectrum to itself (NRP-Sequencing) or by assembling MS^n spectra (NRP-Assembly). The set of residue masses can (†) be discovered by our approach or (‡) considered unknown. The first column indicates the total number of generated de novo sequences; the second column indicates the recovered peptide (modifications are used in place of the unknown residues of the same mass); the third column shows the highest rank of a correct de novo sequence after reranking the generated sequences by their match scores to the MS^n spectra.

returned by NRP-Assembly also has strong support for the sequence recovered by NRP-Sequencing - almost all b-ions are present with intensities well above the background noise. This support is also evidenced by the fact that the top NRP-Sequencing peptide achieves rank 12 when generated by NRP-Assembly. In fact, a closer inspection of the NRP-Assembly consensus spectrum reveals that the discrepancy between the two methods is caused by a single high intensity peak. The first sequence recovered by NRP-Assembly that does not use this peak is exactly the top sequence returned by NRP-Sequencing. Since the converse is not true (i.e. the top NRP-Assembly sequence is not generated by NRP-Sequencing) and high intensity unexplained peaks were also found in the consensus spectrum obtained for Seglitide, one would tend to prefer $T^{+26}GLR^{+5}H^{+4}$ ALNG (NRP-Sequencing) over $T^{+26}GLR^{+5}H^{+4}AMF^{+6}$ (NRP-Assembly). In fact, initial NMR experiments support the result returned by the NRP-Sequencing algorithm. We further note that the analysis (NRP-Sequencing/ NRP-Assembly) of Seglitide and Compound X took only 3-7 minutes on a regular desktop computer (Redhat Linux, Pentium IV, 3.2 GHz, 2 Gb RAM).

3   Discussion

There is a catch-22 when it comes to using mass spectrometry for NRP interpretation. On the one hand, there is hardly any MS data for NRPs because nobody knows how to interpret the spectra automatically, thus providing little incentive for generating large datasets. On the other hand, absence of MS data for NRPs slows down development of algorithms for NRP interpretation because large MS datasets are needed to develop such algorithms. This catch-22 is further complicated by the high cost of many NRPs and by the unavailability of many compounds (often analyzed in the pharmaceutical industry) to academic researchers. This paper presents a collaboration between bioinformatics, mass spectrometry, and NMR researchers that tries to break this vicious cycle. While we acknowledge that the small sample size makes it difficult to estimate how NRP-Sequencing and NRP-Assembly will perform in larger tests, our work represents a first attempt to address the problem and may motivate the natural products community to start generating mass spectrometry data. As has been the case with de novo sequencing of ribosomal peptides, large MS samples can be used to derive elaborate statistical models [31,32,33].

Sequencing NRPs adds two fundamental difficulties to the already challenging task of de novo peptide sequencing: the amino acid masses are not known in advance and the peptides are cyclic. rather than linear. Current de novo sequencing algorithms cannot address these difficulties and standard tandem mass spectrometry (MS^2) provides insufficient information because MS^2 simply results in the linearization of the cyclic peptide. Using additional stages of mass spectrometry leads to spectra containing either a mixture of masses from all linearized versions of the peptide (MS^3 spectrum) or multiple lower-quality spectra from shorter subpeptides (MS^n spectra). Although the theoretical problem of sequencing an MS^3 spectrum is difficult, we have shown that an alignment-based heuristic approach works in practice. An alternative approach that may lead to better results in some cases, is to assemble MS^n spectra and derive the putative peptide sequences from the resulting consensus. Finally, the reliance of both NRP-Sequencing and NRP-Assembly on the alignment of spectrum masses indicates that the accuracy and efficiency of both methods should be greatly increased when applied to high-accuracy mass spectrometry data.

Acknowledgements. We would like to thank Bill Gerwick and Luke Simmons for providing Compound X prior to publication. This project was supported by NIH grant NIGMS 1-R01-RR16522.

References

1. Sieber, S.A., Marahiel, M.A.: Molecular Mechanisms Underlying Nonribosomal Peptide Synthesis: Approaches to New Antibiotics. Chem. Rev. 105, 715–738 (2005)
2. Dorrestein, P.C., Kelleher, N.L.: Dissecting Non-ribosomal and Polyketide Biosynthetic Machineries Using Electrospray Ionization Fourier-Transform Mass Spectrometry. Natural Product Reports 23, 893–918 (2006)
3. Welker, M., Von Doehren, H.: Cyanobacterial Peptides - Nature's Own Combinatorial Biosynthesis. FEMS Microbiology Reviews 30, 530–563 (2006)
4. Butcher, B.G., Helmann, J.D.: Identification of Bacillus subtilis Sigma-dependent Genes that Provide Intrinsic Resistance to Antimicrobial Compounds Produced by Bacilli. Mol. Microbiol. 60, 765–782 (2006)
5. Williams, D., Austin, P., Diaz-Marrero, A., Soest, R., Matainaho, T., Roskelley, C., Roberge, M., Andersen, R · Neopetrosiamides, Peptides from the Marine Sponge Noopetrosia sp. That Inhibit Amoebold Invasion by Human Tumor Cells. Organic Letters 7, 4173–4176 (2005)
6. Luesch, H., Williams, P., Yoshida, W., Moore, R., Paul, V.: Ulongamides A-F, New Beta-Amino Acid-Containing Cyclodepsipeptides from Palauan Collections of the Marine Cyanobacterium Lyngbya sp. Journal of Natural Products 65, 996–1000 (2002)
7. Hamada, T., Matsunaga, S., Yano, G., G nd Fusetani, N.: Polytheonamides A and B, Highly Cytotoxic, Linear Polypeptides with Unprecedented Structural Features, from the Marine Sponge, Theonella swinhoei. J Am. Chem. Soc. 127, 110–118 (2005)
8. Ireland, C.M., Durso, A.R., Newman, R.A., Hacker, M.P.: Antineoplastic Cyclic Peptides from the Marine Tunicate Lissoclinum patella. J. Org. Chem. 47, 360–361 (1982)
9. Kurosawa, K., Matsuura, K., Chida, N.: Total Synthesis of Stevastelins B3 and C3: Structure Confirmation of Stevastelin B3 and Revision of Stevastelin C3. Tetrahedron Letters 46, 389–392 (2005)
10. Li, J., Burgett, A., Esser, L., Amezcua, C., G.Harran, P.: Total synthesis of nominal diazonamides: Part 2. on the true structure and origin of natural isolates. Angew. Chem Intl. Ed. Engl., 4771–4773 (2001)
11. Ikeda, H., Nonomiya, T., Ōmura, S.: Organization of Biosynthetic Gene Cluster for Avermectin in Streptomyces avermitilis: Analysis of Enzymatic Domains in Four Polyketide Synthases. Journal of Industrial Microbiology and Biotechnology 27, 170–176 (2001)
12. Watanabe, K., Hotta, K., Praseuth, A.P., Koketsu, K., Migita, A., Boddy, C.N., Wang, C.C., Oguri, H., Oikawa, H.: Total Biosynthesis of Antitumor Nonribosomal Peptides in Escherichia coli. Nat. Chem. Biol. 2, 423–428 (2006)
13. Barber, M., Bell, D.J., Morris, M.R., Tetler, L.W., Monaghan, J.J., Morden, W.E., Bycroft, B.W., Green, B.N.: An Investigation of the Tyrothricin Complex by Tandem Mass Spectrometry. International Journal of Mass Spectrometry and Ion Processes 122, 143–151 (1992)

14. Hitzeroth, G., Vater, J., Franke, P., Gebhardt, K., Fiedler, H.P.: Whole Cell Matrix-Assisted Laser Desorption/Ionization Time-of-Flight Mass Spectrometry and in situ Structure Analysis of Streptocidins, a Family of Tyrocidine-like Cyclic Peptides. Rapid Communications in Mass Spectrometry 19, 2935–2942 (2005)
15. Gebhardt, K., Pukall, R., Fiedler, H.P.: Streptocidins A-D, Novel Cyclic Decapeptide Antibiotics Produced by Streptomyces sp. Tü 6071. I. Taxonomy, Fermentation, Isolation and Biological Activities. Antibiot. 54, 428–433 (2001)
16. Höltzel, A., Jack, R.W., Nicholson, G.J., Jung, G., Gebhardt, K., Fiedler, H.P., Süssmuth, R.D.: Streptocidins A-D, Novel Cyclic Decapeptide Antibiotics Produced by Streptomyces sp. Tü 6071. II. Structure elucidation. Antibiot. 54, 434–440 (2005)
17. Redman, J., Wilcoxen, K., Ghadiri, M.: Automated Mass Spectrometric Sequence Determination of Cyclic Peptide Library Members. Journal of Combinatorial Chemistry 5, 33–40 (2003)
18. Olsen, J.V., Mann, M.: Improved Peptide Identification in Proteomics by Two Consecutive Stages of Mass Spectrometric Fragmentation. Proc. Natl. Acad. Sci. 101, 13417–13422 (2004)
19. Ulintz, P.J., Bodenmiller, B., Andrews, P.C., Aebersold, R., Nesvizhskii, A.I.: Investigating MS2-MS3 Matching Statistics: A Model for Coupling Consecutive Stage Mass Spectrometry Data for Increased Peptide Identification Confidence. In: Molecular Cellular Proteomics, pp. M700128–MCP200 (2007)
20. Skiena, S.S., Sundaram, G.: A Partial Digest Approach to Restriction Site Mapping. Bulletin of Mathematical Biology 56, 275–294 (1994)
21. Rosenblatt, J., Seymour, P.D.: The Structure of Homometric Sets. SIAM Journal on Algebraic and Discrete Methods 3, 343–350 (1982)
22. Cieliebak, M., Eidenbenz, S., Penna, P.: Partial Digest Problem is Hard to Solve for Erroneous Input Data. Theoretical Computer Science 349, 361–381 (2005)
23. Xu, C., Ma, B.: Complexity and Scoring Function of MS/MS Peptide De Novo Sequencing. Computational Systems Bioinformatics 5, 361–369 (2006)
24. Pevzner, P.A., Dancik, V., Tang, C.: Mutation-Tolerant Protein Identification by Mass Spectrometry. J Comput. Biol. 7, 777–787 (2000)
25. Bandeira, N., Clauser, K.R., Pevzner, P.A.: Shotgun Protein Sequencing: Assembly of Peptide Tandem Mass Spectra from Mixtures of Modified Proteins. Mol. Cell Proteomics 6, 1123–1134 (2007)
26. Bandeira, N., Tsur, D., Frank, A., Pevzner, P.A.: Protein Identification by Spectral Networks Analysis. Proceedings of the National Academy of Sciences 104, 6140–6145 (2007)
27. Bern, M., Goldberg, D.: De Novo Analysis of Peptide Tandem Mass Spectra by Spectral Graph Partitioning. Journal of Computational Biology 13, 364–378 (2006)
28. Chen, T., Kao, M.Y., Tepel, M., Rush, J., Church, G.M.: A Dynamic Programming Approach to De Novo Peptide Sequencing via Tandem Mass Spectrometry. J Comput. Biol. 8, 325–337 (2001)
29. B'Hymer, C., Montes-Bayon, M., Caruso, J.A.: Marfey's Reagent: Past, Present, and Future Uses of 1-Fluoro-2,4-Dinitrophenyl-5-L-Alanine Amide. Journal of Separation Science 26, 7–19 (2003)
30. Bandeira, N., Tang, H., Bafna, V., Pevzner, P.: Shotgun Protein Sequencing by Tandem Mass Spectra Assembly. Analytical Chemistry 76, 7221–7233 (2004)
31. Dancik, V., Addona, T., Clauser, K., Vath, J., Pevzner, P.: De Novo Peptide Sequencing via Tandem Mass Spectrometry. J Comput. Biol. 6, 327–342 (1999)
32. Frank, A.M., Pevzner, P.A.: PepNovo: De Novo Peptide Sequencing via Probabilistic Network Modeling. Anal. Chem. 77, 964–973 (2005)

33. Mo, L., Dutta, D., Wan, Y., Chen, T.: MSNovo: A Dynamic Programming Algorithm for De Novo Peptide Sequencing via Tandem Mass Spectrometry. Anal. Chem. 79, 4870–4878 (2007)

A   Supplementary Materials

A.1   Data Acquisition and Preprocessing

Seglitide was purchased from Aldrich. Compound X was isolated from marine cyanobacteria. Solutions of Seglitide (20 μM in water) and Compound X (50 μg/mL in 50:49:1 water:methanol:acetic acid) underwent nano-electrospray ionization on a Biversa Nanomate (pressure: 0.3 psi, spray voltage: 1.4-1.8 kV) and were analyzed on a Finnigan LTQ-MS by running MS^n ion-tree experiments with Tune Plus and Xcalibur software. [1] The mass spectrometry data for Seglitide contained 400 scans. The scans were grouped by their parent masses and merged into a single spectrum per group. The peaks in the merged spectrum are the average of all peaks in the spectra with the same parent mass. The resulting merged spectra are ranked by their total intensity, defined as the sum of all the averaged peaks. The resulting 41 spectra were selected as the input for the NRP-sequencing algorithm. Among the top 41 spectra, there was one MS^3 spectrum with parent mass of 808 Da and 40 MS^n spectra. The mass spectrometry data for compound X contained 2078 scans, but only 19 different spectra emerged after merging (one MS^3 spectrum with parent mass of 955 Da and 18 MS^n spectra). Only the top 15 MS^n spectra were chosen as the input for the NRP-Assembly algorithm for both Seglitide and Compound X.

A.2   Details on NRP-Sequencing

Below we use Seglitide as an example to illustrate the execution of NRP-Sequencing. We apply the following preprocessing steps before invoking NRP-Sequencing. Let the L-rank of a peak x be the number of peaks in the interval $[x - L, x + L]$ with intensity larger or equal to the intensity of x. We remove all peaks whose L-rank exceeds a threshold γ and make the spectrum S symmetric by combining S with $ParentMass \ominus S = \{ParentMass - s : s \in S\}$. For $L = 25$ and $\gamma = 10$ it results in 140 peaks (as compared to 100 peaks in the original Seglitide MS^3 spectrum).

– **Auto-convolution.** While we implicitly assume that each of the offsets $x_1 \ldots x_k$ corresponds to the mass of a single amino acid, it is straightforward to extend NRP-Sequencing to the case when some of the top peaks corresponds to the mass of 2 or more amino acids. In practice, offsets from

---

[1] Order: breadth-first; max breadth: 25 for Seglitide, 20 for Compound X; max MS^n depth n=5; additional μ-scans: 2, 4, 8 at $n = 3$, 4, 5; normalized energy for Seglitide: 40, 20, 20 at $n = 3$, 4, 5; normalized energy for Compound X: 50, 25, 25 at $n = 3$, 4, 5; isolation width: 3 for Seglitide, 4 for Compound X; exclusion mass width low/high: 2/3.

double/triple masses do not lead to the deterioration of NRP-Sequencing as long as the masses of all single amino acids are present among the top peaks (like in the case of Seglitide). For Seglitide, the maximum peak $x_1 = 85$ corresponds to the mass of A^{+14}.

- **Auto-alignment.** After auto-aligning the MS^3 spectrum with offset x, we construct the consensus spectrum S_x containing only matching masses in the overlapping portion of the auto-alignment. The resulting consensus spectrum is scored with the summed intensities of the corresponding matched masses. For Seglitide, the consensus spectrum S_{x_1} contains all prefix *and* suffix (b/y) ions for the peptide (YWKVF) in the overlapping region.
- **De novo peptide sequencing.** We solve the de novo peptide sequencing problem for the consensus spectrum using the anti-symmetric path algorithm. Since for cyclic peptides the set of amino acid masses is not known in advance we use the top k peaks in auto-convolution as the first approximation for the masses of amino acids.

After selecting the alphabet, NRP-Sequencing generates all de novo peptide reconstructions of S_x with scores above $p \cdot Score(P)$, where p is a parameter and P is the highest scoring peptide representing a solution of the de novo peptide sequencing problem. For Seglitide this approach results in the generation of 290 peptides (in Mode 2 with $p = 0.5$). NRP-Sequencing further scores each candidate peptide by matching all MS^n spectra against it. Finally, NRP-Sequencing reranks cyclic peptides according to their matches to the MS^n spectra.

A.3   Details on NRP-Assembly

The NRP-Assembly algorithm is summarized as follows:

- **Pairwise spectral alignment.** Pairwise spectral alignments are computed for all pairs of preprocessed MS^n spectra by selecting offset maximizing the number of matching peaks.
- **Assembly of pairwise alignments.** Assembly into contigs of size $k + 1$ is done iteratively by adding an additional spectrum to existing contigs of size k. The new contig is formed only if the new spectrum has compatible offsets with the members of the existing contig.
- **Consensus Construction.** A consensus spectrum is derived for each contig using a threshold α to filter the peaks. For instance, for a peak to be in the consensus of Seglitide, we require it to be present in at least α out of all spectra in the contig. The score of a contig is computed by simply summing the percent intensity of all the peaks that are used to construct the consensus. The intensity of each consensus peak is the sum of the peaks that originated it. The peaks at position 0 and $ParentMass$ of each spectrum in the contig are directly imported to the final consensus after the spectrum is aligned.
- **De novo peptide sequencing.** The resulting consensus generally contains more than just b-ion peaks because of neutral loss peaks and occasionally noise peaks. De novo sequencing is used to generate all possible peptide reconstructions from the consensus and these are ranked by matching them

to the MS^n spectra that are not part of the contig, as described in the previous section (NRP-Sequencing).

A.4   Additional Figures

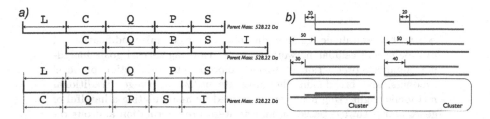

Fig. A-1. a) MS^n spectra may represent a mixture of various subpeptides, when the parent masses of two different linear peptides happen to be the same. In particular, mixed MS^n spectra are common if there are repeated amino acid masses in the cyclic peptide as illustrated by two labeled spectra covering a hypothetical peptide that contains the subpeptide LCQPSI (mass of L equal to the mass of I). This results in an MS^4 spectrum (bottom) that represents a mixture of these two peptides. **b)** The three possible pairwise alignments of three spectra (red, green and blue) in two scenarios. The offsets of the pairwise alignments are compatible (left) and consequently a contig of size three is formed. When the given offsets are incompatible (right), no contig is formed from this triplet.

Systems Metabolic Engineering

Sang Yup Lee*

Korea Advanced Institute of Science and Technology, Daejeon
leesy@kaist.ac.kr

Abstract. Metabolic engineering allows purposeful modification of metabolic and cellular network towards achieving several goals including enhanced production of various bioproducts, production of novel products, and broadening the substrate utilization range. Traditional metabolic engineering has been performed by manipulating a handful of genes and pathways based on known literature information and our rational thinking. Advances in omics technology, computational bioscience, and systems biology are now providing us with new information and knowledge that had not been possible to obtain using traditional approaches. Systems biology is allowing us to elucidate the metabolism and physiology of cells and organisms at the global levels. Metabolic engineering based on the systems-level analysis of cells and organisms, termed systems metabolic engineering, is now offering a new powerful way of designing and developing strains having improved performance. In this lecture, I will present the general strategies of systems metabolic engineering. Also, several examples of applying systems metabolic engineering for the production of amino acids, primary metabolite (succinic acid) and secondary metabolite (lycopene) will be presented.

About the keynote speaker. Dr. Sang Yup Lee is Distinguished Professor and LG Chem Chair Professor at the Department of Chemical and Biomolecular Engineering, KAIST, Korea. He is also the Director of Center for Systems and Synthetic Biotechnology, Director of BioProcess Engineering Research Center, Director of Bioinformatics Research Center, and Co-Director of the Institute for the BioCentury. He has published more than 230 journal papers, 40 books/book chapters, 270 patents, and presented more than 900 papers at conferences. He is currently serving as Senior Editor, Editor, Associate Editor, or Board Member of 12 journals including Biotechnology and Bioengineering.

* This work was supported by Korean Systems Biology Program of the Ministry of Science and Technology, Microsoft, IBM SUR Program and LG Chem Chair Professorship.

Protein Function Prediction Based on Patterns in Biological Networks

Mustafa Kirac and Gultekin Ozsoyoglu

Case Western Reserve University, Cleveland, OH, USA
{kirac,tekin}@case.edu

Abstract. In this paper, we propose a pattern-based protein function annotation framework, employing protein interaction networks, to predict annotation functions of proteins. More specifically, we first detect patterns that appear in the neighborhood of proteins with a particular functionality, and then transfer annotations between two proteins only if they have similar annotation patterns. We show that, in comparison with other techniques, our approach predicts protein annotations more effectively. Our technique (a) produces the highest prediction accuracy of 70-80% precision and recall for different organism specific datasets, and (b) is robust to false positives in protein interaction networks.

1 Introduction

Discovering protein functions is a major task in computational biology. Despite a large number of genome annotation projects, even one of the most well-studied model organisms, *S. cerevisiae*, is reported [SGD] to have more than 3,000 genes, i.e., 40% of its genome, with unknown molecular functions.

Traditionally, *in silico* protein function annotation is achieved through sequence homology, which has the following limitations. First, it requires homologs with known functions in genome databases. Second, transferring functional assignments between proteins with low sequence identity is unreliable [R02, LK03, TS03]. Third, high sequence similarity does not necessarily imply function similarity [V05, F06, BL06]. Thus, a complementary research direction is to assign functional annotations to proteins (and genes) based on biological network information, more specifically, protein interaction network information.

Recently, three distinct network-centric function annotation categories have emerged:

1. Direct annotation schemes [SUS07] infer protein functions based on functionality prevalent across neighbors of proteins. Examples of such schemes include (*a*) assigning the most frequent function among interaction partners [HN+01, SUF00], (*b*) global assignment that minimizes the number of protein interactions between protein pairs that are annotated with different functions [VF+03], (*c*) probabilistic function assignment via graph labeling (binary classification of proteins with respect to existence or non-existence of a function) [LK03, DT+04], and (*d*) function assignment via propagating functions using links of the network [NJ+05].

M. Vingron and L. Wong (Eds.): RECOMB 2008, LNBI 4955, pp. 197–213, 2008.
© Springer-Verlag Berlin Heidelberg 2008

2. Module-assisted schemes [SUS07] first identify clusters of functionally related proteins based on protein connectivity in the network, and then, assign the most recurrent function of a cluster to its members. Major network clustering algorithms [BH06] include Markov Clustering [D00] which simulates the flow among network elements, and separates clusters by no-flow boundaries; Restricted Neighborhood Search Clustering [KPJ04] which minimizes a cost function involving the numbers of intra-cluster and inter-cluster edges; Super Paramagnetic Clustering [BWD96] and Molecular Complex Detection [BH03] which detect and isolate densely connected regions as clusters.

3. Pattern-based approaches [KOY06] learn annotation patterns from a network, and annotate proteins via the patterns found among neighbors of proteins. Techniques such as PST (Probabilistic Suffix Trees) [KOY06] and Correlation Mining [KOY06] are in this category.

Direct and module-assisted annotation schemes have three limitations.

1. Common annotation assumption: Two interacting proteins are assumed likely to share a function. Such an assumption does not hold for all interaction and annotation types. For instance, components of a protein complex probably share the functionality of the complex, whereas receptor and signal transduction proteins have totally different molecular functionalities, despite their physical interactions. To illustrate this, for each Gene Ontology (GO) term [GO], we computed the percentage of interactions that the term annotates both interaction partners over all interactions of proteins that the term annotates. We found that only 18.2%, 18.8%, and 16.5% of GO terms annotate both interaction partners in yeast interactome for molecular function, biological process, and cellular component ontologies, respectively. Moreover, these percentages increase to 25.8%, 39.8%, and 43.6% for the top 100 most frequent GO terms, and to 37.8%, 68.9%, and 83.8% for the top 10 most frequent GO concepts in molecular function, biological process, and cellular component ontologies (after eliminating the root terms), respectively. These percentages indeed explain why direct and module-assisted annotation schemes work well only for "informative", i.e., frequently assigned, function annotations [ZKW02, DT+04], and that accuracy of these techniques are usually presented for only the biological process ontology (vs. molecular function ontology) [SUS07].

2. Connected network assumption: Annotations are transferred between proteins through interaction paths. This implies that annotation transfer is prevented when there are gaps (i.e., disconnected protein pairs) in interaction networks, and the accuracy of such methods reduce dramatically when the network has several disjoint components. Accuracy of many direct and module-assisted annotation schemes are evaluated only on the largest component of the interactome [NJ+05, SUS07].

3. Underutilized cross-species information: Direct annotation methods aim genome-scale functional annotation using network data from single species, and are unable to utilize richly annotated network of a model organism to assign functions to a newly sequenced organism. Module-assisted methods can utilize cross-species data; however, they do not offer any annotation prediction for a protein which is not part of a frequent protein interaction motif conserved among several organisms.

In this paper, we develop and evaluate a new pattern-based function annotation framework. Our goal is to assign GO terms to non-annotated proteins in a given protein interaction network. Proteins within some distance from a given protein U, and their interactions with each other, form a network fragment, namely, the *P-P* (protein-protein) *neighborhood* of protein *U*. When, in a P-P neighborhood, proteins are replaced by their annotations, we have a domain transformation from proteins to functional annotations, resulting in an *annotation neighborhood*. We use the frequently-referred observation [LR+02, MS+02, OM+02, BL04, KGS04, Ki04, TL+04, YL+04, SS+05, PK+07] that proteins that are assigned the same GO term usually have similar annotation neighborhoods as follows. For each GO term *t*, we form a set S_t of annotation neighborhoods of proteins that the term *t* annotates. The set S_t represents consensus information of annotation neighborhoods related to the GO term *t*. Finally, we provide a pairwise graph alignment algorithm that measures the similarity between an annotation neighborhood and neighborhood set of a GO term. We then assign to the non-annotated protein *U* the GO term whose annotation neighborhoods are the most similar to the annotation neighborhood of *U*.

Example 1.1. In Figure 1.1, P_1 to P_6 denote proteins, and lower case letters represent GO terms. Graphs in A and B are protein neighborhoods of proteins P_1 and P_4, respectively. Transforming the protein neighborhoods into (protein) annotation neighborhoods, we obtain the graphs in C and D. E and F are annotation neighborhoods of the term a, and are extracted from C and D by updating their roots.

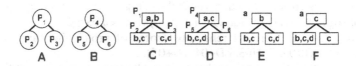

Fig. 1.1. Protein neighborhood and annotation neighborhood examples

Our pattern-based annotation scheme has unique features in comparison with direct and module-assisted annotation schemes.

1. Collective evidence: To assign the functionality *F* to protein *U*, annotation neighborhood of *U* needs to significantly overlap with the annotation neighborhoods of *F* (i.e., annotation neighborhoods of proteins annotated by *F*). In other (direct and module-assisted) schemes, a single interaction is sometimes misleadingly employed as evidence for annotation transfer.

2. Utilization of disjoint network components: Annotations are transferred between proteins through annotation neighborhood similarity; thus two proteins do not need to be connected in a protein-protein (PP) network in order to relate one annotation to another.

3. Utilization of cross-species network information: Cross-species network information can be utilized, since annotation neighborhoods of a GO term may be extracted from a well-known organism to annotate a newly sequenced one.

On the negative side, discovery and storage of patterns can be expensive for some instances. For example, the PST technique [KOY06] involves only sequential patterns (vs. graph patterns), restricts the number of protein functions to a certain number, and is trained via random sampling (vs. employing the complete dataset) since there are exponential numbers of sequential patterns in a network. In this paper, we employ patterns more complex than sequences, namely, annotation graphs. In addition, this new approach can work with a large number of annotations.

The rest of the paper is organized as follows. Section 2 presents details of our methodology. In Section 3, we experimentally evaluate our proposed approach, i.e., the pattern-based function annotation technique. Finally, Section 4 discusses and summarizes our results.

2 Methods

2.1 Protein Annotation Prediction as a Probabilistic Classification Problem

Our goal is to assign GO terms to proteins in protein interaction networks. We formulate our pattern-based protein function prediction as a (multi-class) classification problem: GO terms are classes; proteins are items to be categorized into classes; and network information of proteins corresponds to features of items. We denote network and annotation information of all proteins by single parameter Γ.

Def'n. (*Protein Interaction Network*): $\Gamma = (V, E, \varsigma, W)$ is a vertex-labeled, edge-weighted, undirected graph, called *protein interaction network*, where V is the set of vertices (proteins), E is the set of edges (interactions), ς is a finite alphabet of (annotation) terms (from a function vocabulary, e.g., GO), and $l(v)=\{t| t \in \varsigma\}$, $v \in V$. Weights of edges are reals: $W: <u, v> \rightarrow (0, 1]$, $u, v \in V$, $<u,v> \in E$.

Edge weights in Γ represent reliability scores of protein interactions, obtained computationally [GR03, SSH03, LD+04, NJ+05].

We employ a Bayesian classifier that assigns GO terms to proteins, utilizing annotation and interaction information of other proteins. We are interested in computing the probability that "protein v is associated with GO term t, with respect to given protein interaction network Γ", namely $Prob(t \in l(v) | \Gamma)$. We define a scoring function $f_t(v)$ to be a function of this probability, in order to assign top-k GO terms with highest scores to a given protein. Applying Bayes' rule in log space, we obtain the following:

$$F^1{}_t(v) = \log[Prob(t \in l(v)|\Gamma)] = \log[Prob(\Gamma|t \in l(v))] + \log[Prob(t \in l(v))] - \ln[Prob(\Gamma)]. \quad (1)$$

Since $Prob(\Gamma)$ is independent of t and v, it has the same value for all GO terms and proteins. Thus, we eliminate it from the scoring function:

$$F^2{}_t(v) = \log[Prob(\Gamma|t \in l(v))] + \log[Prob(t \in l(v))]. \quad (2)$$

Thus, the scoring function has been reduced to two factors, namely $Prob(t \in l(v))$ and $Prob(\Gamma|t \in l(v))$. $Prob(t \in l(v))$ depends only on the GO term t, and the protein that the term annotates, namely v. Since v is declared separately from a protein interaction network, v can be considered as an arbitrary protein. A protein having one GO term annotation does not prevent it from having other GO term annotations [KF+03]. Hence, different GO terms independently annotate the same protein. As a result, $Prob(t \in l(v))$ is uniform (for different instances of v), and it can be estimated from a given protein interaction network as follows: $Prob(t \in l(v)) = n_t / n_V$, where n_t is the number of proteins that t annotates, n_V is the number of all proteins in a given protein interaction network.

Prob(Γ|$t \in l(v)$) can be defined as "the likelihood of observing Γ as the protein interaction information of v, given the fact that t annotates v". Direct estimation of this probability from a given protein interaction network is a difficult task due to unknown dependencies between interacting proteins, their annotations, and given protein v. We approximate this probability by incorporating observations from previous work. The function of a protein manifests itself in its neighborhood in the network [SUS07]. In previous protein function prediction models, both direct [LK03] and indirect [CSW07] neighbors of a protein v (within a short distance) are employed to predict functionality of v. On the other hand, distant protein neighbors have little or no effect in determining functionality of a protein [KOY06]. As a result, we approximate *Prob*(Γ|$t \in l(v)$) by employing a subset of the protein interaction network, namely, only the *annotation neighborhood* of v.

Def'n. *(Annotation neighborhood as a rooted graph, rank of vertex):* A *rooted-graph* $\gamma = (U, E', \rho, W')$ is obtained from a subgraph $\Gamma'(V, E, \varsigma, W)$ of Γ, by replacing $v \in V$ with its label set $u = l(v)$, and designating the label set of a node in V as the root ρ, i.e., $U = \{u \mid u = l(v), v \in V\}$, $E' \subseteq U \times U$, W': $<u_1, u_2> \rightarrow (0,1]$, and ρ in U is the *root* of γ. *Rank* of a vertex u in rooted-graph γ, namely *rank*(u, γ), is defined as the distance (i.e., of the length of the path with the highest product of edge weights) between u and the root ρ. *Annotation neighborhood* of a protein P is a rooted-graph with root $l(P)$. An *annotation neighborhood* of a GO term t is a rooted-graph γ with root ρ such that the label set $l(P)$-$\{t\}$ *where* $t \in l(P)$ replaces $l(P)$.

Example 2.1. From Figure 1.1, in protein interaction graphs, vertices have unique labels corresponding to protein identifiers. In comparison, in annotation neighborhood of a protein, vertices are labeled with sets of GO terms, and, a vertex is not necessarily uniquely identified by its label. In annotation neighborhood of a GO term t, the root corresponds to a protein annotated by t, but t is removed from its label set.

Def'n. *(Depth of a graph):* Let γ be an annotation neighborhood graph rooted at ρ, and d be the maximum rank of any node in γ, denoted as γ_ρ^d. Then, d is called the *depth* of the rooted graph γ_ρ^d.

We approximate *Prob*(Γ|$t \in l(v)$) by including only a subset of the protein interaction network (i.e., an annotation neighborhood), since additional information in the protein interaction network other than the annotation pattern of a protein does not significantly change its value for some neighborhood of depth d:

$$Prob(\Gamma | t \in l(v)) = Prob(\gamma_v^d | t \in l(v)) + \varepsilon_{t,v,d} \tag{3}$$

where $\varepsilon_{t,v,d}$ is the residual error of the approximation, and is considered to be negligible. *Prob*(γ_v^d|$t \in l(v)$) can be described as the *likelihood of observing* γ_v^d *as an annotation neighborhood of t*. In the rest of the paper, we employ a global d value for all annotation neighborhoods, and omit d from future definitions.

The core idea of the paper is that *proteins annotated by the same GO term have similar annotation patterns*. When two proteins A and B have a similar annotation, neighboring proteins of A and B, namely N_A and N_B, are also similar in terms of the annotations of proteins in N_A and N_B [LR+02, MS+02, OM+02, BL04, KGS04, Ki04, TL+04, YL+04, SS+05, PK+07]. Given a protein v annotated by GO term t, we want to estimate the "likelihood of observing γ_v as the annotation neighborhood of v, in a situation that the annotation neighborhood information of v is unknown", namely $Prob(\gamma_v |t \in l(v))$. If γ_v is similar to one of the annotation neighborhoods of t, it is indeed likely to observe γ_v as the annotation neighborhood of v, since annotation of v determines its neighborhood. Otherwise, i.e., if γ_v is not similar to any of the annotation neighborhoods of t, probability of observing γ_v is close to observing a random annotation neighborhood, in other words, the probability becomes very small. Therefore, we say that $Prob(\gamma_v |t \in l(v))$ is proportional to the similarity of γ_v to an annotation neighborhood of t.

Annotation neighborhood set S_t of t can be perceived as consensus information of term t. If an annotation neighborhood sub-graph recurs frequently in annotation neighborhoods of t in slightly modified forms, the sub-graph is named a *conserved subgraph*. On the other hand, a sub-graph that occurs only in the annotation neighborhood of a particular protein is named a *discriminative subgraph*. Conserved subgraphs are more likely to represent annotation neighborhoods of a GO term than discriminative subgraphs, therefore $Prob(\gamma_v |t \in l(v))$ should be estimated as the similarity of γ_v to a subgraph conserved among the annotation patterns of t. Let sim be a function that detects regions of annotation neighborhoods conserved in a given set S_t of annotation neighborhoods, and measures similarity between annotation neighborhood γ_v and the set S_t as a real value in the range [0, 1] (i.e., 0 for no similarity, and 1 for exact similarity). Then, we approximate $Prob(\gamma_v |t \in l(v))$ by $sim(\gamma_v, S_t)$. Finally, putting approximated values back to their places in the discriminative function formula, we obtain the following formula:

$$F^3_t(v) = \log[sim(\gamma_v, S_t)] + \log[n_t / n_V] = \log[sim(\gamma_v, S_t)] + \log[n_t] - \log[n_V]. \quad (4)$$

And then, by eliminating the independent term $-\log[n_V]$ from the formula, we obtain the simplified scoring function:

$$f_t(v) = \log[sim(\gamma_v, S_t)] + \log[n_t]. \quad (5)$$

In the next section, we present our protein annotation prediction algorithm that employs $f_t(v)$ as the scoring function.

2.2 Protein Annotation Prediction Algorithm

First, we give a sketch of our prediction algorithm in Figure 2.1.

Algorithm 1: Protein Annotation Prediction

Input: Protein interaction network Γ, target protein v, an integer k to obtain top-k predictions, an integer d as the depth of annotation neighborhoods.

Output: A set T of GO terms as v's annotations.

1. Extract sub-graph Γ' from Γ that consists of v, and v's neighbors within distance d.
2. Construct annotation neighborhood γ_v from the protein graph Γ'.
3. For every GO term t, place annotation neighborhoods of t into set S_t.
4. For each S_t, compute $f_t(v) = \log[sim(\gamma_v, S_t)] + \log[n_t]$, and add $<t, f_t(v)>$ pair in a scoring table S.
5. Sort S by its scores in descending order. Place top k terms in S into set T, and return T.

Fig. 2.1. Sketch of our protein annotation prediction algorithm

The following section describes the construction of S_t and the computation of $sim(\gamma_v, S_t)$.

2.3 Computation of Similarity with an Annotation Neighborhood Set

In this section we describe a procedure to compute the similarity of an annotation neighborhood γ_v to the annotation neighborhood set S_t of a GO term t. Similarity between γ_v and S_t is maximized when a sub-graph of γ_v is found to be conserved (i.e., recurring in a slightly modified form) among the annotation neighborhoods in S_t. Thus, in order to detect whether a sub-graph of γ_v is recurring in S_t, we need to know the correspondence of every vertex and every edge of γ_v to the vertices and edges of annotation neighborhoods in S_t. We represent such correspondences by *pairwise alignments*.

Def'n. *(Pairwise alignment):* Given two annotation patterns γ_1 and γ_2, for every protein u of γ_1, pairwise alignment $PA(u, \gamma_1, \gamma_2)$ maps u to a set of vertices in γ_2.

Initially, suppose that annotation neighborhood set S_t of term t, an annotation neighborhood γ_v, and pairwise alignments between γ_v and every annotation neighborhood in S_t are given. We compute the similarity between γ_v and S_t as the sum of weights of edges that are conserved among annotation neighborhoods of t:

$$sim(\gamma_v, S_t) = \sum_{\langle m,n \rangle \in \gamma_v} tot\_conserve(\langle m, n \rangle, S_t) / \sum_{\langle m,n \rangle \in \gamma_v} W[m, n], \qquad (6)$$

where $\langle m, n \rangle$ is an edge of γ_v, $W[m, n]$ is the weight of the edge $\langle m, n \rangle$, and $tot\_conserve(\langle m, n \rangle, S_t)$ measures how well the edge $\langle m, n \rangle$ is conserved among the annotation neighborhoods in S_t:

$$tot\_conserve(\langle m, n \rangle, S_t) = \sum_{\kappa \in S_t} conserve(\langle m, n \rangle, \kappa) / n_t. \qquad (7)$$

κ is an annotation neighborhood in S_t, and $conserve(\langle m, n \rangle, \kappa)$ measures how well an edge is conserved in a given annotation neighborhood, and n_t is the number of annotation neighborhoods of t. Basically an edge $\langle m, n \rangle$ is conserved in an annotation neighborhood κ if there is an edge $\langle p, q \rangle$ where the labels of p and m, and the labels of q and n are similar, and weights of edges are also arithmetically close. Suppose that $GOSim(l(u), l(u'))$ computes the GO-based similarity between two label sets (i.e., GO terms). Then $conserve(\langle m, n \rangle, \kappa)$ is defined as follows.

$$conserve(\langle m, n \rangle, \kappa) = MAX_{\langle p,q \rangle \in \kappa; p \in PA(m, \gamma_v, \kappa), q \in PA(n, \gamma_v, \kappa)}(edgeSim(\langle m, n \rangle, \langle p, q \rangle)). \qquad (8)$$

Finally, similarity between two edges is defined as follows.

$$edgeSim(\langle m, n \rangle, \langle p, q \rangle) = GOSim(l(m), l(p))GOSim(l(n), l(q))(W[m, n] + W[p, q])/2. \qquad (9)$$

2.4 GO-Based Similarity

Similarity between two GO terms (i.e., vertex label symbols) is more than a simple binary value corresponding to match or mismatch between GO terms. Intuitively, GO terms that annotate a large number of proteins offer little biological insight into the functionality of proteins, whereas GO terms that annotate few proteins are more specific, yet informative. The probability of observing a GO term t can be stated as $Prob(t) = n_t / n_V$, where n_t is the number of proteins that t annotates, n_V is the number of all annotated proteins. Then the *information content* $IC(t)$ of term t is denoted as $IC(t)=-\log(Prob(t))$ [CSC07]. In other words, the more specific a GO term is, the higher IC value it gets.

GO annotations of proteins constitute the nodes of the GO ontology; thus, a protein with more than one GO annotation will have more than one node in the "induced" GO ontology graph.

Def'n *(induced GO ontology graph for a protein)*: Given a protein P with its multiple annotation set S, the *induced GO graph* consists of those nodes in the GO corresponding to the annotations in S, and all of their ancestors (due to the true path rule which states that any protein annotated with the GO term g is also annotated with all the terms that are the ancestors of g in the GO ontology).

The common nodes in two induced GO ontology graphs of two proteins constitute the *annotation intersection graph* of the proteins. The GO ontology graph induced from the union of two proteins' annotations is the *annotation union graph* of the two proteins. Then the GO-based similarity between two annotation neighborhood vertices is expressed as follows.

$$GOSim\ (l(u),\ l(v)) = GOSim\ (T1,\ T2) = \Sigma_{t\in T1\cap T2}\ IC(t)\ /\ \Sigma_{t\in T1\cup T2}\ IC(t), \qquad (10)$$

where $T1$ and $T2$ denote induced GO graphs of the GO term sets $l(u)$ and $l(v)$ of two proteins u and v, respectively. $T1\cap T2$ is the annotation intersection graph, and $T1\cup T2$ is the annotation union graph of $T1$ and $T2$, respectively.

Example 2.2. In Figure 2.2, from left to right, graph (a) illustrates an example GO term hierarchy. Labels of the first graph are GO terms and number of their annotations. Graph (b) is the same with the graph (a), but its labels display information contents of GO terms corresponding to the frequencies in the first graph. Graphs (c) and (d) are annotation intersection and union graphs of annotations $T1$ and $T2$, respectively. As a result, $GOSim(\{c,d\}, \{b,f\}) = GOSim(\{a,b,c,d\}, \{a,b,d,f\}) = .72/2.11 = .34$.

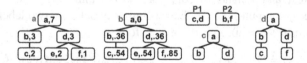

Fig. 2.2. Examples of GO ontology, (a) term frequencies, (b) term information content, (c) annotation intersection graph, and (d) annotation union graph

2.5 Mapping Annotation Neighborhoods Via Pairwise Rooted-Graph Alignment

This section describes our *pairwise (graph) alignment* algorithm that finds a mapping PA from the vertices of one annotation neighborhood to the other. PGA algorithm is employed to provide the pairwise alignment required by annotation neighborhood similarity computation (See Section 2.3).

Given two annotation neighborhoods γ_1 and γ_2, PGA algorithm aims to find the optimal mapping PA from the vertices of γ_1 to the vertices of γ_2, so that (a) the mapping preserves edges as much as possible, i.e., edge $\langle m, n \rangle$ in γ_1 corresponds to an edge $\langle p \in PA(m), q \in PA(n) \rangle$ in γ_2, and (b) the mapping is aware of vertex labels, i.e., value of $edgesim(\langle m,n \rangle, \langle p \in PA(m), q \in PA(n) \rangle)$ is subject to be maximized. Since the graph alignment problem is NP-complete [RC77], we have developed an efficient heuristic that computes an alignment between given two annotation patterns in quadratic time, by ignoring (some) structural properties of annotation neighborhoods. We employ an alignment scoring function $align(u, v)$ to find the best matching vertex v in γ_2 from a given vertex u in γ_1.

Def'n *(relationship between annotation neighborhood vertices):* Suppose that u is a protein in the annotation neighborhood γ. For another protein v in γ; u is an *ancestor* of v if $rank(u, \gamma) < rank(v, \gamma)$, u is a *parent* of v if $rank(u, \gamma) = rank(v, \gamma)-1$, u is a *successor* of v if $rank(u, \gamma) > rank(v, \gamma)$, u is a *child* of v if $rank(u, \gamma) = rank(v, \gamma)+1$, and u is a *sibling* of v if $rank(u, \gamma) = rank(v, \gamma)$. The vertices with highest ranks in an annotation neighborhood are called the *leaves*.

Let v_i and v_j be two vertices in different annotation neighborhoods. Also let v_m be a child of v_i, and v_n be a child of v_j. $W[v_i, v_m]$ is the weight of edge $\langle v_i, v_m \rangle$. We consider three cases for mapping v_i to v_j:

Case 1: v_i is mapped to v_j, and their children are mapped to each other, recursively. In this case, label similarity $GOSim(v_i, v_j)$ is directly added to the alignment score. Every child v_m of v_i is mapped to a child v_n of v_j. For each child v_m of v_i, we find the child v_n of v_j that gives the best alignment score:

Case 2: v_i is mapped to a child v_n of v_j. Since this case skips a vertex (i.e., v_j) and maps v_i to a protein with higher rank, we penalize the alignment score by some factor Φ.

Case 3: v_j is mapped to a child v_m of v_i. This step is the same with Case 2 by exchanging v_i and v_j.

Note that we require two root vertices to be aligned by case 1 only, so that one of them is not aligned on a child of the other even when it would generate a higher alignment score. This constraint ensures that the roots of the annotation neighborhoods of a GO term correspond to each other.

During the alignment of annotation neighborhoods, alignment of the same subgraph pairs may be repeated several times. To prevent re-computations of alignments, we apply a bottom-up dynamic programming approach, and compute $align(u,v)$ without recursion. Our alignment algorithm is shown in Figure 2.3.

The alignment algorithm finds alignment scores for all pairs of vertices in the input annotation neighborhoods. We employ the bottom-up dynamic programming approach, i.e., we start computing the alignment scores of leaf vertices, and then continue iteratively aligning vertices with higher ranks, and finally aligning the roots. During the alignment of two vertices v_i and v_j, only the children of v_i and v_j whose alignment scores are already computed at a previous iteration contribute to the alignment score of v_i and v_j. As a heuristic, we ignore edges between siblings, and do not consider ancestors of vertices while computing the alignment score. Edges between siblings do not connect any vertex to the root via some shortest path. Employing only the shortest path edges to represent functional relationship between genes and proteins is a common approach [Wag01, ZCA07]. As a result, our algorithm requires $O(|\gamma_1| \times |\gamma_2| \times \varepsilon^2)$ (where ε is the expected number of a vertex's children) space and time to find an optimal mapping producing the highest alignment score. Note that there may be different alignments with the same maximum score. Our algorithm arbitrarily generates one of the optimal alignments.

Algorithm 2: Pairwise Annotation Neighborhood Alignment
Input: Two annotation neighborhoods γ_1 and γ_2.
Output: Two mappings $PA(v_m, \gamma_1, \gamma_2)$ and $PA(v_n, \gamma_2, \gamma_1)$ between vertices of γ_1 and γ_2, and γ_2 and γ_1, respectively.

1. Prepare three matrices M, C, and D of size $|\gamma_1| \times |\gamma_2|$, where $|\gamma_1|$ and $|\gamma_2|$ are number of vertices in input annotation patterns γ_1 and γ_2, respectively. Each entry of M, C, and D is a real number corresponding to alignment scores, a binary value corresponds whether the alignment is done via case 1, and a list of integer pairs that correspond to mapped children. Rows and columns of these matrices correspond to vertices of annotation neighborhoods ordered by their distances to the roots.

2. Fill matrices by starting from the lower right hand corner in the matrix:
   for i = $|\gamma_1|$ downto 1 do
      for j = $|\gamma_2|$ downto 1 do
      $align_1(v_i, v_j) = GOSim(l(v_i), l(v_j)) + \sum_{v_m} MAX_{v_n} M_{m,n} W[v_i, v_m] W[v_j, v_n]$, //case 1
         where v_m is a child of v_i and v_n is a child of v_j
      $align_2(v_i, v_j) = MAX_{v_n}(M_{i,n} W[v_i, v_n]/\Phi)$, // case 2 (or 3)
         where v_m is a child of v_i and v_n is a child of v_j
      $align_3(v_i, v_j) = align_2(v_j, v_i)$
      if (i=1, j=1) then $align(v_i, v_j) = align_1(v_i, v_j)$. // roots are being aligned.
      else $align(v_i, v_j) = MAX[align_1(v_i, v_j), align_2(v_i, v_j), align_3(v_i, v_j)]$.
      $M_{i,j} = align(v_i, v_j)$
      if $align(v_i, v_j) = align_1(v_i, v_j)$ then $C_{i,j}$ = true //if case 1 alignment
      if ($C_{i,j}$=true) then $D_{i,j}$ is the list of mappings between children of v_i and v_j.
      else $D_{i,j}$ is assigned the following: //case 2 or case 3 alignments
         if $align(v_i, v_j) = align_2(v_i, v_j)$, $D_{i,j} = \{<i,m>\}$ (i.e., v_i is mapped to child v_m of v_j)
         else if $align(v_i, v_j) = align_3(v_i, v_j)$, $D_{i,j} = \{<n,j>\}$ (i.e., v_j is mapped to child v_n of v_i)
      end for // j loop
   end for // i loop

3. Trace back the matrices, and find vertex mappings that result in the highest alignment scores.
   Push vertex pair <1,1>, i.e., mapping of roots, to stack Q.
   while Q is not empty do
      Pop a vertex mapping <m,n> from Q.
      if ($C_{m,n}$) then add mappings $PA(v_m, \gamma_1, \gamma_2) \rightarrow v_n$, and $PA(v_n, \gamma_2, \gamma_1) \rightarrow v_m$.
      Push all vertex pairs in $D_{i,j}$ to Q.
   end while // while loop

4. Return $PA(v_m, \gamma_1, \gamma_2)$, and $PA(v_n, \gamma_2, \gamma_1)$.

Fig. 2.3. Sketch of our pairwise annotation neighborhood alignment algorithm

3 Results

3.1 Data Preparation

For our experiments, we built protein interaction networks from organism-specific protein interaction information provided by the BioGrid database [SB+06]. We employed three datasets, namely YEAST (*Saccharomyces cerevisiae*), WORM (*Caenorhabditis elegans*), and FLY (*Drosophila melanogaster*). YEAST, WORM, and FLY datasets contain 5197, 2778, and 7542 proteins, and 70772, 4351, and 25325 interactions, respectively. In addition, we employed the dataset of Lee et al. [LR+02] to assign weights to interactions in the YEAST dataset. We downloaded GO annotations of proteins from the gene ontology website [GOW].

3.2 Measuring Prediction Accuracy

Having prepared the datasets, we evaluated our pattern-based annotation prediction (PAP) method. We compare our method to the correlation mining (CM) technique [KOY06] and the neighbor counting (NC) technique [SUF00]. CM method is chosen to illustrate how much improvement is gained by the utilization of additional neighborhood information, since CM employs only direct neighbors of proteins, and is shown to have reasonable accuracy in comparison with other methodologies [KOY06]. NC method is chosen as a baseline in order to contrast with its assumption that interacting protein pairs have common annotations [SUF00]. Given a protein interaction network, and annotations of proteins, these techniques generate a prediction set of GO terms that are likely to annotate a target protein. We evaluate the prediction accuracy of each technique via cross-validation. We remove annotations of each protein (with known annotations), and then run one of these techniques on the protein interaction network to obtain a prediction set P for the protein. Then, we compare P with the true annotation set T of the protein, and compute the precision and recall of the predictions. To achieve high prediction accuracy, the technique should have high precision and recall values. Since there is a tradeoff between having high precision and high recall, we evaluate the accuracy of different techniques by the F-values of predictions, instead. We employ the standard definitions of precision and recall [KOY06], and F-value is defined [SBH97] as the harmonic mean of precision and recall of a prediction set.

For NC, CM, and our technique PAP, we computed the F-values of GO term predictions on FLY, WORM, and YEAST datasets. For this experiment, we generate annotation neighborhoods of depth 1, i.e., we consider only the first level neighborhood of proteins. Impact of different depth values on the prediction accuracy are evaluated in Section 3.5.

We computed the F-value for each k value in top-k prediction tests. To sum up the prediction results, for each individual protein, we picked the k value that produces the highest F-value for that protein. Therefore F-values of techniques represent the highest possible accuracy of the technique, rather than the accuracy specific to the value of k. In table 3.1, we list the F-values of NC, CM, and PAP techniques on three different datasets, employing the molecular functionality GO annotations of proteins. Our method PAP performs better than NC and CM, in FLY and WORM datasets. In the YEAST dataset, although F-values are very close to each other, best prediction accuracy is obtained by the NC method.

In this experiment, we also merged data from the three species to observe how the prediction accuracy changes. We tested each technique on FLY+YEAST and FLY+YEAST+WORM (ALL) combinations, using molecular function annotations only. See Table 3.2 for the results of this experiment. The accuracy of our method does not decrease by the integration of cross-species information. In comparison, accuracies of NC and CM techniques decrease. Next, we employ the FLY dataset, and test the three techniques for different k values (picking the top k GO terms with highest scores). See Figure 3.3 for a plot of F-values against increasing values of k. Our method generates F-values higher than NC and CM techniques for every value of k, except k =1.

3.3 Ontology Comparison

Next, we test the accuracy of PAP, NC, and CM techniques on FLY and WORM datasets for Biological Process (BP), Molecular Function (MF), and Cellular Component (CC) sub-ontologies of GO. Tables 3.4(a-c) display the results of this experiment. We find that all three techniques produce best results on the CC ontology. We explain this observation as follows. Protein-protein interactions usually occur in the same cellular location; therefore protein interaction partners are usually annotated by the same CC annotations. Direct annotation methods (e.g., NC), can correctly transfer a CC term from an interaction partner of a protein. Similarly, our technique also produces its most accurate annotations for the CC ontology. An exception is the FLY dataset, where our method performs better with MF ontology in comparison with BP and CC ontologies. In FLY and WORM datasets, our method annotates MF terms better than NC and CM techniques. In these datasets, MF terms are not frequently annotated to both partners of interactions (low accuracy of NC) and the only way of correctly predicting an MF annotation is to propagate it from a distant (or disconnected) protein via annotation neighborhood similarity (see Section 2.1). An exception to this is the YEAST dataset, which has the largest number of interactions per protein (in comparison with FLY and WORM datasets); therefore annotation neighborhoods of sticky proteins (i.e., proteins with a large number of interactions) generate noise.

Table 3.1. Comparison of NC, CM and PAP on different organism specific interaction datasets

	FLY	WORM	YEAST
NC	39.6%	41.3%	**80.7%**
CM	54.2%	55.8%	76.3%
PAP	**70.7%**	**79.2%**	78.3%

Table 3.2. Accuracy of NC, CM, and PAP in cross-species interaction datasets.

	FLY+YEAST	ALL
NC	74.7%	70.0%
CM	71.9%	68.6%
PAP	**78.3%**	**78.9%**

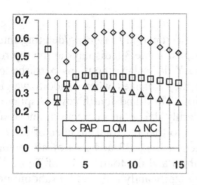

Fig. 3.3. Accuracy of NC, CM, and PAP for different values of k in top-k prediction experiments

Table 3.4(a-c). Ontology comparison on FLY (a), WORM (b), and YEAST (c) datasets

FLY	MF	BP	CC
NC	39.6%	41.5%	46.2%
CM	54.2%	50.2%	58.6%
PAP	**70.7%**	**62.7%**	**60.6%**

WORM	MF	BP	CC
NC	41.3%	44.3%	64.7%
CM	55.8%	59.2%	80.4%
PAP	**79.2%**	**84.6%**	**93.0%**

YEAST	MF	BP	CC
NC	**80.7%**	79.0%	89.7%
CM	76.3%	67.6%	86.6%
PAP	78.3%	**91.2%**	**92.4%**

3.4 Pattern Size and Gap Penalty Factor

In this experiment we observe how our method performs when we limit the size of annotation neighborhoods. We employed the FLY dataset, and MF annotations only. When the depths of annotation neighborhoods are limited, we employ only the nodes whose ranks are lower than or equal to the depth threshold (i.e., parameter d in Algorithm 1). We limit the maximum number of children that a vertex can have during pairwise alignment. When the number of children is limited to a constant number, say n, we compute the alignment scores of two nodes by employing their only top-n best aligning children. Our results show that increasing the number of children increases the accuracy of the method (See Table 3.5). In addition, we observe that increasing the depth improves the prediction accuracy (See Table 3.6). Finally when we test the impact of the value of Φ, namely the gap penalty factor (see Section 2.5), we find that the optimal value of Φ is between 2 and 3 (See Table 3.7).

Table 3.5. Accuracy of PAP by fixed depth (i.e., 1) and varying number of vertex children

#Children	Accuracy
1	59.8%
4	70.7%
10	71.0%
20	71.2%

Table 3.6. Accuracy of PAP by fixed number of vertex children (i.e., 1) and varying depth

Depth	Accuracy
1	59.8%
2	60.3%
3	62.2%
4	63.3%

Table 3.7. Accuracy of PAP by different Φ values, when depth = 4, #children =1

Φ	Accuracy
1	60.5%
2	63.3%
3	62.5%
5	60.4%

Table 3.8. Robustness of PAP in different perturbation conditions

	WORM	FLY
Original	79.2%	70.7%
Weights	64.4%	67.4%
33% pert.	77.1%	72.8%
67% pert.	66.4%	65.0%

3.5 Robustness Test

Next, we test the robustness of our approach by (a) randomly inserting/deleting inter-actions, and (b) assigning random weights on edges. Table 3.8 displays at the top row the accuracy of our technique on WORM and FLY datasets without any perturbation. Second row of this table shows how the accuracy changes when random weights between 0 and 1 are assigned to each interaction. Third and fourth rows display re-sults of our perturbation experiments. We randomly delete an interaction, and insert another one between a pair of random proteins (so that the total number of interac-tions remain unchanged). Perturbation of a fraction (i.e., .33 and .67) of all interac-tions changes the accuracy as displayed in the third and the fourth rows of Table 3.8. We find that the accuracy of our method is not significantly affected by random per-turbations. Thus, we conclude that our method is robust to false positives in the inter-action data. More interestingly, the accuracy of our method using the FLY dataset is actually improved by perturbation, similar to the results of Chua et al. [CSW07].

4 Discussion

In this paper, we have proposed a pattern-based function annotation framework which formulates protein function prediction as a (multi-class) classification problem. Es-sentially, we define a classification function via probabilistic reasoning. Then, we translate the protein function prediction task to pattern-detection and pattern-matching tasks. We detect annotation neighborhoods, i.e., patterns that appear in the protein interaction neighborhood of proteins with a particular functionality. Then we match the annotation patterns to neighborhoods of proteins to be annotated. We provide a pairwise graph alignment algorithm that ignores some structural elements of annota-tion neighborhoods in order to reduce the computational complexity of pattern match-ing. In addition, since we allow a vertex (i.e., a protein) to have more than one label (i.e., GO term), pattern matching task gains an additional level of complexity. We reduce this complexity by exploiting the GO hierarchy and computing the similarity between two label sets in a single operation, rather than comparing every possible subset of vertex labels.

In our experiments, we have applied our methodology to predict annotations of proteins from yeast, fly, and worm protein interaction networks. We have shown that our approach effectively predicts protein annotations, and performs with better accu-racy in comparison to the NC and the CM techniques, despite the fact that the accu-racy of our method reduces in a densely connected network, e.g., the yeast protein interaction network. In addition, we have shown that our approach is robust to false positives in protein interaction networks.

5 Related Work

Our work differs from the previous work in two aspects. First, the previous research on protein function prediction focuses on a particular protein function set, and builds models based on the direct interactions of proteins [TD+03, SL03, DT+04, SUF00, HN+01, VF+03]. In comparison, we process protein interaction network as a whole in order to locate patterns of annotation. We assign a GO term annotation to a protein P if the annotation is implied by the existing annotation patterns of proteins that interact with P. Our prediction of a GO term requires a statistically significant usage of that GO term in a particular pattern, reducing the effects of false interactions/false annotations as long as the corrupt data does not span a major portion of the interaction data, causing our framework to generate fake patterns.

Our method is closely related to alignment-based graph classification approaches [WH+07]. Other graph classification methods include kernel-based [KI02], and frequent pattern-based [CK+07] classification. Weskamp *et al.* [WH+07] provides an alignment-based graph classification methodology to classify a given graph with respect to its similarity to consensus of a group of graphs of the same class. Our approach is not classifying graphs as a whole, but prediction of vertex labels based on the similarity of a sub-graph that contain an unclassified vertex, to another sub-graph where labels of vertices are known. In addition, we provide heuristics to reduce the complexity of pairwise graph alignments, and we consider multiple vertex labels (i.e., GO terms).

Acknowledgement

We would like to thank Dr Mehmet Koyuturk for helpful discussions. This research is supported by the US National Science Foundation grants DBI-0218061 and CNS-0551603.

References

[BH02]   Bader, G.D., Hogue, C.W.: Analyzing yeast protein–protein interaction data obtained from different sources. Nat. Biotechnol. 20, 991–997 (2002)

[BH03]   Bader, G.D., Hogue, C.W.V.: An automated method for finding molecular complexes in large protein interaction networks. BMC Bioinformatics 4, 2 (2003)

[BH06]   Brohée, S., van Helden, J.: Evaluation of clustering algorithms for protein-protein interaction networks. BMC Bioinformatics 7, 488 (2006)

[BL04]   Berg, J., Lässig, M.: Local graph alignment and motif search in biological networks. PNAS 101, 14689–14694 (2004)

[BL06]   Berg, J., Lässig, M.: Cross-species analysis of biological networks by Bayesian alignment. PNAS 103, 10967–10972 (2006)

[BWD96] Blatt, M., Wiseman, S., Domany, E.: Superparamagnetic clustering of data. Phys. Rev. Lett. 76(18), 3251–3254 (1996)

[CK+07]   Cakmak, A., Kirac, M., Reynolds, M.R., Ozsoyoglu, Z.M., Ozsoyoglu, G.: Gene Ontology-Based Annotation Analysis and Categorization of Metabolic Pathways. SSDBM 33 (2007)

[CSC07]   Couto, F., Silva, M., Coutinho, P.: Measuring Semantic Similarity between Gene Ontology Terms. DKE 61, 137–152 (2007)

[CSW07]   Chua, H.N., Sung, W.K., Wong, L.: Using indirect protein interactions for the prediction of Gene Ontology functions. BMC Bioinformatics 8(Suppl 4), 8 (2007)

[D00]   Van Dongen, S.: Graph clustering by flow simulation. PhD thesis Centers for mathematics and computer science (CWI), University of Utrecht (2000)

[DT+04]   Deng, M., Tu, Z., Sun, F., Chen, T.: Mapping gene ontology to proteins based on protein–protein interaction data. Bioinformatics 20, 895–902 (2004)

[F06]   Friedberg, I.: Automated protein function prediction—the genomic challenge. Briefings in Bioinformatics 7(3), 225–242 (2006)

[GO]   Consortium, Gene Ontology: The GO database and informatics resource. Nucleic Acids Res, 32, D258-D261 (2004)

[GOW]   Gene Ontology Annotations Database, http://www.geneontology.org/GO.current.annotations.shtml

[GR03]   Goldberg, D.S., Roth, F.: Assessing experimentally derived interactions in a small world. PNAS 100(8), 4372–4376 (2003)

[HN+01]   Hishigaki, H., Nakai, K., Ono, T., Tanigami, A., Takagi, T.: Assessment of prediction accuracy of protein function from protein–protein interaction data. Yeast 18, 523–531 (2001)

[HY+05]   Hu, H., Yan, X., Huang, Y., Han, J., Zhou, X.J.: Mining coherent dense subgraphs across massive biological networks for functional discovery. Bioinformatics 21(Suppl 1), i213–i221 (2005)

[KF+03]   King, O.D., Foulger, R.E., Dwight, S.S., White, J.V., Roth, F.P.: Predicting gene function from patterns of annotation. Genome Res 13(5), 896–904 (2003)

[Ki04]   Kitano, H.: Biological Robustness. Nat Genet 5, 826–838 (2004)

[KI02]   Kashima, H., Inokuchi, A.: Kernels for Graph Classification. In: ICDM 2002 (AM-2002) (2002)

[KGS04]   Koyutürk, M., Grama, A., Szpankowski, W.: An efficient algorithm for detecting frequent subgraphs in biological networks. Bioinformatics 20(Suppl 1), i200–i207 (2004)

[KOY06]   Kirac, M., Ozsoyoglu, G., Yang, J.: Annotating proteins by mining protein interaction networks. Bioinformatics 22, e260–e270 (2006)

[KPJ04]   King, A.D., Przulj, N., Jurisica, I.: Protein complex prediction via cost-based clustering. Bioinformatics 20(17), 3013–3020 (2004)

[LD+04]   Lee, I., Date, S.V., Adai, A.T., Marcotte, E.M.: A Probabilistic Functional Network of Yeast Genes. Science 306(5701), 1555–1558 (2004)

[LK03]   Letovsky, S., Kasif, S.: Predicting protein function from protein–protein interaction data: a probabilistic approach. Bioinformatics 19, i197–i204 (2003)

[LR+02]   Lee, et al.: Transcriptional Regulatory Networks in Saccharomyces cerevisiae. Science 298(5594), 799–804 (2002)

[MK+02]   von Mering, C., Krause, R., Snel, B., Cornell, M., Oliver, S.G., Fields, S., Bork, P.: Comparative assessment of large-scale data sets of protein–protein interactions. Nature 417, 399–403 (2002)

[MS+02]   Milo, R., Shen-Orr, S., Itzkovitz, S., Kashtan, N., Chklovskii, D., Alon, U.: Network Motifs: Simple Building Blocks of Complex Networks. Science 298, 824–827 (2002)

[NJ+05]   Nabieva, E., Jim, K., Agarwal, A., Chazelle, B.: Whole-proteome prediction of protein function via graphtheoretic analysis of interaction maps. Bioinformatics 21(Suppl. 1), i302–i310 (2005)

[OM+02]   Orr, S.S., Milo, R., Mangan, S., Alon, U.: Network motifs in the transcriptional regulation network of Escherichia coli. Nat Genet. 31, 64–68 (2002)

[PK+07]   Pandey, J., Koyuturk, M., Kim, Y., Szpankowski, W., Subramaniam, S., Grama, A.: Functional annotation of regulatory pathways. Bioinformatics 23(13), i377–i386 (2007)

[R02]   Rost, B.: Enzyme function less conserved than anticipated. J Mol. Biol. 318, 595–608 (2002)

[RC77]   Ronald, C.R., Corneil, D.G.: The graph isomorphism disease. Journal of Graph Theory 1(4), 339–363 (1977)

[SB+06]   Stark, C., Breitkreutz, B.J., Reguly, T., Boucher, L., Breitkreutz, A., Tyers, M.: BioGRID: a general repository for interaction datasets. Nucleic Acids Res. 34, D535–D539 (2006)

[SBH97]   Shaw, W.M.J., Burgin, R., Howell, P.: Performance standards and evaluations in IR test collections: Vector-space and other retrieval models. Info Proc. Manag. 33(1), 15–36 (1997)

[SGD]   Saccharomyces Genome Database (SGD), http://www.yeastgenome.org/

[SL03]   Samanta, M.P., Liang, S.: Predicting protein functions from redundancies in large-scale protein interaction networks. PNAS 100, 12579–12583 (2003)

[SS+05]   Sharan, R., Suthram, S., Kelley, R.M., Kuhn, T., McCuine, S., Uetz, P., Sittler, T., Karp, R., Ideker, T.: Conserved patterns of protein interaction in multiple species. PNAS 102, 1974–1979 (2005)

[SSH03]   Saito, R., Suzuki, H., Hayashizaki, Y.: Construction of reliable protein–protein interaction networks with a new interaction generality measure. Bioinformatics 19(6), 756–763 (2003)

[SUF00]   Schwikowski, B., Uetz, P., Fields, S.: A network of protein–protein interactions in yeast Nat Biotechnol. 18, 1257–1261 (2000)

[SUS07]   Sharan, R., Ulitsky, I., Shamir, R.: Network-based prediction of protein function. Mol. Sys. Bio. 3, 88 (2007)

[TD+03]   Troyanskaya, O.G., Dolinski, K., Owen, A.B., Altman, R.B., Botstein, D.: A Bayesian framework for combining heterogeneous data sources for gene function prediction (in Saccharomyces cerevisiae). PNAS 100(14), 8348–8353 (2003)

[TL+04]   Tong, et al.: Global Mapping of the Yeast Genetic Interaction Network. Science 303(5659), 808–813 (2004)

[TS03]   Tian, W., Skolnick, J.: How well is enzyme function conserved as a function of pairwise sequence identity? J Mol. Biol. 333, 863–882 (2003)

[V05]   Valencia, A.: Automatic annotation of protein function. Curr. Opin. Struct. Biol. 15, 267–274 (2005)

[VF+03]   Vazquez, A., Flammini, A., Maritan, A., Vespignani, A.: Global protein function prediction from protein–protein interaction networks. Nat. Biotechnol. 21, 697–700 (2003)

[Wag01]   Wagner, A.: The Yeast Protein Interaction Network Evolves Rapidly and Contains Few Redundant Duplicate Genes. Mol. Biol. Evol. 18(7), 1283–1292 (2001)

[WH+07]   Weskamp, N., Hüllermeier, E., Kuhn, D., Klebe, G.: Multiple graph alignment for the structural analysis of protein active sites. IEEE/ACM Trans. Comput. Biol. Bioinform. 4(2), 310–320 (2007)

[YL+04]   Yu, H., Luscombe, N.M., Lu, H.X., Zhu, X., Xia, Y., Han, J.J., Bertin, N., Chung, S., Vidal, M., Gerstein, M.: Annotation Transfer Between Genomes: Protein–Protein Interologs and Protein–DNA Regulogs. Genome Res. 14, 1107–1118 (2004)

[ZCA07]   Zhao, X., Chen, L., Aihara, K.: Gene Function Prediction with the Shortest Path in Functional Linkage Graph. OSB, 68–74 (2007)

[ZKW02]   Zhou, X., Kao, M.C.J., Wong, W.H.: From the Cover: Transitive functional annotation by shortest-path analysis of gene expression data. PNAS 99, 12783–12788 (2002)

Automatic Parameter Learning for Multiple Network Alignment

Jason Flannick[1], Antal Novak[1], Chuong B. Do[1], Balaji S. Srinivasan[2], and Serafim Batzoglou[1]

[1] Department of Computer Science, Stanford University, Stanford, CA 94305, USA
flannick@cs.stanford.edu
[2] Department of Statistics, Stanford University, Stanford, CA 94305, USA

Abstract. We developed Græmlin 2.0, a new multiple network aligner with (1) a novel scoring function that can use arbitrary features of a multiple network alignment, such as protein deletions, protein duplications, protein mutations, and interaction losses; (2) a parameter learning algorithm that uses a training set of known network alignments to learn parameters for our scoring function and thereby adapt it to any set of networks; and (3) an algorithm that uses our scoring function to find approximate multiple network alignments in linear time.

We tested Græmlin 2.0's accuracy on protein interaction networks from IntAct, DIP, and the Stanford Network Database. We show that, on each of these datasets, Græmlin 2.0 has higher sensitivity and specificity than existing network aligners. Græmlin 2.0 is available under the GNU public license at http://graemlin.stanford.edu.

1 Introduction

This paper describes Græmlin 2.0, a multiple network aligner with a novel scoring function, a fully automatic algorithm that learns the scoring function's parameters, and an algorithm that uses the scoring function to align multiple networks in linear time. Græmlin 2.0 significantly increases accuracy when aligning protein interaction networks and aids network alignment users by automatically adapting alignment algorithms to any network dataset.

Network alignment compares interaction networks of different species [1]. An interaction network contains nodes, which represent genes, proteins, or other molecules, as well as edges between nodes, which represent interactions. By comparing networks, network alignment finds conserved biological modules or pathways [2,3]. Because conserved modules are usually functionally important, network alignment research growth [1] has paralleled interaction network dataset growth [4,5].

Network alignment algorithms use a scoring function and a search algorithm. The scoring function assigns a numerical value to network alignments—high values indicate conservation. The search algorithm searches the set of possible network alignments for the highest scoring network alignment.

M. Vingron and L. Wong (Eds.): RECOMB 2008, LNBI 4955, pp. 214–231, 2008.

Most network alignment research has focused on pairwise network alignment search algorithms. PathBLAST uses a randomized dynamic programming algorithm to find conserved pathways [6] and uses a greedy algorithm to find conserved protein complexes [7]. MaWISh formulates network alignment as a maximum weight induced subgraph problem [8]. MetaPathwayHunter uses a graph matching algorithm to find inexact matches to a query pathway in a network database [9], and QNet exactly aligns query networks with bounded tree width [10]. Other network alignment algorithms use ideas behind Google's PageRank algorithm [11] or cast network alignment as an Integer Quadratic Programming problem [12]. Two network aligners can perform multiple network alignment. NetworkBLAST extends PathBLAST to align three species simultaneously [13]. Græmlin 1.0 can align more than 10 species at once [14].

Scoring function research has focused on various models of network evolution. MaWISh [8] scores alignments with a duplication-divergence model for protein evolution. Berg et. al. [15] perform Bayesian network alignment and model network evolution with interaction gains and losses as well as protein sequence divergences. Hirsh et. al. [16] model protein complex evolution with interaction gains and losses as well as protein duplications.

Despite these advances, scoring functions still have several limitations. First, existing scoring functions cannot automatically adapt to multiple network datasets. Because networks have different edge densities and noise levels, which depend on the experiments or integration methods used to obtain the networks, parameters that align one set of networks accurately might align another set of networks inaccurately.

Second, existing scoring functions use only sequence similarity, interaction conservation, and protein duplications to compute scores. As scoring functions use additional features such as protein deletions and paralog interaction conservation, parameters become harder to hand-tune.

Finally, existing evolutionary scoring functions do not apply to multiple network alignment. Existing multiple network aligners either have no evolutionary model (NetworkBLAST) or use heuristic parameter choices with no evolutionary basis (Græmlin 1.0).

In this paper, we first present a scoring function that addresses these limitations. We next present an algorithm that uses a training set of known alignments to automatically learn parameters for our scoring function. We then present an algorithm that uses our scoring function to perform approximate global network alignment in linear time. Finally, we present benchmarks comparing Græmlin 2.0, a new multiple network aligner that includes these three pieces, to existing network aligners.

2   Methods

2.1   Network Alignment Formulation

The input to multiple network alignment is d networks $G_1, \ldots, G_d$. Each network represents a different species and contains a set of nodes V_i and a set of edges

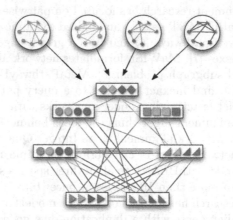

Fig. 1. A network alignment is an equivalence relation. In this example, four protein interaction networks are input to multiple alignment. A network alignment partitions proteins into equivalence classes (indicated by boxes).

E_i linking pairs of nodes. One common type of network is a protein interaction network, in which nodes represent proteins and edges represent interactions, either direct or indirect, between proteins.

A *multiple network alignment* is an equivalence relation a over the nodes $V = V_1 \cup \cdots \cup V_d$. An equivalence relation is transitive and partitions V into a set of disjoint equivalence classes [14]. A *local alignment* is a relation over a subset of the nodes in V; a *global alignment* [11] is a relation over all nodes in V. Figure 1 shows an example of an alignment of four protein interaction networks.

Network alignments have a biological interpretation. Nodes in the same equivalence class are functionally orthologous [17]. The subset of nodes in a local alignment represents a conserved module [2] or pathway.

A *scoring function* for network alignment is a map $s : \mathcal{A} \to \mathbb{R}$, where $\mathcal{A}$ is the set of potential network alignments of $G_1, \ldots, G_d$. The *global network alignment problem* is to find the highest-scoring global network alignment. The *local network alignment problem* is to find a set of maximally-scoring local network alignments.

In this paper, we restrict attention to global network alignment. Many ideas that apply to global network alignment also apply to local alignment. In addition, a local alignment algorithm can use global network alignment as a first step and then segment the global alignment into a set of local alignments [6,7].

2.2 Scoring Function

General Definition. Our scoring function computes "features" [18,19] of a network alignment. Formally, we define a vector-valued *feature function* $\mathbf{f} : \mathcal{A} \to \mathbb{R}^n$, which maps a global alignment to a numerical *feature vector*. More specifically, we define a node feature function $\mathbf{f}^N$ that maps equivalence classes to a feature vector and an edge feature function $\mathbf{f}^E$ that maps pairs of equivalence classes to a feature vector. We then define

$$\mathbf{f}(a) = \begin{bmatrix} \sum\limits_{[x] \in a} \mathbf{f}^N([x]) \\ \sum\limits_{\substack{[x],[y] \in a \\ [x] \neq [y]}} \mathbf{f}^E([x],[y]) \end{bmatrix} \qquad (1)$$

with the first sum over all equivalence classes in the alignment a and the second sum over all pairs of equivalence classes in a.

Given a numerical *parameter vector* $\mathbf{w}$, the score of an alignment a is $s(a) = \mathbf{w} \cdot \mathbf{f}(a)$. The *parameter learning problem* is to find $\mathbf{w}$. We discuss our parameter learning algorithm below.

The feature function isolates the biological meaning of network alignment. Our learning and alignment algorithms make no further biological assumptions. Furthermore, one can define a feature function for any kind of network. Our scoring function therefore applies to any set of networks, regardless of the meaning of nodes and edges.

Implementation for Protein Interaction Networks. We implemented a feature function that computes evolutionary events. We first describe our feature function for the special case of pairwise network alignment (the alignment of two networks), and we then generalize our feature function to multiple

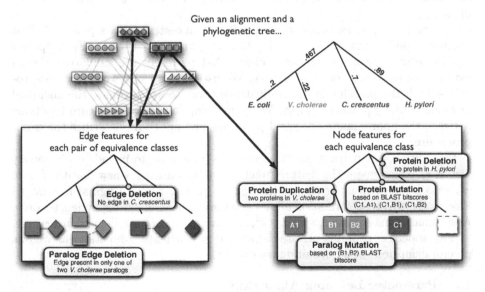

Fig. 2. Alignment feature functions compute evolutionary events. This figure shows the set of evolutionary events that our node and edge feature functions compute. We use a phylogenetic tree with branch lengths to determine the events. The appendix gives precise definitions of the evolutionary events.

network alignment. Figure 2 illustrates the evolutionary events our feature function computes.

Our pairwise node feature function computes the occurrence of the following four evolutionary events between the species in an equivalence class:

- *Protein deletion* is the loss of a protein in one of the two species.
- *Protein duplication* is the duplication of a protein in one of the two species.
- *Protein mutation* is the divergence in sequence of two proteins in different species.
- *Paralog mutation* is the divergence in sequence of two proteins in the same species.

Our pairwise edge feature function computes the occurrence of the following two evolutionary events between the species in a pair of equivalence classes:

- *Edge deletion* is the loss of an interaction between two pairs of proteins in different species.
- *Paralog edge deletion* is the loss of an interaction between two pairs of proteins in the same species.

The value of each event is one if the event occurs and zero if it does not. The entries in the feature vector are the values of the events.

We take two steps to generalize these pairwise feature functions to multiple network alignment. First, we use a phylogenetic tree to relate species and then sum pairwise feature functions over pairs of species adjacent in the tree, including ancestral species. Second, we modify the feature functions to include evolutionary distance.

Our pairwise feature functions generalize to ancestral species pairs. We first compute species weight vectors [20] for each ancestral species. Each species weight vector contains numerical weights that represent the similarity of each extant species to the ancestral species. We use these species weight vectors, together with the proteins in the equivalence class, to approximate the ancestral proteins in the equivalence class. We then compute pairwise feature functions between the approximate ancestral proteins. The appendix describes the exact procedure.

In addition, our pairwise feature functions generalize to include evolutionary distance. We augment the feature function by introducing a new feature $f_i \times b$, where b is the distance between the species pair, for each original feature f_i. Effectively, this transformation allows features to have linear dependencies on b. Additional terms such as $f_i \times b^2, f_i \times b^3, \ldots$ have more complex dependencies on b.

The appendix contains precise definitions of our feature function, as well as precise definitions of all evolutionary events.

2.3   Parameter Learning Algorithm

Inputs. Our algorithm to find **w** requires a *training set* of known alignments. The training set is a collection of m *training examples*; each training example i specifies a set of networks $\{G^{(i)} = G_1^{(i)}, \ldots G_d^{(i)}\}$ and their correct alignment $a^{(i)}$.

Our learning algorithm requires a *loss function* $\Delta : \mathcal{A} \times \mathcal{A} \to \mathbb{R}^+$. By definition, $\Delta(a^{(i)}, a)$ must be 0 when $a^{(i)} = a$ and positive when $a^{(i)} \neq a$ [21]. Intuitively, $\Delta(a^{(i)}, a)$ measures the distance of an alignment a from the training alignment $a^{(i)}$; the learned parameter vector should therefore assign higher scores to alignments with smaller loss function values.

To train parameters for our feature function, we used a training set of KEGG Ortholog (KO) groups [22]. Each training example contained the networks from a set of species, with nodes removed that did not have a KO group. The correct alignment contained an equivalence class for each KO group.

We also defined a loss function that grows as alignments diverge from the correct alignment $a^{(i)}$. More specifically, let $[x]_{a^{(i)}}$ denote the equivalence class of $x \in V^{(i)} = \bigcup_j V_j^{(i)}$ in $a^{(i)}$ and $[x]_a$ denote the equivalence class of x in a. We define $\Delta(a^{(i)}, a) = \sum_{x \in V^{(i)}} |[x]_a \setminus [x]_{a^{(i)}}|$, where $A \setminus B$ denotes the set difference between A and B. This loss function is proportional to the number of nodes aligned in a that are not aligned in the correct alignment $a^{(i)}$.

We experimented with the natural opposite of this loss function – the number of nodes aligned in the correct alignment $a^{(i)}$ that are not aligned in a. As expected, this alternate loss function resulted in a scoring function that aligned more nodes. We found empirically, however, that our original loss function was more accurate.

Theory. We pose parameter learning as a maximum margin structured learning problem. We find a parameter vector that solves the following convex program [21]:

$$\min_{\mathbf{w}, \xi_1, \ldots, \xi_m} \frac{\lambda}{2} \|\mathbf{w}\|^2 + \frac{1}{m} \sum_{i=1}^{m} \xi_i$$

$$\text{s.t. } \forall i, a \in \mathcal{A}^{(i)}, \mathbf{w} \cdot \mathbf{f}(a^{(i)}) + \xi_i \geq \mathbf{w} \cdot \mathbf{f}(a) + \Delta(a^{(i)}, a).$$

The constraints in this convex program encourage the learned $\mathbf{w}$ to satisfy a set of conditions: each training alignment $a^{(i)}$ should score higher than all other alignments a by at least $\Delta(a^{(i)}, a)$. The slack variables ξ_i are penalties for each unsatisfied condition. The objective function is the sum of the penalties with a regularization term that prevents overfitting. Given the low risk of overfitting the few free parameters in our model, we set $\lambda = 0$ for convenience. In more complex models with richer feature sets, overfitting can be substantially more severe when the amount of training data is limited; employing effective regularization techniques in such cases is a topic for future research.

We can show [21] that this constrained convex program is equivalent to the unconstrained minimization problem

$$c(\mathbf{w}) = \frac{1}{m} \sum_{i=1}^{m} r^{(i)}(\mathbf{w}) + \frac{\lambda}{2} \|\mathbf{w}\|^2, \tag{2}$$

where $r^{(i)}(\mathbf{w}) = \max_{a \in \mathcal{A}^{(i)}} \left(\mathbf{w} \cdot \mathbf{f}(a) + \Delta(a^{(i)}, a) \right) - \mathbf{w} \cdot \mathbf{f}(a^{(i)})$.

This objective function is convex but nondifferentiable [21]. We can therefore minimize it with subgradient descent [23], an extension of gradient descent to nondifferentiable objective functions.

A subgradient of equation (2) is [21]

$$\lambda \mathbf{w} + \frac{1}{m} \sum_{i=1}^{m} \left(\mathbf{f}(a_*^{(i)}) - \mathbf{f}(a^{(i)}) \right),$$

where $a_*^{(i)} = \arg \max_{a \in \mathcal{A}^{(i)}} \mathbf{w} \cdot \mathbf{f}(a) + \Delta(a^{(i)}, a)$ is the optimal alignment, determined by the loss function $\Delta(a^{(i)}, a)$ and current $\mathbf{w}$, of $G^{(i)}$.

Algorithm. Based on these ideas, our learning algorithm performs subgradient descent. It starts with $\mathbf{w} = 0$. Then, it iteratively computes the subgradient $\mathbf{g}$ of equation (2) at the current parameter vector $\mathbf{w}$ and updates $\mathbf{w} \leftarrow \mathbf{w} - \alpha \mathbf{g}$, where α is the *learning rate*. The algorithm stops when it performs 100 iterations that do not reduce the objective function. We set the learning rate to a small constant ($\alpha = 0.05$).

The algorithm for finding $\arg \max_{a \in \mathcal{A}^{(i)}} \mathbf{w} \cdot \mathbf{f}(a) + \Delta(a^{(i)}, a)$ is the inference algorithm. It is a global alignment algorithm with a scoring function augmented by Δ. Below we present an efficient approximate global alignment algorithm that we use as an approximate inference algorithm.

Our learning algorithm has an intuitive interpretation. At each iteration it uses the loss function Δ and the current $\mathbf{w}$ to compute the optimal alignment. It then decreases the score of features with higher values in the optimal alignment than in the training example and increases the score of features with lower values in

LEARN($\{G_1^{(i)}, \ldots, G_d^{(i)}, a^{(i)}\}_{i=1}^{m}$: training set , α : learning rate , λ : regularization)

```
1   var w ← 0 // the current parameter vector
2   var c* ← ∞ // a measure of progress
3   var w* ← w // the best parameter vector so far
4   while c* updated in last 100 iterations
5   do
6       var g ← 0 // the current subgradient
7       var c = 0 // the current objective function
8       for i = 1 : m
9       do // sum over all training examples
10          var a*⁽ⁱ⁾ = ALIGN(G₁⁽ⁱ⁾, ..., G_d⁽ⁱ⁾, w, Δ)
11          g ← g + f(a*⁽ⁱ⁾) − f(a⁽ⁱ⁾) // update the subgradient
12          c ← c + w · f(a*⁽ⁱ⁾) + Δ(a⁽ⁱ⁾, a*⁽ⁱ⁾) − w · f(a⁽ⁱ⁾) // update the margin
13       g ← (1/m)g − λw; c ← (1/m)c + (1/2)||w||² // add in regularization
14       if c < c*
15       then
16              c* ← c; w* = w // update the best parameter vector so far
17       w ← w − αg // update current parameter vector
18   return w*
```

Fig. 3. Our parameter learning algorithm

the optimal alignment than in the training example. Figure 3 shows our learning algorithm.

Our learning algorithm also has performance guarantees. If the inference algorithm is exact, and if the learning rate is constant, our learning algorithm converges at a linear rate to a small region surrounding the optimal **w** [24,21]. A bound on convergence with an approximate inference algorithm is a topic for further research.

2.4 Global Alignment Algorithm

Our global alignment algorithm serves two roles. It finds the highest scoring global alignment once the optimal parameter vector has been learned, and it performs inference as part of our learning algorithm.

We implemented a local hillclimbing algorithm for global alignment [25]. Our alignment algorithm is approximate but efficient in practice. It requires that the alignment feature function decomposes into node and edge feature functions as in equation (1).

Our alignment algorithm (Figure 4) iteratively performs updates of a current alignment. The initial alignment contains every node in a separate equivalence class. Our algorithm then proceeds in a series of iterations. During each iteration, it processes each node and evaluates a series of moves for each node:

- Leave the node alone.
- Create a new equivalence class with only the node.
- Move the node to another equivalence class.
- Merge the entire equivalence class of the node with another equivalence class.

For each move, our algorithm computes the alignment score before and after the move and performs the move that increases the score the most. Once our algorithm has processed each node, it begins a new iteration. It stops when an iteration does not increase the alignment score.

Our alignment algorithm performs inference as part of our learning algorithm. It can use any scoring function that decomposes as in equation (1). Therefore, to perform inference, we need only augment the scoring function with a loss function Δ that also decomposes into node and edge feature functions. The loss function presented above has this property.

Our alignment algorithm depends on the set of candidate equivalence classes to which processed nodes can move. As a heuristic, it considers as candidates only equivalence classes with a node that has homology (BLAST [26] e-value $< 10^{-5}$) to the processed node.

Our alignment algorithm also depends on the order in which it processes nodes. As a heuristic, it uses node scores—the scoring function with the edge feature function set to zero—to order nodes. For each node, our algorithm computes the node score change when it moves the node to each candidate equivalence class. It saves the maximum node score change for each node and then considers nodes in order of decreasing maximum node score change.

In practice, our alignment algorithm runs in linear time. To align networks with n total nodes and m total edges, our algorithm has b iterations that each

ALIGN($G_1, \ldots, G_d$: set of networks , w : parameter vector , Δ : optional loss function)
```
 1    var a ← an alignment with one equivalence class per node
 2    while true
 3    do
 4        var δ_t = 0 // the total change in score of this iteration
 5        for  each node p ∈ ∪_i G_i
 6        do
 7            var δ* ← 0 // best score
 8            var o* ← undef // best move
 9            for  each move o of node p
10            do
11                var a_t ← o(a) // alignment after move o
12                δ ← w · f(a_t) + Δ(a_t) − (w · f(a) + Δ(a)) // change in score after move o
13                if δ > δ*
14                    then
15                        δ* = δ; o* = o // new best move
16                a ← o*(a) // do best move on alignment
17                δ_t ← δ_t + δ* // update total change in score of this iteration
18        if δ_t = 0
19            then break
20    return w
```

Fig. 4. Our global alignment algorithm

process n nodes. For each node our algorithm computes the change in score when it moves the node to, on average, C candidate classes. Because the feature function decomposes as in equation (1), to perform each score computation our algorithm needs only to examine the candidate class, the node's old class, and the two classes' neighbors. Its running time is therefore $O(bC(n + m))$. Empirically, b is usually a small constant (less than 10). While C can be large, our algorithm runs faster if it only considers candidate classes with high homology to the processed node (BLAST e-value $\ll 10^{-5}$.)

3   Results

Experimental Setup. We tested our aligner on three different network datasets: IntAct [27], DIP [28], and the Stanford Network Database [29] (SNDB). We ran pairwise alignments of the human and mouse IntAct networks, yeast and fly DIP networks, *Escherichia coli* K12 and *Salmonella typhimurium* LT2 SNDB networks, and *E. coli* and *Caulobacter crescentus* SNDB networks. We also ran a three-way alignment of the yeast, worm, and fly DIP networks, and a six-way alignment of *E. coli*, *S. typhimurium*, *Vibrio cholerae*, *Campylobacter jejuni* NCTC 11168, *Helicobacter pylori* 26695, and *C. crescentus* SNDB networks.

We used KO groups [22] for our alignment comparison metrics. To compute each metric, we first removed all nodes in the alignment without a KO group and we then removed all equivalence classes with only one node. We then defined an equivalence class as *correct* if every node in it had the same KO group.

To measure specificity, we computed two metrics:

1. the fraction of equivalence classes that were correct (C_{eq})
2. the fraction of nodes that were in correct equivalence classes (C_{node})

To measure sensitivity, we computed two metrics:

1. the total number of nodes that were in correct equivalence classes (Cor)
2. the number of equivalence classes that contained k species, for $k = 2, \ldots, n$

We used cross validation to test Græmlin 2.0. For each set of networks, we partitioned the KO groups into ten equal sized test sets. For each test set, we trained Græmlin 2.0 on the KO groups not in the test set as described in the Methods section. We then aligned the networks and computed our metrics on only the KO groups in the test set. Our final numbers for a set of networks were the average of our metrics over the ten test sets.

To limit biases we used cross validation to test all aligners. For aligners other than Græmlin 2.0 we aligned the networks only one time. However, we did not compute our metrics on all KO groups at once; instead, we computed our metrics separately for each test set and then averaged the numbers.

As a final check that our test and training sets were independent, we computed similar metrics using Gene Ontology (GO) categories [30,13] instead of KO groups. We do not report the results of these tests because they showed no change in the relative performance of the aligners.

We compared Græmlin 2.0 to the local aligners NetworkBLAST[1] [13], MaW-ISh [8], and Græmlin 1.0 [14], as well as the global aligner IsoRank [11] and a global aligner (Græmlin-global) that used our new alignment algorithm with Græmlin 1.0's scoring function.

While we simultaneously compared Græmlin 2.0 to IsoRank and Græmlin-global, we compared Græmlin 2.0 to each local aligner separately. Local aligners may have lower sensitivity than global aligners simply because local aligners only consider nodes in conserved modules while global aligners consider all nodes. Therefore, for each comparison to a local aligner, we removed equivalence classes in Græmlin 2.0's output that did not contain a node in the local aligner's output.

Performance Comparisons. Table 1 shows that Græmlin 2.0 is the most specific aligner. Across all datasets, it produces both the highest fraction of correct equivalence classes as well as the highest fraction of nodes in correct equivalence classes.

Table 2 shows that Græmlin 2.0 is also the most sensitive aligner. In the SNDB pairwise alignments, Græmlin 2.0 and IsoRank produce the most number of nodes in correct equivalence classes. In the other tests, Græmlin 2.0 produces the most number of nodes in correct equivalence classes.

Figure 5 shows that Græmlin 2.0 also finds more cross-species conservation than Græmlin 1.0 and Græmlin-global. Relative to Græmlin 1.0 and Græmlin-global, Græmlin 2.0 produces two to five times as many equivalence classes with four, five, and six species.

---

[1] We used the latest C++ version of NetworkBLAST available at the time of writing, dated Dec. 1, 2007. For the eukaryotic networks, the number of homologs was too large for this version, so we used an older Java implementation, NBlast-0.5. On the SNDB data, the two versions produced virtually identical results.

Table 1. Græmlin 2.0 has higher specificity. As described in the text, we measured the fraction of correct equivalence classes (C_{eq}) and the fraction of nodes in correct equivalence classes (C_{node}). We compared Græmlin 2.0 (Gr2.0) to NetworkBLAST (NB), MaWISh (MW), Græmlin 1.0 (Gr), IsoRank (Iso), and Græmlin-global (GrG). Abbreviations: eco = *E. coli*; stm = *S. typhimurium*; cce = *C. crescentus*; hsa = human; mmu = mouse; sce = yeast; dme = fly.

	SNDB						IntAct		DIP			
	eco/stm		eco/cce		6-way		hsa/mmu		sce/dme		3-way	
	C_{eq}	C_{node}	C_{eq}	C_{node}	C_{eq}	C_{node}	C_{eq}	C_{node}	C_{eq}	C_{node}	C_{eq}	C_{node}
Local aligner comparisons												
NB	0.77	0.49	0.78	0.50	–	–	0.33	0.06	0.39	0.14	–	–
Gr2.0	**0.95**	**0.94**	**0.79**	**0.78**	–	–	**0.83**	**0.81**	**0.58**	**0.58**	–	–
MW	0.84	0.64	**0.77**	0.54	–	–	0.59	0.36	0.45	0.37	–	–
Gr2.0	**0.97**	**0.96**	**0.77**	**0.76**	–	–	**0.88**	**0.86**	**0.90**	**0.91**	–	–
Gr	0.80	0.77	0.69	0.64	0.76	0.67	0.59	0.53	0.33	0.29	0.23	0.15
Gr2.0	**0.96**	**0.95**	**0.82**	**0.81**	**0.86**	**0.85**	**0.86**	**0.84**	**0.61**	**0.61**	**0.57**	**0.57**
Global aligner comparisons												
GrG	0.86	0.86	0.72	0.72	0.80	0.81	0.64	0.64	0.68	0.68	0.71	0.71
Iso	0.91	0.91	0.65	0.65	–	–	0.62	0.62	0.63	0.63	–	–
Gr2.0	**0.96**	**0.96**	**0.78**	**0.78**	**0.87**	**0.87**	**0.81**	**0.80**	**0.73**	**0.73**	**0.76**	**0.76**

Table 2. Græmlin 2.0 has higher sensitivity. We measured the number of nodes in correct equivalence classes (Cor), as described in the text. To show the number of nodes considered in each local aligner comparison, we also measured the number of nodes aligned by each local aligner (Tot). Methodology and abbreviations are the same as in Table 1.

	SNDB						IntAct		DIP			
	eco/stm		eco/cce		6-way		hsa/mmu		sce/dme		3-way	
	Cor	Tot	Cor	Tot	Cor	Tot	Cor	Tot	Cor	Tot	Cor	Tot
Local aligner comparisons												
NB	457	1016	346	697	–	–	65	1010	43	306	–	–
Gr2.0	**627**		**447**		–		**258**		**155**		–	
MW	1309	2050	458	841	–	–	87	241	10	27	–	–
Gr2.0	**1611**		**553**		–		**181**		**20**		–	
Gr	985	1286	546	847	1524	2287	108	203	35	122	27	180
Gr2.0	**1157**		**608**		**2216**		**151**		**75**		**86**	
Global aligner comparisons												
GrG	1496		720		2388		268		384		564	
Iso	**2026**	–	**1014**	–	–	–	306	–	534	–	–	–
Gr2.0	2024		1012		**3578**		**350**		**637**		**827**	

These results suggest that a network aligner's scoring function is more important than its search algorithm. Græmlin 2.0 performs better than existing aligners, despite its simple search algorithm, because of its accurate scoring function.

Number of species per equivalence class

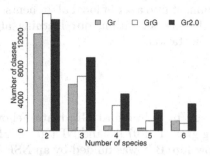

Fig. 5. Græmlin 2.0 finds more cross-species conservation. We counted the number of equivalence classes that contained k species for $k = 2, 3, 4, 5, 6$ as described in the text. We compared Græmlin 2.0 (Gr2.0) to Græmlin 1.0 (Gr) and a global aligner (GrG) that used our new alignment algorithm with Græmlin 1.0's scoring function. We ran the six-way alignment described in the text.

For pairwise alignment, Græmlin 2.0, MaWISh, Græmlin 1.0, and Græmlin-global each ran for less than a minute, while NetworkBLAST and IsoRank ran for over an hour. For each pairwise alignment training run, Græmlin 2.0 ran for under ten minutes. On the six-way alignment, Græmlin 2.0, Græmlin 1.0, and Græmlin-global each ran for under three minutes, and Græmlin 2.0 trained in under forty-five minutes.

4 Discussion

In this paper we presented Græmlin 2.0, a multiple network aligner with a new feature-based scoring function, an algorithm that automatically learns the scoring function's parameters, and an algorithm that uses the scoring function to approximately align multiple networks in linear time. We implemented Græmlin 2.0 for protein interaction network alignment, with a feature function that computes evolutionary events. Græmlin 2.0 has higher accuracy than existing network alignment algorithms across multiple network datasets.

Græmlin 2.0 allows users to easily apply network alignment to their network datasets. Our learning algorithm automatically learns parameters specific to any set of networks. In contrast, existing alignment algorithms require manual recalibration to adjust parameters to different datasets.

Græmlin 2.0 also extends in principle beyond protein interaction network alignment. As more experimental data gathers and network integration algorithms improve, network datasets with multiple data types will appear, such as regulatory networks with directed edges and metabolic networks with chemical compounds [31]. With redefined feature functions, our scoring function and parameter learning algorithm apply to these kinds of networks.

Future research can analyze our learning algorithm. In particular, Græmlin 2.0 might yield better results with a different learning rate or more robust convergence criteria.

Future research can also extend our approach to local alignment. One option is to segment a global alignment into a set of local alignments. With an appropriate feature function and inference algorithm, our learning algorithm can learn a scoring function for segmentation.

Acknowledgments

JF was supported in part by a Stanford Graduate Fellowship. AN was supported by NLM training grant LM-07033 and NIH grant UHG003162. CBD was funded by an NSF Fellowship. BSS was funded by an NSF VIGRE postdoctoral fellowship (NSF grant EMSW21-VIGRE 0502385).

References

1. Sharan, R., Ideker, T.: Modeling cellular machinery through biological network comparison. Nat. Biotechnol. 24, 427–433 (2006)
2. Hartwell, L.H., Hopfield, J.J., Leibler, S., Murray, A.W.: From molecular to modular cell biology. Nature 402, 47–52 (1999)
3. Pereira-Leal, J.B., Levy, E.D., Teichmann, S.A.: The origins and evolution of functional modules: lessons from protein complexes. Philos. Trans. R. Soc. Lond. B. Biol. Sci. 361, 507–517 (2006)
4. Uetz, P., Finley Jr., R.L.: From protein networks to biological systems. FEBS Lett. 579, 1821–1827 (2005)
5. Cusick, M.E., Klitgord, N., Vidal, M., Hill, D.E.: Interactome: gateway into systems biology. Hum. Mol. Genet. 14(2), 171–181 (2005)
6. Kelley, B.P., Sharan, R., Karp, R.M., Sittler, T., Root, D.E., Stockwell, B.R., Ideker, T.: Conserved pathways within bacteria and yeast as revealed by global protein network alignment. Proc. Natl. Acad. Sci. USA 100, 11394–11399 (2003)
7. Sharan, R., Ideker, T., Kelley, B., Shamir, R., Karp, R.M.: Identification of protein complexes by comparative analysis of yeast and bacterial protein interaction data. J Comput. Biol. 12, 835–846 (2005)
8. Koyuturk, M., Kim, Y., Topkara, U., Subramaniam, S., Szpankowski, W., Grama, A.: Pairwise alignment of protein interaction networks. J Comput. Biol. 13, 182–199 (2006)
9. Pinter, R.Y., Rokhlenko, O., Yeger-Lotem, E., Ziv-Ukelson, M.: Alignment of metabolic pathways. Bioinformatics 21, 3401–3408 (2005)
10. Dost, B., Shlomi, T., Gupta, N., Ruppin, E., Bafna, V., Sharan, R.: QNet: A Tool for Querying Protein Interaction Networks. In: Speed, T., Huang, H. (eds.) RECOMB 2007. LNCS (LNBI), vol. 4453, pp. 1–15. Springer, Heidelberg (2007)
11. Singh, R., Xu, J., Berger, B.: Pairwise global alignment of protein interaction networks by matching neighborhood topology. In: Speed, T., Huang, H. (eds.) RECOMB 2007. LNCS (LNBI), vol. 4453, pp. 16–31. Springer, Heidelberg (2007)
12. Zhenping, L., Zhang, S., Wang, Y., Zhang, X.-S., Chen, L.: Alignment of molecular networks by integer quadratic programming. Bioinformatics 23, 1631–1639 (2007)
13. Sharan, R., Suthram, S., Kelley, R.M., Kuhn, T., McCuine, S., Uetz, P., Sittler, T., Karp, R.M., Ideker, T.: Conserved patterns of protein interaction in multiple species. Proc. Natl. Acad. Sci. USA 102, 1974–1979 (2005)

14. Flannick, J., Novak, A., Srinivasan, B.S., Batzoglou, S., McAdams, H.H.: Graemlin: General and Robust Alignment of Multiple Large Interaction Networks. Genome Res. 16 (2006)
15. Berg, J., Lassig, M.: Cross-species analysis of biological networks by Bayesian alignment. Proc. Natl. Acad Sci. USA 103, 10967–10972 (2006)
16. Hirsh, E., Sharan, R.: Identification of conserved protein complexes based on a model of protein network evolution. Bioinformatics 23, 170–176 (2007)
17. Remm, M., Storm, C.E., Sonnhammer, E.L.: Automatic clustering of orthologs and in-paralogs from pairwise species comparisons. J Mol. Biol. 314, 1041–1052 (2001)
18. Do, C.B., Gross, S.S., Batzoglou, S.: Contralign: Discriminative training for protein sequence alignment. In: Apostolico, A., Guerra, C., Istrail, S., Pevzner, P.A., Waterman, M. (eds.) RECOMB 2006. LNCS (LNBI), vol. 3909, pp. 160–174. Springer, Heidelberg (2006)
19. Do, C.B., Woods, D.A., Batzoglou, S.: CONTRAfold: RNA secondary structure prediction without physics-based models. Bioinformatics 22, 90–98 (2006)
20. Felsenstein, J.: Maximum-likelihood estimation of evolutionary trees from continuous characters. Am. J. Hum. Genet. 25, 471–492 (1973)
21. Ratliff, N., Bagnell, J., Zinkevich, M. (online) subgradient methods for structured prediction. In: Eleventh International Conference on Artificial Intelligence and Statistics (AIStats) (2007)
22. Kanehisa, M., Goto, S.: KEGG: kyoto encyclopedia of genes and genomes. Nucleic. Acids. Res. 28, 27–30 (2000)
23. Shor, N.Z., Kiwiel, K.C., Ruszcayński, A.: Minimization methods for non-differentiable functions. Springer, New York (1985)
24. Nedic, A., Bertsekas, D.: Convergence rate of incremental subgradient algorithms (2000)
25. Russell, S., Norvig, P.: Artificial Intelligence: A Modern Approach, 2nd edn. Prentice-Hall, Englewood Cliffs (2003)
26. Altschul, S.F., Madden, T.L., Schaffer, A.A., Zhang, J., Zhang, Z., Miller, W., Lipman, D.J.: Gapped BLAST and PSI-BLAST: a new generation of protein database search programs. Nucleic Acids Res. 25, 3389–3402 (1997)
27. Kerrien, S., Alam-Faruque, Y., Aranda, B., Bancarz, I., Bridge, A., Derow, C., Dimmer, E., Feuermann, M., Friedrichsen, A., Huntley, R., Kohler, C., Khadake, J., Leroy, C., Liban, A., Lieftink, C., Montecchi-Palazzi, L., Orchard, S., Risse, J., Robbe, K., Roechert, B., Thorneycroft, D., Zhang, Y., Apweiler, R., Hermjakob, H.: IntAct–open source resource for molecular interaction data. Nucleic Acids Res. 35, 561–565 (2007)
28. Xenarios, I., Salwinski, L., Duan, X.J., Higney, P., Kim, S.-M., Eisenberg, D.: DIP, the Database of Interacting Proteins: a research tool for studying cellular networks of protein interactions. Nucleic Acids Res. 30, 303–305 (2002)
29. Srinivasan, B.S., Novak, A.F., Flannick, J.A., Batzoglou, S., McAdams, H.H.: Integrated protein interaction networks for 11 microbes. In: Apostolico, A., Guerra, C., Istrail, S., Pevzner, P.A., Waterman, M. (eds.) RECOMB 2006. LNCS (LNBI), vol. 3909, pp. 1–14. Springer, Heidelberg (2006)
30. Ashburner, M., Ball, C.A., Blake, J.A., Botstein, D., Butler, H., Cherry, J.M., Davis, A.P., Dolinski, K., Dwight, S.S., Eppig, J.T., Harris, M.A., Hill, D.P., Issel-Tarver, L., Kasarskis, A., Lewis, S., Matese, J.C., Richardson, J.E., Ringwald, M., Rubin, G.M., Sherlock, G.: Gene ontology: tool for the unification of biology. The Gene Ontology Consortium. Nat. Genet. 25, 25–29 (2000)

31. Srinivasan, B.S., Shah, N.H., Flannick, J.A., Abeliuk, E., Novak, A.F., Batzoglou, S.: Current progress in network research: toward reference networks for key model organisms. Brief Bioinform (2007)
32. Altschul, S.F., Carroll, R.J., Lipman, D.J.: Weights for data related by a tree. J Mol. Biol. 207, 647–653 (1989)

A   Feature Function Definition

This section presents precise definitions of our feature function and the evolutionary events that our feature function computes.

We define evolutionary events for possibly ancestral species. We assume that we have n extant species $1, \ldots, n$ and m ancestral species $n+1, \ldots, n+m$,[2] all related by a phylogenetic tree.

Each species $i \in [1 : n+m]$ is represented by a species weight vector $\mathbf{s}^i \in \mathbb{R}^n$, where $\sum_{j=1}^{n} s_j^i = 1$ and s_j^i represents the similarity of species $j \in [1 : n]$ to species i. We can use a phylogenetic tree to compute the weight vectors efficiently [20,32]. Each extant species $j \in [1 : n]$ has a species weight vector $[s_1^j = 0, \ldots, s_{j-1}^j = 0, s_j^j = 1, s_{j+1}^j = 0, \ldots, s_n^j = 0]$.

We denote an equivalence class $[x]$ as a set of proteins $\bigcup_{i=1}^{n} \Pi_i^{[x]}$, where $\Pi_i^{[x]}$ is the projection of $[x]$ to species i.

A.1   Node Feature Function

We compute the node feature function $\mathbf{f}^N$ for an equivalence class $[x]$ as follows. First, we compute events for species r at the phylogenetic tree root.

Protein Present. We define $\mathbf{p} \in \mathbb{R}^n$ as $p_i = 1$ if $\Pi_i^{[x]} \neq \varnothing$ and 0 otherwise.

- $f_1^N = \mathbf{s}^r \cdot \mathbf{p}$ is the probability that species r has a protein in $[x]$.
- $f_2^N = 1 - \mathbf{s}^r \cdot \mathbf{p}$ is the probability that species r does not have a protein in $[x]$.

Protein Count. We define $\mathbf{c} \in \mathbb{R}^n$ as $c_i = |\Pi_i^{[x]}|$, the number of proteins that species i has in $[x]$.

- $f_3^N = \frac{\mathbf{s}^r \cdot \mathbf{c}}{\mathbf{s}^r \cdot \mathbf{p}}$ is the expected number of proteins species r has in $[x]$, given that r has a protein.
- $f_4^N = (f_3^N)^2$

The protein present and protein count features describe the most recent common ancestor of the extant species in the equivalence class.

Next, we compute events for all pairs of species $i, j \in [1 : n+m], i \neq j$ adjacent in the tree.

---

[2] In the appendix, the symbols n and m have different meanings than in the main text.

Protein Deletion. We define $p(k) = \mathbf{s}^k \cdot \mathbf{p}$ as the probability that species k has a protein in $[x]$.

- $f_5^N(i,j) = p(i) \times (1 - p(j)) + (1 - p(i)) \times p(j)$ is the probability a protein deletion occurs between species i and j.
- $f_6^N(i,j) = p(i) \times p(j)$ is the probability a protein deletion does not occur between species i and j.

Protein Duplication. We define $c(k) = \frac{\mathbf{s}^k \cdot \mathbf{c}}{\mathbf{s}^k \cdot \mathbf{p}}$ as the expected numbers of proteins that species k has in $[x]$.

- $f_7^N(i,j) = |c(i) - c(j)|$ is the expected number of proteins gained between species i and j.

Protein Mutation. We define a species pair weight matrix $\mathbf{S}^{ij} \in \mathbb{R}^{n \times n}$ as $S_{kl}^{ij} = s_k^i s_l^j$. We define $\mathbf{B} \in \mathbb{R}^{n \times n}$ as

$$B_{kl} = \frac{1}{|\Pi_k^{[x]}||\Pi_l^{[x]}|} \sum_{p \in \Pi_k^{[x]}} \sum_{q \in \Pi_l^{[x]}} b(p,q)$$

where $b(p,q)$ is the BLAST bitscore [26] of proteins p and q. B_{kl} is the average bitscore among the proteins in species k and l. B_{kl} equals 0 if either species k or l has no proteins in $[x]$.

- $f_8^N(i,j) = \mathrm{tr}(\mathbf{S}^{ij^T}\mathbf{B})$, the sum of entry-wise products, is the expected bitscore between the proteins in species i and j.
- $f_9^N(i,j) = (f_8^N)^2$
- $f_{10}^N(i,j) = (f_8^N)^{-1}$
- $f_{11}^N(i,j) = (f_8^N)^{-2}$

Features f_9^N through $f_1^N 1$ allow our scoring function to include nonlinear dependencies on the BLAST bitscore of the proteins.

Finally, we compute events for all extant species $i \in [1:n]$.

Paralog Mutation

- $f_{12}^N(i) = \mathbf{B}_{ii}$ is the expected average bitscore between a protein in species i and its paralogs.
- $f_{13}^N(i,j) = (f_{12}^N)^2$
- $f_{14}^N(i,j) = (f_{12}^N)^{-1}$
- $f_{14}^N(i,j) = (f_{12}^N)^{-2}$

A.2   Edge Feature Function

We compute the edge feature function $\mathbf{f}^E$ for equivalence classes $[x]$ and $[y]$ as follows. First, we compute events for all pairs of species $i, j \in [1:n+m], i \neq j$ adjacent in the tree.

Edge Deletion. For $k \in [1:n], p \in \Pi_k^{[x]}, q \in \Pi_k^{[y]}$, we define $e(k,p,q) = 1$ if there is an edge between p and q and 0 otherwise. We then define $\mathbf{e} \in \mathbb{R}^n$ as

$$e_k = \frac{1}{|\Pi_k^{[x]}||\Pi_k^{[y]}|} \sum_{p \in \Pi_k^{[x]}} \sum_{q \in \Pi_k^{[y]}} e(k,p,q)$$

which represents the average probability that species k has an edge. We define e_k as NULL if $\Pi_k^{[x]}$ or $\Pi_k^{[y]}$ is empty. We define

$$e(l) = \left(\frac{1}{\displaystyle\sum_{k:e_k \neq \text{NULL}} e_k} \right) \sum_{k:e_k \neq \text{NULL}} e_k s_k^l \qquad l \in \{i,j\}$$

which represent the probabilities that species i and j have edges.

- $f_1^E(i,j) = e(i) \times (1 - e(j)) + (1 - e(i)) \times e(j)$ is the probability that an edge is lost between species i and j.
- $f_2^E(i,j) = e(i) * e(j)$ is the probability that an edge is not lost between i and j.

Next, we compute events for all extant species $i \in [1:n]$.

Paralog Edge Deletion. We define $\tilde{e}(k,p,q) = 1$, for $k \in [1:n], p \in \Pi_k^{[x]}, q \in \Pi_k^{[y]}$ as

$$\tilde{e}(k,p,q) = \frac{1}{|\Pi_k^{[x]}||\Pi_k^{[y]}|} \sum_{\substack{p' \in \Pi_k^{[x]} \\ q' \in \Pi_k^{[y]} \\ (p',q') \neq (p,q)}} e(k,p',q')$$

which represents the probability, ignoring p and q, that species k has an edge.

- $f_3^E(i) = \sum_{p \in \Pi_k^{[x]}} \sum_{q \in \Pi_k^{[y]}} \left(e(i,p,q) \times (1 - \tilde{e}(i,p,q)) + (1 - e(i,p,q)) \times \tilde{e}(i,p,q) \right)$ is the average probability an edge is lost between a pair of proteins in species i and all other pairs of proteins in species i.
- $f_4^E(i) = \sum_{p \in \Pi_k^{[x]}} \sum_{q \in \Pi_k^{[y]}} e(i,p,q) \times \tilde{e}(i,p,q)$ is the average probability an edge is not lost between a pair of proteins in species i and all other pairs of proteins in species i.

For pairwise alignment of two species s and t, the final node feature function is

$$\mathbf{f}^N([x]) =$$
$$[f_1^N, f_2^N, f_3^N, f_4^N, f_5^N(s,t), f_6^N(s,t), f_7^N(s,t), f_8^N(s,t), f_9^N(s,t), f_{10}^N(s,t),$$
$$f_{11}^N(s,t), f_{12}^N(s) + f_{12}^N(t), f_{13}^N(s) + f_{13}^N(t), f_{14}^N(s) + f_{14}^N(t), f_{15}^N(s) + f_{15}^N(t)]$$

and the final edge feature function is

$$\mathbf{f}^E([x],[y]) = \left[f_1^E(s,t), f_2^E(s,t), f_3^E(s) + f_3^E(t), f_4^E(s) + f_4^E(t)\right]$$

For multiple alignment, the final node feature function is

$$\mathbf{f}^N([x]) =$$

$$\left[f_1^N, f_2^N, f_3^N, f_4^N, \sum_{(i,j)} f_5^N(i,j), \sum_{(i,j)} f_5^N(i,j) \times b, \right.$$

$$\sum_{(i,j)} f_6^N(i,j), \sum_{(i,j)} f_6^N(i,j) \times b, \sum_{(i,j)} f_7^N(i,j), \sum_{(i,j)} f_7^N(i,j) \times b,$$

$$\sum_{(i,j)} f_8^N(i,j), \sum_{(i,j)} f_8^N(i,j) \times b, \sum_{(i,j)} f_8^N(i,j) \times b^2, \sum_{(i,j)} f_8^N(i,j) \times b^3,$$

$$\sum_{(i,j)} f_9^N(i,j), \sum_{(i,j)} f_9^N(i,j) \times b, \sum_{(i,j)} f_9^N(i,j) \times b^2, \sum_{(i,j)} f_9^N(i,j) \times b^3,$$

$$\sum_{(i,j)} f_{10}^N(i,j), \sum_{(i,j)} f_{10}^N(i,j) \times b, \sum_{(i,j)} f_{10}^N(i,j) \times b^2, \sum_{(i,j)} f_{10}^N(i,j) \times b^3,$$

$$\sum_{(i,j)} f_{11}^N(i,j), \sum_{(i,j)} f_{11}^N(i,j) \times b, \sum_{(i,j)} f_{11}^N(i,j) \times b^2, \sum_{(i,j)} f_{11}^N(i,j) \times b^3,$$

$$\left. \sum_{i=1}^{n} f_{12}^N(i), \sum_{i=1}^{n} f_{13}^N(i), \sum_{i=1}^{n} f_{14}^N(i), \sum_{i=1}^{n} f_{15}^N(i) \right]$$

and the final edge feature function is

$$\mathbf{f}^E([x],[y]) = \left[\sum_{(i,j)} f_1^E(i,j), \sum_{(i,j)} f_1^E(i,j) \times b, \sum_{(i,j)} f_2^E(i,j), \sum_{(i,j)} f_2^E(i,j) \times b, \sum_{i=1}^{n} f_3^E(i), \sum_{i=1}^{n} f_4^E(i) \right]$$

where the sums over (i, j) are taken over branches of the phylogenetic tree and the sums i are taken over the leaves of the tree.

An Integrative Network Approach to Map the Transcriptome to the Phenome

Michael R. Mehan[1], Juan Nunez-Iglesias[1], Mrinal Kalakrishnan[1], Michael S. Waterman[1], and Xianghong Jasmine Zhou[1,2]

[1] Program in Computational Biology, Department of Biological Sciences
University of Southern California, Los Angeles CA 90089, USA
[2] To whom correspondence should be addressed: xjzhou@usc.edu

Abstract. Although many studies have been successful in the discovery of cooperating groups of genes, mapping these groups to phenotypes has proved a much more challenging task. In this paper, we present the first genome-wide mapping of gene coexpression modules onto the phenome. We annotated coexpression networks from 136 microarray datasets with phenotypes from the Unified Medical Language System (UMLS). We then designed an efficient graph-based simulated annealing approach to identify coexpression modules frequently and specifically occurring in datasets related to individual phenotypes. By requiring phenotype-specific recurrence, we ensure the robustness of our findings. We discovered 9,183 modules specific to 47 phenotypes, and developed validation tests combining Gene Ontology, GeneRIF and UMLS. Our method is generally applicable to any kind of abundant network data with defined phenotype association, and thus paves the way for genome-wide, gene network-phenotype maps.

1 Introduction

The fundamental aim of genetics is to link phenotype to genotype, and traditional genetic studies have sought to associate single genes to a particular phenotypic trait. However, it has become clear that complex diseases, such as cancer, autoimmune disease, or heart disease, are effected by the interaction of many different genes. For this problem, genetic association studies lack power. Locus heterogeneity, epistasis, low penetrance, and pleiotropy all contribute to mask or reduce the detectable signal [1,2].

In recent years, high-throughput approaches have been used to study the interaction of groups of genes. In a gene network, nodes represent genes (or gene products), and links between nodes represent functional relationship between the nodes. Examples include protein-protein interaction networks, genetic interaction networks, and gene coexpression networks. Borrowing or expanding tools from the fields of network analysis and graph theory, researchers have devised numerous ways to use these networks to determine which genes work together [3,4,5,6,7,8]. However, virtually all of this work fails to complete the link between genotype and phenotype. Genes and gene products are grouped into

M. Vingron and L. Wong (Eds.): RECOMB 2008, LNBI 4955, pp. 232–245, 2008.

modules and complexes, but these are not linked to phenotypes. We note two remarkable exceptions: Butte and Kohane used differential expression analysis [9] to systematically associate genes with specific phenotypes and environments, using data from the Gene Expression Omnibus [10]; and Lage and colleagues [11] used OMIM protein annotations to associate protein complexes with disease phenotypes. However, the former approach does not consider genes in a network context, while the latter approach only considers annotated nodes in a single static network. Neither approach, nor any other, has systematically mapped gene networks to the experimental phenotype conditions under which they are activated.

In this paper, we introduce the first approach to explicitly bridge this gap. Like Butte and Kohane, we used the large amount of microarray gene expression data from the Gene Expression Omnibus. Here, instead of gene-phenotype associations, we used integrative network analysis to infer network module to phenotype associations. A series of microarray datasets can be modeled as a series of coexpression networks as follows: each node represents a gene, and a link is placed between two nodes if their expression profiles in that dataset are highly similar. The crucial advantage of this approach is that each generated network can be labeled with the phenotypic information of that dataset, such as the type of biological sample, the disease state, drug treatment, etc. The Unified Medical Language System (UMLS) [12] provides an extensive catalog of medical concepts and their relationships, as well as language processing tools that enable the automated mapping of text onto UMLS concepts. This allowed us to automatically annotate each microarray dataset with UMLS phenotype classes by using the associated MEDLINE reference.

For each phenotype, we partitioned the datasets into a *phenotype class*, consisting of datasets annotated with that phenotype, and a *background class*, consisting of the rest of the datasets. We designed a graph-based simulated annealing [13] approach to efficiently identify groups of genes which form dense subnetworks preferentially and repeatedly in the phenotype class. Note that a dense subnetwork in a coexpression graph represents a coexpression cluster. Although microarray data is noisy, we have shown in our previous work [14,15] that coexpression clusters recurrent across multiple datasets represent true functional or transcriptional modules with high probability. Here, we further show that if a frequent coexpression cluster additionally is specific to a phenotype class, it is likely to effect that phenotype.

We applied our approach to the analysis of 136 microarray datasets, covering 60 phenotype conditions. We discovered approximately 9,000 modules specific to 47 of these phenotypes, and developed a novel way to validate this specificity by integrating gene and dataset annotations from Gene Ontology [16], Gene Reference Into Function (GeneRIF) [17], and UMLS. Our method lays the foundation for a genome-wide, gene network-phenotype map, which will benefit our understanding of complex diseases and their treatment. Our present map of network patterns to phenotypes has many applications, such as predicting the phenotypic effects of multiple interacting genetic perturbations, *in silico* testing of genetically complex hypotheses, and prioritization of candidate genes for targeted intervention. Furthermore, the concept of our approach is general, and can be easily extended to incorporate any standardized phenomic procedures, as suggested, for example, by the Human Phenome Project [18].

2 Methods

2.1 Dataset Preparation

Dataset Selection. We selected every microarray dataset from NCBI's Gene Expression Omnibus that met the following criteria: all samples were of human origin; the dataset had at least 8 samples (a minimum for accurate correlation estimation); and the platform was either GPL91 (corresponding to Affymetrix HG-U95A) or GPL96 (Affymetrix HG-U133A). Throughout this study, we only considered the genes shared by the two platforms (and therefore all datasets), of which there are 8,635. The 136 datasets that met these criteria on 28 Feb 2007 were used for the analysis described herein.

Dataset Annotation. We determined the phenotypic context of a microarray dataset based on the Medical Subject Headings (MeSH) of its corresponding PubMed record, mapped to UMLS concepts. This is more refined than attempting to scan the abstract or full text of the paper, and in practice it results in much cleaner and more reliable annotations [9,19]. UMLS is the largest available compendium of biomedical vocabularies, spanning approximately one million interrelated concepts, including diseases, treatments, and phenotypic concepts at different levels of resolution (molecules, cells, tissues and whole organisms). In order to infer higher-order links between datasets, we annotated datasets with the matched UMLS concept and, in addition, all its ancestor concepts. This resulted in a total of 467 annotations, of which 80 mapped to more than 5 datasets, or 60 after merging annotations that mapped onto identical sets of datasets.

Correlation Estimation and Graph Generation. For each dataset, we used the Jackknife Pearson correlation as a measure of similarity between two genes (the minimum of the leave-one-out Pearson correlations). To determine the coexpression network, we selected a cutoff corresponding to the top-ranking 150,000 correlations of the total $\binom{8635}{2} \approx 3.73 \times 10^7$ gene pairs (0.4%). The cutoff was generated by exploring the statistical distribution of pairwise correlations, which we do not detail here.

Once a cutoff has been determined, we defined that dataset's coexpression network as the graph $G_i = (V, E_i)$, where V corresponds to the set of genes being investigated, and $(g_a, g_b) \in E_i$ if the correlation between g_a and g_b is higher than the cutoff.

Differential Coexpression Graphs. To dramatically increase the probability of finding optimal modules across the many massive networks, we wished narrow down the search space. We therefore constructed a weighted *differential* *coexpression graph* for each phenotype, which summarizes the differences between the gene coexpression networks in the phenotype class and those in the background class. This graph was used by the simulated annealing algorithm to create neighboring states (see "Neighbor Selection" under Section 2.2). We describe it formally as follows.

To begin, we define $\mathcal{G}$ as the set of all graphs constructed from the microarray datasets. For each phenotype $\mathcal{P}$, we partition $\mathcal{G}$ into the *phenotype* graphs $\mathcal{G}_\mathcal{P}$, corresponding to datasets annotated with $\mathcal{P}$, and the *background* graphs $\mathcal{G}_\mathcal{P}^c = \mathcal{G} \setminus \mathcal{G}_\mathcal{P}$, corresponding to the rest of the datasets.

We then construct a weighted differential coexpression graph $G_\Delta = (V, E_\Delta)$ to reflect edges (coexpression relationships) that are present frequently in $\mathcal{G}_\mathcal{P}$ but not in $\mathcal{G}_\mathcal{P}^c$. This specificity can be measured by the significance p of a hypergeometric test, assessing the abundance of an edge in $\mathcal{G}_\mathcal{P}$ relative to its overall abundance in $\mathcal{G}$. In G_Δ, the vertex set V is the same as in every graph in $\mathcal{G}$, and the weight associated with (g_a, g_b) is then $w_\Delta(g_a, g_b) = -\log(p)$. Edges of weight 0 are not in E_Δ. In this way, heavier edges in this graph represent pairs of genes that exhibit elevated coexpression highly specific to $\mathcal{G}_\mathcal{P}$.

2.2 Simulated Annealing Design

Goal and Rationale. Our aim was to find sets of genes that satisfy three criteria: first, the genes must be coexpressed in multiple datasets; second, the annotations of these datasets must be enriched in some specific phenotype; and third, the gene set must be maximally large while meeting the first two criteria.

As explained in section 2.1, from each annotated dataset we derived a coexpression graph. For a set of vertices $V' \subset V$ having m edges between them, the *density* is $\delta(V') = m/\binom{|V'|}{2} = 2m/(|V'|(|V'|-1))$. This is exactly the proportion of gene pairs from V' that are coexpressed, taken over all possible pairs $\{(u, v) : u \in V', v \in V'\}$. We say that a vertex set is *dense* if δ is large (typically greater than 0.6). Then, for each phenotype, we wanted to find a set of vertices that is dense in a large proportion of datasets annotated with that phenotype, and that is *not* dense in datasets not annotated with it.

As we demonstrated in our earlier work [15], the problem of identifying frequent dense vertex sets is NP-complete. It is easy to show that the additional requirement of phenotype specificity does not decrease the complexity of the problem. Hence, we decided to use simulated annealing, a well-established stochastic algorithm with successful application in other NP-complete problems [20]. Our design for the simulated annealing (SA) algorithm follows.

Search Space. A state in our SA design is defined as a set of vertices, and the search space is the set of all sets of vertices, although for simplicity and for computational considerations we limited ourselves to sets smaller than 50 vertices. We believe this to be an ample margin for phenotypically relevant gene sets. Formally, we define the search space as $\mathcal{S} = \{x : x \subset V, |x| \leq 50, |x| \geq 3\}$.

Objective Functions. Recall from "Goal and Rationale" in this section that we needed to maximize three different objectives: size, density, and specificity.

Therefore, we used three objective functions, which we combined into a single function using a weighted sum. Much work has been done to generalize the simulated annealing process to multiple objectives, collectively known as MOSA (Multiple Objective Simulated Annealing). The simplest strategy is to create an energy function f_i for each objective i, and then combine them into a single energy function by using a weighted sum $f(x) = \sum_{i=1}^{k} w_i f_i(x)$. The key difficulty with this approach is determining an appropriate set of weights. In previous studies, this has been accomplished empirically [21], and this is the approach that we take for the following reasons: we were interested in a single optimal combination of objective functions, rather than exploring the extremes of each; our design for individual functions was such that overall effectiveness of the algorithm was consistent throughout a range of weights; and the parameters we chose based on performance on simulated data behaved well on the real data. The weights we chose for size, density and specificity were 1, 0.5, and 0.25 respectively.

For the three objectives, we knew hard thresholds that we had determined in past studies: size of 5 or more, specificity p-value of less than 0.001, and density greater than 0.6. For simulated annealing, however, we want to accept intermediate states that may be unfavorable. We therefore designed the energy functions to have soft thresholds, by using an exponential increase in energy for unfavorable values of the three objectives. Since we combine the functions using a linear weighted sum, for any arbitrary set of non-zero weights, extreme solutions will be rejected by the exponential energy increase. The individual energy functions that we designed take the following forms:

$$f_{size}(x) = \left(1 - \frac{|x|}{50}\right) \cdot \left(1 + \exp\left\{-\gamma\left(\frac{|x|}{50} - o_s\right)\right\}\right) \tag{1}$$

$$f_{dens}(x) = \exp\left\{-\alpha\left(\min_{i \in \mathcal{G}_A}\left(\delta_i(x)\right) - o_\delta\right)\right\} \tag{2}$$

$$f_{spec}(x) = \log\left(\mathbb{P}\left(Y \geq |\mathcal{G}_A \cap \mathcal{G}_P|\right)\right) \tag{3}$$

where

$\mathcal{G}_P$ is the set of datasets annotated with the current phenotype,

$\mathcal{G}_A$ is the set of datasets in which the gene cluster is dense,

and $Y \sim hypergeometric\left(|\mathcal{G}_A|, |\mathcal{G}_P|, |\mathcal{G}_P^c|\right)$.

Equation 1 shows the size objective function, which contains both a linear component (first expression) and an exponential component (second expression): this sets a soft threshold at low sizes (4-5 vertices), but continues to reward size increases above the threshold, due to the linear component. Equation 2 shows that the density objective function consists only of the soft threshold. Finally, equation 3 shows the specificity objective function, which favors low p-values of specificity to the phenotype at an exponential rate.

Because of the exponential component of these functions, extreme cases (such as a single triangle that is very dense and specific, but very small) are rejected on the basis of just one of the energy functions, but improvements to existing cases are rewarded, and intermediate cases are accepted with good probability. We selected the parameters $\gamma = 15$, $o_s = -0.5$, $\alpha = 15$ and $o_\delta = 0.5$ based on our experience with biologically validated clusters compared with clusters arising from random chance.

Initial State. A SA approach aims to find a global optimum during each run. Therefore, if we were to use random initial states and run the algorithm for a long enough time, we will always find approximately the same set of vertices, representing the largest set having the most evidence for coexpression and phenotype specificity. We were, however, interested in a large number of vertex sets showing evidence for coexpression and phenotype specificity. To this end, we designed a systematic way of generating initial states, or seeds, and we restricted the SA search space to clusters containing these seeds.

We define a *triangle* as a set of three vertices that is fully connected in at least one dataset. The hypothesis underlying our strategy is that, if a set of genes is coexpressed specifically in datasets annotated with the phenotype of interest, then at least one recurrent triangle will appear in the phenotype datasets and it is unlikely to appear in many of the background datasets.

Therefore, for each phenotype, we tested every possible gene triplet for enrichment (using the hypergeometric test) of triangles in the phenotype datasets with respect to the background datasets. For each seed having a hypergeometric p-value less than 0.01, we ran the SA algorithm, with the constraint that states in that run must be supersets of the initial triplet.

Selection of Neighboring States. We defined a neighbor as a state that contains either one more or one less vertex than the current state. We created neighboring states by first determining whether to add to or remove a vertex, then choosing the vertex based on the appropriate probability distribution.

If a cluster has size 3, it consists only of the initial seed, and so a vertex must be added. Conversely, if a cluster has size 50, it has reached the maximum size, and a vertex must be removed. For intermediate values, we proceeded as follows.

Let x be the current cluster. We narrowed the search space of vertices to be added by considering only vertices that have at least one edge to a vertex in x in at least one of the phenotype datasets. This is easily justified because vertices not meeting this criterion could not possibly contribute to x as a dense, phenotype-specific cluster, even as an intermediate step. It can be shown that this

set corresponds exactly to $\mathcal{N}_x = \left\{ g : g \notin x, \sum_{h \in x} w_\Delta(g, h) > 0 \right\}$ (See "Differential

Coexpression Graphs" under section 2.1).

The probability of removing a vertex is then given by $p_{rem} = s_0 / |\mathcal{N}_x|$, where s_0 is an estimate of how many vertices will improve the cluster. This is to allow the SA process ample time to consider many neighbors before attempting to remove a vertex, since the number of neighboring vertices vastly outnumbers the number of vertices in a cluster. We heuristically used $s_0 = 10$ as an appropriate average number. In the future, an iterative estimation of s_0 as the average size of the returned clusters might improve the performance of the algorithm.

In the event that a gene is to be removed, it is chosen uniformly from the cluster. When adding a gene, however, we made the probability that a vertex $g \in \mathcal{N}_x$ is added proportional to the sum of the weights of edges from g to the members of x in the differential coexpression graph. Formally, we have: $\mathbb{P}\left(g_a \text{ is added}\right) = \sum_{a \in x} w_\Delta\left(g_a, a\right) / \sum_{b \in \mathcal{N}_x} \sum_{a \in x} w_\Delta\left(a, b\right)$.

Annealing Schedule. We used the schedule $T_k = T_{max}/\log\left(k+1\right)$, where k is the iteration number and T_k is the temperature at that iteration, as suggested by Geman and Geman [22]. The initial temperature was 10. This schedule form guarantees optimality for long running times. Although it can be argued that the exponential running times required makes this schedule impractical, we found that for an identical number of SA iterations, it resulted in lower-energy clusters than the often-used exponential schedule, $T_{k+1} = \alpha T_k = \alpha^k T_{max}$. We ran the algorithm for 700,000 iterations per run.

Post-filtering. Recall that we enforced the inclusion of the initial seed triangle in the final result. Clearly, some seeds will result from noise alone, and therefore the final output will not be biologically significant. To remove these clusters, we filtered the SA output clusters by discarding any vertex set not meeting the following criteria: size greater than 6; density greater than 0.66; phenotype-specificity p-value less than 0.0066; and dense in at least 3 datasets related to the target phenotype. After filtering, we merged redundant clusters, defined as pairs clusters for which intersection/union was higher than 0.8.

3 Results

3.1 Functional Homogeneity

We applied our simulated annealing approach to the 136 microarray datasets covering 60 phenotype classes. These included a range of diseases (e.g. leukemia, myopathy, and nervous system disorders) and tissues (e.g. brain, lung, and muscle). Starting from the recurrent triangle seeds for each of the 60 distinct phenotypes, we identified 9,183 clusters that satisfied our criteria for a concept-specific coexpression cluster. The number of clusters we found for a given phenotype increased with the number of datasets annotated with it: most of the phenotypes with only a few associated datasets yielded few clusters. The most represented phenotype we studied was "Hemic and Lymphatic Diseases," which had 19 associated datasets and a total of 322 clusters.

We used two different methods to evaluate cluster quality. First, we assessed cluster functional homogeneity by testing for enrichment for specific Gene Ontology [16] biological process terms. If a cluster is enriched in a GO term with a hypergeometric p-value less than 0.01, we declare the cluster functionally homogeneous. Of the 9,183 clusters derived from all phenotypes, 74.8% were functionally homogenous by this measure. An advantage of our approach is demonstrated by this validation: since we identified clusters specific to only subsets of all our datasets, we were less likely to detect constitutively expressed clusters, such as those consisting of ribosomal genes or genes involved in protein synthesis.

While the GO approach provides information about gene function, it fails to describe its phenotypic implications. To map individual genes to phenotypes, we used GeneRIF [17]. The GeneRIF database contains short statements derived directly from publications describing functions, processes, and diseases in which a gene is implicated. We annotated genes with phenotypes by mapping the GeneRIF notes to UMLS metathesaurus terms as we did with the dataset MeSH headings (see Section 2.1). Similar to GO annotations, we assessed the *conceptual homogeneity* of gene clusters in specific UMLS keywords with the hypergeometric test, enforcing a minimum p-value of 0.01. The proportion of modules that were conceptually homogeneous was 48.3%. Clusters usually show less conceptual homogeneity than functional homogeneity, which is likely due to the sparsity of GeneRIF annotations. There are cases, however, in which GeneRIF performs better. For example, many of the cancer related phenotypes, such as "Carcinoma," "Neoplasm Metastasis," and "Neoplastic Processes," show higher GeneRIF homogeneity, which could be attributed to the abundance of related literature. The functional and conceptual homogeneity of clusters derived from different phenotype classes is summarized in Figure 1.

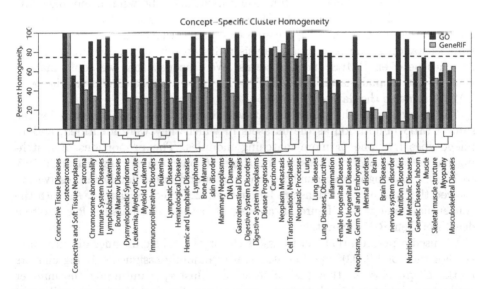

Fig. 1. Functional homogeneity of clusters by phenotype. The dotted lines show the overall homogeneity across all phenotypes. The dendrogram shows the distance between phenotypes in dataset overlap.

3.2 Phenotype Specificity

In addition to testing for functional and conceptual homogeneity, we assessed whether the returned clusters were involved in the phenotype condition in which they were found. Again, we used both GO and GeneRIF independently for this.

Recall that each functionally homogeneous module is associated with one or more GO biological functions, and that it is also associated with the phenotype in which it was found. We summarized these GO functions by mapping them to "informative nodes," which we introduced in our earlier work [3], and then tested them for overrepresentation in that phenotype class. This gave us, for each of 31 phenotypes (out of 47 phenotypes having at least one module), a list of gene module functions that are active in that phenotype more often than expected by chance. Many of these GO functions are clearly related to the phenotype in which they were found. For example, the phenotype "Lymphoma," which is a cancer of immune system cells, has 3 coupled GO biological processes related to cell proliferation – "cell cycle process" (7e-36), "cell cycle" (3e-27), and "cell division" (9e-17) – as well as "immune system process" (5e-6). Furthermore, our approach identified biological processes related not only to disease phenotypes, but also to tissue phenotypes. For example, the "Muscle" phenotype is significantly enriched with modules homogeneous for the biological function "muscle contraction."

We show in Table 1 the full list of phenotypes and the functions overrepresented in their modules. The functional association between a module's GO function and the phenotype in which it is active suggests that our clusters are indeed linked to the phenotype conditions under which they were identified. Notwithstanding, the preceding analysis relies on our subjective evaluation of matches between UMLS and GO terms. We reasoned that we could make a more objective analysis with GeneRIF, as it can be mapped directly to the same UMLS terms as the dataset phenotypes.

We thus counted the modules that were conceptually homogeneous for the same UMLS annotation as the datasets in which they were identified, and 16 phenotypes had one or more matching modules. The proportions of matching modules to total modules among these 16 phenotypes ranged from 0.75% to 7.88%. Although these numbers may not sound immediately impressive, we showed that these proportions are larger than expected by chance. As with the conceptual homogeneity figures, it appears that these numbers are low due to a dearth of GeneRIF annotations.

We used a permutation test to assess the statistical significance of our analysis. For each of 1,000,000 permutations, we randomly assigned existing clusters to the 47 phenotypes that had at least one cluster, maintaining the number of clusters assigned to each phenotype constant. The five phenotypes with statistical significance after Bonferonni correction were "Lymphoma" (<4.7e-5), "Lymphoproliferative Disorders" (<4.7e-5), "Musculoskeletal Diseases" (4.7e-5), "Hematological Disease" (9.2e-3), and "Hemic and Lymphatic Diseases" (0.01).

Although these five phenotypes do not represent the majority of the phenotypes, they show that our results are statistically significant for some phenotypes. The lack of consistent validation of phenotype specificity can be attributed to limitations in the GeneRIF-UMLS mapping as well as lack of gene phenotype annotations.

Table 1. Gene Ontology biological processes that are overrepresented in dataset UMLS phenotypes (hypergeometric enrichment p-value < 0.001 after Bonferroni correction). The concept name appears as in the UMLS metathesaurus. The number of Functionally Homogeneous Clusters (FHC) is listed along with the total number of clusters (TC) identified for that phenotype.

Concept Name	FHC (TC)	Overrepresented GO biological functions (Bonferroni-corrected p-value)
Bone Marrow	406 (417)	biosynthetic process(4e-30), cell cycle(3e-20), cell cycle process(9e-17), response to biotic stimulus(5e-15), cell division(2e-14), multi-organism process(2e-14), DNA metabolic process(3e-4)
Bone Marrow Diseases	40 (51)	generation of precursor metabolites and energy(1e-7)
Carcinoma	65 (78)	ion transport(8e-16), biological adhesion(4e-10), anatomical structure development(2e-5), immune system process(1e-4)
Chromosome abnormality	132 (145)	biosynthetic process(4e-21)
Connective and Soft Tissue Neoplasm	25 (45)	RNA metabolic process(1e-4)
Disease Progression	169 (176)	DNA metabolic process(1e-132), response to endogenous stimulus(8e-65), cell cycle process(4e-48), response to stress(5e-43), cell cycle(2e-34), cell division(2e-13)
DNA Damage	42 (46)	biosynthetic process(1e-4)
Dysmyelopoietic Syndromes	228 (277)	RNA metabolic process(5e-6)
Hemic and Lymphatic Diseases	204 (322)	response to external stimulus(4e-9), defense response(2e-5)
Immune System Diseases	441 (474)	biosynthetic process(1e-124)
Immunoproliferative Disorders	517 (702)	defense response(2e-24), multi-organism process(2e-23), response to biotic stimulus(2e-23), immune system process(1e-12), response to external stimulus(1e-11), catabolic process(2e-10), protein folding(2e-4), carbohydrate metabolic process(3e-4)
Inflammation	18 (23)	anatomical structure morphogenesis(2e-4)
leukemia	438 (592)	RNA metabolic process(7e-23), protein folding(5e-16), defense response(8e-9), immune system process(6e-8), response to external stimulus(1e-9)
Leukemia, Myeloid, Acute	110 (159)	biosynthetic process(1e-11), RNA metabolic process(2e-5)
Lung	64 (69)	response to biotic stimulus(1e-20), immune system process(2e-20), multi-organism process(7e-18), cell-cell signaling(2e-5), regulation of transcription from RNA polymerase II promoter(3e-4)
Lung diseases	159 (186)	biosynthetic process(4e-20), establishment of protein localization(6e-9), establishment of cellular localization(7e-6)
Lymphatic Diseases	229 (322)	RNA metabolic process(5e-8)
Lymphoblastic Leukemia	107 (113)	biosynthetic process(2e-43)
Lymphoma	363 (381)	cell cycle process(7e-36), cell cycle(3e-27), cell division(9e-17), intracellular signaling cascade(7e-15), multi-organism process(8e-10), response to biotic stimulus(9e-10), DNA metabolic process(7e-9), biosynthetic process(1e-6), organelle organization and biogenesis(1e-6), immune system process(5e-6), phosphorus metabolic process(5e-5), response to stress(2e-4)
Muscle	58 (79)	muscle contraction(6e-6), biosynthetic process(7e-5)
Musculoskeletal Diseases	410 (685)	negative regulation of biological process(4e-5), cell motility(2e-4)
Myeloid Leukemia	279 (334)	biosynthetic process(2e-5), DNA metabolic process(3e-5), cell cycle(8e-4)
Myopathy	387 (668)	proteolysis(1e-7), ion transport(2e-6), cell motility(8e-6)
Neoplasm Metastasis	118 (150)	cellular developmental process(5e-35), regulation of multicellular organismal process(3e-26), keratinization(2e-23), biopolymer modification(1e-18), muscle contraction(2e-14), anatomical structure development(4e-7)
Neoplasms, Germ Cell and Embryonal	35 (37)	cell cycle process(4e-40), cell cycle(1e-30), cell division(2e-28), organelle organization and biogenesis(1e-11), DNA metabolic process(4e-7)
Neoplastic Processes	484 (668)	cellular developmental process(3e-144), keratinization(1e-91), biopolymer modification(8e-83), anatomical structure development(4e-71), biological adhesion(5e-40), ion transport(9e-18), regulation of multicellular organismal process(7e-4)
nervous system disorder	197 (336)	hydrogen transport(4e-4), cofactor metabolic process(4e-4)
Nutrition Disorders	27 (27)	biosynthetic process(3e-11)
Nutritional and Metabolic Diseases	47 (51)	biosynthetic process(5e-8)
sarcoma	34 (52)	cell cycle process(5e-6)
Skeletal muscle structure	600 (871)	ion transport(2e-5), cell motility(6e-5)

3.3 Example Modules Identified by Our Algorithm

Below we illustrate two examples of identified phenotype-specific modules, one from a disease phenotype and another from a tissue phenotype.

The first example is a 7-gene module (CSF3R, CD14, ITGB2, FCGR3B, LST1, S100A9, S100A12) which is specific to the phenotypes "Immunoproliferative Disorders" and "Lymphoproliferative Disorders", which annotate the same set of datasets (Figure 2a). The module has density higher than 0.66 in 5 out of the total 136 datasets. Of those 5 datasets, 4 are annotated with the phenotype "Lymphoproliferative Disorders" (GDS1067, GDS1284, GDS1388, GDS1454), which gives a specificity p-value of 6.2e-4. Two of these datasets study B-cell chronic lymphocytic leukemia and the other two study multiple myeloma. The fifth dataset, GDS1021, is not annotated with that phenotype, but, somewhat consistently with the four other datasets, it studies gene expression in peripheral blood mononuclear cells in renal cancer patients.

Strikingly, all 7 genes are annotated as "defense response," "immune system process," or both, in the GO biological process database. Additionally, 3 of the genes are associated with the UMLS concept "Lymphoproliferative Disorders," including LST1 and ITGB2, which have been used as diagnostic predictors of lymphoproliferative diseases. Further evidence of the cluster's validity as a biological module comes in the form of a direct protein-protein interaction between ITGB2 and CD14 *in vitro*. Multiple genes used for lymphoproliferative disorder diagnosis, interactions between module members, and complete immune system functional homogeneity all suggest a role for this module in lymphoproliferative diseases. Knowledge of this module can guide further experiments in the study of these diseases, as well as to elucidate the module's regulators.

The second example module consists of 9 genes, and is specific to the "Skeletal muscle structure" phenotype (Figure 2b). Three of the five active datasets study expression in muscle tissue (GDS268, GDS563, GDS1259). This cluster is highly functionally enriched, containing six genes that are annotated with a GO biological process related to muscle contraction. Specifically, in muscle fibers, troponin genes (TNNI1, TNNT1) along with tropomyosin (TPM3) associate with actin (ACTA1) to regulate muscle contraction via binding to the myosin complex (MYL2, MYH2, MYLPF). The module also contains SLN, which catalyzes the ATP-dependent transport of Ca^{2+} in muscle cells. In total, eight of the nine genes in the cluster are related to muscle, providing strong evidence for its phenotypic specificity.

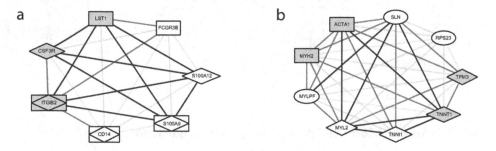

Fig. 2. Two examples of phenotype-specific modules. The opacity of an edge is proportional to the recurrence of the edge in the active datasets. a) A module specific to "Lymphoproliferative Disorders" datasets. Genes represented as rectangles are annotated with the GO term "immune system process." Gene represented as diamonds are annotated with the GO term "defense response." Shaded nodes represent genes implicated in "Lymphoproliferative Disorders" b) A module specific to the "Skeletal muscle structure" datasets. Genes represented as diamonds and rectangles are annotated with GO terms "regulation of muscle contraction" and "muscle contraction" respectively. The shaded genes are implicated in "Musculoskeletal Diseases."

Availability. The complete catalog of phenotype-specific gene clusters can be found at our website: http://zhoulab.usc.edu/Phenotype/

4 Discussion

The importance of considering the phenotypic context of gene modules cannot be overstated. Ultimately, molecular understanding is most useful when its macroscopic effects are well understood. In this paper, we described a graph-based approach integrating many microarray datasets to derive a genome-wide mapping of coexpression modules to phenotypes.

The provable computational complexity of this problem drew us to stochastic algorithms, and as a result we developed a number of useful graph-mining optimizations to the simulated annealing method. Firstly, we devised a strategy to divide the search space effectively by defining fully connected triplet (triangle) seeds. Secondly, we designed highly robust energy functions that could be linearly combined over a range of weights. And thirdly, we designed a method to prioritize neighbor searching. Overall, we have demonstrated that simulated annealing is a highly effective and adaptable strategy for pattern-mining in graphs.

We associated gene modules with human diseases on a genome-wide scale. The resulting map emphasizes that multiple genes must act together to effect phenotype, and, more specifically, that a gene in different contexts may participate in the manifestation different phenotypes. It has not escaped our notice that our map may represent the largest collection of examples of genetic pleiotropy to date. We reserve the results of this analysis for a future work. [23,24,25]

In this study, we applied our method to microarray data, which is so far the most abundant data measuring the genome-wide molecular activity under different phenotype conditions. We are well aware that microarray data has limitations, and that not all module activities can be assessed with expression profiling. We emphasize, however, that our method is generally applicable to any kind of abundant network data having clearly defined phenotype annotations. One possibility is a dynamically-annotated protein-protein interaction network consisting of conditional interactions. [26] Given the current unrelenting pace of technological innovation in the biological sciences, we envision that a vast amount of genome-wide, phenome-annotated profiling data will soon complement our current view of the genome-phenome association, for not only mRNA but also other molecules, such as protein and miRNA.

References

1. Lander, E.S., Schork, N.J.: Genetic dissection of complex traits. Science 265(5181), 2037–2048 (1994)
2. Risch, N.J.: Searching for genetic determinants in the new millennium. Nature 405(6788), 847–856 (2000)
3. Zhou, X., Kao, M.C.J., Wong, W.H.: Transitive functional annotation by shortest-path analysis of gene expression data. Proc. Natl. Acad. Sci. USA 99(20), 12783–12788 (2002)

4. Bader, G.D., Hogue, C.W.V.: An automated method for finding molecular complexes in large protein interaction networks. BMC Bioinformatics 4, 2 (2003)
5. Spirin, V., Mirny, L.A.: Protein complexes and functional modules in molecular networks. Proc. Natl. Acad. Sci. USA 100(21), 12123–12128 (2003)
6. Kelley, B.P., Sharan, R., Karp, R.M., Sittler, T., Root, D.E., Stockwell, B.R., Ideker, T.: Conserved pathways within bacteria and yeast as revealed by global protein network alignment. Proc. Natl. Acad. Sci. U.S.A. 100(20), 11394–11399 (2003)
7. Hu, H., Yan, X., Huang, Y., Han, J., Zhou, X.J.: Mining coherent dense subgraphs across massive biological networks for functional discovery. Bioinformatics 21(Suppl 1), i213–i221 (2005)
8. Yip, A.M., Horvath, S.: Gene network interconnectedness and the generalized topological overlap measure. BMC Bioinformatics 8, 22 (2007)
9. Butte, A.J., Kohane, I.S.: Creation and implications of a phenome-genome network. Nat. Biotechnol. 24(1), 55–62 (2006)
10. Barrett, T., Troup, D.B., Wilhite, S.E., Ledoux, P., Rudnev, D., Evangelista, C., Kim, I.F., Soboleva, A., Tomashevsky, M., Edgar, R.: Ncbi geo: mining tens of millions of expression profiles–database and tools update. Nucleic Acids Res. 35(Database issue), D760–D765 (2007)
11. Lage, K., Karlberg, E.O., Størling, Z.M., Olason, P.I., Pedersen, A.G., Rigina, O., Hinsby, A.M., Tümer, Z., Pociot, F., Tommerup, N., Moreau, Y., Brunak, S.: A human phenome-interactome network of protein complexes implicated in genetic disorders. Nat. Biotechnol. 25(3), 309–316 (2007)
12. Bodenreider, O.: The unified medical language system (umls): integrating biomedical terminology. Nucleic Acids Res. 32(Database issue), D267–D270 (2004)
13. Kirkpatrick, S., Gelatt, C., Vecchi, M.: Optimization by simulated annealing. Science 220(4598), 671–680 (1983)
14. Zhou, X.J., Kao, M.C.J., Huang, H., Wong, A., Nunez-Iglesias, J., Primig, M., Aparicio, O.M., Finch, C.E., Morgan, T.E., Wong, W.H.: Functional annotation and network reconstruction through cross-platform integration of microarray data. Nat. Biotechnol. 23(2), 238–243 (2005)
15. Yan, X., Mehan, M.R., Huang, Y., Waterman, M.S., Yu, P.S., Zhou, X.J.: A graph-based approach to systematically reconstruct human transcriptional regulatory modules. Bioinformatics 23(13), i577–586 (2007)
16. Consortium, G.O.: The gene ontology (go) project in 2006. Nucleic Acids Res 34(Database issue), D322–D326 (2006)
17. Mitchell, J.A., Aronson, A.R., Mork, J.G., Folk, L.C., Humphrey, S.M., Ward, J.M.: Gene indexing: characterization and analysis of nlm's generifs. In: AMIA Annual Symposium proceedings / AMIA Symposium AMIA Symposium, January 2003, pp. 460–464 (2003)
18. Freimer, N., Sabatti, C.: The human phenome project. Nat. Genet. 34(1), 15–21 (2003)
19. Butte, A.J., Chen, R.: Finding disease-related genomic experiments within an international repository: first steps in translational bioinformatics. In: AMIA Annual Symposium proceedings / AMIA Symposium AMIA Symposium, pp. 106–110 (2006)
20. Suman, B., Kumar, P.: A survey of simulated annealing as a tool for single and multiobjective optimization. Journal of the Operational Research Society 57(10), 1143–1160 (2006)
21. Collette, Y., Siarry, P.: Multiobjective Optimization: Principles and Case Studies, 2nd edn., pp. 45–51. Springer, Heidelberg (2004)

22. Geman, S., Geman, D.: Stochastic relaxation, gibbs distributions, and the bayesian restoration of images. IEEE-PAMI 6, 721–741 (1984)
23. Jeffery, C.J.: Multifunctional proteins: examples of gene sharing. Ann. Med. 35(1), 28–35 (2003)
24. Jeffery, C.J.: Moonlighting proteins: old proteins learning new tricks. Trends Genet. 19(8), 415–417 (2003)
25. Zhang, M.: Multiple functions of maspin in tumor progression and mouse development. Front. Biosci. 9, 2218–2226 (2004)
26. Han, J.D.J., Bertin, N., Hao, T., Goldberg, D.S., Berriz, G.F., Zhang, L.V., Dupuy, D., Walhout, A.J.M., Cusick, M.E., Roth, F.P., Vidal, M.: Evidence for dynamically organized modularity in the yeast protein-protein interaction network. Nature 430(6995), 88–93 (2004)

Fast and Accurate Alignment of Multiple Protein Networks

Maxim Kalaev[1], Vineet Bafna[2], and Roded Sharan[1]

[1] School of Computer Science, Tel Aviv University, Tel Aviv 69978, Israel
{kalaevma,roded}@post.tau.ac.il
[2] CSE, University of California San Diego, USA
vbafna@cs.ucsd.edu

Abstract. Comparative analysis of protein networks has proven to be a powerful approach for elucidating network structure and predicting protein function and interaction. A fundamental challenge for the successful application of this approach is to devise an efficient multiple network alignment algorithm. Here we present a novel framework for the problem. At the heart of the framework is a novel representation of multiple networks that is only linear in their size as opposed to current exponential representations. Our alignment algorithm is very efficient, being capable of aligning 10 networks with tens of thousands of proteins each in minutes. We show that our algorithm outperforms a previous strategy for the problem that is based on progressive alignment, and produces results that are more in line with current biological knowledge.

1 Introduction

Recent technological advances enable the systematic characterization of protein-protein interaction (PPI) networks across multiple species. Procedures such as yeast two-hybrid ([1]) and protein co-immunoprecipitation ([2]) are routinely employed nowadays to generate large-scale protein interaction networks for human and most model species ([3,4,5,6,7]). Key to interpreting these data is the inference of cellular machineries. As in other biological domains, a comparative approach provides a powerful basis for addressing this challenge, calling for algorithms for protein network alignment.

In the network alignment problem one has to identify network regions that are conserved in their sequence and interaction pattern across two or more species. While the general problem is hard, generalizing subgraph isomorphism, heuristic methods have been devised to tackle it. One heuristic approach for the problem creates a merged representation of the networks being compared, called a *network alignment graph*, facilitating the search for conserved subnetworks. In a network alignment graph, the nodes represent sets of proteins, one from each species, and the edges represent conserved PPIs across the investigated species.

The network alignment paradigm has been applied successfully by a number of authors to search for conserved pathways [8] and complexes [9,10,11]. However,

M. Vingron and L. Wong (Eds.): RECOMB 2008, LNBI 4955, pp. 246–256, 2008.

its extension to more than a few (3) networks proved difficult due to the exponential growth of the alignment graph with the number of species. Recently, an algorithm was suggested to overcome this difficulty, proposing the idea of imitating progressive sequence alignment techniques [12]. The latter algorithm was successfully applied to align up to 10 microbial networks. Very recently, Dutkowsky and Tiuryn [13] proposed another framework for efficient alignment of multiple networks, but this approach was applied to date to three networks only.

Here we propose a new algorithm for multiple network alignment that is based on a novel representation of the network data. The algorithm allows avoiding the explicit representation of every set of potentially orthologous proteins (which form a node in the network alignment graph), thereby achieving dramatic reduction in time and memory requirements. We compare our algorithm to previous approaches using various data sets, showing that it is extremely fast and accurate, outperforming the progressive alignment approach. For lack of space, some proofs are shortened or omitted.

2   Methods

2.1   Data Representation

Given k protein-protein interaction networks, we represent them using a k-layer graph, which we call a *layered alignment graph*. Each layer corresponds to a species and contains the corresponding network. Additional edges connect proteins from different layers if they are sequence similar. Formally, layer i has a set V_i of vertices and a set E_i of edges. For exposition purposes, assume that $|V_i| = n$ for all i. Additionally, we have a set of *inter-layer* denoted by E_H. Let $G_H = (\cup_i V_i, E_H)$ denote the graph restricted to the inter-layer edges. Let δ be the largest degree in G_H. The relation between an alignment graph and a layered alignment graph should be clear: while in the former every set of potentially orthologous proteins is represented by a vertex; in the latter such a set is represented by a subgraph of size k which includes a vertex from each of the layers. We call such a subgraph a k-*spine*. Key to the algorithmic approach presented below is the assumption that a k-spine corresponding to a set of truly orthologous proteins must be connected and, hence, admits a spanning tree. Thus, we can identify all potential vertex sets inducing k-spines by looking for trees instead.

A collection of (connected) k-spines induces a candidate conserved subnetwork. We score it using a likelihood ratio score as described in [11]. The score evaluates the fit of the protein-protein interactions within this subnetwork to a conserved subnetwork model versus the chance that they arise at random. The conserved subnetwork model assumes that each pair of proteins from the same species in the subnetwork should interact, independently of all other pairs, with high probability β. The random model assumes that each species' network was chosen uniformly at random from the collection of all graphs with the same vertex degrees as the ones observed. This random model induces a probability of occurrence p_{uv} for each edge (u, v) of the graph. To accommodate for information on the reliability of interactions, the interaction status of every vertex pair is treated as a noisy observation, and its reliability is combined into the likelihood score. Overall, for a subnetwork with vertex set U, the likelihood ratio score factors over the vertex pairs in it: $\mathcal{L}(U) = \sum_{(u,v) \in U \times U} w(u, v)$ where $w(v, v) = 0$ and for $u \neq v$,

$$w(u, v) = \log \frac{\beta Pr(O_{uv}|T_{uv}) + (1 - \beta) Pr(O_{uv}|F_{uv})}{p_{uv} Pr(O_{uv}|T_{uv}) + (1 - p_{uv}) Pr(O_{uv}|F_{uv})},$$

Here O_{uv} denotes the set of experimental observations on the interaction status of u and v, T_{uv} denotes the event that u and v truly interact, and F_{uv} denotes the event the u and v do not interact. The computation of $Pr(O_{uv}|T_{uv})$ and $Pr(O_{uv}|F_{uv})$ is based on the reliability assigned to the interaction between u and v (see [11] for further details).

This notion of a conserved subnetwork is extended easily to a layered alignment graph. If we considered every k-spine to be a (super-)node in a graph, then an m-node subgraph is a subgraph of m k-spines, with a dense interconnection of PPI edges. Formally, define an *m-subnet* as a collection U of k multi-sets $U_i = \{u_i[1], \ldots, u_i[m]\}$ with the following properties:

- For all $1 \leq i \leq k$ and $1 \leq j \leq m$, $u_i[j] \in V_i$.
- For all $1 \leq j \leq m$, the set $U[j] = \{u_1[j], u_2[j], \ldots, u_k[j]\}$ is a k-spine.

The score $\mathcal{S}(U)$ of the m-subnet is given by $\mathcal{S}(U) = \sum_{i=1}^{k} \mathcal{L}(U_i)$.

2.2 The Search Algorithm

The main algorithmic task is to look for high scoring m-subnets, for a fixed m. This problem is computationally hard even when there is only a single network, and edge-weights are restricted to +1 for all edges, and −1 for all non-edges [14]. Thus, we resort to a greedy heuristic which starts from high weight seeds and expands them using local search. Such greedy heuristics have been successfully applied to search for conserved subnetworks in a network alignment graph [11].

There are two sub-tasks we need to tackle: (i) computing high weight seeds; and (ii) extending a seed. We provide algorithmic solutions for both tasks below.

Computing seeds: We start by computing d-subnets as *seeds*, where $d << m$. Notably, even when $d = 2$, we do not know of any algorithm better than the naive approach, which involves looking at all pairs of k-spines. This $O(n^{dk})$ time algorithm is intractable for typical sized networks, so we consider two assumptions on the inter-layer edges that reduce the computational complexity while retaining sensitivity.

The first assumption asserts that the k-spines of a seed support the same topology of inter-connections. This is motivated by the observation that proteins within the same pathway or complex are typically present or absent in the genome as a group [15]. Thus, we consider the following problem:

Problem 1. d-identical-spine-subnet : Compute a set of d k-spines with identical topologies and maximum score.

Theorem 1. *The d-identical-spine-subnet problem admits an $O((n\delta)^d k 3^k)$ solution.*

Proof. Recall that a d-subnet can be described as a collection U of size d multisets $U_1, U_2, \ldots U_k$. Let $(U_{i_1}, U_{i_2}) \in E_H$ iff $(u_{i_1}[j], u_{i_2}[j]) \in E_H$ for all $1 \le j \le d$.

First, consider the case where each of the d k-spines is restricted to be a path (Figure 1). This implies that the d-subnet itself can be considered as a path $U_{i_1}, U_{i_2}, \ldots, U_{i_k}$. For a subset of species S, let $\mathscr{S}(U, S)$ denote the score of the best d-subnet that uses only species in S, and consists of a path that ends with U. Let $s(U)$ be the species corresponding to U. To compute $\mathscr{S}(U, S)$, note that we only need to recurse using the predecessor of U in the path. Formally:

$$\mathscr{S}(U, S) = \begin{cases} \max_{\substack{(U,W) \in E_H \\ s(W) \in S \setminus \{s(U)\}}} \mathscr{S}(W, S \setminus \{s(U)\}) + \mathcal{L}(U) & \text{if } |S| > 1 \\ \\ \mathcal{L}(U) & \text{if } |S| = 1 \end{cases}$$

Thus, for paths, the overall complexity is $O((n\delta)^d k 2^k)$.

A similar recursion can be applied when searching for k-spines that are trees with identical topology. For a subset of species S, let $\mathscr{S}(U, S)$ denote the score of the best d-subnet that uses only the species in S, and consists of a tree rooted at U. Then for $|S| > 1$:

$$\mathscr{S}(U, S) = \max_{\substack{(U,W) \in E_H, S_1 \subset S \\ s(U) \in S_1, s(W) \in S \setminus S_1}} \mathscr{S}(U, S_1) + \mathscr{S}(W, S \setminus S_1)$$

The overall complexity is $O((n\delta)^d k 3^k)$.   ♣

A second, slightly different assumption is based on the phylogeny (described as a rooted, binary tree T) of the investigated species. Consider a set of nodes

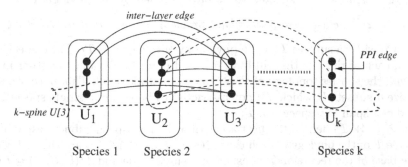

Fig. 1. A seed defined by a d-identical-spine subnet, where the k-spines are restricted to be paths with identical topology. The dashed line encloses one of the three k-spines.

a, b, c whose underlying species follow the phylogenetic triple $(s(a), (s(b), s(c)))$. We make the following *phylogenetic* assumption: if a, b, c are connected via inter-layer edges, then b and c must be connected. This implies that we can restrict our attention to k-spines that are *guided* by the phylogeny T in the following sense: any restriction of the k-spine to species that form a clade in T is a subtree of the k-spine. Note that two guided spines can have very different topologies (see Figure 2).

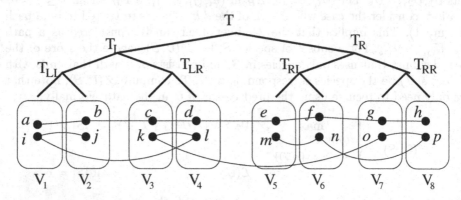

Fig. 2. Sketch of a 2-*guided-spine-subnet* . Note that while the paths of the two k-spines have different topologies, they are both guided by the underlying tree. Following the notation in the proof of Theorem 2, let $U = \{a, j\}, W = \{h, m\}$, and consider two possible distant sets $X = \{d, k\}$ and $Y = \{e, p\}$. By definition, $T_{LL}(U \cup X) = \{a, j\}$, $T_{LR}(U \cup X) = \{d, k\}$, $T_{RL}(Y \cup W) = \{e, m\}$, $T_{RR}(Y \cup W) = \{h, p\}$. Hence, $\mathscr{S}(U, W, T) \geq \mathscr{S}(\{a, j\}, \{d, k\}, T_L) + \mathscr{S}(\{e, m\}, \{h, p\}, T_R) \geq \mathscr{S}(\{a, i\}, \{b, j\}, T_{LL}) + \mathscr{S}(\{c, k\}, \{d, l\}, T_{LR}) + \mathscr{S}(\{e, m\}, \{f, n\}, T_{RL}) + \mathscr{S}(\{g, o\}, \{h, p\}, T_{RR}) \geq \mathcal{L}(a, i) + \mathcal{L}(b, j) + \mathcal{L}(c, k) + \mathcal{L}(d, l) + \mathcal{L}(e, m) + \mathcal{L}(f, n) + \mathcal{L}(g, o) + \mathcal{L}(h, p)$.

Problem 2. **The d-*guided-spine-subnet*** problem: Compute a set of d k-spines guided by the underlying phylogeny, with maximum score.

Unfortunately, we do not know of any efficient algorithm better than the naive $O(n^{kd})$ for this problem. However, we show a better solution for d-*guided-paths*, where the k-spines are restricted to be paths guided by the phylogeny.

Theorem 2. *The d-guided-path-subnet problem can be solved in $O(k^3(n^3\delta)^d)$.*

Proof. Consider a subtree T of the phylogeny with subtrees T_L, T_R, respectively. Clearly, each of the d paths will have one end-point in T_L, and the other in T_R. However, the species topology of these paths is not identical. Therefore, we work with size d subsets U which are not restricted to be within a single species, but instead can span any species in T.

Let $\mathscr{S}(U, W, T)$ denote the best score of a d-guided-path-subnet restricted to a subtree T of the phylogeny such that $s(U) \subseteq T_L, s(W) \subseteq T_R$ are the end nodes. At the base of the recursion T consists of a single node and $\mathscr{S}(U, U, T) = \mathcal{L}(U)$. Otherwise, let $U = \langle u[1], u[2] \ldots u[d] \rangle \in T_L$, and $W = \langle w[1], w[2] \ldots w[d] \rangle \in T_R$. Denote the root of T by $root(T)$.

For a node u, s.t. $s(u) \in T$, define its *distant set* $\mathcal{D}_T(u) = \{x | LCA_T(s(u),$ $s(x)) = \text{root}(T)\}$, where $LCA_T(a,b)$ is the least common ancestor of a and b in T. Extend this to d elements by defining $\mathcal{D}_T(U) = \{X \mid LCA_T(s(u[j]), s(x[j])) = \text{root}(T) \; \forall j\}$ The key idea to note is that if $X \in \mathcal{D}_T(U)$, then for all j $s(x[j]) \in T_L, s(u[j]) \in T_R$ or $s(x[j]) \in T_R, s(u[j]) \in T_L$. Define $T_L(U \cup X) \; (T_R(U \cup X))$ as the set of all vertices in $U \cup X$ with species in $T_L \; (T_R)$. Then,

$$\mathscr{S}(U, W, T) = \max_{\substack{X \in \mathcal{D}_{T_L}(U) \\ Y \in \mathcal{D}_{T_R}(W) \\ (X,Y) \in E_H}} (\mathscr{S}(T_{LL}(U \cup X), T_{LR}(U \cup X), T_L) + \mathscr{S}(T_{RL}(Y \cup W), T_{RR}(Y \cup W), T_R))$$

For an example see Figure 2. For the running time, note that there are $k^2 n^{2d}$ cells in the table $\mathscr{S}$. For each cell, there are kn^d choices for the set X and for each there are δ^d choices for a set Y s.t. $(X,Y) \in E_H$. The total time is therefore $O(k^3(n^3\delta)^d)$. ♣

In fact, we can improve the running time to $O((k^2 n^2 \delta)^d)$ (the proof will appear in the full version of the paper), but this is still not practical for reasonable values of n.

Extending a seed: The next phase of the algorithm is performing an iterative expansion of the seed by adding, in each iteration, the k-spine that contributes the most to the score. Let us denote by $H = (V', E')$ the current seed, and by $\mathscr{S}(v, S)$ the score of the best partial extension of H by a subtree that is rooted at vertex v and visits the species in S. Further denote by $s(v)$ the species corresponding to vertex v, and let $W(v) = \sum_{u \in V'} w(u, v)$. Then $\mathscr{S}(v, S)$ can be computed using the following recursive relation:

$$\mathscr{S}(v, S) = \begin{cases} \max\limits_{\substack{(v,w) \in E_H, S_1 \subset S \\ s(v) \in S_1, s(w) \in S \setminus S_1}} \mathscr{S}(v, S_1) + \mathscr{S}(w, S \setminus S_1) & \text{if } |S| > 1 \\[2ex] \mathcal{L}(v) & \text{if } |S| = 1 \end{cases}$$

The overall complexity is $O(n\delta k 3^k)$.

There are two speedups one can introduce to this basic extension scheme. The first is to constrain k-spines to paths (rather than trees), obtaining an $O(n\delta k 2^k)$ time algorithm. The second is to set in advance the order of the species along the tree, eliminating the 3^k factor. We term this variant *restricted order* as opposed to the previous *relaxed order* variant.

2.3 Implementation Notes

We have designed a software package, *NetworkBLAST-M*, implementing the multiple network alignment approach outlined above. The implementation allows looking for 2-identical-spine seeds with spines constrained to trees with relaxed and restricted topologies. For efficiency reasons, we restricted the seed vertices in each network to be of distance at most 2 from one another.

To verify that using 2-identical-spines is adequate for our problem, we analyzed alignment nodes within conserved network regions output by NetworkBlast [11] for different networks sets. When aligning yeast, worm and fly networks, in 85% of the cases, the pertaining alignment nodes respected the yeast-worm-fly phylogeny-based orientation. In two additional microbial network sets (C. jejuni, E. coli, H. pylori and C. crescentus, V. cholerae and H. pylori) more than 95% of the alignment nodes respected the same phylogeny-based orientation. Moreover, 72% of the alignment nodes actually formed cliques in G_H.

The final collection of conserved subnetworks was filtered to remove redundant solutions. This was done using an iterative greedy procedure that selects each time the highest scoring subgraph and removes all subgraphs intersecting it by more than 50%. For two conserved subnetworks A and B, containing $|A|$ and $|B|$ proteins, respectively, the intersection rate is computed as the number of common proteins over $\min\{|A|, |B|\}$.

3 Results

We applied our algorithm to eukaryotic and microbial PPI networks, summarized in Table 1. The three eukaryotic networks were taken from [11] and the microbial networks were taken from [12]. As in [11], we used a BLAST E-value threshold of 10^{-7} for sequence similarity, ensuring a corrected significance value of 0.01.

We evaluated the identified conserved subnetworks by computing the functional coherency of their member proteins with respect to the biological process annotation of the gene ontology (GO) [16], for each species separately. To this end, we used the GO TermFinder tool [17] to compute empirical enrichment p-values, and corrected for multiple testing using the false discovery rate procedure [18]. For each species we report the percent of process coherent subnetworks discovered, and the number of distinct GO categories they cover. The

Table 1. A summary of the PPI networks analyzed in this study

Species (tax id)	#Proteins	#PPIs
S. coelicolor (100226)	6678	230409
E. coli E12 (83333)	4087	216326
M. tuberculosis (83332)	3457	128932
S. typhimurium (99287)	4239	94609
C. crescentus (190650)	3341	40524
V. cholerae (243277)	2948	36038
S. pneumoniae (170187)	1843	25726
C. jejuni (192222)	1442	22116
H. pylori (85962)	1070	12943
Synechocystis sp. (1148)	2371	69439
S. cerevisiae (4932)	4738	15147
C. elegans (6239	2853	4472
D. melanogaster (7227)	7165	23484

first measure quantifies the specificity of the method, and the second provides an indication on the sensitivity of the method.

To establish the validity of our method, we first compared it to Network-BLAST [11]. NetworkBLAST is an exhaustive approach that relies on explicitly constructing a network alignment graph and, hence, is limited in application to the alignment of up to 3 networks. Both methods use same scoring function and scoring parameters were set equal for both methods for fair comparison. The results in Table 2 show that the performance of NetworkBLAST-M is comparable to that of NetworkBLAST. The latter has higher specificity, but fewer GO categories enriched. The sensitivity of NetworkBLAST-M further improves when using the relaxed-order variant. Notably, the application of NetworkBLAST-M took less than 30 seconds in both configurations, while NetworkBLAST's run took more than six hours.

Table 2. A comparison of NetworkBLAST-M and NetworkBLAST on three eukaryotic networks. For these networks NetworkBLAST produced 59 conserved regions, while NetworkBLAST-M identified 64 regions in the restricted-order variant and 92 in the relaxed-order variant,

Species	Specificity (%)	# GO categories enriched
NetworkBLAST		
S. cerevisiae	100.0	14
C. elegans	88.0	13
D. melanogaster	94.9	16
NetworkBLAST-M restricted order		
S. cerevisiae	100.0	29
C. elegans	68.8	32
D. melanogaster	98.4	37
NetworkBLAST-M relaxed order		
S. cerevisiae	94.6	45
C. elegans	67.0	29
D. melanogaster	90.1	41

Next, we compared the performance of NetworkBLAST-M to that of Graemlin [12] on a set of 10 microbial networks. Graemlin's results were taken from the original publication, considering only alignments which contain all 10 species (a total of 21 conserved regions). NetworkBLAST-M was applied only in the restricted-order variant due to the high computation burden. The algorithm detected a total of 33 conserved network regions. As summarized in Table 3, NetworkBLAST-M outperforms Graemlin, providing uniformly higher specificity and sensitivity.

Statistics on the running times of NetworkBLAST-M on different sets of microbial networks with 3-10 species are given in Table 4. As evident, the restricted-order variant is considerably faster and can process up to 10 networks in minutes.

Table 3. A comparison of NetworkBLAST-M and Graemlin on 10 microbial networks. Results are provided for nine of the ten species for which we had gene ontology information (for Synechocystis we did not have functional information readily available).

Species	Specificity (%)	# GO categories enriched
NetworkBLAST-M restricted order		
S. coelicolor	100	17
E. coli E12	90	16
M. tuberculosis	87.9	17
S. typhimurium	93.1	14
C. crescentus	84.8	15
V. cholerae	90.6	16
S. pneumoniae	97.0	14
C. jejuni	96.2	12
H. pylori	92.3	13
Synechocystis	N/A	N/A
Graemlin		
S. coelicolor	71.4	12
E. coli E12	76.5	10
M. tuberculosis	76.9	8
S. typhimurium	81.3	10
C. crescentus	86.7	11
V. cholerae	80.0	9
S. pneumoniae	71.4	8
C. jejuni	76.9	9
H. pylori	56.3	8
Synechocystis	N/A	N/A

Table 4. NetworkBLAST-M run-time as a function of the number of species and the size of the layered alignment graph. All the tests were performed on Intel Xeon 3.06GHz 3GB memory machine.

#Species	#Nodes	#PPI edges	#Sequence similarity edges	Restricted order run time (sec)	Relaxed order run time (sec)
3	8132	102288	26834	40	44
5	11945	193843	57142	72	1587
7	17236	301365	103887	83	46686
10	31458	877032	327219	140	N/A

4   Conclusions

We have provided a fast and accurate framework for multiple network alignment. Our framework is based on a novel representation of multiple protein-protein interaction networks and the orthology relations among their proteins. The framework performs comparably to an exhaustive approach while allowing dramatic reduction in running time and memory requirements. It is shown to outperform a previous approach based on progressive alignment ideas.

Future research includes a more extensive comparison of the different seed computation variants presented here. Our initial experiments in this regard indicate that the relaxed-order yields higher sensitivity on eukaryotic data sets, while the two perform similarly on microbial networks (data not shown). This may reflect the fact that sequence similarity among the pertaining microbial proteins tends to be transitive and, hence, any order of the species will form a tree in G_H. The development of efficient network alignment techniques, such as the one described here, is crucial to the study of protein network evolution and is expected to become increasingly important as protein-protein interaction databases continue to grow in size and species coverage.

Acknowledgments

VB was supported in part by a research gift from Glaxo SmithKline. This research was supported by the Israel Science Foundation (grant no. 385/06).

References

1. Ito, T., Chiba, T., Yoshida, M.: Exploring the yeast protein interactome using comprehensive two-hybrid projects. Trends Biotechnology 19, 23–27 (2001)
2. Aebersold, R., Mann, M.: Mass spectrometry-based proteomics. Nature 422, 198–207 (2003)
3. Uetz, P., et al.: A comprehensive analysis of protein-protein interactions in Saccharomyces cerevisiae. Nature 403, 623–627 (2000)
4. Ito, T., et al.: A comprehensive two-hybrid analysis to explore the yeast protein interactome. Proc. Natl. Acad. Sci. USA 98, 4569–4574 (2001)
5. Ho, Y., et al.: Systematic identification of protein complexes in Saccharomyces cerevisiae by mass spectrometry. Nature 415, 180–183 (2002)
6. Gavin, A., et al.: Functional organization of the yeast proteome by systematic analysis of protein complexes. Nature 415, 141–147 (2002)
7. Stelzl, U., et al.: A human protein-protein interaction network: a resource for annotating the proteome. Cell 122, 830–832 (2005)
8. Kelley, B., et al.: Conserved pathways within bacteria and yeast as revealed by global protein network alignment. Proc. Natl. Acad. Sci. 100, 11394–11399 (2003)
9. Sharan, R., Ideker, T., Kelley, B., Shamir, R., Karp, R.: Identification of protein complexes by comparative analysis of yeast and bacterial protein interaction data. Journal of Computational Biology 12, 835–846 (2005)
10. Koyuturk, M., et al.: Pairwise local alignment of protein interaction networks guided by models of evolution. Journal of Computational Biology 13, 182–199 (2006)
11. Sharan, R., et al.: Conserved patterns of protein interaction in multiple species. Proc. Natl. Acad. Sci. 102, 1974–1979 (2005)
12. Flannick, J., Novak, A., Srinivasan, B., McAdams, H., Batzoglou, S.: Graemlin:general and robust alignment of multiple large interaction networks. Genome Research 16, 1169–1181 (2006)
13. Dutkowsky, J., Tiuryn, J.: Identification of functional modules from conserved ancestral protein-protein interactions. Bioinformatics 23, 149–158 (2007)

14. Shamir, R., Sharan, R., Tsur, D.: Cluster graph modification problems. Discrete Applied Mathematics 144, 173–182 (2004)
15. Pellegrini, M., Marcotte, E.M., Thompson, M.J., Eisenberg, D., Yeates, T.O.: Assigning protein functions by comparative genome analysis: Protein phylogenetic profiles. PNAS 96, 4285–4288 (1999)
16. Ashburner, M., et al.: The gene ontology consortium. gene ontology: Tool for the unification of biology 25, 25–29 (2000)
17. Boyle, E., Weng, S., Gollub, J., Jin, H., Botstein, D., Cherry, J., Sherlock, G.: Go:termfinder–open source software for accessing gene ontology information and finding significantly enriched gene ontology terms associated with a list of genes. Bioinformatics 20, 3710–3715 (2004)
18. Benjamini, Y., Hochberg, Y.: Controlling the false discovery rate: a practical and powerful approach to multiple testing. Journal of the Royal Statistical Society 57 (1), 289–300 (1995)

High-Resolution Modeling of Cellular Signaling Networks

Michael Baym[1,2,*], Chris Bakal[3,4,*], Norbert Perrimon[3,4],
and Bonnie Berger[1,2,**]

[1] Department of Mathematics, MIT, Cambridge, MA 02139
[2] Computer Science and Artificial Intelligence Laboratory, MIT, 02139
[3] Department of Genetics, Harvard Medical School, Boston, MA 02115
[4] Howard Hughes Medical Institute, Boston MA 02215
bab@mit.edu

Abstract. A central challenge in systems biology is the reconstruction
of biological networks from high-throughput data sets. A particularly difficult case of this is the inference of dynamic cellular signaling networks.
Within signaling networks, a common motif is that of many activators
and inhibitors acting upon a small set of substrates. Here we present a
novel technique for high-resolution inference of signalling networks from
perturbation data based on parameterized modeling of biochemical rates.
We also introduce a powerful new signal-processing method for reduction
of batch effects in microarray data. We demonstrate the efficacy of these
techniques on data from experiments we performed on the *Drosophila*
Rho-signaling network, correctly identifying many known features of the
network. In comparison to existing techniques, we are able to provide significantly improved prediction of signaling networks on simulated data,
and higher robustness to the noise inherent in all high-throughput experiments. While previous methods have been effective at inferring biological
networks in broad statistical strokes, this work takes the further step of
modeling both specific interactions and correlations in the background
to increase the resolution. The generality of our techniques should allow
them to be applied to a wide variety of networks.

1 Introduction

Biological signaling networks regulate a host of cellular processes in response
to environmental cues. Due to the complexity of the networks and the lack of
effective experimental and computational tools, there are still few biological signaling networks for which a systems-level, yet detailed, description is known [1].
Substantial evidence now exists that the architecture of these networks is highly
complex, consisting in large part of enzymes that act as molecular switches to
activate and inhibit downstream substrates via post-translational modification.
These substrates are often themselves enzymes, acting in similar fashion.

* These authors contributed equally to this work.
** To whom correspondence should be addressed.

M. Vingron and L. Wong (Eds.): RECOMB 2008, LNBI 4955, pp. 257–271, 2008.
© Springer-Verlag Berlin Heidelberg 2008

In experiments, we are able to genetically inhibit or over-express the levels of activators, inhibitors and the substrates themselves, but rarely are able to directly observe the levels of active substrate in cells. Without the ability to directly observe the biochemical repercussions of inhibiting an enzyme in real-time, determining the true in vivo targets of these enzymes requires indirect observation of genetic perturbation and inference of enzyme-substrate relationships. For example, it is possible to observe downstream transcription levels which are affected in an unknown way by the level of active substrate [2].

The specific problem we address is the reconstruction of cellular signaling networks studied by perturbing components of the network, and reading the results via microarrays. We take a model-based approach to the problem of reconstructing network topology. For every pair of proteins in the network, we predict the most likely strength of interaction based on the data, and from this predict the topology of the network. This is computationally feasible as we are considering a subset of proteins for which we know the general network motif.

We demonstrate the efficacy of this approach by inferring from experiments the Rho-signaling network in *Drosophila*, in which some 40 enzymes activate and inhibit a set of approximately seven substrates. This network plays a critical role in cell adhesion and motility, and disruptions in the orthologous network in humans have been implicated in a number of different forms of cancer [3]. This structure, with many enzymes and few substrates (Fig. 1), is a common motif in signaling networks [4, 5].

To complicate the inference of the Rho-signaling network further, not every enzyme-substrate interaction predicted *in vitro* is reflected *in vivo* [6]. As such, we need more subtle information than is provided by current high-throughput protein-protein interaction techniques such as yeast two-hybrid screening [7, 8].

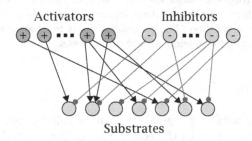

Fig. 1. The many enzyme-few substrate motif. A triangular arrowhead represents activation, a circular arrowhead inhibition.

To probe this network, we have carried out and analyzed a series of knockout and overexpression experiments in the *Drosophila* S2R+ cell line. We measure the regulatory effects of these changes using DNA microarrays. It is important to note that microarrays measure the relative abundance of the gene transcript, which can be used as a rough proxy for the total concentration of gene product. What they do not elucidate, however, is the relative fraction of an enzyme in an active or inactive state, which is crucial to the behavior of signaling networks.

To reconstruct the network from measurement, rather than directly use the microarray features corresponding to the proteins of interest, we instead use correlations in observations of the affected downstream gene products.

A number of related techniques for inferring global patterns based on high-throughput data exist. Many of these utilize the technique of probabilistic graphical models [9, 10, 11, 12, 13]. While these techniques are effective for inferring networks in broad statistical strokes, we increase the resolution and model the rate coefficients of individual reactions. The mathematics of our methodology is in fact isomorphic to a probabilistic graphical model approach; however as our parameters correspond directly to physical quantities or coefficients, we are able to dramatically narrow our model space when compared to a more general technique such as Bayesian or Markov networks [9]. In doing so we are able to gain both greater sensitivity, specificity, and robustness to noise. A related technique, based on modeling of rate kinetics in the framework of Dynamic Bayesian Networks has been effective in modeling genetic regulatory networks [14]. Techniques from information theory, such as ARACNE (Algorithm for the Reconstruction of Accurate Cellular Networks) [15, 16] and nonparameteric statistics, such as GSEA (Gene Set Enrichment Analysis) [17] have also been used to infer connections in high-throughput experiments. While not generally used for signaling network reconstruction, GSEA notably has been popular recently [18, 19], in part for its efficacy in overcoming batch effect noise.

We take the novel approach of constructing and optimizing a detailed parameterized model, based on the biochemistry of the network we aim to reconstruct. For the first part of the network model, namely the connections of the enzymes to substrates, we know the specific rate equations for substrate activation and inhibition. By modeling the individual interactions in like manner to the well-established Michaelis-Mentin rate kinetics [20, 21, 14], we are able to construct a model of the effects of knockout experiments on the level of active substrate. Lacking prior information, we model the effect of the level of active substrate on the microarray data by a linear function. If the only source of error were uncorrelated Gaussian noise in the measurements, we could then simply fit the parameters of this model to the data to obtain a best guess at the model's topology.

However, noise and "batch effects" [18] in microarray data are a real-world complication for most inference methods, which we address in a novel way. Noise in microarrays is seemingly paradoxical. On one hand, identical samples plated onto two different microarrays will yield almost identical results [22, 23]. On the other hand, with many microarray data sets, when one simply clusters experiments by similarity of features, the strongest predictor of the results is to group by the day on which the experiment was performed. We hypothesize, in this analysis, that the batch effects in microarrays are in fact other cellular processes in the sample unrelated to the experimental state. Properly filtering the ever-present batch effects in microarray data requires more than simply considering them to be background noise. Specifically, instead of the standard approach of fitting the data to our signal and assuming noise cancels, we consider the data to be a combination of the signal we are interested in and a second, structured signal of the batch effects.

Fitting this many-parameter model with physical constraints to the actual data optimizes our prediction for the signaling network, with remarkably good results.

To test this method we have constructed random networks with structure similar to the expected biology, and used these to generate data in simulated experiments. We find that when compared to reconstructions based on naïve correlation, GSEA, and ARACNE, we were able to obtain significantly more accurate network reconstructions. That is to say, at every specificity we obtained better sensitivity and vice-versa. The details of how GSEA and ARACNE were used in this manner can be found in Sec. 3.1.

We have also reconstructed the Rho-signaling network in *Drosophila* S2R+ cells from a series of RNAi and overexpression experiments we performed. While very little is experimentally known about this network, of the 40 pairs for which we have any biological evidence, we were able to predict 26 correctly, considerably better than chance (a p-value of 0.0079). It is important to remember that this standard is far from certain, and the known data represents a small fraction of the over 180 connections we aim to predict. Notably, many of the global features of the predicted network are in line with what is believed from biological experiments. While there is little doubt that with further experiments we will predict a more accurate network, this is the first detailed systems-level model of the *Drosophila* Rho-signaling network.

Contributions. We have introduced a novel parameterized model-based approach to signaling network inference from high-throughput data. We use this to provide testable predictions for connections in the *Drosophila* Rho-signaling network. Large-scale general statistical techniques have painted networks in broad strokes. Given the broad generality of such modeling, and the prevalence of similar motifs to the example studied here, the present approach is a crucial step in the program of systems biology.

Additionally we have developed a method for incorporating a noise model into this fit so as to greatly reduce the impact of batch effects in microarray data. This approach to noise in microarrays is widely applicable.

2   Models and Algorithms

In broad terms, we first aim to derive a model of the effects of our perturbations on the data whose parameters correspond to the edge weights of the cellular signaling network we wish to reconstruct. We first model how the level of active substrate changes in response to perturbations of the activators or inhibitors. To do this we derive an equilibrium condition based on well-known biochemical rate kinetics. We then make a linear model of how this affects the experimental data.

To fully understand the data, however, requires more than simply a model of the network. We need, as pointed out earlier, to model the noise, in order

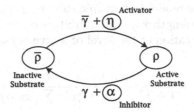

Fig. 2. The dynamics of an activator-inhibitor-substrate trio. The circled variables are proportional to protein concentrations.

to account for correlations in the background levels on unperturbed repeat experiments; we take a low-dimensional linear approximation of the batch effects present in microarray data. By fitting the parameters of the resultant model to the experimental data, we are able to predict both the topology and edge weights of the signaling network.

2.1   Biochemical Model

We first illustrate our approach for a single activator-inhibitor-substrate trio before extending to the many-node case. We start by deriving the time dependence of the concentration[1] ρ of active substrate in terms of the concentrations $\bar{\rho}$ of inactive substrate, η of activator, α of inhibitor, and the base rates $\bar{\gamma}$ of activation and γ of de-activation. Fig. 2 depicts these kinetics. As the rate at which inactive substrate becomes active is proportional to its concentration times the rate of activation and vice-versa,

$$\frac{d\rho}{dt} = -\frac{d\bar{\rho}}{dt} = \bar{\rho}(\bar{\gamma} + \eta) - \rho(\gamma + \alpha). \tag{1}$$

We are primarily interested in ρ, the level of active substrate, as the downstream effects of the substrate are dependent on this concentration. As the measurements are taken several days after perturbation and are an average over the expression levels of many individual cells, by ergodicity we expect to find approximately the equilibrium $(d\rho/dt = 0)$ concentration of substrate.

Solving for ρ at equilibrium yields:

$$\rho = \frac{\kappa(\bar{\gamma} + \eta)}{\bar{\gamma} + \eta + \gamma + \alpha}. \tag{2}$$

where $\kappa = \rho + \bar{\rho}$ is total concentration of the substrate, approximately available from the microarray data. By choice of time units we can let $\bar{\gamma} = 1$. This result, by no coincidence, is similar to the familiar Michaelis-Mentin rate kinetics.

---

[1] Choice of units of concentration is absorbed by scalar factors of the fit once the x_{jk} and y_{jk} coefficients are added; see Eq. 3.

We now generalize the model to multiple substrates κ_k, interchangeable activators η_j with relative strength x_{kj}, and inhibitors α_j with relative strength y_{kj}. The equilibrium concentration of the level of active substrate ρ_k then becomes:

$$\rho_k = \frac{\kappa_k \left(1 + \sum_j x_{kj}\eta_j\right)}{1 + \sum_j x_{kj}\eta_j + \gamma_k + \sum_j y_{kj}\alpha_j}. \tag{3}$$

Lacking more detailed biological information, and aiming to avoid the introduction of unnecessary parameters, we assume a linear response from features in the microarray. Specifically, for a vector of microarray feature data φ, we model the effect as a general linear function of the levels of active substrate, of the form $a\rho + r$. Additionally we introduce a superscripted index z for those variables which vary by experiment. The level, φ_i^z, of the ith feature in microarray z is in our model:

$$\varphi_i^z = \sum_k a_{ik} \left(\frac{\kappa_k^z \left(1 + \sum_j x_{kj}\eta_j^z\right)}{1 + \sum_j x_{kj}\eta_j^z + \gamma_k + \sum_j y_{kj}\alpha_j^z} \right) + r_i + \beta_i^z + \epsilon_i^z, \tag{4}$$

where the batch effects β and noise ϵ are considered additively.

2.2   Noise Filtration

As batch effects in microarrays are highly correlated, our approach is to construct a linear model of their structure. Empirically, batch effects tend to have a small number, s, of significant singular values (from empirical data $s \simeq 4$). In the singular vector basis, we can model the batch effects as a (features $\times$ s) matrix c. To determine the background as a function of experiment batch, we rotate by an ($s \times$ batches) rotation matrix u. Thus $cu = \sum_j c_{ij}u_{jd}$ is a (features $\times$ batches) matrix whose columns are the background signal by batch. Finally to extract the batch effect for a given experiment z, we multiply by the characteristic function of experiments by batches, χ, where $\chi_d^z = 1$ if experiment z happened in batch d and is 0 otherwise. Our model of batch effects is then:

$$\beta_i = \sum_{l,d} c_{il}u_{ld}\chi_d^z. \tag{5}$$

All together, our detailed model for experimental data based on the network, experiments, and noise becomes:

$$\varphi_i^z = \sum_k a_{ik} \left(\frac{\kappa_k^z \left(1 + \sum_j x_{kj}\eta_j^z\right)}{1 + \sum_j x_{kj}\eta_j^z + \gamma_k + \sum_j y_{kj}\alpha_j^z} \right) + r_i + \sum_{l,d} c_{il}u_{ld}\chi_d^z + \epsilon_i. \tag{6}$$

2.3   Model Fitting

Having now constructed a model of our system, we minimize the least-squares difference between the model predictions and observed data (detailed in Sec. 3.2), to obtain optimal model parameters. The resultant values of x and y predict the relative strengths of the activator-substrate interactions.

It is important to keep in mind which parameters are known and which we must fit. We know s and χ from experiment. In lieu of detailed knowledge of the activity levels of the activator and inhibitor, we take κ_k^z, η_j^z and α_j^z to be 1 normally, 0 on those experiments for which the gene is silenced, and 2 for those in which it is overexpressed. The remaining fitting parameters of our model are x, y, a, γ, r, c, and u.

For a vector of experimental data d, we construct, as above, a model for the predicted data φ. Fitting the model to data is done by minimizing:

$$f(x, y, a, \gamma, r, c, u) = \sum_{i,z} (d_i^z - \varphi_i^z)^2, \tag{7}$$

where φ_i^z is given in Eq. 6, subject to the constraints

$$x_{kj}, y_{kj}, \delta_k, \kappa_k \geq 0 \tag{8}$$

and the additional constraint that u is a rotation matrix. As the solution space is non-convex and likely has local minima, we use a general trust-regions [24] method for minimization starting at multiple starting points. The fit with lowest objective value is taken to be the best predictor of the network.

To verify that we have more data than parameters, we consider a microarray with Φ features and a network model with a total of θ activators and inhibitors and σ substrates. Additionally we consider a 4-dimensional noise model for λ batches. Then for ζ experiments, we have more data than parameters precisely when:

$$\zeta > \sigma + 4 + \frac{(\theta + 3)\sigma + 4\lambda - 10}{\Phi} \tag{9}$$

In a realistic setting, for 26 enzymes, six substrates, with on average six experiments per batch, and assuming each experiment has at least 50 features, then we need to perform at least 14 experiments in order to have more data than parameters. As the batch effect model has substantially lower rank than the number of batches, as long as there are at least five batches, over-fitting is unlikely.

In the above setting with 70 experiments, network optimization takes approximately 8 hours on a Powerbook G4 using an off-the-shelf constrained local nonlinear optimization routine in the MATLAB Optimization Toolbox [25] to a convergence tolerance of $1e-6$. While we aim to find the network which globally minimizes f, this trust-regions based local search technique occasionally reaches the convergence threshold at a demonstrably sub-optimal value. Continuing to optimize on a subset of the variables followed by repeated total optimization is often sufficient to pass these obstacles. Nevertheless, this still yields a good network prediction (see below). With more refined optimization tools, we will likely make even more accurate predictions. While we find that in very noisy cases the global minimum of f is smaller than that predicted by the actual connections, an overfit of the data, in practice this is a good guess.

3   Results

3.1   Simulations

We have generated simulated data on randomly created networks. The density of activator-substrate and inhibitor-substrate connections was chosen to reflect what is expected in the Rho-signaling network described in Sec. 3.2. From this, we have generated model experiment sets consisting of one knockout twice of each of the substrates and a single knockout of each activator and inhibitor in batches in random order. To further mimic our biological data set we included at least one baseline experiment in each batch. From this model we simulated experimental data with both noise and a batch-effect signal and attempted to fit the generated data.

To test against other techniques, we applied the statistics used by GSEA and ARACNE, modified for use on our model data sets. While GSEA is not typically used for signaling network reconstruct, its general usefulness in microarray analysis necessitates the comparison. ARACNE, on the other hand, while designed for a similar situation, does not directly apply, and so needs to be modified to make a direct comparison. As a baseline, we also computed the naïve (Pearson) correlation of experimental states.

GSEA starts by constructing, for each experimental condition, two subsets ("gene sets") of the features, one positive and one negative, which are used as indicators of the condition. To test whether a specific state is represented in a new experiment, the Kolmogorov-Smirnov enrichment score of those subsets in the new data is calculated (for details, see [17]). If the positive set is positively enriched and the negative set negatively enriched, the test state is said to be represented in the data. Likewise if the reverse occurs, the state is said to be negatively represented. If both are positively or negatively enriched, GSEA does not make a prediction. We are able to apply GSEA by computing positive and negative gene sets based on perturbation data for the substrates and then testing for enrichment in each of states in which we perturb an activator or inhibitor.

ARACNE, on the other hand, begins by computing the kernel-smoothed approximate mutual information (AMI) of every pair of features (for details, see [15]). In order to remove transitive effects, for every trio of features A, B, C, the pair with the smallest mutual information is marked to not be an edge. The remaining set of all unmarked edges is then a prediction of the network. As already discussed, we do not have features in our experiment that correspond directly to the levels we wish to measure. However, treating each experimental state as a feature, we are able to apply the AMI metric to obtain the relative efficacies of the activator and inhibitor perturbation experiments as predictors of the substrate perturbations. We know from the outset that the network we are trying to predict has no induced triangles, and so ARACNE would not remove any of the edges. However, the relative strengths of these predictions yield a predicted network topology.

On noiseless data, with only a minimal set of experiments and batch effects of comparable size to the perturbation signal, we are able to achieve a perfect network reconstruction which was not achieved by any of the other methods we consider. On highly noisy data, we cannot reconstruct the network perfectly; however we consistently outperform the other methods in both specificity and sensitivity (Fig. 3). Moreover, we find that while the model alone out-performs other techniques (comparably to AMI), the batch effect fit is of crucial importance. While this is clearly a biased result, as the simulated data is generated by the same model we assume in the fit, it does show that we are able to obtain a partial reconstruction even under high noise conditions. As this is a best-guess model from prior biological knowledge, the assumptions are far from unreasonable.

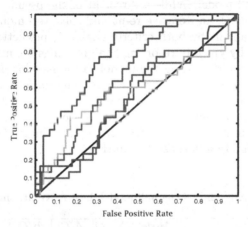

Fig. 3. Typical ROC curve for highly noisy simulated data. Our model (dark blue) is closest to the actual network, which would be a point at $[0, 1]$. Model fitting without batch effects (purple) is also considered. The other lines represent the predictions obtained by a GSEA-derived metric (red), an ARACNE-derived metric (light blue), and naïve correlation (green). The diagonal black line is the expected performance of random guessing. This particular set of simulated data has no repeat experiments for GAPs or GEFs, a batch signal of half the intensity of the perturbations, and an approximate total signal-to-noise ratio of 1.5.

3.2 Biological Data

We used our method, discussed above, on forthcoming microarray data collected from RNAi and overexpression experiments to predict the structure of the Rho-signaling network in *Drosophila* S2R+ cells. This network consists of approximately 47 proteins, divided roughly as 7 GTPases, 20 Guanine Nucleotide Exchange Factors (GEFs) and 20 GTPase Activating Proteins (GAPs). Importantly, we have the additional information that, despite their misleading names, the GEFs serve to activate certain GTPases and the GAPs serve to inhibit them. The exact connections, however, are for the vast majority, unknown.

Labeled aRNA, transcribed from cDNA, was prepared from S2R+ Drosophila cells following five days incubation with dsRNA or post-transfection of overexpression constructs. The aRNA was then hybridized to CombiMatrix 4x2k CustomArrays designed to include those genes most likely to yield a regulatory effect from a perturbation to the Rho-signaling network. After standard spatial and consensus Lowess [26] normalization, we k-means clustered [27] the data into 50 pseudo-features to capture only the large-scale variation in the data.[2]

After fitting, we have computed the significance of our fit using the Akaike and Bayesian Information Criteria (AIC and BIC) [28, 29]. These measure parameter fit quality as a function of the number of parameters, with smaller numbers being better. AIC tends to under-penalize free parameters while BIC tends to over-penalize, thus we computed both. As a baseline, we computed the AIC/BIC of the null model. While a direct fit of the pseudo-features yielded a lower AIC but not BIC, an iterative re-fit and solve technique, not unlike EM, produced a significant fit by both criteria (Table 1, prediction in Fig 5). This re-fitting was done by greedily resorting the groupings for meta-features based on the model fitness and refitting the model to the new meta-features. As each step strictly increases fit quality, and there are only finitely many sets of meta-features, this is naïvely guaranteed to converge in $O(n^k)$ iterations for n features and k meta-features. We find, however that the convergences generally to happens in around 5 iterations, leaving feature variance intact (an indication that this is not converging to a degenerate solution).

Table 1. AIC/BIC of the null model, best naïve fit, and best fit

Model	Fit (f)	AIC	BIC
Null Model ($\varphi_i^z = 0$)	0.9885	-8.389	-8.387
Best Fit	0.2342	-9.480	-8.366
Adapted Features	0.0328	-11.446	-10.332

To further test the accuracy of our model, we fit the model to four subsets of the 87 experiments and tested the prediction quality on the remaining experiments. The prediction error is calculated as the mean squared error of the predicted values divided by the mean standard deviation by feature. We tested on four sets: Sets 1 and 2 were chosen randomly to have nine (10.3% of experiments) and seventeen (19.5% of experiments) elements respectively, of which four of each are unduplicated experiments. Sets 3c and 4c were chosen randomly to have nine elements but were constrained not to have two elements from the same batch or experimental condition. We find that the model accurately predicts test set data (Table 2) for repeated experiments. Note that in Set 1, when

---

[2] The fact there are fewer than 50 significant singular values in the data and the linearity of a, r and β, indicates that we can not get more information from more clusters.

Table 2. Prediction error on test data.

Test Set	Size	# Unduplicated	Total Fit (f)	Test Set Fit	Error
1	9	4	0.0280	0.1307	14.6%
2	17	4	0.0288	0.0632	6.10%
3c	9	0	0.0302	0.0371	3.13%
4c	9	0	0.0301	0.0517	4.06%

44% of the experiments in the test set are non-duplicated, the prediction error is significantly higher. This indicates the necessity of both the batch and network components of the model.

While very little is known about the actual structure of the network, our reconstruction performed well when compared to previous biological data from in vivo experiments [30, 31, 32, 33, 34, 35, 36, 37] or mammalian homology, [38, 39, 40, 41, 42, 43, 44, 45, 46]. We predicted the existence of 57 of the 156 possible connections. Of the 23 known connections, both from in vivo experiments and inferred by orthology, we successfully predicted 11. Of the 17 pairs of proteins for which there is evidence they do not interact, we correctly predicted 15. This compares quite favorably to the predictions of other methods (Fig. 4). On this set of known interactions and non-interactions, the probability that our set of predicted connections overlapped correctly at least 26 times by chance is 0.0079. It is important to keep in mind that the known data represents less than a quarter of the testable connections predicted by our method.

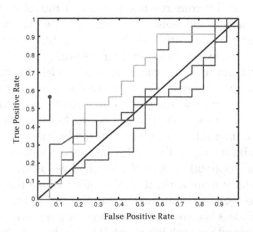

Fig. 4. ROC curve of network predictions vs. known data. Our model (dark blue) is closest, the curve discontinuity is on account of many of the predictions being zero. The other lines represent the predictions obtained by our model without a batch effect model (green), a GSEA-derived metric (purple), an ARACNE-derived metric (red), and naïve correlation (light blue).

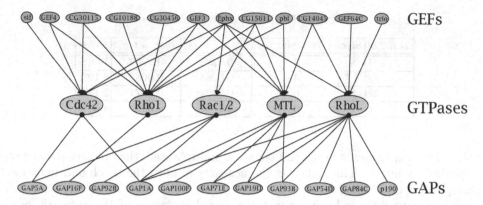

Fig. 5. The predicted Rho-signaling network in *Drosophila*

Two global network features of note, that the GTPase Rho1 is more highly connected than either Rac1/2 or Cdc42, and that the GEF Ephx has broad specificity, were reflected in our predictions as well. We also note that the prediction quality is not substantially different for GEFs (7 of 12 positives and 8 of 10 negatives) or GAPs (4 of 7 positives and 7 of 7 negatives).

4   Conclusion

In this paper we infer a signaling network from microarray data on perturbation experiments. We do so by constructing a detailed model of both the network and experimental background noise. We demonstrate the effectiveness of this technique on simulated data, and use it to make testable predictions of the connections in the *Drosophila* Rho-signaling network.

There are several natural extensions to our model. First, it is possible to backtrack errors in prediction in order to guide future experiments. We can also obtain a better fit on the unknown connections by incorporating further biological knowledge. For example, if it is known that a given enzyme-substrate pair does or does not interact, we can limit our model space to reflect this with an appropriate constraint on x_{jk} in Eq. 8. Recent advances in optimization promise greater efficiency and scalability than the method we used.

Our approaches have more general applicability. Since the many enzyme-few substrate motif is so common, we can use similar techniques to elucidate more networks as the data sets become available. Furthermore, microarray data is used in many contexts beyond network inference. The method of filtering batch effects proposed here will provide a potentially very useful tool for future exploration.

Acknowledgements. We are grateful to Jonathan Kelner, Kenneth Kamrin, and Nathan Palmer for helpful input. M.B. gratefully acknowledges support from the Fannie and John Hertz Foundation, and the National Defense Science and Engineering Program. C.B is a Fellow of the Leukemia and Lymphoma Society.

References

1. Friedman, A., Perrimon, N.: Genetic screening for signal transduction in the era of network biology. Cell 128, 225–231 (2007)
2. Hughes, T.R., Marton, M.J., Jones, A.R., Roberts, C.J., Stoughton, R., Armour, C.D., Bennett, H.A., Coffey, E., Dai, H., He, Y.D., Kidd, M.J., King, A.M., Meyer, M.R., Slade, D., Lum, P.Y., Stepaniants, S.B., Shoemaker, D.D., Gachotte, D., Chakraburtty, K., Simon, J., Bard, M., Friend, S.H.: Functional discovery via a compendium of expression profiles. Cell 102, 109–126 (2000)
3. Sahai, E., Marshall, C.J.: Rho-gtpases and cancer. Nat. Rev. Cancer 2(2), 133–142 (2002)
4. Albert, R.: Scale-free networks in cell biology. J. Cell Sci. 118(21), 4947–4957 (2005)
5. Csete, M., Doyle, J.: Bow ties, metabolism and disease. Trends in Biotechnology 22(9), 446–450 (2004)
6. Michiels, F., Habets, G.G.M., Stam, J.C., van der Kammen, R.A., Collard, J.G.: A role for rac in tiaml-induced membrane ruffling and invasion. Nature 375, 338–340 (1995)
7. Fields, S., Song, O.-K.: A novel genetic system to detect protein-protein interactions. Nature 340, 245–246 (1989)
8. Giot, L., Bader, J.S., Brouwer, C., Chaudhuri, A., Kuang, B., Li, Y., Hao, Y.L., Ooi, C.E., Godwin, B., Vitols, E., Vijayadamodar, G., Pochart, P., Machineni, H., Welsh, M., Kong, Y., Zerhusen, B., Malcolm, R., Varrone, Z., Collis, A., Minto, M., Burgess, S., McDaniel, L., Stimpson, E., Spriggs, F., Williams, J., Neurath, K., Ioime, N., Agee, M., Voss, E., Furtak, K., Renzulli, R., Aanensen, N., Carrolla, S., Bickelhaupt, E., Lazovatsky, Y., DaSilva, A., Zhong, J., Stanyon, C.A., Finley, R.L.: A protein interaction map of drosophila melanogaster. Science 302(5651), 1727–1736 (2003)
9. Friedman, N.: Inferring cellular networks using probabilistic graphical models. Science 303(5659), 799–805 (2004)
10. Sachs, K., Perez, O., Pe'er, D., Lauffenburger, D.A., Nolan, G.P.: Causal protein-signaling networks derived from multiparameter single-cell data. Science 308(5721), 523–529 (2005)
11. Friedman, N., Linial, M., Nachman, I., Pe'er, D.: Using Bayesian networks to analyze expression data. J. of Computational Biology 7(3-4), 601–620 (2000)
12. Peõer, D., Regev, A., Elidan, G., Friedman, N.: Inferring subnetworks from perturbed expression profiles. Bioinformatics 17, S214–S224 (2001)
13. Li, C., Suzuki, S., Ge, Q.-W., Nakata, M., Matsuno, H., Miyano, S.: Structural modeling and analysis of signaling pathways based on petri nets. J. Bioinformatics and Computational Biology 4(5), 1119–1140 (2006)
14. Nachman, I., Regev, A., Friedman, N.: Inferring quantitative models of regulatory networks from expression data. Bioinformatics 20(suppl. 1), i248–i256 (2004)
15. Margolin, A.A., Wang, K., Lim, W.K., Kustagi, M., Nemenman, I., Califano, A.: Reverse engineering cellular networks. Nature Protocols 1, 662–671 (2006)
16. Basso, K., Margolin, A.A., Stolovitzky, G., Klein, U., Dalla-Favera, R., Califano, A.: Reverse engineering of regulatory networks in human B cells. Nature Genetics 37(4), 382–390 (2005)
17. Subramanian, A., Tamayo, P., Mootha, V.K., Mukherjee, S., Ebert, B.L., Gillette, M.A., Paulovich, A., Pomeroy, S.L., Golub, T.R., Lander, E.S., Mesirov, J.P.: Gene set enrichment analysis: A knowledge-based approach for interpreting genome-wide expression profiles. Proc. Natl. Acad. Sci. USA 102(43), 15545–15550 (2005)

18. Lamb, J., Crawford, E.D., Peck, D., Modell, J.W., Blat, I.C., Wrobel, M.J., Lerner, J., Brunet, J.-P., Subramanian, A., Ross, K.N., Reich, M., Hieronymus, H., Wei, G., Armstrong, S.A., Haggarty, S.J., Clemons, P.A., Wei, R., Carr, S.A., Lander, E.S., Golub, T.R.: The Connectivity Map: Using Gene-Expression Signatures to Connect Small Molecules, Genes, and Disease. Science 313(5795), 1929–1935 (2006)

19. Baur, J.A., Pearson, K.J., Price, N.L., Jamieson, H.A., Lerin, C., Kalra, A., Prabhu, V.V., Allard, J.S., Lopez-Lluch, G., Lewis, K., Pistell, P.J., Poosala, S., Becker, K.G., Boss, O., Gwinn, D., Wang, M., Ramaswamy, S., Fishbein, K.W., Spencer, R.G., Lakatta, E.G., Couteur, D.L., Shaw, R.J., Navas, P., Puigserver, P., Ingram, D.K., de Cabo, R., Sinclair, D.A.: Resveratrol improves health and survival of mice on a high-calorie diet. Nature 444(7117), 337–342 (2006)

20. Michaelis, L., Menten, M.: Die kinetik der invertinwirkung. Biochem. Z. 49, 333–369 (1913)

21. Briggs, G.E., Haldane, J.B.S.: A note on the kinetics of enzyme action. Biochem. J. 19, 339–339 (1925)

22. Borup, R., Zhao, P., Nagaraju, K., Bakay, E.P.H.M., Chen, Y.-W.: Sources of variability and effect of experimental approach on expression profiling data interpretation. BMC Bioinformatics 3(4) (2002)

23. Larkin, J.E., Frank, B.C., Gavras, H., Sultana, R., Quackenbush, J.: Independence and reproducibility across microarray platforms. Nature Methods 2(5), 337–344 (2005)

24. Coleman, T., Li, Y.: An Interior, Trust Region Approach for Nonlinear Minimization Subject to Bounds. SIAM Journal on Optimization 6, 418–445 (1996)

25. The Mathworks: Optimization toolbox 3.1.2 (2007), http://www.mathworks.com/products/optimization/

26. Cleveland, W.S.: Robust locally weighted regression and smoothing scatterplots. J. Amer. Stat. Assoc. 74, 829–836 (1979)

27. Macqueen, J.B.: Some methods of classification and analysis of multivariate observations. In: Proceedings of the Fifth Berkeley Symposium on Mathemtical Statistics and Probability, pp. 281–297 (1967)

28. Akaike, H.: A new look at the statistical model identification. IEEE Transactions on Automatic Control 19(6), 716–723 (1974)

29. Schwarz, G.: Estimating the dimension of a model. Annals of Statistics 6(2), 461–464 (1978)

30. Sone, M., Hoshino, M., Suzuki, E., Kuroda, S., Kaibuchi, K., Nakagoshi, H., Saigo, K., Nabeshima, Y.-i., Hama, C.: Still life, a protein in synaptic terminals of drosophila homologous to gdp-gtp exchangers. Science 275(5299), 543–547 (1997)

31. Newsome, T.P., Schmidt, S., Dietzl, G., Keleman, K., Asling, B., Debant, A., Dickson, B.J.: Trio combines with dock to regulate pak activity during photoreceptor axon pathfinding in drosophila. Cell 101(3), 283–294 (2000)

32. Billuart, P., Winter, C.G., Maresh, A., Zhao, X., Luo, L.: Regulating axon branch stability: the role of p190 rhogap in repressing a retraction signaling pathway (2001)

33. Bashaw, G.J., Hu, H., Nobes, C.D., Goodman, C.S.: A novel dbl family rhogef promotes rho-dependent axon attraction to the central nervous system midline in drosophila and overcomes robo repulsion (2001)

34. Gonzalez, C.: Cell division: The place and time of cytokinesis. Current Biology 13(9), R363–R365 (2003)

35. Rossman, K.L., Der, C.J., Sondek, J.: Gef means go: turning on rho gtpases with guanine nucleotide-exchange factors. Nat. Rev. Mol. Cell Biol. 6(2), 167–180 (2005)

36. Hu, H., Li, M., Labrador, J.P., McEwen, J., Lai, E.C., Goodman, C.S., Bashaw, G.J.: Cross gtpase-activating protein (crossgap)/vilse links the roundabout receptor to rac to regulate midline repulsion (2005)
37. Nahm, M., Lee, M., Baek, S.-H., Yoon, J.-H., Kim, H.-H., Lee, Z.H., Lee, S.: Drosophila rhogef4 encodes a novel rhoa-specific guanine exchange factor that is highly expressed in the embryonic central nervous system. Gene 384, 139–144 (2006)
38. Reid, T., Bathoorn, A., Ahmadian, M.R., Collard, J.G.: Identification and characterization of hpem-2, a guanine nucleotide exchange factor specific for cdc42. J. Biol. Chem. 274(47), 33587–33593 (1999)
39. Shamah, S.M., Lin, M.Z., Goldberg, J.L., Estrach, S., Sahin, M., Hu, L., Bazalakova, M., Neve, R.L., Corfas, G., Debant, A., Greenberg, M.E.: Epha receptors regulate growth cone dynamics through the novel guanine nucleotide exchange factor ephexin. Cell 105(2), 233–244 (2001)
40. Hall, C., Michael, G.J., Cann, N., Ferrari, G., Teo, M., Jacobs, T., Monfries, C., Lim, L.: alpha2-chimaerin, a cdc42/rac1 regulator, is selectively expressed in the rat embryonic nervous system and is involved in neuritogenesis in n1e-115 neuroblastoma cells (2001)
41. Niu, J., Profirovic, J., Pan, H., Vaiskunaite, R., Voyno-Yasenetskaya, T.: G protein βγ subunits stimulate p114rhogef, a guanine nucleotide exchange factor for rhoa and rac1: Regulation of cell shape and reactive oxygen species production. Circ. Res. 93(9), 848–856 (2003)
42. Nagata, K.-I., Inagaki, M.: Cytoskeletal modification of rho guanine nucleotide exchange factor activity: identification of a rho guanine nucleotide exchange factor as a binding partner for sept9b, a mammalian septin. Oncogene 24(1), 65–76 (2004)
43. Wells, C.D., Fawcett, J.P., Traweger, A., Yamanaka, Y., Goudreault, M., Elder, K., Kulkarni, S., Gish, G., Virag, C., Lim, C., Colwill, K., Starostine, A., Metalnikov, P., Pawson, T.: A rich1/amot complex regulates the cdc42 gtpase and apical-polarity proteins in epithelial cells. cell 125(3), 535–548 (2006)
44. Cho, Y.J., Cunnick, J.M., Yi, S.J., Kaartinen, V., Groffen, J., Heisterkamp, N.: Abr and bcr, two homologous rac gtpase-activating proteins, control multiple cellular functions of murine macrophages (2007)
45. Dalva, M.B.: There's more than one way to skin a chimaerin (2007)
46. Mitin, N., Betts, L., Yohe, M.E., Der, C.J., Sondek, J., Rossman, K.L.: Release of autoinhibition of asef by apc leads to cdc42 activation and tumor suppression. Nat. Struct. Mol. Biol. 14(9), 814–823 (2007)

At the Origin of Life:
How Did Folded Proteins Evolve?

Andrei Lupas

Max-Planck-Institute for Developmental Biology, Tuebingen
andrei.lupas@tuebingen.mpg.de

Abstract. Proteins are essential building blocks of living cells; indeed, life can be viewed as resulting substantially from the chemical activity of proteins. Because of their importance, it is hardly surprising that ancestors for most proteins observed today were already present at the time of the 'last common ancestor', a primordial organism from which all life on Earth is descended. Yet folded proteins are too complex to have arisen de novo. How then did they evolve? We are pursuing the hypothesis that folded proteins evolved by fusion and recombination from an ancestral set of peptides, which emerged in the context of RNA-dependent replication and catalysis (the "RNA world"). Systematic studies should allow a description of this ancient peptide set in the same way in which ancient vocabularies have been reconstructed from the comparative study of modern languages.

About the Keynote Speaker. Born on September 6, 1963 in Bucharest (Romania). Studies in Biology at the Technical University Munich (1982–1985) and in Molecular Biology at Princeton University (1985–1990); PhD with Jeff Stock on the mechanism of signal transduction in bacterial chemotaxis (1991). Postdoctoral fellow with Andreas Plueckthun at the Gene Center of the University, Munich (1992–1993), working on antibody engineering. Research assistant with Wolfgang Baumeister at the Max-Planck-Institute for Biochemistry, Martinsried (1992–1997), working on the development and application of sequence analysis tools, and on the structure and function of the proteasome. Senior Computational Biologist, later Assistant Director of Bioinformatics, at SmithKline Beecham Pharmaceuticals, Collegeville, USA (1997–2001). Since 2001, director of the department of protein evolution at the Max-Planck-Institute for Developmental Biology, Tuebingen.

M. Vingron and L. Wong (Eds.): RECOMB 2008, LNBI 4955, p. 272, 2008.
© Springer-Verlag Berlin Heidelberg 2008

Locating Multiple Gene Duplications through Reconciled Trees

J. Gordon Burleigh[1], Mukul S. Bansal[2], Andre Wehe[2], and Oliver Eulenstein[2]

[1] National Evolutionary Synthesis Center, Durham, NC, USA
jgb12@duke.edu
[2] Department of Computer Science, Iowa State University, Ames, IA, USA
{bansal, awehe, oeulenst}@cs.iastate.edu

Abstract. We introduce the first exact and efficient algorithm for Guigó et al.'s problem that given a collection of rooted, binary gene trees and a rooted, binary species tree, determines a minimum number of locations for gene duplication events from the gene trees on the species tree. We examined the performance of our algorithm using a set of 85 genes trees that contain genes from a total of 136 plant taxa. There was evidence of large-scale gene duplication events in *Populus, Gossypium,* Poaceae, Asteraceae, Brassicaceae, Solanaceae, Fabaceae, and near the root of the eudicot clade. However, error in gene trees can produce erroneous evidence of large-scale duplication events, especially near the root of the species tree. Our algorithm can provide hypotheses for precise locations of large-scale gene duplication events with data from relatively few gene trees and can complement other genomic approaches to provide a more comprehensive view of ancient large-scale gene duplication events.

1 Introduction

Polyploidy is a major component of plant genome evolution [27,14]. Analyses of genomic data from numerous plants such as grasses [16,24], *Arabidopsis* or Brassicaceae [30,26,3], poplar [28], cotton [4], *Physcomitrella* [25], and *Vitis* [10] have revealed evidence of ancient genome duplications. Yet the number of ancient genome duplications and their precise location in the evolutionary history of plants is still unclear. We describe the first exact polynomial time algorithm for Guigó et al.'s problem [15] that maps large-scale gene duplications, such as polyploidy, on a species tree, and we demonstrate its ability to identify and place ancient polyploidy events in plants.

The presence of large, duplicated chromosomal segments within a genome provided the first evidence of ancient polyploidy (e.g. [30,26,3,5,16,24,10]). These duplications can be dated based on the sequence divergence between paralogous genes on duplicated blocks. However, rapid gene loss and gene rearrangements after a polyploidy event can make it difficult or impossible to detect ancient duplicated chromosomal segments [20,26], and few plant taxa have adequate gene mapping data. It is also possible to detect ancient polyploidy based solely on the age distributions of pairs of duplicated (paralogous) genes (e.g. [20,30,4,28,8,25]). The date of the inferred duplications is estimated from amino acid or, more commonly, silent (synonymous) substitution rates, using molecular clock assumptions. Examining genomic data from multiple taxa in a

M. Vingron and L. Wong (Eds.): RECOMB 2008, LNBI 4955, pp. 273–284, 2008.
© Springer-Verlag Berlin Heidelberg 2008

comparative phylogentic context has the potential to improve estimates of the timing of large-scale duplication events (e.g. [5,7]). In the simplest approach, a phylogenetic tree is constructed with a pair of paralogous genes from one taxon, and the best homolog from a second taxon and from an outgroup taxon [5,7]. This allows one to date the duplication from the first taxon relative to the divergence with the second taxon. Yet placing a duplication event relative to a single taxonomic divergence is not very specific.

Guigó et al. [15] first addressed a more comprehensive phylogenetic approach that maps duplication events from a collection of rooted, binary gene trees onto a rooted, binary species tree. Later on, Page and Cotton [22] refined this problem and used it to examine gene duplication events in vertebrates. We refer to the refined problem as the Episode Clustering problem. An alternative version of this problem was introduced by Fellows et al., which they proved to be intrinsically difficult [9]. Hence, we direct the focus of this work to the Episode Clustering problem. This problem determines duplication events using the Gene Duplication Model from Goodman et al. [13]. Each duplication can be placed on any species on a path between the two (not necessarily distinct) most recent species that could have contained the duplication and its parent respectively. In case the parent does not exist, the path runs between the most recent species for the duplication and the root of the species tree. An example is depicted in Fig. 1. The duplications in gene tree G are represented by the three bold nodes. Associated with each bold node is its path represented by an interval. For example, the interval $[5, 3]$ represents the path $5, 4, 3$ in the species trees S. Let g denote the node corresponding to the interval $[5, 3]$. Species 5 is the most recent species that could have contained g and the parent of species 3, i.e. 2, is the most recent species that could have contained the parent of g. The *Episode Clustering (EC)* problem is, given a collection of gene trees and a species tree, find a minimum number of locations in the species tree where all duplications in the gene trees can be placed. For example, all three duplications in Fig. 1 can be placed on species nodes 2 and 3. Page and Cotton [22] observed that the EC problem can be efficiently reduced to the set-cover problem [11]. They approach the EC problem using a heuristic for the intrinsically difficult set-cover problem. In this paper we present an efficient and exact solution for the EC problem, which is based on established graph theoretical results. Note, that the gene duplications and the paths where duplications can be placed are computable in linear time using efficient least common ancestor computations [2,31].

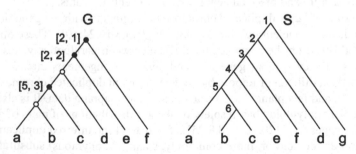

Fig. 1. A gene tree G and a comparable species tree S is depicted. The bold nodes in G are duplications and their intervals represent their allowed locations in the species tree S.

2 Methods

2.1 Basic Definitions, Notation, and Preliminaries

In this section we first introduce basic definitions and notation that we will be dealing with and then define preliminaries required for this work.

Basic Definitions and Notation. A *tree* T is a connected graph with no cycles, consisting of a node set $V(T)$ and an edge set $E(T)$. T is *rooted* if it has exactly one distinguished node called the *root* which we denote by $\text{Ro}(T)$. Let T be a rooted tree. We define $\leq_T$ to be the partial order on $V(T)$ where $x \leq_T y$ if y is a node on the path between $\text{Ro}(T)$ and x. We denote by $x \smallfrown_T y$ that x, y are related by $\leq_T$, and by $<_T$ the strict counterpart of the relation $\leq_T$. The set of minima under $\leq_T$ is denoted by $\text{Le}(T)$ and its elements are called *leaves*. If $x \leq_T y$ and $\{x, y\} \in E(T)$, then we call y the *parent* of x denoted by $\text{Pa}(x)$ and we call x a *child* of y. The set of all children of y is denoted by $\text{Ch}_T(y)$. The *least common ancestor (lca)* of a non-empty subset $L \subseteq V(T)$ denoted as $\text{lca}(L)$, is the unique smallest upper bound of L under $\leq_T$. A subtree of T rooted at node $y \in V(T)$, denoted by T_y, is the tree induced by $\{x \in V(T) \colon x <_T y\}$, T is called (fully) binary if every node has either zero or two children.

The *interval* for $a <_T b$ is defined as $[a, b] = \{x \subset V(T) \mid a <_T x \leq_T b\}$. Let $\mathcal{I}$ be a collection of intervals in $\leq_T$. The *node cover* of a node $v \in V(T)$ is defined as $cover(v) := \{I \in \mathcal{I} \mid v \in I\}$ and the *node cover* of a node set $V \subseteq V(T)$ is defined as $cover(V) = \bigcup_{v \in V} cover(v)$. A set $V \subseteq V(T)$ is called a *cover* of $\mathcal{I}$, if $cover(V) = \mathcal{I}$. If V is a cover of minimum cardinality, we call V a *minimum cover* of $\mathcal{I}$.

The *intersection graph* of a collection of intervals $\mathcal{I}$, denoted $int(\mathcal{I})$, is the graph $(\mathcal{I}, E)$ where $\{I, I'\} \in E$ precisely if $I \cap I' \neq \emptyset$. Let $G = (V, E)$ be a graph, then $V(G) = V$ and $E(G) = E$. A *clique* in G is a set $C \subseteq V$ which induces a completely connected subgraph in G. A *clique cover* of a G is a set of cliques $\mathcal{C}$ in G such that $\bigcup_{C \in \mathcal{C}} C = V$. A *minimum clique cover* is a clique cover of minimum size.

Problem 1. *Tree Interval Cover (TIC)*
   Instance: *A collection of intervals $\mathcal{I}$ in the order $\leq_T$.*
   Find: *A minimum cover of $\mathcal{I}$.*

The Episode Clustering problem is a special case of the TIC problem.

The Episode-Clustering (EC) Problem. The EC problem is to place duplications onto a minimum number of species in a species tree, where each duplication is associated with an interval in the species tree describing the locations where that duplication can be placed. The definition of duplication and its associated interval are based on the Gene Duplication (GD) model [23] introduced by Goodman et al. [13]. Here we only provide definitions necessary to state the EC problem.

   The GD model is based on a gene and species tree from which gene duplications and their associated intervals can be derived. A *species tree* is a tree that depicts the evolutionary relationships of a set of species. Given a gene family for a set of species, a *gene tree* is a tree that depicts the evolutionary relationships among the sequences encoding only that gene family in the given species. Thus the nodes in a gene tree represent genes. To compare a gene tree G with a species tree S a mapping from each gene $g \in V(G)$ to the most recent species in S that could have contained g is required.

Definition 1 (Mapping). *A leaf-mapping $\mathcal{L}_{G,S}$: $\text{Le}(G) \rightarrow \text{Le}(S)$ specifies, for each gene g the species from which it was sampled. The extension $\mathcal{M}_{G,S}$: $V(G) \rightarrow V(S)$ of $\mathcal{L}_{G,S}$ is the mapping defined by $\mathcal{M}_{G,S}(g) = \text{lca}(\mathcal{L}_{G,S}(\text{Le}(G_g)))$.*

Definition 2 (Comparability). *The trees G and S are comparable if there exists a leaf-mapping $\mathcal{L}_{G,S}$. A set of gene trees $\mathcal{G}$ and S are comparable if each gene tree in $\mathcal{G}$ is comparable with S.*

Throughout the remainder of this paper, $\mathcal{G}$ denotes a collection of input gene trees, S a comparable species tree, and G denotes an arbitrary gene tree in $\mathcal{G}$.

Definition 3 (Duplication). *A node $v \in V(G)$ is a (gene) duplication if $\mathcal{M}_{G,S}(v) = \mathcal{M}_{G,S}(u)$ for some $u \in \text{Ch}(v)$ and we define $\text{Dup}(G, S) = \{g \in V(G) \mid g$ is a duplication$\}$.*

Definition 4. *For every $g \in V(G)$ we define the interval*

$$I(g) = \begin{cases} [\mathcal{M}(g), \text{Ro}(S)], & \text{if } g = \text{Ro}(G), \\ [\mathcal{M}(g), \mathcal{M}(g)], & \text{if } \mathcal{M}(g) = \mathcal{M}(\text{Pa}(g)), \quad (1) \\ [\mathcal{M}(g), \mathcal{M}(\text{Pa}(g))] - \{\mathcal{M}(\text{Pa}(g))\}, & \text{otherwise.} \end{cases}$$

Problem 2. *Episode Clustering (EC)*
   Instance: *A collection of gene trees $\mathcal{G}$ and a comparable species tree S.*
   Find: *A solution to the TIC instance $\bigcup_{g \in \text{Dup}(\mathcal{G},S)} \{I(g)\}$ in the order $\leq_S$.*

The TIC instance $\bigcup_{g \in \text{Dup}(\mathcal{G},S)} \{I(g)\}$ can be computed in linear time [31] using efficient lca computation (e.g. [2]). To solve the EC problem we give an efficient solution for the TIC problem in the following section.

2.2 Solving the TIC Problem

Let $\mathcal{I}$ be a collection of intervals in the order $\leq_T$. In the interest of brevity, proofs for Lemmas 1 and 2, Theorems 1 and 2, and Corollary 1 appear in the Appendix.

Lemma 1. *Let C be a clique in the intersection graph $int(\mathcal{I})$. Then, $\bigcap_{I \in C} I$ is an interval in the order $\leq_T$. In particular $\bigcap_{I \in C} I = [a, b]$ where $a = \text{lca}(\bigcup_{[x,y] \in C} x)$ and $b = \min(\bigcup_{[x,y] \in C} y)$.*

Lemma 2. *Let $\mathcal{I}$ be a collection of intervals over $\leq_T$ and $V \subseteq V(T)$ covers $\mathcal{I}$. Then, $\mathcal{C} := \bigcup_{v \in V} \{cover(v)\}$ forms a clique cover of the intersection graph $int(\mathcal{I})$.*

Theorem 1. *Let $\mathcal{I}$ be a collection of intervals over $\leq_T$, and $\mathcal{C}$ be a minimum clique cover of the intersection graph $int(\mathcal{I})$. Define the function $f \colon \mathcal{C} \rightarrow V(T)$ that maps $f(C)$ to some element in $\bigcap_{I \in C} I$. Note, f is well defined by Lemma 1. Then, the node set $f(\mathcal{C})$ is a minimum interval cover of $\mathcal{I}$.*

The following two results are well known (see [21], and [12]).

Lemma 3. *If G is the intersection graph of a family of paths on a tree, then G is triangulated.*

Every interval in $\leq_T$ is equivalent to a path on T. Thus, the intersection graph $int(\mathcal{I})$ is triangulated.

Lemma 4. *Given a triangulated graph G with n nodes and m edges, a minimum clique cover for G can be computed in $O(n + m)$ time.*

Theorem 2. *Given a collection of intervals $\mathcal{I}$ in $\leq_T$ that are presented through paths on the tree T. Then, the TIC problem can be solved in $O(n^2 + nm + l)$ where $n = |V(int(\mathcal{I}))|$, $m = |E(int(\mathcal{I}))|$ and $l = |\text{Le}(T)|$.*

Corollary 1. *Let $\mathcal{G}$ be a collection of gene trees and S a comparable species tree, where $k = \Sigma_{G \in \mathcal{G}} |\text{Le}(G)|$ and $l = |\text{Le}(S)|$. Then, the EC problem for the instance $\mathcal{G}$ and S can be solved in $O(k^2 + km + l)$ time, where m is the number of intersecting intervals that are associated with the duplications in the collection of gene trees $\mathcal{G}$*

2.3 Plant Gene Analysis

We tested our algorithm using a set of plant gene family trees made from alignments obtained from Phytome, an online comparative genomics database for plants [18]. We selected the masked amino acid alignments from all 85 gene families in Phytome that contain sequences from at least 100 of the 136 total taxa. The gene trees were inferred with maximum likelihood (ML) phylogenetic analyses using RAxML-VI-HPC version 2.2.3. The ML analyses used the JTT amino acid substitution model [19] with the PROTMIX option for modeling rate variation among sites. The ML gene trees were first rooted using mid-point rooting. However, if any alternate rootings of the gene trees decreased the minimum number of gene duplications needed to reconcile the gene trees with the species tree, we chose a rooting that minimizes the number of duplications. Finally, since it is difficult to distinguish allelic variants of a single gene from paralogs, if a gene tree had any clades that contain only sequences from a single taxon, we removed all but a single leaf from the clade. We used a species tree based on currently accepted plant phylogenetic hypotheses (e.g. [1]).

Inferring Gene Duplications Events. We used our EC algorithm to infer the minimum number of duplication locations for the set of ML gene trees on the specified species tree. Our algorithm provides a solution for the minimum number of duplication locations that also includes the total number of duplications at each node, the number of duplication episodes at each node, and the number of genes with duplications at each node. In order to examine the performance of our algorithm in the absence of phylogenetic signal, we also performed 10 replicates our analysis after randomly permuting the leaf labels from each of the gene trees. This experiment will provide an expectation of the results of our algorithm if there was no phylogenetic signal in the gene trees, or if the gene trees were essentially random.

3 Results

Plant Duplication Analysis. We found that gene duplication events involving at least one of the 85 gene trees occur on a minimum of 119 of the 135 internal nodes. While some nodes show evidence of many duplications, others have evidence of very few duplications. For example, 51 nodes have evidence of ≤ 10 duplications, and 4 nodes have evidence of ≥ 1000 duplications. Since we are most interested in identifying large-scale duplications, we focus on the 25 nodes with duplications involving at least half (≥ 43) of the gene trees (Table 1 and Fig. 2). These are especially abundant among the root nodes (Fig. 2). However, they are also common at the base of major clades including Poaceae, Solanaceae, Asteraceae, Brassicaceae, as well as *Populus* and *Gossypium* (Table 1 and Fig. 2). Each analysis of the 85 gene data set took approximately 15 minutes on a Macintosh Power PC laptop computer with a 1.5 GHz G4 processor and Mac OSX 10.4 operating system.

Random Leaves Analysis. The 10 analyses using gene trees with randomly permuted leaf labels found evidence for gene duplication events on only between 25 and 33 (ave. 28.3) internal nodes. In all replicates there was evidence for gene duplications involving many if not all genes in the root nodes (A-C, F-I in Fig. 2) of the species tree as well as the root nodes of the eudicots (nodes L, M, N, and R in Fig. 2), but generally few genes in the other nodes of the species tree (Table 1 and Fig. 2).

4 Discussion

Gene and Genome Duplications in Plants. Our analyses first emphasize the ubiquity of gene duplications throughout the evolutionary history of plants. While we examined only 85 gene families with incomplete sampling, there is evidence of gene duplications on nearly 90% of the internal nodes. Our analyses also provide a hypothesis for the history of large-scale gene duplications in plants that is generally consistent with previous hypotheses (e.g., [8]). Our focus on the 25 nodes with evidence of duplications in at least half of the gene families identified many previously hypothesized ancient polyploidy events. These include events at the base of the Poaceae (node J [16,24]), Brassicaceae (node T [30,26]), and Asteraceae (node Q [8]), within Solanaceae (nodes O and P [8]) and Fabaceae (node W [6]), and in *Populus* (nodes X and Y [28]) and *Gossypium* (node V [4] Fig. 2). In some cases, our analyses provide more precise hypotheses of the phylogenetic location of these duplications because of our higher taxon sampling. For example, while there has been evidence of a large-scale gene duplication common to many grasses (e.g. [29,24]), our analysis places it between the divergence of *Ananas* and the Poaceae (node J, Fig. 2). There is little evidence for large-scale duplications at the root nodes (nodes A-C, F-I Fig. 2), and at most of the early eudicot nodes (nodes L-N, R-S; Fig. 2); yet, these also are the nodes where large numbers of duplications map in our analysis of the randomly permuted gene trees (Table 1; Fig. 2). When mapping duplications from a single gene tree to a species tree, error in the gene trees erroneously places duplications towards the root of the species tree [17]. Our results suggest that erroneously placed genes in gene trees also provide erroneous evidence of large-scale duplications at the root nodes. Thus, we advise interpreting evidence of large-scale

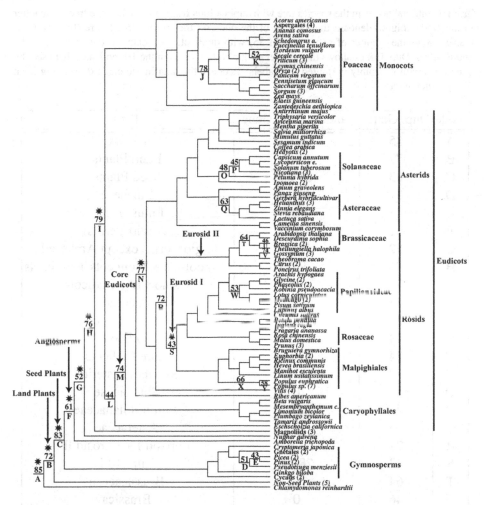

Fig. 2. Species tree with potential locations of large-scale gene duplication events. The species tree used in the analysis contains 136 taxa, and in some cases, multiple (usually congeneric) species in a clade were combined into a single taxon for this figure. In these cases, the total number of species in the combined group is written in parentheses beside the leaf name. The internal nodes with duplications from ≥ 43 of the 85 gene trees have letters under the branch leading to the node, and the number of gene trees with duplications on top of the branch. Stars on top of the branch denote nodes where the analyses using gene trees with randomly permuted leaf labels identified gene duplications from as many gene trees as the analysis with ML gene trees. In other words, the estimated number of duplicated genes at the nodes with stars may be greatly influenced by, if not totally due to, error in the gene trees.

duplications near the root of a tree with great caution. If we disregard the potentially erroneous events at the root nodes, our analysis provides an overall picture of ancient polyploidy in angiosperms that is largely consistent with the recent data from the *Vitis* genome [10]. We hypothesize that the two genome duplications in *Arabidopsis* since its

Table 1. Internal nodes in the species tree with duplications from at least 43 gene trees. The letter in the Node column denoted the location of the node on the species tree figure (Fig. 2). Dup. Genes shows the number of genes (out of 85) with duplications located at the specified node, and Random Dup. Genes shows the number of duplicated genes in the 10 replicates that used the gene trees with randomly permuted leaf labels. Taxa are the taxa in the clade descending from the specified node.

Node	Dup. Genes	Random Dup. Genes	Taxa
A	85	85	All Taxa
B	72	84-85	Land Plants
C	83	84-85	Seed Plants
D	51	0	Pinaceae
E	43	0	Pinus, Abies
F	61	45-65	Angiosperms
G	52	49-58	Angiosperms except Amborella
H	76	79-83	Magnoliids + Monocots + Eudicots
I	79	85	Monocots + Eudicots
J	78	0-20	Poaceae
K	52	0	Secale + Triticum
L	44	26-36	Eudicots
M	74	55-69	Core Eudicots
N	77	84-85	Rosids + Asterids
O	48	0	Solanaceae
P	45	0	within Solanaceae
Q	63	0	Asteraceae
R	72	62-68	Eurosid I + Eurosid II
S	43	28-44	Eurosid I
T	64	0-5	Brassicaceae
U	46	0	Brassica
V	74	0	Gossypium
W	53	0-18	within Fabaceae
X	66	0-8	Populus
Y	58	0	within Populus

common ancestor with *Vitis* occurred at the base of the Brassicaceae (node T; Fig. 2) and at the base of the eurosid I + eurosid II clade (node R; Fig. 2). The ancestral hexiploidization of the *Vitis* and *Arbaidopsis* genomes occured at nodes L and/or M (Fig. 2), after the divergence of eudicots and monocots.

Algorithm Performance and Limitations. The results of analysis of plant gene trees also suggest some weaknesses in our approach and directions for future research. First, though our analysis uses only 85 gene trees, we find evidence of duplications on nearly all of the internal nodes. With more gene trees, there will doubtlessly be evidence for duplications on every node of the tree. In this case, an algorithm that seeks to find the minimum number of nodes with duplications will cease to be informative. It may be more informative to find the duplication mappings that minimize the overall number of duplication episodes. The randomized leaf analysis also suggests that gene tree error can produce evidence of apparently anomalous large-scale gene duplication events. Unfortunately, some error is likely inherent in any gene tree inference. Even if the unrooted gene tree topology is correct, it is extremely difficult to determine the correct rooting when there is a history of duplications. It may be useful to develop methods for mapping large-gene duplication events that can account for possible error in the gene trees, either by utilizing unresolved or unrooted gene trees or by allowing small changes in the topology of the gene trees if they will lead to better solutions.

5 Conclusion

We introduce a new exact algorithm that solves a biological problem: how can we reconstruct the history of gene duplications across a phylogeny in a way that minimizes the locations of the duplications. By placing large-scale duplication events in such a phylogenetic context, we can help specify the precise location and timing of the duplications. Unlike other methods, our approach does not require gene map data and does not rely on molecular clock assumptions. Furthermore, it can be used with relatively few gene family trees. However, error in the gene trees, and possibly the species tree, can confound the results from our approach, creating evidence for apparently anomalous large-scale duplication events. Thus, our approach may be most effective as a complement to other methods for detecting large-scale duplications from genomic data of one or few taxa.

References

1. APG II: An update of the angiosperm phylogeny group classification for the orders and families of flowering plants: APG II. Bot. J. Linn. Soc. 141, 399–436 (2000)
2. Bender, M.A., Farach-Colton, M.: The LCA problem revisited. LATIN, pp. 88–94 (2000)
3. Blanc, G., Hokamp, K., Wolfe, K.H.: A recent polyploidy superimposed on older large-scale duplications in the Arabidopsis genome. Genome Res. 13, 137–144 (2003)
4. Blanc, G., Wolfe, K.H.: Widespread paleopolyploidy in model plant species inferred from age distributions of duplicate genes. Plant Cell 16, 1093–1101 (2004)
5. Bowers, J.E., Chapman, B.A., Rong, J., Paterson, A.H.: Unravelling angiosperm genome eolution by phylogenetic analysis of chromosomal duplication events. Nature 422, 433–438 (2003)
6. Cannon, S.B., et al.: Legume genome evolution viewed through the Medicago truncatula and Lotus japonicus genomes. Proc. Natl. Acad. Sci. 103, 14959–14964 (2006)
7. Chapman, B.A., Bowers, J.E., Schulze, S.R., Paterson, A.H.: A comparative phylogenetic approach for dating whole genome duplication events. Bioinformatics 20, 180–185 (2004)
8. Cui, L., et al.: Widespread genome duplications throughout the history of flowering plants. Genome Res. 16, 738–749 (2006)
9. Fellows, M., Hallet, M., Stege, U.: On the multiple gene duplication problem. ISAAC, pp. 347–356 (1998)

10. F.-I.P.C.: for Grapevine Genome Characterization: The grapevine genome sequence suggests ancestral hexaploidization in major angiosperm phyla. Nature 449, 463–467 (2007)
11. Garey, M.R., Johnson, D.S.: Computers and Intractability: A guide to the theory of NP-completeness. W. H. Freeman, New York (1979)
12. Golumbic, M.R.: Algorithmic Graph Theory and Perfect Graphs, Annals of Discrete Mathematics, 2nd edn., vol. 57. Academic Press, London (2004)
13. Goodman, M., Czelusniak, J., Moore, G.W., Romero-Herrera, A.E., Matsuda, G.: Fitting the gene lineage into its species lineage, a parsimony strategy illustrated by cladograms constructed from globin sequences. Systematic Zoology 28, 132–163 (1979)
14. Grant, V.: Plant speciation, 2nd edn. Columbia University Press (1981)
15. Guigó, R., Muchnik, I., Smith, T.F.: Reconstruction of ancient molecular phylogeny. Molecular Phylogenetics and Evolution 6(2), 189–213 (1996)
16. Guyot, Keller: Ancestral genome duplication in rice. Genome 47, 610–614 (2004)
17. Hahn, M.: Bias in phylogenetic tree reconciliation methods: implications for vertebrate genome evolution. Genome Biol. 8, R141 (2007)
18. Hartmann, S., Lu, D., Phillips, J., Vision, T.J.: Phytome: A platform for plant comparative genomics. Nucleic Acids Research 34, D724–D730 (2006)
19. Jones, D.T., Taylor, W.R., Thornton, J.M.: The rapid generation of mutation data matrices from protein sequences. Comp. Appl. Biosci. 8, 25–282 (1992)
20. Lynch, M., Conery, J.S.: The evolutionary fate and consequence of duplicate genes. Science 290, 1151–1155 (2000)
21. Monma, C.L., Wei, V.K.: Intersection graphs of paths in a tree. Journal of Combinatorial Theory 41, 141–181 (1985)
22. Page, R.D.M., Cotton, J.A.: Vertebrate phylogenomics: reconciled trees and gene duplications. In: Pacific Symposium on Biocomputing, pp. 536–547 (2002)
23. Page, R.D.M., Holmes, E.C.: Molecular evolution: a phylogenetic approach. Blackwell Science, Malden (1998)
24. Paterson, A.H., Bowers, J.E., Chapman, B.A.: Ancient polyploidization predating divergence of the cereals, and its consequences for comparative genomics. Proc. Natl. Acad. Sci. 101, 9903–9908 (2004)
25. Rensing, S.A., Ick, J., Fawcett, J.A., Lang, D., Zimmer, A., Van de Peer, Y., Reski, R.: An ancient genome duplication contributed to the abundance of metabolic genes in the moss *Physcomitrella patens*. BMC Evol. Biol. 7, 130 (2007)
26. Simillion, C., Vandepoele, K., Van Montagu, M.C.E., Zabeau, M., Van de Peer, Y.: The hidden duplication past of *Arabidopsis thaliana*. Proc. Natl. Acad. Sci. 99, 13627–13632 (2002)
27. Stebbins, G.: Variation and evolution in plants. Columbia Univ. Press (1950)
28. Sterck, L., Rombauts, S., Jansson, S., Sterky, F., Rouzé, P., Van de Peer, Y.: EST data suggest that poplar is an ancient polyploidy. New Phytologist 167, 165–170 (2005)
29. Vandepoele, K., Simillion, C., van de Peer, Y.: Evidence that rice and other cereals are ancient aneuploids. Plant Cell 15, 2192–2202 (2003)
30. Vision, T.J., Brown, D.G., Tanksley, S.: The origins of genome duplications in *Arabidopsis*. Science 290, 2114–2117 (2000)
31. Zhang, L.: On a Mirkin-Muchnik-Smith conjecture for comparing molecular phylogenies. Journal of Computational Biology 4(2), 177–187 (1997)

A   Appendix

Proof (Lemma 1). The proof is by induction on $|C|$. Clearly, the result holds for $|C| \leq 1$. Now, assume that $|C| \geq 2$ and that the result holds for all cliques with fewer nodes.

Let $V = [v, v']$ be an interval in C. Then, for $C' = C - \{V\}$ it holds by the inductive assumption that $\bigcap_{I \in C'} I$ is an interval, say $U = [u, u']$ where $u = \mathrm{lca}(\bigcup_{[x,y] \in C'} x)$ and $u' = \min(\bigcup_{[x,y] \in C'} y)$.

We first show that $u' \smallfrown_T v'$. Any interval $W \in C'$ intersects with V since $V, W \in C$, and thus there exists $x \in V \cap W$ where $x \leq v'$. The interval W also contains the interval U and especially the element u', since $U = \bigcap_{I \in C'} I$. Since $x, u' \in W$ it follows $x \smallfrown_T u'$. Thus either $x \leq_T u'$ or $x >_T u'$. In the first case x is a lower bound on u' and a lower bound on v', since $x \leq_T v'$. Thus $v' \smallfrown_T u'$. In the latter case it follows $v' \leq_T u'$ from $x >_T u'$ and $v' \geq_T x$.

Now, consider the following two cases:

Case $V \cap U \neq \emptyset$ We show that $\bigcap_{I \in C} I$ is an interval in $\leq_T$. From $V \cap U \neq \emptyset$ and $u' \smallfrown_T v'$ follows that $V \cap U = [\mathrm{lca}(u, v), \min(u', v')]$. With our hypothesis $u = \mathrm{lca}(\bigcup_{[x,y] \in C'} x)$ and $u' = \min(\bigcup_{[x,y] \in C'} y)$, the desired statement follows.

Case $V \cap U = \emptyset$ We show that this case is not possible. Consider the two possible cases for $u' \smallfrown_T v'$:

Case $u' \leq_T v'$ Thus $[u', v']$ is an interval, and $[u', v'] \cap V$ is an interval with the minimum element $v'' = \mathrm{lca}(u', v)$. With $U \cap V = \emptyset$ follows that $u' < v''$ and further that $v'' \notin U$. We show that v'' is an element in every $W \in C'$ and thus $v'' \in U$, a contradiction. Consider any $W \in C'$, then $u' \in W$, and there exists $x \in W \cap V$, since $W, V \in C$. With $u' < v''_T$ we follow that $w \leq u' <_T v'' \leq_T x \leq_T w'$ and further $v'' \in W$ as desired.

Case $v' <_T u'$ Thus $[v', u']$ is an interval. We show that v' is an element in every $W \in C'$ and thus $v' \in U$, a contradiction to $V \cap U = \emptyset$. Consider any $W \in C'$ we have $u' \in W$, and there exists $x \in V \cap W$ where $x \leq_T v'$. Therefore we have $w \leq_T x \leq_T v' < u'' \leq_T u' \leq_T w'$ from which follows that $v' \in W$ as desired. $\qquad\square$

Proof (Lemma 2). We first show that $cover(v)$ forms a clique in the intersection graph $int(\mathcal{I})$ for any $v \in V$. Let U, V be distinct intervals in $cover(v)$, then $v \in (U \cap V)$. Thus $\{U, V\} \in E(int(\mathcal{I}))$ and it follows that $int(\mathcal{I})$ is a clique.

From the proven statement above follows that $\mathcal{C}$ is a collection of cliques in $int(\mathcal{I})$. To show that $\mathcal{C}$ covers $int(\mathcal{I})$ consider an interval $I \in V(int(\mathcal{I}))$. Since V covers $\mathcal{I}$, there exists an element $v \in V$ such that $I \in cover(v)$. We have shown that $cover(v)$ is a clique in $\mathcal{C}$. Hence, $\mathcal{C}$ covers $int(\mathcal{I})$. $\qquad\square$

Proof (Theorem 1). We first show that $f(\mathcal{C})$ is an interval cover of $\mathcal{I}$, and then we show the minimality of the interval cover $f(\mathcal{C})$.

$f(\mathcal{C})$ **is an interval cover for $\mathcal{I}$:** Let $I \in \mathcal{I}$. Since $\mathcal{C}$ is a clique cover of $int(\mathcal{I})$, there exists a clique $C \in \mathcal{C}$ where $I \in C$. Thus $f(\mathcal{C})$ is an element in I and therefore covers I. Hence, every interval $I \in \mathcal{I}$ is covered by $f(\mathcal{C})$.

$f(\mathcal{C})$ **is a minimum interval cover for $\mathcal{I}$:** We first prove that $|f(\mathcal{C})| = |\mathcal{C}|$ by showing that f is injective. Suppose that there exist distinct cliques $C, C' \in \mathcal{C}$ such that $f(C) = f(C')$. Then, $f(C) \in I$ for every interval $I \in (C \cup C')$. Therefore, $C \cup C'$ forms a clique in $int(\mathcal{I})$, and $\mathcal{C}' = \mathcal{C} - \{C, C'\} \cup \{C \cup C'\}$ is a clique cover of $int(\mathcal{I})$ where $|\mathcal{C}'| < |\mathcal{C}|$. Hence, $\mathcal{C}$ is not a minimum clique cover of $int(\mathcal{I})$, a contradiction.

Now, suppose for the purpose of a contradiction that there exists an interval cover $V \subseteq V(T)$ such that $|V| < |f(\mathcal{C})|$. By Lemma 2, $\mathcal{C}' := \bigcup_{v \in V}\{cover(v)\}$ is a clique cover and $|\mathcal{C}'| \leq |V| < |f(\mathcal{C})| = |\mathcal{C}|$. Hence, $\mathcal{C}$ is not a minimum clique cover, a contradiction.   □

Proof (Theorem 2). Theorem 1 states that the TIC problem for an instance $\mathcal{I}$ can be solved by finding a minimum clique cover $\mathcal{C}$ in the intersection graph $int(\mathcal{I})$ and then constructing an interval cover by selecting for every clique $C \in \mathcal{C}$ a node $v \in [a, b]$ where $a = \mathrm{lca}(\bigcup_{[x,y] \in C} x)$ and $b = \min_{[x,y] \in C} y$.

The intersection graph $int(\mathcal{I})$ can be constructed naively through a tree traversal of T in time $O(n^2 + l)$. A minimum clique cover $\mathcal{C}$ of $int(\mathcal{I})$ can be found in $O(n + m)$ by Lemma 4. Also naively the node a (using [2] for the lca computation) or b can be computed in $O(n)$ time for each clique in $\mathcal{C}$. This results in $O(nm)$ time to construct an interval cover from $\mathcal{C}$. In summary the TIC problem can be solved in time $O(n^2 + nm + l)$.   □

Proof (Corollary 1). The EC problem for the instance $(\mathcal{G}, S)$ is the TIC problem for the instance $\mathcal{I} = \bigcup_{g \in \mathrm{Dup}(\mathcal{G}, S)} I(g)$. Therefore, the overall time to solve the EC problem is the time to compute the instance $\mathcal{I}$ in addition to the running time to solve the TIC problem for the instance $\mathcal{I}$.

After $O(l)$ preprocessing time, the mapping $\mathcal{M}$ for all gene trees in $\mathcal{G}$ can be computed in $O(k)$ time [31] (using [2]). Traversing all trees $G \in \mathcal{G}$ the gene duplications and their intervals can computed in $O(k)$ time. Hence, the desired TIC problem instance can be computed in $O(k + l)$ time. The TIC problem for the $O(k)$ intervals over $\leq_S$ can be solved in time $O(k^2 + km + l)$ by Theorem 1. In summary the EC problem can be solved in time $O(k^2 + km + l)$.   □

Rapid and Accurate Protein Side Chain Prediction with Local Backbone Information

Jing Zhang[1,2,*], Xin Gao[1,*], Jinbo Xu[3,**], and Ming Li[1,**]

[1] David R. Cheriton School of Computer Science, University of Waterloo,
Waterloo, Ontario, Canada N2L 6P7
[2] The Institute for Theoretical Computer Science,
Department of Computer Science and Technology, Tsinghua University,
Beijing 100084, China
[3] Toyota Technological Institute at Chicago, Chicago, IL, USA, 60637
{j2zhang, x4gao, mli}@cs.uwaterloo.ca, j3xu@tti-c.org

Abstract. High-accuracy protein structure modeling demands accurate and very fast side chain prediction since such a procedure must be repeatedly called at each step of structure refinement. Many known side chain prediction programs, such as SCWRL and TreePack, depend on the philosophy that global information and pairwise energy function must be used to achieve high accuracy. These programs are too slow to be used in the case when side chain packing has to be used thousands of times, such as protein structure refinement and protein design.

We present an unexpected study that local backbone information can determine side chain conformations accurately. LocalPack, our side chain packing program which is based on only local information, achieves equal accuracy as SCWRL and TreePack, yet runs 4-14 times faster, hence providing a key missing piece in our efforts to high-accuracy protein structure modeling.

Keywords: side chain prediction, local backbone features, multiclass Support Vector Machines.

1 Introduction

Protein side chain packing is a key step towards accurate protein structure modeling and has been studied for three decades [1, 2, 3, 4]. Given the backbone conformation of a protein, side chain prediction determines the coordinates of all the side chain atoms. Accurate and very fast side chain prediction is vital to accurate protein structure modeling since such a procedure needs to be repeatedly called at each step of the entire protein structure refinement process, which usually samples a very large number of backbone conformations. Protein side chain packing is also an indispensable component of protein design, which finds a protein sequence that can fold into a given three-dimensional protein

---

* The first two authors contributed equally to this paper.
** Corresponding authors.

M. Vingron and L. Wong (Eds.): RECOMB 2008, LNBI 4955, pp. 285–299, 2008.

structure [5, 6]. Whenever a protein backbone conformation (in protein structure modeling) or its primary sequence (in protein design) is changed, side chain packing has to be conducted to re-determine the coordinates of the affected side chain atoms or even all the side chain atoms. Many known side chain prediction programs, such as SCWRL [7] and TreePack [8,9], predict the positions of side chain atoms using global information and pairwise energy function, in order to achieve high accuracy. Thus these programs are too slow to be called tens of thousands of times in high-accuracy protein structure modeling or protein design. Therefore, an ultra-fast side chain prediction method is urgently needed.

An important discovery on side chain conformation is that the side chains have a few frequently occurred conformations (referred to as rotamers) [1,3,7,10,11]. Thus, most side chain prediction methods assume side chains can only take several highly probable rotamers, while others consider conformations sampled around rotamers.

Problem Description. Given a finite set of side chain rotamers for each amino acid type, and a backbone conformation b. Let p denote a possible side chain conformation vector indicating the rotamer choice for each residue position. Traditional side chain prediction problem can be formulated as a combinatorial search problem:

$$p^* = \arg\min_p[E_{SS}(p,p) + E_{SB}(p,b) + E_{BB}(b,b)] \tag{1}$$

where p^* denotes the optimal side chain conformation, $E_{SS}(p,p)$ is a pairwise energy item representing interactions among side chain atoms, $E_{SB}(p,b)$ represents interaction energy between side chain atoms and backbone atoms, and $E_{BB}(b,b)$ represents backbone-backbone interaction energy. Among them, $E_{BB}(b,b)$ can be considered as a constant since the backbone conformation is fixed.

Following this formulation, almost all side chain prediction methods employ a pairwise energy function and a rotamer library, then apply a global or local search strategy to find the optimal solution for this combinatorial problem.

Rotamer Libraries. A rotamer library is a finite set of rotamers, each of which has an occurring probability. Rotamer libraries can be either backbone-independent [2, 12, 13, 14, 15, 16] or backbone-dependent [1, 3, 7, 10, 17, 18, 19], according to whether the occurring probability of a rotamer is estimated based on backbone information. Chandrasekaran *et al.* developed the first backbone-independent library [12]. Janin *et al.* [1] and McGregor *et al.* [3] examined the relationship between side chain conformation and secondary structure, and then developed a secondary-structure-dependent rotamer library. Dunbrack *et al.* developed the first backbone dihedral angle based rotamer library [17] and refined it by Bayesian statistical analysis [10].

Backbone-dependent rotamer library is widely used to predict side chain conformations [8,9,20,21,22,23,24,25,26]. Rotamer library not only can make side chain prediction a discrete-optimization problem, but also can provide the probability of each rotamer in energy function calculation. However, since many side chain prediction methods use rotamer probabilities in their energy functions, their performance is sensitive to these values which are hard to be estimated accurately.

Energy Functions. The energy functions are considered to be a bottleneck of existing side chain prediction methods. Although many studies aim to improve the accuracy of side chain packing energy functions [20, 26, 27, 28, 29], all side chain predictors claim that their methods can perform much better if the energy function is more accurate. As mentioned above, energy functions used in side chain prediction contain both side chain-backbone interaction energy and side chain-side chain interaction energy.

Roitberg *et al.* [27] used a mean field approximation, which probably has the same global minimum as the original system, to direct their search strategy. A much more accurate energy function was developed by Liang *et al* [20]. Their energy function contains contact surface, volume overlap, backbone dependency, electrostatic interactions, and desolvation energy. In [26], ROSETTA's energy function [30], which is the sum of Lennard-Jones potential, rotamer energy, atomic clash penalty, and hydrogen-bonding potential, was improved by the tree-reweighted belief propagation (TRBP) technique.

Search Methods. A large number of search methods have been developed to optimize the energy function and find the side chain conformation with the minimum energy, such as Metropolis Monte Carlo [31], Gibbs sampling Monte Carlo [32], genetic algorithm [33], dead-end elimination (DEE) [16, 34], neural networks [35], simulated annealing [35, 36], graph theory methods [8, 9, 21], semidefinite programming [23], and integer linear programming [24, 37].

Besides the energy function, search strategy is another bottleneck for side chain prediction. The side chain prediction problem has been proved to be NP-hard [38, 39] if pairwise or multi-body energy function is used. Heuristics such as Monte Carlo or genetic algorithm can find local minimum of an energy function relatively quickly, but cannot guarantee to find the optimal solution of the energy function. On the other hand, some global search methods can find the global optimal solution at the cost of running time. For example, the widely-used program, SCWRL3.0 [21], can optimize its energy function to its global optimum by first decomposing a protein backbone structure into some substructures and then employing a divide-and-conquer strategy to determine the positions of side chain atoms. SCWRL is not fast enough to be used for iterative refinements and protein design. Another global search method, TreePack [8, 9], achieves similar accuracy as SCWRL3.0, but runs much faster. In contrast to SCWRL, TreePack can decompose a protein structure into much smaller substructures without losing accuracy, and thus reduce running time dramatically. However, both SCWRL and TreePack are likely to fail in the case when the backbone conformation implies heavy steric atomic clashes and thus cannot be cut into small substructures without losing accuracy.

In this paper, we present a study on the relationship between local backbone information and side chain conformations, and develop a side chain packing program LocalPack. LocalPack predicts the side chain conformations using local backbone information only and is as accurate as SCWRL, a program that uses pairwise energy function and global search method. We first reformulate side chain packing problem and then solve it using multi-class Support Vector

Machines (multi-class SVM). Our method has the following three features: 1) Instead of using the occurring probabilities contained in a rotamer library, our method only uses the angle values of rotamer candidates. 2) Our method does not use any pairwise energy function. Instead, only local backbone information is employed to predict side chain positions. Furthermore, these local backbone features can be calculated extremely fast. 3) We do not need to optimize an energy function. By contrast, our method generates a set of linear classifiers based on local backbone features and then use these classifiers to predict side chain positions.

The rest of this paper is organized as follows: In Section 2, we introduce our new formulation of side chain prediction problem. Section 3 describes our multi-class SVM model and the features used to construct the classification rule. A cutting plane algorithm is proposed to obtain solutions to the multi-class SVM. In Section 4, we present some experimental results and compare our method to existing methods on both native and nonnative backbones. We also analyze the relative importance of the features in our model. Finally, Section 5 discusses potential applications and future development of our method.

2   New Formulation for Side Chain Prediction

Given a position on a protein backbone sequence, we can calculate a set of backbone related local features on this position. Starting from a rotamer library, our basic assumption is that a certain set of local features can determine the correct rotamer of the side chain on this position.

Table 1. An example of the basic assumption of this paper: a backbone related feature vector A can determine the rotamer choice. Except for the last column, the first 6 columns show examples of possible backbone related feature vectors. The last column shows χ_1 rotamer values corresponding to the feature vectors.

Residue Type	ϕ	ψ	Secondary Stru.	Solvent Access.	# Contacts	χ_1 Rotamer
ARG	60°	45°	Helix	82.75%	11	63°
PHE	112°	42°	Helix	10.23%	4	114°
GLN	34°	16°	Loop	8.65%	6	125°
MET	156°	107°	Sheet	65.22%	19	178°

Let $\mathcal{A} = \{A_1, A_2, \ldots, A_n\}$ denote the set of feature vectors for a given protein with length n, where vector $A_j = \{a_1^j, a_2^j, \ldots, a_k^j\}$ denote the set of backbone related features on the j-th position, either continuous values, such as solvent accessibility, or discrete values, such as secondary structure and amino acid type. Let $\mathcal{R} = \{r_1, r_2, \ldots, r_m\}$ denote an arbitrary rotamer set. Table 1 shows some examples of feature vectors, according to which the rotamer choice for each residue position is determined.

Based on our assumption, given a rotamer set $\mathcal{R}$, we can consider side chain predictor as a function $f(A_j)$ that maps from a given feature vector A_j to a rotamer. $f(A_j)$ is defined to be

$$f(A_j) = \arg \max_{i, r_i \in R} h(A_j, r_i), j = 1, \ldots, m \qquad (2)$$

where $h(A_j, r_i)$ is a scoring function that evaluates the score of assigning the rotamer r_i to the j-th position with feature vector $A_j \in \mathcal{A}$. We aim to find a function $h(A_j, r_i)$ such that $f(A_j)$ matches the correct rotamer choices as well as possible for all the position j.

The formulation 2 is based on a general rotamer library $\mathcal{R}$. Studies on backbone-dependent rotamer libraries [7, 10, 17, 19] show that side chains do prefer some rotamers for a fixed amino acid type and a fixed pair of ϕ, ψ backbone dihedral angles. This kind of rotamer libraries can also fit into our model easily by removing the features (amino acid type, ϕ, ψ) from vector A_j and finding h on a rotamer library which is a (amino acid type, ϕ, ψ)-dependent subset of the original rotamer library $\mathcal{R}$. We will introduce how to find the scoring function h in the next section.

3 A Multi-class SVM Model for the Side Chain Prediction Problem

3.1 A Multi-class SVM Model

In this paper, we consider side chain prediction problem as described in formulation 2 that is a linear function on feature vector A. That is, $h(A_j, r_i) = w_i \cdot A_j$, where w_i is a parameter vector for rotamer i that we want to learn. Thus, according to formulation 2, side chain prediction problem can be formulated as a classification problem:

$$f(A_j) = \arg \max_{i, r_i \in R} w_i \cdot A_j, \quad j = 1, \ldots, n, \qquad (3)$$

in which we want to find such a f that matches correct rotamer choices as well as possible.

To learn the parameter vectors w_i from a training example set $S = \{(A^1, r^1), \ldots, (A^p, r^p)\}$ with size p, where A^j is the feature vector of a residue and r^j is the experimentally determined rotamer of this residue, we applied a multi-class Support Vector Machine (multi-class SVM) model. Multi-class Support Vector Machines provide powerful approaches to deal with the general problem of learning a mapping from a high dimensional feature space to a discrete set [40]. However, traditional multi-class SVM do not directly fit into the side chain prediction problem. The reason is that the number of rotamer labels is usually very large in the real world, which will result in a large number of constraints in multi-class SVM. This will make the traditional quadratic programming based algorithm unfeasible to solve the side chain prediction problem.

To solve this large class problem, we applied the idea of loss function $\triangle$ from structured SVM [41,42], a generalized version of multi-class SVM. Different from multi-class SVM, which were developed to solve classification problems on discrete set $\mathcal{Y} = \{1, \ldots, k\}$, structured SVM were developed to solve classification problems that involve features extracted jointly from the inputs and the outputs, such as sequences, strings, graphs, or labeled trees. Loss function $\triangle$ is widely used in structured SVM [41, 42] to deal with the case in which $|\mathcal{Y}|$ is large. In our side chain prediction problem, we used the concept of loss function and defined it to be: $\triangle : \mathcal{R} \times \mathcal{R} \to \{0, 1\}$, where $\triangle(y', y)$ returns 0 if $y' = y$, and 1 otherwise. $\triangle(y', y)$ quantifies how "bad" it is to predict y' when y is the correct label.

Here we use the loss function $\triangle$ to re-scale the margin as proposed by Taskar *et al.* [43] and formulate the problem of finding parameter vectors w_i, $i = 1, \ldots, m$ in the form of the following optimization problem:

$$\min_{w_i, \xi_j} \frac{1}{2} \sum_{i=1}^{m} \|w_i\|^2 + \frac{C}{p} \sum_{j=1}^{p} \xi_j \qquad (4)$$

$$\forall j, l \quad w_{r^j} \cdot A_j - w_l \cdot A_j \geq \triangle(l, r^j) - \xi_j$$

where m is the size of rotamer library, p is the size of training set, $\xi_j \geq 0$ are called *slack variables*. $\|w_i\|$ is the norm of vector w_i, which determines the size of margin in SVM. $C > 0$ is a tradeoff between training error minimization and margin maximization.

We then apply a cutting plane algorithm described in [41] to solve this optimization problem. The basic idea of the algorithm is to find a relatively small set of constraints without losing too much accuracy. They achieved this goal by building a nested sequence which successively tights relaxations of the original problem. It can be proved that:

- Accuracy: the cutting plane algorithm can compute arbitrarily close approximation to the optimal solution.
- Efficiency: the number of steps that the cutting plane algorithm needs to converge is polynomial on the number of data points.

In practice, the cutting plane algorithm works very well on solving our side chain prediction problem, which we will show later. For more details about the algorithm, please refer to [41].

3.2   Model Features

The relationship between side chain conformations and backbone dihedral angles (ϕ, ψ) has been well studied. Many side chain prediction programs use a backbone-dependent or backbone-independent rotamer library. This work uses the backbone-dependent rotamer library [7, 10] developed by Dunbrack *et al.*. The major problem to be addressed is what kind of backbone structure features a side chain conformation depends on. Many works [3, 19, 44] have been done to analyze the relationship between side chain dihedral angles and local backbone features, such as backbone dihedral angles, secondary structure and solvent accessibility. Here we introduce the local structure features used in our prediction and show how to use them in training and testing.

backbone dihedral angles. Given an amino acid and a pair of (ϕ, ψ) angles, the backbone-dependent rotamer library can provide a set of candidate side chain conformations. We do not use backbone dihedral angles as features in the training. Instead, we divide training data point into many groups according to the amino acid types and ϕ, ψ angles, and develop a classifier for each group based on its corresponding rotamer subset.

secondary structure. Secondary structure is local conformation of a protein backbone. Previous works [3] have shown that secondary structure is highly relevant to the distribution of side chain dihedral angles. We use P-SEA [45] to calculate the secondary structure of a given protein backbone. P-SEA can generate the secondary structure type for each backbone position. Since SVM can only take numerical values as input, we use the expected occurring probability of each secondary structure type as its feature value. Let $N(\alpha)$, $N(\beta)$, $N(loop)$ denote the numbers of residues in α-helices, β-sheets and loops in a training data group, and N denote their sum. The expected occurring probabilities are calculated as $N(\alpha)/N$, $N(\beta)/N$ and $N(loop)/N$, respectively.

solvent accessibility. The accessible surface area is the area of a biomolecule's surface that is accessible to a solvent. It can be calculated by using a sphere of a certain radius to probe the surface of the molecule. A typical radius value is 1.4Å, which approximates the radius of a water molecule. Solvent-accessible surface of atoms have been used to predict conformations of side chains in [44], where they added this term into the energy function during the global optimization and calculated it iteratively. Their results show that the prediction accuracy can be significantly improved by adding the solvent term. This implies the importance of solvent accessibility in modeling side chain conformations. We use Naccess [46] to calculate the backbone solvent accessibility. The output of Naccess is normalized value and we use it as one of our features directly.

contact number. The contact number of a residue in a protein structure is a quantity similar to, but different from solvent accessible surface area. The contact number of a given residue is defined as the number of C_α atoms within a predefined distance $D(= 8\text{Å})$ to the C_α atom of this given residue. The contact numbers are scaled to values between 0 and 1 using a standard max-min normalization method, such that the smallest contact number becomes zero and the largest number becomes one.

4 Results

4.1 Implementation Details

We implemented LocalPack with C++. To improve the efficiency of feature calculation, we used a quick K-nearest-neighbor (KNN) algorithm [47,48] to calculate contact numbers. After extracting backbone related features, such as solvent accessibility, secondary structure, and contact number, we encoded these features into a multi-class SVM model as described in Section 3.1. The SVM model is trained using $SVM^{multiclass}$ [49] with linear kernel function, a program that solves multi-class SVM problem by applying cutting plane algorithm described in [41].

We applied 10-fold cross-validation on our training set to estimate the best C (see Equation 4), a tradeoff between model parameter complexity and tolerable model training errors. A big C indicates that a small training error is tolerated but a big model parameter complexity allowed. A model trained using such a C may not generalize well to the test data. Hsu *et al.* showed in [50] that by testing on a sequence of exponentially growing C values, a good model can be identified in practice. Thus, we tried $C = 2^{-5}, 2^{-4}, ..., 2^{20}$ for each training case, and determined its best C value.

4.2 Training and Test Set

Selecting reasonable training and test sets is very important for fairly evaluating the performance of machine learning methods. We used PDB20 as our training set, in which any two proteins do not share more than 20% sequence identity. We also removed those proteins in this set with resolution worse than 2.Å. This results in a data set of 3060 proteins. For test set, we used Dunbrack's benchmark set [7], which consists of 180 proteins. Since we also used the rotamer library extracted from a set of 800 proteins [10], we examined the overlap among PDB20, the set of 800 proteins for rotamer library generation, and Dunbrack's benchmark set. It turns out that Dunbrack's benchmark set contains 87 proteins in PDB20 and 102 in the set of proteins for rotamer library generation. Thus, we removed all the overlapping proteins from Dunbrack's benchmark set and obtain a reduced benchmark set of 78 proteins. It can be seen from Fig.1 that both our PDB20 training set and the reduced test set are good samples of real world proteins in terms of amino acid composition.

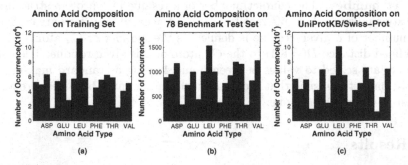

Fig. 1. The amino acid compositions on PDB20 training set(a), reduced 78 benchmark test set(b), and the UniProtKB/Swiss-Prot protein knowledgebase(c), respectively. UniProtKB/Swiss-Prot protein knowledgebase [51] is one of the largest protein sequence databases. The statistics of UniProtKB/Swiss-Prot was taken on 283,454 protein sequences on Sep.11, 2007.

We evaluated the performance of our method on both this reduced benchmark set and Dunbrack's original benchmark set which has overlapping proteins to our training set. Not surprisingly, the accuracy of our method is approximately 8% higher on the Dunbrack's benchmark set than on the reduced set, while the accuracy of SCWRL3.0 is consistent on the two benchmark sets. Thus, in the following experimental studies, we will only evaluate our method on this reduced benchmark set.

4.3 Prediction Accuracy on Native Backbones

We compared the accuracy of our method to the most widely used program SCWRL3.0 in terms of χ_1 and χ_{1+2} accuracy. Other widely used programs, such as Modeller [52], SCAP [11], and TreePack [8,9], performs no better than SCWRL3.0 on both the 180 benchmark set and the 78 benchmark set. Due to the page limitations, we only show the comparison between our method and SCWRL3.0 in Table 2. A prediction is considered to be correct if its value is within 40° from its experimental value. The prediction accuracy of one amino acid is calculated as the ratio of the number of correctly predicted side chains to the total number of side chains of this amino acid type.

As shown in Table 2, the overall accuracy of our method is very close to that of SCWRL3.0. In fact, the χ_1 accuracy of our method is only 0.61% lower than that of SCWRL3.0, while the χ_{1+2} accuracy is 0.51% lower. Although our method is based on local backbone information only, it does not lose any accuracy while is much more computationally efficient, which we will show later. In fact, the χ_1 accuracy of our method is higher than SCWRL3.0 on nine out of the eighteen amino acids, especially LYS, SER and THR. However, our method is much worse than SCWRL3.0 on CYS, LEU, PHE and TRP. Meanwhile, the χ_{1+2} accuracy of our method is higher than SCWRL3.0 on eight out of the eighteen amino acids. This means local backbone information can also determine χ_2 conformation accurately. On the other hand, results shown in Table 2 also demonstrate that the accuracy of our method is not worse than any global optimization methods.

We further examined the eight amino acids on which our method did not perform well (with χ_1 accuracy $\leq$ 82%). They are ARG, ASN, GLN, GLU, LEU, LYS, MET and SER. Except for SER, all the other seven amino acids have large side chain groups as shown in Fig. 2. This result is consistent with the model on which our method is built. Our method assumes that local backbone information can determine side chain conformations. However, if a side chain group is large, its position will be more likely to be impacted by other side chain groups around it and thus cannot be completely determined using only local information. Thus, for such cases, we probably need more information to determine side chain conformations. Interestingly, the global optimization method, SCWRL3.0, which considers all side chain and backbone atoms around one side chain, did worse than our method on four out of these seven amino acids as shown in red boxes in Fig. 2.

Table 2. Prediction accuracy of LocalPack and SCWRL 3.0 on the 78 benchmark set. A prediction of a side chain is correct if its deviation from the experimental value is no more than 40°. χ_1 accuracy of one amino acid is the ratio of the number of correctly predicted χ_1 angles to the total number of this amino acid type, while χ_{1+2} accuracy of one amino acid is the ratio of the number of side chains with both χ_1 and χ_2 being predicted correctly to the total number of this amino acid type.

amino acid	LocalPack		SCWRL 3.0	
	χ_1 accuracy	χ_{1+2} accuracy	χ_1 accuracy	χ_{1+2} accuracy
ARG	0.7701	0.6060	0.7558	0.6226
ASN	0.7888	0.7011	0.7956	0.6882
ASP	0.8322	0.7337	0.8218	0.6974
CYS	0.8497	0.8497	0.8915	0.8915
GLN	0.7493	0.5416	0.7449	0.5319
GLU	0.6841	0.5077	0.7084	0.5128
HIS	0.8226	0.7551	0.8382	0.7745
ILE	0.9172	0.7884	0.9114	0.8060
LEU	0.7851	0.7321	0.8996	0.8142
LYS	0.7678	0.5768	0.7199	0.5444
MET	0.8169	0.6097	0.8160	0.6720
PHE	0.8410	0.7740	0.9361	0.8774
PRO	0.8426	0.7701	0.8517	0.7879
SER	0.7556	0.7556	0.6883	0.6883
THR	0.9193	0.9193	0.8855	0.8855
TRP	0.8328	0.6851	0.8843	0.6688
TYR	0.9239	0.8616	0.9171	0.8615
VAL	0.8922	0.8922	0.9075	0.9075
overall	0.8205	0.7314	0.8266	0.7365

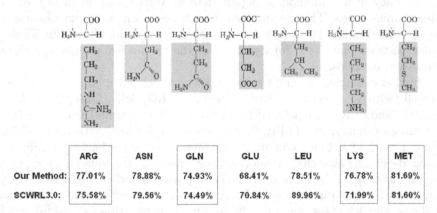

	ARG	ASN	GLN	GLU	LEU	LYS	MET
Our Method:	77.01%	78.88%	74.93%	68.41%	78.51%	76.78%	81.69%
SCWRL3.0:	75.58%	79.56%	74.49%	70.84%	89.96%	71.99%	81.60%

Fig. 2. The χ_1 accuracy of LocalPack on amino acid types ARG, ASN, GLN, GLU, LEU, LYS, and MET. The four amino acids on which the accuracy of LocalPack is higher than that of SCWRL3.0 are marked in red boxes.

4.4 Feature Importance Analysis

A key step in feature based machine learning study is to evaluate the importance of each feature encoded. We evaluated the importance of each feature by removing it from the whole set of features, and testing the accuracy on the rest feature set. Table 3 shows the χ_1 accuracy on different feature sets on amino acid arginine (ARG). The comparisons on other amino acids or on χ_{1+2} are similar. Due to the page limits, we only show the results on χ_1 accuracy of ARG here.

Table 3. Feature importance analysis on ARG. The 1st column is the χ_1 accuracy of LocalPack with all 3 features. Starting from the 2nd column, the χ_1 accuracy on feature sets without solvent accessibility, without secondary structure, and without contact number are listed, respectively.

	with all 3 features	without solvent accessibility	without secondary structure	without contact number
χ_1 Accuracy	0.7701	0.7226	0.7352	0.7320

It can be seen from Table 3 that all of the three features are important to our method. More specifically, removing solvent accessibility feature will reduce the accuracy by 4.8%, while removing secondary structure and contact number will reduce the accuracy by 3.5% and 3.8%, respectively. This means that solvent accessibility is the most important feature in our method, while secondary structure is the least. This makes sense becuase the backbone-dependent rotamer library [10] we used has already partially encoded secondary structure information by considering backbone ϕ, ψ angles in their statistics.

4.5 Performance on Non-native Backbones

We further evaluated the accuracy of our method on nonnative backbones. We compared the χ_1 accuracy of our method to four commonly used side chain prediction methods: MODELLER, TreePack, SCWRL3.0, and SCAP, on a nonnative backbone test set provide by Xu et al. in [9]. The test set contains prediction models generated by a protein threading program, RAPTOR [53], on 24 CASP6 test proteins [54]. RAPTOR generated good alignments for most of these targets. MODELLER [52] was called by RAPTOR to generate model backbones according to the alignments. Besides, MODELLER is also able to predict side chains based on a statistical method. SCAP was tested using the CHARMM force field with the heavy atom model and the largest rotamer library available to SCAP.

The overall χ_1 accuracy is shown in Table 4. The prediction accuracy of our method is the same as TreePack, and slightly worse than SCWRL3.0, while much better than MODELLER and SCAP. This indicates that our method also works well on nonnative backbones.

4.6 Computational Efficiency

Since our method is based on only local backbone features, it can be expected that our method is much more computationally efficient. TreePack has been

Table 4. The overall χ_1 accuracy of MODELLER, TreePack, SCWRL3.0, SCAP, and LocalPack on the 24 nonnative test proteins

	MODELLER	TreePack	SCWRL3.0	SCAP	LocalPack
χ_1 Accuracy	0.428	0.520	0.530	0.488	0.520

Table 5. CPU time comparison of TreePack, SCWRL3.0, and LocalPack on the 78 protein benchmark set

	TreePack	SCWRL3.0	LocalPack
Time	186 seconds	657 seconds	46 seconds

reported as one of the fastest methods for side chain prediction. Table 5 shows the total CPU time comparison of TreePack, SCWRL3.0, and our method on the 78 benchmark set. All three programs are tested on a Debian Linux box with a 1.7GHz CPU.

From Table 5, it is clear that our method is much faster than both TreePack and SCWRL3.0. In fact, we are more than 14 times faster than SCWRL3.0, and more than 4 times faster than TreePack. The average CPU time of our method on one test protein is 0.58 seconds. We also tested the CPU time of our method on the original 180 benchmark set, the results are consistent with the 78 benchmark set.

5   Discussions

This paper formulated protein side chain packing as a classification problem and developed a multi-class SVM method for protein side chain prediction. As far as we know, this is the first attempt to apply multi-class SVM method to the side chain prediction problem. Our experimental results demonstrate that this new method works very well.

This paper demonstrated that protein side chain positions can be predicted using local backbone information to the same accuracy as those programs employing pairwise energy functions and computationally-intensive optimization algorithms, such as SCWRL and TreePack. We hope our discovery will change the way researchers look at this problem and lead to rapid and accurate protein side chain packing programs, which are indispensable in high-accuracy protein structure modeling.

One of the major bottlenecks in protein structure refinement is how to quickly generate a huge number of possible full-atom conformations so that a full-atom energy function can be used to pick up the energetically most favorable conformations. Our method enables us to generate a good side chain packing extremely fast after a change of backbone conformation. Since our method depends on local backbone information only, our method can be made even much more faster when only a local part of a protein structure is refined. This allows us to do side chain packing at each step of protein structure refinement and thus makes it feasible to apply an accurate full-atom energy function to each generated conformation.

We plan to further examine the features used in our method to see if more improvement can be achieved. For example, we only used a feature "contact number" to describe how many residues are in contact with a given residue. This feature does not capture the types of amino acids that are in contact with this given residue. We can extend this single "contact number" to a vector of twenty contact numbers, each of which is the number of residues, of the same amino acid type, in contact with this given residue. We only used three types of secondary structure in our model. This may be enriched by eight types of secondary structure.

Acknowledgements

We are grateful to Dongbo Bu and Bo Jiang for the thought provoking discussion and comments. We also want to thank Shuai Cheng Li for helping us prepare the PDB20 data set. This work is supported by NSERC RGPIN46506 and Canada Research Chair Program.

References

1. Janin, J., Wodak, S., Levitt, M., Maigret, B.: The conformation of amino acid side chains in proteins. J. Mol. Biol. 125, 357–386 (1978)
2. Bhat, T.N., Sasisekharan, V., Vijayan, M.: An analysis of side-chain conformation in proteins. Int. J. Pept. Protein Res. 14, 170–184 (1979)
3. McGregor, M., Islam, S., Sternberg, M.: Analysis of the relationship between side-chain conformation and secondary structure in globular proteins. J. Mol. Biol. 198, 295–310 (1987)
4. Summers, N.L., Karplus, M.: Construction of side-chains in homology modeling: Application to the c-terminal lobe of rhizopuspepsin. J. Mol. Biol. 210, 785–810 (1989)
5. Desjarlais, J., Handel, T.: De novo design of the hydrophobic cores of proteins. Protein Science 4, 2006–2018 (1995)
6. Dahiyat, B., Mayo, S.: Protein design automation. Protein Science 5, 895–903 (1996)
7. Dunbrack, R.: Rotamer libraries in the 21st century. Curr. Opin. Struct. Biol. 12, 431–440 (2002)
8. Xu, J.: Rapid Protein Side-Chain Packing via Tree Decomposition. In: Miyano, S., Mesirov, J., Kasif, S., Istrail, S., Pevzner, P.A., Waterman, M. (eds.) RECOMB 2005. LNCS (LNBI), vol. 3500, pp. 423–439. Springer, Heidelberg (2005)
9. Xu, J., Berger, B.: Fast and accurate algorithms for protein side-chain packing. Journal of ACM 53, 533–557 (2006)
10. Dunbrack, R., Cohen, F.: Bayesian statistical analysis of protein side-chain rotamer preferences. Protein Science 6, 1661–1681 (1997)
11. Xiang, Z., Honig, B.: Extending the accuracy limits of prediction for side-chain conformations. J. Mol. Biol. 311, 421–430 (2001)

12. Chandrasekaran, R., Ramachandran, G.: Studies on the conformation of amino acids. XI. Analysis of the observed side group conformations in proteins. Int. J. Protein Research 2, 223–233 (1994)
13. Benedetti, E., Morelli, G., Nemethy, G., Scheraga, H.: Statistical and energetic analysis of sidechain conformations in oligopeptides. Int. J. Peptide Protein Res. 22, 1–15 (1983)
14. Ponder, J., Richards, F.: Tertiary templates for proteins. use of packing criteria in the enumeration of allowed sequences for different structural classes. J. Mol. Biol. 193, 775–791 (1987)
15. Kono, H., Doi, J.: A new method for side-chain conformation prediction using a hopfield network and reproduced rotamers. J. Comp. Chem. 17, 1667–1683 (1996)
16. Maeyer, M., Desmet, J., Lasters, I.: All in one: a highly detailed rotamer library improves both accuracy and speed in the modelling of sidechains by dead-end elimination. Fold Des. 2, 53–66 (1997)
17. Dunbrack, R., Karplus, M.: Backbone-dependent rotamer library for proteins: Application to side-chain prediction. J. Mol. Biol. 230, 543–574 (1993)
18. Schrauber, H., Eisenhaber, F., Argos, P.: Rotamers: To be or not to be? An analysis of amino acid sidechain conformations in globular proteins. J. Mol. Biol. 230, 592–612 (1993)
19. Dunbrack, R., Karplus, M.: Conformational analysis of the backbone-dependent rotamer preferences of protein sidechains. Nature Struct. Biol. 1, 334–340 (1994)
20. Liang, S., Grishin, N.: Side-chain modeling with an optimized scoring function. Protein Science 11, 322–331 (2002)
21. Canutescu, A., Shelenkov, A., Dunbrack, R.: A graph-theory algorithm for rapid protein side-chain prediction. Protein Science 12, 2001–2014 (2003)
22. Peterson, R., Dutton, P., Wand, A.: Improved side-chain prediction accuracy using an ab initio potential energy function and a very large rotamer library. Protein Science 13, 735–751 (2004)
23. Chazelle, B., Kingsford, C., Singh, M.: A semidefinite programming approach to side chain positioning with new rounding strategies. Informs Journal on Computing 16, 380–392 (2004)
24. Kingsford, C., Chazelle, B., Singh, M.: Solving and analyzing side-chain positioning problems using linear and integer programming. Bioinformatics 21, 1028–1036 (2005)
25. Jain, T., Cerutti, D., McCammon, J.: Configurational-bias sampling techinique for predicting side-chain conformations in proteins. Protein Science 15, 2029–2039 (2007)
26. Yanover, C., Schueler-Furman, O., Weiss, Y.: Minimizing and learning energy functions for side-chain prediction. In: Speed, T., Huang, H. (eds.) RECOMB 2007. LNCS (LNBI), vol. 4453, pp. 381–395. Springer, Heidelberg (2007)
27. Roitberg, A., Elber, R.: Modeling side chains in peptides and proteins: Application of the locally enhanced sampling and the simulated annealing methods to find minimum energy functions. Chem. Phys. 95, 9277–9287 (1991)
28. Street, A., Mayo, S.: Intrinsic beta-sheet propensities result from van der waals interactions between side chains and the local backbone. PNAS 96, 9074–9076 (1999)
29. Mendes, J., Nagarajaram, H., Soares, C., Blundell, T., Carrondo, M.: Incorporating knowledge-based biases into an energy-based side-chain modeling method: Application to comparative modeling of protein structure. Biopolymers 59, 72–86 (2001)
30. Rohl, C., Strauss, C., Chivian, D., Baker, D.: Modeling structurally variable regions in homologous proteins with rosetta. Proteins: Structure, Function, and Bioinformatics 55, 656–677 (2004)

31. Holm, L., Sander, C.: Fast and simple monte carlo algorithm for side chain optimization in proteins: Application to model building by homology. Proteins: Structure, Function and Genetics 14, 213–223 (1992)
32. Vasquez, M.: An evaluation of discrete and continuum search techniques for conformational analysis of side-chains in proteins. Biopolymers 36, 53–70 (1995)
33. Tuffery, P., Etchebest, C., Hazout, S., Lavery, R.: A new approach to the rapid determination of protein side chain conformations. J. Biomol. Struct. Dyn. 8, 1267–1289 (1991)
34. Desmet, J., Maeyer, M., Hazes, B., Laster, I.: The dead-end elimination theorem and its use in protein side-chain positioning. Nature 356, 539–542 (1992)
35. Hwang, J., Liao, W.: Side-chain prediction by neural networks and simulated annealing optimization. Protein Eng. 8, 363–370 (1995)
36. Lee, C., Subbiah, S.: Prediction of protein side-chain conformation by packing optimization. J. Mol. Biol. 217, 373–388 (1991)
37. Eriksson, O., Zhou, Y., Elofsson, A.: Side chain-positioning as an integer programming problem. In: Gascuel, O., Moret, B.M.E. (eds.) WABI 2001. LNCS, vol. 2149, pp. 128–141. Springer, Heidelberg (2001)
38. Akutsu, T.: NP-hardness results for protein side-chain packing. Genome Informatics 8, 180–186 (1997)
39. Pierce, N., Winfree, E.: Protein design is NP-hard. Protein Eng 15, 770–700 (2002)
40. Crammer, K., Singer, Y.: On the algorithmic implementation of multiclass kernel-based vector machines. Journal of Machine Learning Research 2, 265–292 (2001)
41. Tsochantaridis, I., Hofmann, T., Joachims, T., Altun, Y.: Support vector machine learning for interdependent and structured output spaces. In: The 21st International Conference on Machine Learning, vol. 69, pp. 104–111 (2004)
42. Tsochantaridis, I., Joachims, T., Hofmann, T., Altun, Y.: Large margin methods for structured and interdependent output variables. Journal of Machine Learning Research 6, 1453–1484 (2005)
43. Taskar, B., Guestrin, C., Koller, D.: Max-margin markov networks. NIPS 16 (2004)
44. Eyal, E., Najmanovich, R., Mcconkey, R.J., Enelman, M., Sobolev, V.: Importance of solvent accessibility and contact surfaces in modeling side-chain conformations in proteins. J. Comput. Chem. 25, 712–724 (2004)
45. Labesse, G., Colloc'h, N., Pothier, J., Mornon, J.P.: P-SEA, a new efficient assignment of secondary structure from C_α trace of proteins. CABIOS 13, 291–295 (1997)
46. Hubbard, S.J., Thornton, J.M.: 'NACCESS', Computer Program, Department of Biochemistry and Molecular Biology, University College London (1993)
47. Dasarathy, B.V.: Nearest neighbor (NN) norms: NN pattern classification techniques. IEEE Computer Society Press, Los Alamitos (1990)
48. Shakhnarovich, G., Darrell, T., Indyk, P.: Nearest-Neighbor Methods in Learning and Vision: Theory and Practice (Neural Information Processing). The MIT Press, Cambridge (2006)
49. http://svmlight.joachims.org/svm_multiclass.html
50. Hsu, C.W., Chang, C.C., Lin, C.J.: A practical guide to support vector classification. Technical report, Taipei (2003)
51. http://ca.expasy.org/sprot/relnotes/relstat.html
52. Sali, A., Blundell, T.L.: Comparative protein modelling by satisfaction of spatial restraints. J. Mol. Biol. 234, 779–815 (1993)
53. Xu, J., Li, M., Kim, D., Xu, Y.: RAPTOR: optimal protein threading by linear programming. Journal of Bioinformatics and Computational Biology 1, 95–117 (2003)
54. http://predictioncenter.org/casp6/Casp6.html

Algorithms for Joint Optimization of Stability and Diversity in Planning Combinatorial Libraries of Chimeric Proteins

Wei Zheng[1], Alan M. Friedman[2], and Chris Bailey-Kellogg[1]

[1] Department of Computer Science, Dartmouth College
6211 Sudikoff Laboratory, Hanover NH 03755, USA
wei.zheng@dartmouth.edu, cbk@cs.dartmouth.edu
[2] Department of Biological Sciences and Purdue Cancer Center
Purdue University, West Lafayette, IN 47907, USA
afried@purdue.edu

Abstract. In engineering protein variants by constructing and screening combinatorial libraries of chimeric proteins, two complementary and competing goals are desired: the new proteins must be similar enough to the evolutionarily-selected wild-type proteins to be stably folded, and they must be different enough to display functional variation. We present here the first method, STAVERSITY, to simultaneously optimize stability and diversity in selecting sets of breakpoint locations for site-directed recombination. Our goal is to uncover all "undominated" breakpoint sets, for which no other breakpoint set is better in both factors. Our first algorithm finds the undominated sets serving as the vertices of the lower envelope of the two-dimensional (stability and diversity) convex hull containing all possible breakpoint sets. Our second algorithm identifies additional breakpoint sets in the concavities that are either undominated or dominated only by undiscovered breakpoint sets within a distance bound computed by the algorithm. Both algorithms are efficient, requiring only time polynomial in the numbers of residues and breakpoints, while characterizing a space defined by an exponential number of possible breakpoint sets. We applied STAVERSITY to identify 2–10 breakpoint sets for three different sets of parent proteins from the purE family of biosynthetic enzymes. The average normalized distance between our plans and the lower bound for optimal plans is around 1 percent. Our plans dominate most (60–90% on average for each parent set) of the plans found by other possible approaches, random sampling or explicit optimization for stability with implicit optimization for diversity. The identified breakpoint sets provide a compact representation of good plans, enabling a protein engineer to understand and account for the trade-offs between two key considerations in combinatorial chimeragenesis.

1 Introduction

Protein engineering by site-directed recombination (Fig. 1(a)) generates libraries of hybrid proteins (or "chimeras") by mimicking the mixing and inheritance that

M. Vingron and L. Wong (Eds.): RECOMB 2008, LNBI 4955, pp. 300–314, 2008.
© Springer-Verlag Berlin Heidelberg 2008

occur in natural reproduction. A set of homologous parent genes are recombined at defined breakpoint locations, yielding a combinatorial set of hybrids [1,2,3,4]. In contrast to stochastic library construction methods (e.g., [5,6,7]), site-directed approaches explicitly choose breakpoint locations to optimize expected library quality (e.g., predicted disruption [2,8,9] or library diversity [10]). In contrast to mutagenesis, the mutations introduced by site-directed recombination are known to be compatible with each other in parent proteins with a similar structural context (due to homology), and are thus expected to be less disruptive. Without requiring precise modeling or prediction of the effects of mutation, site-directed recombination can produce variant proteins with improved properties and activities [2,3,11,12].

There are two competing goals in recombination experiment planning. We want the resulting hybrids to be stably folded, which is easiest to achieve if they are just like wild-type proteins. At the same time, we want the hybrids to have different activity, which of course requires that they be different from wild-type. By construction, site-directed recombination preserves single-position conservation statistics, since each residue position in the hybrid library is simply taken from one of the parents, and all parents are equally represented within the combinatorial library. Thus evaluation of stability typically focuses on correlation statistics between interacting residues [1,13,14,15,0]. The key insight (middle of Fig. 1(a)) is that recombination "perturbs" the distributions of amino acid types for interacting residues, thereby potentially disrupting the interactions underlying stable folding. Models of residue correlation have been shown to capture important information in a number of applications, including prediction of free energy changes caused by hydrophobic core mutations [16], prediction and recognition of native-like protein structure [17], and functional classification of members of protein families [18]. Pairwise [1] and higher-order [9] models have been used in algorithms to plan site-directed recombination experiments minimizing perturbation (and thereby maximizing expected stability), and have led to the creation of variant proteins with improved or novel activities [2,3,11,12].

In addition to stable hybrids, we also want a diverse hybrid library in order to obtain hybrids with improved or novel activities. Under various methodologies and on a number of systems, including cytochromes P450 [14], β-lactamases [19], and single chain Fv antibodies [20], functional change from wild-type has been correlated with the number of mutations in protein variants. Earlier work on site-directed recombination optimizes for stability while indirectly forcing diversity by constraining the minimum fragment length [14]. Our recent work developed the first approach to explicitly optimize for diversity, by finding breakpoint locations that sample protein sequence space relatively uniformly [10] (right of Fig. 1(a)). However, since diversity competes with stability, it is desirable to explicitly consider both criteria simultaneously.

This paper presents STAVERSITY (a hybrid word with both "stability" and "diversity"), the first method to explicitly optimize both stability and diversity in planning site-directed recombination experiments. A set of breakpoints defines a hybrid library that can be evaluated by metrics we call "perturbation"

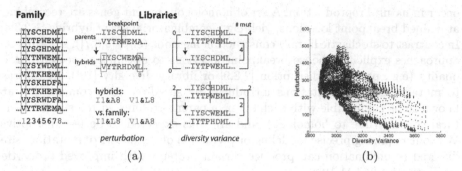

Fig. 1. Stability and diversity are two competing criteria for selecting breakpoint locations for site-directed recombination. (a) Given a family of homologous proteins (here just a cartoon fragment of a multiple sequence alignment), we select a small number of parents to be recombined at specific locations, generating a library of hybrids. Middle: recombination may perturb the previously observed correlations between amino acid types for interacting residues, here at positions 1 and 8, thereby possibly affecting stability. Right: the resulting library may have more or less diversity; the choice of breakpoint location on the top yields hybrids that are identical to the parents, while that on the bottom results in sequence space being sampled relatively uniformly, with an equal number of mutations between each hybrid and each parent. (b) Diversity variance (x-axis) and perturbation (y) for completely enumerated 2-breakpoint sets for three purE proteins, from *E. coli*, *G. gallus*, and *M. thermautotrophicus*. Blue dots are all possible breakpoint sets (for larger numbers of breakpoints, we would not be able to enumerate them all); red diamonds (to the lower left, minimizing one objective for a fixed value of the other) are the undominated ones.

and "diversity variance" (Fig. 1(a)). We seek to minimize perturbation as a way to ensure stable hybrids, and we seek to minimize diversity variance as a way to evenly spread out hybrids in sequence space. Using diversity variance and perturbation values as two dimensions, we can consider possible breakpoint sets as points in a two-dimensional space (Fig. 1(b)). Since it is difficult for an experimenter to decide *a priori* upon the "best" combination of these two incommensurate factors, our methods provide insights into the trade-offs by finding *undominated* sets of breakpoints—those for which no other set of breakpoints is better for both factors ("Pareto optimal", in economics jargon). Our goal is to find the undominated sets efficiently, without explicitly enumerating the exponential number of possible plans.

The problem of finding optimal trade-offs between competing desired criteria is a common one. For example, in considering how to segment records in a large shared database, Eisner and Severance studied the optimal trade-off between the cost of storage and the benefit of retrieval [21]. The goal was assumed to be either a linear or non-linear combination of cost and benefit, and parametric analysis was applied to find the optimal trade-off under all possible parameter values. In computational biology, such ideas are also at the heart of a parametric approach to sequence alignment, e.g., trading off match scores and gap penalties [22]. A comprehensive analysis of parameteric sequence alignment [23] showed that for both global and local alignment, the number of parametric regions to be considered is bounded, and that fast algorithms [24] can be employed to perform the alignment.

In the present case, there is no underlying notion of an optimal trade-off; instead, we want to provide the experimenter with an overview of all possibilities worth considering (because they are undominated). We prove that the results of convex optimization (breakpoint sets on the lower convex hull) are undominated, and develop the natural polynomial-time algorithm to find those breakpoint sets (similar to the parametric analysis in [21,22,23]). We also develop a polynomial-time algorithm to uncover many of the breakpoint sets in the concavities which are either undominated or can be shown by the algorithm to be within a small distance of any undiscovered set that would dominate them.

We present planning results for cases with from 2 to 10 breakpoints and three different sets of parents from the purE family of biosynthetic enzymes that we are currently studying by site-directed recombination. Overall, our plans can be proved to be quite good—the average normalized distance between our plans and the lower bound on optimal plans is around 1 percent. Other possible methods (either sampling breakpoint sets randomly, or explicitly optimizing for stability while implicitly optimizing for diversity) don't do nearly as well—on average for each parent set, our plans dominate 60–90% of those.

2   Methods

Let $\mathcal{P} = \{P_1, P_2, \ldots, P_n\}$ represent a multiple sequence alignment of n parent proteins, with each sequence of length l including residues and gaps. A recombination experiment with λ breakpoints is defined by a set of breakpoint locations $X = \{x_1, x_2, \ldots, x_\lambda \mid 1 \leq x_1 < x_2 < \ldots < x_\lambda < l\}$. The breakpoints partition each parent P_a into $\lambda + 1$ fragments with sequences $P_a[1, x_1], P_a[x_1 + 1, x_2], \ldots, P_a[x_\lambda, l]$, where in general we use $P_a[r, r']$ to denote the amino acid string from position r to r' in sequence P_a, and $P_a[r]$ to denote the single amino acid at position r. A hybrid protein H_i is a concatenation of chosen parental fragments, assembled in the original order. Thus it is also of length l. Then a hybrid library $\mathcal{H}(\mathcal{P}, X) = \{H_1, H_2, \ldots, H_{n^{\lambda+1}}\}$ includes all combinations.

Given a breakpoint set X, we can evaluate the perturbation and diversity variance of the resulting hybrid library with metrics $v_p(X)$ and $v_d(X)$ (see below). We assume, without loss of generality, that both v_d and v_p are to be minimized. For two breakpoint sets X and X', if $v_p(X) \leq v_p(X')$ and $v_d(X) \leq v_d(X')$, and one of the two inequalities is strict, we say that X' is *dominated* by X. Let $\mathcal{X}_\lambda$ be the set of all possible λ-breakpoint sets. If for some breakpoint set X there is no $X' \in \mathcal{X}_\lambda$ that dominates it, we say that X is *undominated*. If X is not dominated by any $X' \in \mathcal{X}'_\lambda$ for some subset $\mathcal{X}'_\lambda \subset \mathcal{X}_\lambda$ of possible breakpoint sets, we say that X is *locally undominated*. Our goal is then:

Goal: Given parent proteins $\mathcal{P}$ and number of breakpoints λ, find the set $\mathcal{U}_\lambda$ of undominated λ-breakpoint sets.

Once undominated breakpoint sets have been computed, the experimenter can readily evaluate trade-offs between diversity variance and perturbation. For example, the minimal perturbation experiment X_* for a given maximum diversity variance threshold θ_d is readily found as $X_* = \arg\min_{X \in \mathcal{U}_\lambda \,:\, v_d(X) \leq \theta_d} v_p(X)$ $= \arg\max_{X \in \mathcal{U}_\lambda \,:\, v_d(X) \leq \theta_d} v_d(X)$. If desired, appropriate data structures can be established to efficiently support such queries.

2.1 Metrics

To evaluate diversity and perturbation, we adopt here the metrics from our previous work [9,10]. However, the method presented below is generic enough to support other metrics, including the perturbation scores of Arnold and co-workers [1] or Moore and Maranas [13].

Perturbation v_p is computed according the hypergraph model of pairwise and higher-order interactions, developed to characterize stability of hybrid libraries [9]. A hyperedge e is defined for each set of residue positions that are in mutual contact. A "hyperresidue" R represents a tuple of amino acids for the residues. An edge-specific potential score $\Phi_e(R)$ is calculated for each hyperresidue for each hyperedge, based on occurrence statistics in a multiple sequence alignment of the specific protein family, as well as in proteins in general. The potential score captures the degree of "hyperconservation" for the edge—how important it appears to be to preserve the combination of amino acid types. Then, given a set of parent proteins $\mathcal{P}$ and a breakpoint set X defining a hybrid library $\mathcal{H}$, we can compute the perturbation as the difference in amino acid distributions (see again Fig. 1(a)), weighted by the potentials:

$$v_p(X) = \sum_e \left(\frac{f_{e,\mathcal{P}}(R)}{|\mathcal{P}|} \cdot \Phi_e(R) \right) - \sum_e \left(\frac{f_{e,\mathcal{H}}(R)}{|\mathcal{H}|} \cdot \Phi_e(R) \right), \quad (1)$$

where $f_{e,\mathcal{P}}(R)$ and $f_{e,\mathcal{H}}(R)$ are the number of occurrences of R at e in the parent proteins and hybrid library, respectively.

We introduced the idea of evaluating diversity in a library according to the variance in the number of mutations between each hybrid–parent pair (illustrated in Fig. 1(a)) [10]. (Hybrid–hybrid diversity variance can likewise be calculated, and is highly correlated with the hybrid–parent metric.) We have shown that the total number of mutations is a constant determined only by the parents, but that by assessing the squared-differences in the numbers, we are optimizing for a relatively uniform sampling of sequence space. We use here the average diversity variance, the original metric divided by the number of hybrids (of course these have the same minima):

$$v_d(X) = 1/n^{\lambda+2} \cdot \sum_{a=1}^{n} \sum_{i=1}^{n^{\lambda+1}} m(H_i, P_a)^2 \quad (2)$$

where $m(H_i, P_a) = \sum_{1 \leq r \leq l} I\{H_i[r] \neq P_a[r]\}$ is the number of positions at which hybrid H_i and parent P_a have different residues. To ignore conservative substitutions, we test "equality" according to standard sets of amino acid classes $\{\{C\},\{F,Y,W\},\{H,R,K\},\{N,D,Q,E\},\{S,T,P,A,G\},\{M,I,L,V\}\}$.

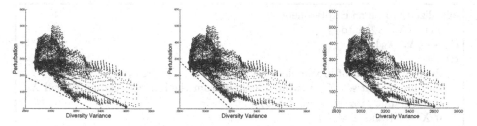

Fig. 2. Finding undominated breakpoint sets on the convex hull, for the completely enumerated 2-breakpoint test system of Fig. 1(b). In practice, we would not enumerate all the points within the hull. (Left) Undominated breakpoint set X below and farthest from the line connecting X_p and X_d, the sets optimizing perturbation alone and diversity variance alone. (Middle) Undominated breakpoint set between X_d and X. (Right) All undominated breakpoint sets on the hull.

2.2   Finding Undominated Breakpoint Sets on the Convex Hull

Based on v_p and v_d, we equate breakpoint set X with its location in the two-dimensional space with axes for perturbation and diversity variance. If breakpoint set X' is dominated by X, then for any line passing through X with a negative slope, X' must be above the line. Thus we know X is undominated if we can find a negative-slope line through it such that all other breakpoint sets are on or above the line.

This insight leads us to the basis for our first algorithm (see Fig. 2), which constructs the lower convex hull connecting the breakpoint sets X_p and X_d minimizing perturbation and diversity variance alone, respectively. For simplicity, we assume throughout the paper that there is a unique X_p minimizing perturbation and a unique X_d minimizing diversity. The extensions to handle non-unique minima are straightforward, requiring us to consider inequalities that aren't strict.

Claim. Let X_p and X_d be the (unique) breakpoint sets minimizing perturbation alone and diversity variance alone, respectively. Then any breakpoint set X on the lower envelope of the convex hull of all breakpoint sets, below the line connecting X_p and X_d, is undominated.

Proof. Consider such an X, and let X' be an adjacent breakpoint set on the convex hull (X' could be X_p or X_d). By the definition of convex hull, all other breakpoint sets must be on one side of the line connecting X and X'. In fact, they must be above the line since otherwise X_p or X_d would be below the line, contradicting the definition. The line connecting X_p and X_d must have a negative slope. Otherwise, since $v_d(X_d) < v_d(X_p)$, we would also have $v_p(X_d) < v_p(X_p)$, contradicting X_p's optimality. It similarly follows that $v_d(X_d) < v_d(X) < v_d(X_p)$ and that the line connecting X and X' has a negative slope. Thus, all breakpoint sets lie on or above a negative-slope line through X, so X is undominated.   □

Of course we want to find the breakpoint sets on the convex hull without enumerating the exponential number of breakpoint sets inside the hull. Our algorithm,

initialize Q to be an empty queue
enqueue (X_p, X_d) into Q
$\mathcal{B}_H \leftarrow \{X_p, X_d\}$
repeat
    dequeue one pair of breakpoint sets (X_1, X_2) $(v_d(X_1) < v_d(X_2))$ from Q
    find breakpoint set X below and farthest from the line connecting X_1 and X_2
    (by dynamic programming, Eq. 3)
    if $X \neq X_1$ and $X \neq X_2$
        $\mathcal{B}_H \leftarrow \mathcal{B}_H + \{X\}$
        enqueue (X_1, X) or (X, X_2) into Q
    end if
until Q is empty
return $\mathcal{B}_H$

Fig. 3. Algorithm for finding undominated breakpoint sets on the convex hull

illustrated in Fig. 2 and described in Fig. 3, is similar to the quickhull algorithm [25] but efficiently finds the hull points without knowing the interior points. The algorithm starts with X_p and X_d, and recursively finds hull points between an existing pair of hull points. The key is finding the intermediate hull breakpoint set X below and farthest from the line connecting hull breakpoint sets X_1 and X_2. (The same method can find the initial X_p and X_d as special cases.) Let $\alpha = v_p(X_1) - v_p(X_2)$ and $\beta = v_d(X_1) - v_d(X_2)$, so that α/β is the slope of the line connecting X_1 and X_2. For the X we seek, all other breakpoint sets must be above the line passing through X with slope α/β. Thus X is the breakpoint set minimizing the value of $\alpha v_p(X) + \beta v_d(X)$.

To find X, we adopt the dynamic programming frameworks from our earlier methods for perturbation alone and diversity variance alone, to handle convex combinations. The idea is to add breakpoints one-by-one from left to right in the sequence (N- to C-terminus), at each point considering the change to $\alpha v_p + \beta v_d$ for this breakpoint given previous breakpoints. Optimal substructure holds since a hybrid library with breakpoints $X_k = \{x_1, \ldots, x_{k-1} = r', x_k = r\}$ extends a hybrid library with breakpoints $X_{k-1} = \{x_1, \ldots, x_{k-1} = r'\}$ by concatenating each of the hybrids with each parent fragment $P_a[r' + 1, r]$. The best choice for x_k depends only on the best choice for x_{k-1}.

Let $d_{pd}(r, k)$ be the optimal value for the linear combination $\alpha v_p + \beta v_d$ with k breakpoints, with the last breakpoint at residue position r. The structure of the recurrence to compute $d_{pd}(r, k)$ is as follows:

$$d_{pd}(r, k) = \begin{cases} C_{pd}(r) & \text{if } k = 1, \\ \min_{r' < r}\{d_{pd}(r', k-1) + \Delta d_{pd}(r', r)\} & \text{if } k > 1. \end{cases} \quad (3)$$

where C_{pd} is the initialization value for only one breakpoint and $\Delta d_{pd}(r', r)$ is the increment when one more breakpoint is put after residue position r. Straightforward algebraic manipulations to derive C_{pd} and $\Delta d_{pd}(r', r)$ have been omitted due to lack of space; the resulting formulas are as follows:

$$C_{pd}(r) = \alpha v_p(\{r\}) + \frac{\beta}{n^2} \cdot \sum_{a=1}^{n}\sum_{b=1}^{n} m(P_a[1,r], P_b[1,r])^2, \tag{4}$$

$$\Delta d_{pd}(r',r) = \alpha \left(v_p\left(\{r',r\}\right) - v_p\left(\{r'\}\right)\right) + \frac{\beta}{n^2} \cdot \sum_{a=1}^{n}\sum_{b=1}^{n} m(P_a[r'+1,r], P_b[r'+1,r])^2 +$$

$$\frac{\beta}{n^3} \cdot \sum_{a=1}^{n}\left(\sum_{b=1}^{n} m(P_a[1,r'], P_b[1,r']) \cdot \sum_{b=1}^{n} m(P_a[r'+1,r], P_b[r'+1,r])\right) \tag{5}$$

To compute this recurrence by dynamic programming requires a table of size λl (recall that λ is the number of breakpoints and l is the sequence length) and each entry depends on $O(l)$ previous entries in computing the minimum. Based on previous derivations [9,10], the complexity of calculating $\Delta d_{pd}(r',r)$ (done in a preprocessing step, for look up during the dynamic programming) includes $O(lE)$ for the increment in perturbation (where E is the number of hyperedges) and $O(n^2 l^2)$ for the increment in diversity variance (where we have n sequences). Thus the complexity for dynamic programming is $O(lE + n^2 l^2 + \lambda l^2)$. We run this algorithm once to find each undominated breakpoint set on the convex hull, so to compute the whole set $\mathcal{B}_{II}$ requires $O(\mathcal{B}_H(lE + n^2 l^3 + \lambda l^2))$—polynomial in each of the input variables and output size.

2.3 Finding Locally Undominated Breakpoint Sets in Concavities

The algorithm of Fig. 3 finds all undominated breakpoint sets on the convex hull, but as Fig. 1 illustrates, many undominated breakpoint sets (45/59 in that example) lie in the concavities. Since our underlying dynamic programming framework (Eq. 3) is limited to convex combinations of v_p and v_d, in order to use it we must focus on smaller regions whose convex hulls intersect the concavities. We can then find breakpoint sets that are locally undominated with respect to the various regions. While these breakpoint sets are not necessarily undominated globally, in the next section we develop an approach to evaluate their optimality.

Let us consider how to constrain our optimization to regions within the perturbation-diversity space. Consider the effect of moving from a breakpoint set X with breakpoint i fixed to residue position r, to breakpoint set X' with i fixed to $r + 1$. The contribution to the perturbation score v_p is changed only for those edges incident on $r + 1$. If we assume a constant degree in the contact graph (since physically each residue can only contact a limited number of other residues), then the expected change in perturbation, $|v_p(X') - v_p(X)|$, is bounded by a constant fraction of the overall perturbation range. Similarly, the contribution to diversity variance v_d is changed only for the fragments from position X_{i-1} to X_i and from X_i to X_{i+1}. While we omit the details, which aren't essential here, it follows that the expected difference $|v_d(X') - v_d(X)|$ is bounded by a linear function of the total number of mutations in those fragments, a small amount compared to the range of diversity variance. Thus each time we advance a single breakpoint location, we take a small step in perturbation-diversity space.

Based on this insight, our algorithm for exploring the concavities iterates over all possible (breakpoint, position) pairs. With a breakpoint fixed to a position, we apply a variant of our dynamic programming algorithm, changing the formulas appropriately for v_p and v_d to account for the fixed breakpoint. After obtaining the locally undominated breakpoint sets for each (breakpoint, location) pair, we take the union of the sets and eliminate those that are dominated. The dynamic programming framework is used to find each point on the local lower convex hull. Thus to find a multiset (including duplicates) $\mathcal{B}_L$ of locally undominated breakpoint sets, the total complexity is $O(\mathcal{B}_L(lE + n^2l^2 + \lambda l^2))$. By fixing more breakpoint locations (e.g., pairs), we would explore the concavities even better, but of course at increased cost. We could also consider variations, such as sampling positions rather than trying each one. However, our results show that fixing each breakpoint at each position is fast enough and yields high quality results.

2.4 Optimality Guarantees

Suppose that for a particular experiment, we want to find the optimal breakpoint set X minimizing perturbation such that $v_d(X) \leq \theta_d$, for some diversity variance threshold θ_d (minimizing diversity subject to a perturbation threshold can be handled similarly). In our concavity-exploring algorithm, when we fix breakpoint i to be at residue position r, we obtain a "local" convex hull. This hull may have a lower convex chord $\overline{X_1 X_2}$ (connecting consecutive points on the lower hull) that intersects the θ_d line (i.e., the vertical line $v_d = \theta_d$) at some point with perturbation value p (see Fig. 4(left)). We represent this convex chord as $c = (i, r, p)$. We can use the set C of all convex chords that intersect the θ_d line over the various local hulls, to bound the best possible perturbation for X.

Claim. Given diversity variance threshold θ_d and set C of all convex chords intersecting the line for θ_d, let $S_i = \{(r, p) \mid (i, r, p) \in C\}$ $(1 \leq i \leq \lambda)$ and let $\mathcal{T} = \{\{(r_1, p_1), (r_2, p_2), \ldots, (r_\lambda, p_\lambda)\} \in S_1 \times S_2 \times \ldots \times S_\lambda \mid r_1 < r_2 < \ldots r_\lambda\}$. If $\mathcal{T}$ is empty, the experiment plan provided by STAVERSITY is optimal. Otherwise, the undiscovered optimal undominated breakpoint set $X_o = \{x_{o,1}, x_{o,2}, \ldots, x_{o,\lambda}\}$ has $v_p(X_o) \geq \min_{T \in \mathcal{T}} \max_{(r,p) \in T} p$.

Proof. Suppose X_o is the optimal undominated breakpoint set and is not found by STAVERSITY. Then for each local hull fixing breakpoint i at residue position $x_{o,i}$, X_o is not found, implying that it is inside the hull and above a lower convex chord. Let X_1 and X_2 be the breakpoint sets at the left and right ends, respectively, of the convex chord below X_o (i.e., $v_d(X_1) < v_d(X_o) < v_d(X_2)$), as in Fig. 4(left). Since the line through X_1 and X_2 has a negative slope (as in the proof in Sec. 2.2), we have $v_p(X_1) > v_p(X_2)$ and $v_p(X_o) > v_p(X_2)$. Thus the line of θ_d must intersect the convex chord $\overline{X_1 X_2}$, since otherwise either $v_d(X_1) > \theta_d$ and $v_d(X_o) > \theta_d$ (contradicting its satisfaction of the threshold), or $v_d(X_2) < \theta_d$ and $v_p(X_2) < v_p(X_o)$ (contradicting its optimality). Furthermore, the perturbation value of the intersection is less than $v_p(X_o)$ as the convex chord has a negative slope and $v_d(X_o) \leq \theta_d$.

Thus, if X_o exists, for each local hull fixing breakpoint i at residue position $x_{o,i}$, we have a convex chord $c = (i, x_{o,i}, p_i)$ such that c is below X_o

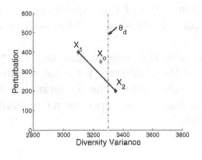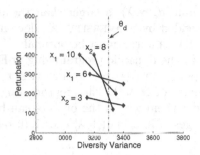

Fig. 4. Illustration of optimality guarantees. (Left) The dashed line representing a diversity threshold θ_d intersects convex chords below X_o. (Right) Bounding the optimal perturbation for a 2-breakpoint set with respect to the θ_d threshold. Each (breakpoint, position) pair generates one convex chord (blue line segment) intersecting the θ_d line. The lower bound for perturbation is the maximum perturbation from a set of convex chords for a consistent breakpoint set, here $x_1 = 6$ and $x_2 = 8$.

and intersects the line of θ_d at perturbation p_i. So $v_p(X_o) > \max\{p_i\}$ and $\{(x_{o,1}, p_1), (x_{o,2}, p_2), ..., (x_{o,\lambda}, p_\lambda)\} \in \mathcal{T}$. As a result, if $\mathcal{T}$ is empty, no X_o exists, and the plan is optimal. Otherwise, following the argument above, we can bound the perturbation for any missed X_o by $v_p(X_o) \geq \min_{T \in \mathcal{T}} \max_{(r,p) \in T} p$. $\square$

This claim suggests an approach for computing the perturbation bound: consider convex chords for the local hulls in order of perturbation, moving up the line of θ_d. When we have found a set of chords, one for each breakpoint, such that the corresponding breakpoint locations are in increasing order, then we have the best possible perturbation value. Fig. 4(right) gives an example.

To efficiently compute the lower bound of $v_p(X_o)$, we develop another dynamic programming algorithm. Let $\mathcal{T}_{k,\gamma}$ be the valid breakpoint sets from S_i with breakpoint k ($\leq \lambda$) at position γ, i.e., $\mathcal{T}_{k,\gamma} = \{(r_1, p_1), (r_2, p_2), \ldots, (r_k, p_k) \in S_1 \times S_2 \times \ldots \times S_k \mid r_1 < r_2 < \ldots < r_k = \gamma\}$. We can then define the minimum perturbation with breakpoint k at position γ as:

$$d_e(\gamma, k) = \min_{T \in \mathcal{T}_{k,\gamma}} \max_{(r,p) \in T} p. \tag{6}$$

If $\mathcal{T}_{k,\gamma}$ is empty, then the k-breakpoint set cannot be constructed, and $d_e(\gamma, k) = \infty$. Otherwise, we can form a k-breakpoint set by extending a valid $k - 1$-breakpoint set ending at residue position $\tau < \gamma$. Optimal substructure holds, and to compute the perturbation we have the recurrence:

$$d_e(\gamma, k) = \begin{cases} \max\{p, \min_{\tau < \gamma} d_e(\tau, k-1)\} & \text{if } \exists (r, p) \in S_k \text{ with } r = \gamma, \\ \infty & \text{otherwise.} \end{cases} \tag{7}$$

And we get the lower bound of $v_p(X_o)$ from the final column:

$$v_p(X_o) \geq \min_{T \in \mathcal{T}_\lambda} \max_{(r,p) \in T} \{p\} = \min_\gamma d_e(\gamma, \lambda). \tag{8}$$

If $\min_\gamma d_e(\gamma, \lambda)$ is larger than the perturbation value of the experiment plan provided by STAVERSITY, X_o does not exist and the experiment plan provided by STAVERSITY must be optimal.

In the dynamic programming of Eq. 7, the table is of size λl, and each entry depends on $O(l)$ previous entries in computing the minimum, for a total complexity of $O(\lambda l^2)$. The preprocessing to put the chords into the S_i buckets and order them within the buckets can be done in linear time, since we have small ranges of integers ($[1, \lambda]$ and $[1, l]$, respectively).

3   Results and Discussion

We have been studying by site-directed recombination homologous proteins of the purE family (COG 41 and pfam 731), which catalyze steps in the *de novo* synthesis of purines. While clear homologs, purE proteins carry out substantially different enzymatic activities in different organisms: in eubacteria, fungi and plants (as well as probably most archaebacteria), the purE product functions as a mutase in the second step of a two-step reaction, while in metazoans and methanogenic archaebacteria, the purE product functions as a carboxylase in a single-step reaction that yields the same product [26,18]. This striking difference in activity makes the purE family a valuable target in protein engineering—by exploring sequence space through site-directed recombination, we seek to find the features of the "boundaries" enclosing the distinct activities.

To identify a set of possible purE parents, we created a multiple sequence alignment of the purE family, then eliminated columns not mapped to the structure of *E. coli* purE (PDB id: 1qcz) and eliminated sequences with more than 20% gaps. This yielded a diverse set of 367 sequences of 162 residues each, including 28 of the rarer class of metazoans and methanogens with inferred carboxylase activity. The average pairwise sequence identity is 65.8%. We selected three parent sets, each consisting of three purE parents with varying diversity—**medium diversity** (average identity 55%): *Escherichia coli, Gallus gallus, Methanothermobacter thermautotrophicus;* **high diversity** (31%): *Drosophila melanogaster, Bdellovibrio bacteriovorus, Treponema denticola;* **low diversity** (80%): *Gibberella zeae, Magnaporthe grisea, Saccharomyces cerevisiae.*

For each parent set, and for 2 to 10 breakpoints, we applied STAVERSITY to find breakpoint sets. On average, it took around 5 minutes for 2 breakpoints. The running time increased according to the number of breakpoints, and it took around 2 hours for 10 breakpoints.

To assess the completeness of STAVERSITY, we enumerated for the medium diversity parents all 2-breakpoint sets (plotted in Fig. 1) and all 3-breakpoint sets, deeming it impractical to enumerate plans with more breakpoints. STAVERSITY finds 55 of the 59 undominated 2-breakpoint sets in the enumeration and 77 of the 115 undominated 3-breakpoint sets. In both cases, the breakpoint sets that STAVERSITY missed were quite close to others that it found. For missed set X and found sets $\mathcal{B}$, we compute the distance as $\min_{X' \in \mathcal{B}\,:\,v_d(X') < v_d(X)} v_p(X') - v_p(X)$, divided by the range of perturbation values over all breakpoint sets. The

Table 1. Comparison of STAVERSITY with RAND and IMPLICIT for three different parent sets and from 2 to 10 breakpoints

	2	3	4	5	6	7	8	9	10
	Medium diversity								
STAVERSITY	55	89	112	134	180	215	229	263	284
RAND	36	44	39	36	45	37	37	24	33
IMPLICIT	43	63	68	78	71	65	61	47	64
STAVERSITY dom. RAND	72.2%	90.9%	100.0%	100.0%	100.0%	100.0%	100.0%	100.0%	100.0%
STAVERSITY dom. IMPLICIT	14.0%	41.3%	58.8%	76.9%	85.9%	92.3%	90.2%	87.2%	89.1%
RAND dom. STAVERSITY	0.0%	3.4%	0.0%	0.0%	0.0%	0.0%	0.0%	0.0%	0.0%
IMPLICIT dom. STAVERSITY	0.0%	9.0%	2.7%	0.7%	0.0%	0.5%	0.0%	0.0%	0.0%
	High diversity								
STAVERSITY	79	104	160	179	230	263	300	352	369
RAND	63	62	55	55	49	50	52	43	45
IMPLICIT	86	82	106	107	106	120	98	122	109
STAVERSITY dom. RAND	82.5%	100.0%	100.0%	100.0%	100.0%	100.0%	100.0%	100.0%	100.0%
STAVERSITY dom. IMPLICIT	4.7%	23.2%	49.1%	55.1%	70.8%	83.3%	81.6%	85.2%	85.3%
RAND dom. STAVERSITY	0.0%	0.0%	0.0%	0.0%	0.0%	0.0%	0.0%	0.0%	0.0%
IMPLICIT dom. STAVERSITY	0.0%	1.9%	0.6%	0.0%	0.0%	0.0%	0.0%	0.0%	0.0%
	Low diversity								
STAVERSITY	20	26	34	38	62	68	79	87	106
RAND	14	12	21	25	25	29	29	31	30
IMPLICIT	6	6	10	11	17	20	23	24	27
STAVERSITY dom. RAND	35.7%	83.3%	81.0%	100.0%	96.0%	100.0%	100.0%	100.0%	100.0%
STAVERSITY dom. IMPLICIT	33.3%	33.3%	70.0%	78.6%	82.4%	85.0%	87.0%	87.5%	88.9%
RAND dom. STAVERSITY	5.0%	0.0%	5.9%	0.0%	0.0%	0.0%	0.0%	0.0%	0.0%
IMPLICIT dom. STAVERSITY	0.0%	0.0%	0.0%	0.0%	0.0%	0.0%	0.0%	0.0%	0.0%

average value for the 4 missed 2-breakpoint sets is 0.5%, as is that for the 38 missed 3-breakpoint sets.

The only other method available for optimizing stability and diversity, based on RASPP [8], does so implicitly while optimizing perturbation. The lengths of the fragments to be recombined are constrained to lie between minimum and maximum values; a perturbation-optimal library is generated for each minimum-maximum pair. This restriction does provide some sampling of various levels of diversity, since larger fragments generally lead to greater diversity. For comparison, we implemented a version of this approach (called IMPLICIT below) using our metrics and returning only the locally undominated breakpoint sets (i.e., not dominated by any others in the set). For a baseline for comparison, we also applied a simple random selection method (called RAND below), in which we randomly sample sets of breakpoints and return the locally undominated ones.

Tab. 1 summarizes the results on the different tests. To put the methods on a relatively equal footing and avoid saturation by random sampling (which happens with small numbers of breakpoints), the number of random samples for each test case was set as the total number of breakpoint sets found by STAVERSITY in the local convex hulls. (We also tested a large number of random samples; see below.) In each table, the rows of "STAVERSITY", "RAND" and "IMPLICIT" give the numbers of breakpoint sets found by each method. The rows of the form "STAVERSITY dom. RAND" give the percentage of breakpoint sets found by the second method that are dominated by those found by the first method (not counting the breakpoint sets common to both methods).

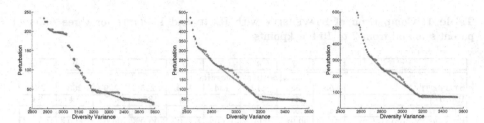

Fig. 5. Optimality guarantees: breakpoint sets found by STAVERSITY (green diamonds) compared to lower bounds on optimal perturbation for 100 different diversity variance values (red crosses), for (left) 2, (middle) 6, and (right) 10 internal breakpoint sets

On average, for the medium diversity parents, RAND finds only 28 percent as many breakpoint sets as STAVERSITY does, and 96 percent of the RAND ones are dominated by STAVERSITY ones. IMPLICIT finds 45 percent as many, of which 71 percent are dominated. RAND performs similarly badly on the high diversity parents, finding 32 percent as many with 98 percent dominated, and improves a little on the low diversity set, at 48 percent with 88 percent dominated. IMPLICIT improves a little on the high diversity set, finding 56 percent with 60 percent dominated, but for the low diversity parents finds only 29 percent with 72 percent dominated. STAVERSITY always finds more and better breakpoint sets than RAND and IMPLICIT. One possible explanation for the variation of IMPLICIT's performance with parent diversity level is that when parents become more diverse (in the limit, being entirely different), the fragment length (the implicit diversity control) is increasingly important for generating diversity. When parents become less diverse, longer fragment length does not necessarily mean more diversity, so the impact of fragment length on diversity is not so significant. The performance of random selection is clearly subject to the curse of dimensionality.

We tried using a large number of samples ($> 10^7$) for the different parent sets with 10 breakpoints. Even with this large number of samples, STAVERSITY still significantly outperforms random selection. For the medium diversity parents, RAND finds only 83 breakpoints, all of which are dominated by STAVERSITY ones, while with the high diversity parents, it finds only 117, again all dominated. It does relatively better for the low diversity parents, finding 101 breakpoints of which 94% are dominated, and it dominates 2 of the STAVERSITY breakpoints.

In addition to finding more, better breakpoint sets, STAVERSITY can provide optimality guarantees. To evaluate how close our results are to optimal perturbation-diversity trade-offs, we tested 100 diversity variance thresholds θ_d for the bound on the optimal perturbation value. As Fig. 5 illustrates, our plans are very close to optimal. Quantitatively, we can compute the average distance between the vector of perturbation bounds and corresponding actual perturbation values, normalized by the range of perturbation values as above. For the medium diversity parents, the difference is 0.9% for 2 breakpoints, 1.0% for 6, and 0.9% for 10. The results are similar for the high diversity set (1.2%, 1.4%, and 1.3%) and low diversity set (3.4%, 1.1%, and 1.4%). The one outlier, 3.4% for low-diversity parents with 2 breakpoints, comes mainly from the extreme ends of the diversity range. Overall (and even in that case), the breakpoint sets are provably close to providing optimal trade-offs.

4 Conclusion

We present a method to optimize breakpoint selection for site-directed protein recombination considering stability and diversity simultaneously. Our method is generic in metrics for stability and diversity, and by finding undominated breakpoint sets, it allows the experimenter to assess the trade-offs between these factors. Unlike other methods, we can provide optimality guarantees on identified breakpoint sets. In practice, our method significantly outperforms existing alternatives, finding more and better breakpoint sets. STAVERSITY should be a valuable tool enabling protein engineers to choose experiments that better explore sequence space, improving the "hit-rate" of finding proteins that are stably folded and have novel activity.

Acknowledgments. This work was supported in part by an NSF CAREER award to CBK (IIS-0444544) and an NSF SEIII grant to CDK, AMF, and Bruce Craig (IIS-0502801).

References

1. Voigt, C., Martinez, C., Wang, Z., Mayo, S., Arnold, F.: Protein building blocks preserved by recombination. Nat. Struct. Biol. 9(7), 553–558 (2002)
2. Meyer, M., Silberg, J., Voigt, C., Endelman, J., Mayo, S., Wang, Z., Arnold, F.: Library analysis of SCHEMA-guided protein recombination. Protein Sci. 12, 1686–1693 (2003)
3. Otey, C., Landwehr, M., Endelman, J., Hiraga, K., Bloom, J., Arnold, F.: Structure-guided recombination creates an artificial family of cytochromes P450. PLoS Biol. 4(5), e112 (2006) **PLoS Biol. 4(5), e112**
4. Saftalov, L., Smith, P., Friedman, A., Bailey-Kellogg, C.: Site-directed combinatorial construction of chimaeric genes: general method for optimizing assembly of gene fragments. Proteins 64(3), 629–642 (2006)
5. Stemmer, W.: Rapid evolution of a protein in vitro by DNA shuffling. Nature 370(6488), 389–391 (1994)
6. Aguinaldo, A., Arnold, F.: Staggered extension process (StEP) in vitro recombination. Methods Mol. Biol. 231, 105–110 (2003)
7. Coco, W.: RACHITT: Gene family shuffling by random chimeragenesis on transient templates. Methods Mol. Biol. 231, 111–127 (2003)
8. Endelman, J., Silberg, J., Wang, Z., Arnold, F.: Site-directed protein recombination as a shortest-path problem. Protein Eng. Des. Sel. 17, 589–594 (2004)
9. Ye, X., Friedman, A., Bailey-Kellogg, C.: Hypergraph model of multi-residue interactions in proteins: sequentially-constrained partitioning algorithms for optimization of site-directed protein recombination. In: Apostolico, A., Guerra, C., Istrail, S., Pevzner, P.A., Waterman, M. (eds.) RECOMB 2006. LNCS (LNBI), vol. 3909, pp. 777–790. Springer, Heidelberg (2006)

10. Zheng, W., Ye, X., Friedman, A., Bailey-Kellogg, C.: Algorithms for selecting breakpoint locations to optimize diversity in protein engineering by site-directed protein recombination. In: Proc. CSB, pp. 31–40 (2007)
11. Meyer, M., Hochrein, L., Arnold, F.: Structure-guided SCHEMA recombination of distantly related beta-lactamases. Protein Engineering, Design & Selection 19, 563–570 (2006)
12. Landwehr, M., Carbone, M., Otey, C., Li, Y., Arnold, F.: Diversification of catalytic function in a synthetic family of chimeric cytochrome P450s. Chemistry & Biology 14, 269–278 (2007)
13. Moore, G., Maranas, C.: Identifying residue-residue clashes in protein hybrids by using a second-order mean-field approach. PNAS 100(9), 5091–5096 (2003)
14. Otey, C., Silberg, J., Voigt, C., Endelman, J., Bandara, G., Arnold, F.: Functional evolution and structural conservation in chimeric cytochromes P450: calibrating a structure-guided approach. Chem. Biol. 11(3), 309–318 (2004)
15. Saraf, M.C., Gupta, A., Maranas, C.: Design of combinatorial protein libraries of optimal size. Proteins 60(4), 769–777 (2005)
16. Carter Jr., C., LeFebvre, B., Cammer, S., Tropsha, A., Edgell, M.: Four-body potentials reveal protein-specific correlations to stability changes caused by hydrophobic core mutations. J. Mol. Biol. 311, 621–638 (2001)
17. Krishnamoorthy, B., Tropsha, A.: Development of a four-body statistical pseudo-potential to discriminate native from non-native protein conformations. Bioinformatics 19, 1540–1548 (2003)
18. Thomas, J., Ramakrishnan, N., Bailey-Kellogg, C.: Graphical models of residue coupling in protein families. IEEE/ACM Transactions on Computational Biology and Bioinformatics (in press, 2007) (Preprint),
http://www.cs.dartmouth.edu/~cbk/papers/tcbb07.pdf
19. Zaccolo, M., Gherardi, E.: The effect of high-frequency random mutagenesis on in vitro protein evolution: a study on TEM-1 beta-lactamase. J. Mol. Biol. 285, 775–783 (1999)
20. Daugherty, P., Chen, G., Iverson, B., Georgiou, G.: Quantitative analysis of the effect of the mutation frequency on the affinity maturation of single chain Fv antibodies. PNAS 97, 2029–2034 (2000)
21. Eisner, M., Severance, D.: Mathematical techniques for efficient record segmentation in large shared databases. Journal of the ACM 23, 619–635 (1976)
22. Waterman, M.S., Eggert, M., Lander, E.: Parametric sequence comparisons. Proc. Natl. Acad. Sci. USA 89, 6090–6093 (1992)
23. Gusfield, D., Balasubramanian, K., Naor, D.: Parametric optimization of sequence alignment. Algorithmica 12, 312–326 (1994)
24. Gusfield, D.: Parametric combinatorial computing and a problem of program module distribution. Journal of the ACM 30, 551–563 (1983)
25. Bykat, A.: Convex hull of a finite set of points in two dimensions. Info. Proc. Letters 7, 296–298 (1978)
26. Firestine, S., Poon, S., Mueller, E., Stubbe, J., Davisson, V.: Reactions catalyzed by 5-aminoimidazole ribonucleotide carboxylases from Escherichia coli and Gallus gallus: a case for divergent catalytic mechanisms. Biochemistry 33, 11927–11934 (1994)

DLIGHT – Lateral Gene Transfer Detection Using Pairwise Evolutionary Distances in a Statistical Framework

Christophe Dessimoz*, Daniel Margadant, and Gaston H. Gonnet

ETH Zurich, Institute of Computational Science,
CH-8092 Zurich and
Swiss Institute of Bioinformatics,
cdessimoz@inf.ethz.ch

Abstract. This paper presents an algorithm to detect lateral gene transfer (LGT) on the basis of pairwise evolutionary distances. The prediction is made from a likelihood ratio derived from hypotheses of LGT versus no LGT, using multivariate normal theory. In contrast to approaches based on explicit phylogenetic LGT detection, it avoids the high computational cost and pitfalls associated with gene tree inference, while maintaining the high level of characterization obtainable from such methods (species involved in LGT, direction, distance to the LGT event in the past). We validate the algorithm empirically using both simulation and real data, and compare its predictions with standard methods and other studies.

1 Introduction

Lateral gene transfer (LGT), or horizontal gene transfer (HGT), is widely recognized as a major force in prokaryotic genome evolution, but the study of its nature and extent is constrained by the limitations of current methods for LGT detection [1,2]. These methods can be divided in two broad categories: parametric methods and phylogenetic methods. In parametric methods, sequence properties such as nucleotide composition [3,4], dinucleotide frequencies [5], codon usage biases [6,7,8], or, more recently, nucleotide substitution matrices [9] are calculated for a specific gene and compared with the rest of the genome. A transferred gene has parameter values typical for its donor genome, which makes it distinguishable from the recipient genome. For this reason, the method can only detect LGT events taking place between organisms with significantly different patterns of evolution. Furthermore, parametric methods are limited to recent LGT transfers because the transferred sequences adapt to their new host relatively rapidly [3]. Lastly, some native genes may have atypical nucleotide composition for reasons other than LGT.

Phylogenetic methods identify LGT events by analyzing the discrepancy between the phylogeny of laterally transferred genes and their host genomes. Therefore, most phylogenetic methods consist of inference of gene and species trees,

---

* Corresponding author.

M. Vingron and L. Wong (Eds.): RECOMB 2008, LNBI 4955, pp. 315–330, 2008.

and their reconciliation [10,11]. Other methods, such as Lawrence's rank correlation test [12] or Clarke's phylogenetic discordance test [13] use unexpected sequence similarity scores to detect LGT, and do not require the inference of gene trees. To distinguish between the two types, we refer to the former by *explicit*, the latter by *implicit* phylogenetic methods. Explicit methods have the potential of describing in detail LGT events (involved species, direction of transfer, time of the transfer), but suffer from the difficulties associated with the inference of gene trees, a task both computationally expensive and error-prone. On the other hand, the two implicit phylogenetic methods mentioned here are fast and robust, though limited by their reliance on similarity scores, which do not always reflect phylogeny [14] in the first place, and by the relative coarseness of their underlying models, which limits their detection power.

In this manuscript, we introduce a new phylogenetic method for LGT detection, which we call DLIGHT (Distance Likelihood based Inference of Genes Horizontally Transfered). Based on evolutionary distances and applied in a probabilistic framework, it combines the speed, the lack of gene tree requirement, and the robustness of implicit methods with the high level of details obtained by explicit methods. The next section presents the algorithm, and is followed by validation using simulation and biological data.

2   Method

2.1   Preliminaries

Definition (family of orthologs). *A set of sequences (genes or proteins[1]) $f = \{x_1, x_2, ...\}$ is a family of orthologs if all pairs of sequences (x_i, x_j) in f are either orthologs or xenologs through orthologous replacement. We denote the set of all such families by F.*

DLIGHT's objective is to detect LGT in such families of orthologs. In the above definition, we require that the families have no paralogs (paralogy detection is beyond the scope of this article). This also ensures that there is at most one sequence per species in any family of orthologs. Thus, a sequence is also uniquely referenced by the pair (f, g), where f is a family that contains the sequence and g the species it belongs to (or the genome – the two terms are used here interchangeably). We denote by $G(f)$ the set of species of sequences of f. We denote the evolutionary distance between sequences of species i and j in family f by $d_f(i, j)$.

Assumption 1 (interspecies distance, family-specific rates). *We assume that, in the absence of LGT, all distances between orthologs of species i and j are proportional to an interspecies distance $d(i, j)$, with a family-specific proportionality constant τ_f. Formally, $d_f(i, j) = \tau_f \cdot d(i, j)$. Furthermore, we require that on average, the proportionality constant be one ($\frac{1}{|F|} \sum_{f \in F} \tau_f = 1$). This model is refered to as proportional branch lengths by [15].*

---

[1] In this work, we consider at most one protein sequence per gene.

Estimator $\hat{d}_f(i, j)$. The evolutionary distance $d_f(i, j)$ can be estimated from a pairwise alignment by maximum likelihood (ML) under a model of amino-acid substitution. We call this estimator $\hat{d}_f(i, j)$. The ML estimator is asymptotically unbiased and asymptotically normally distributed. The ML procedure also provides an estimate of its variance $\sigma^2(\hat{d}_f(i, j))$. Furthermore, covariance estimation is shown in [16].

Estimator $\hat{d}(i, j)$. We estimate the interspecies distance $d(i, j)$ using the unweighted sample average over all $|F|$ families of orthologs:

$$\hat{d}(i, j) = \frac{1}{|F|} \sum_{f \in F} \hat{d}_f(i, j)$$

The estimator is unbiased, because:

$$\mathbb{E}(\hat{d}(i, j)) = \frac{1}{|F|} \sum_{f \in F} \mathbb{E}(\hat{d}_f(i, j)) = \frac{1}{|F|} \sum_{f \in F} \tau_f \cdot d(i, j) = d(i, j) \underbrace{\frac{1}{|F|} \sum_{f \in F} \tau_f}_{=1} = d(i, j)$$

Assumption 2. *In the following, we will consider $\hat{d}(i, j)$ to be a point estimate, that is, we assume that $\sigma^2(\hat{d}(i, j)) = 0$.*

This assumption may appear to be quite strong, especially if the number of families under consideration is small. In most cases, however, the number of families is relatively large (larger than the size of a typical family), and the variances of interspecies distances are much smaller than those of the other estimators under consideration here. In terms of computation, the assumption considerably reduces the time complexity of our approach.

Estimator $\hat{\tau}_f$. We estimate the rate τ_f of family f using the following estimator:

$$\hat{\tau}_f = \frac{\frac{1}{n_f(n_f-1)} \sum_{i,j \in G(f), i \neq j} \hat{d}_f(i, j)}{\frac{1}{n_f(n_f-1)} \sum_{i,j \in G(f), i \neq j} \hat{d}(i, j)} = \frac{\sum_{i,j \in G(f), i \neq j} \hat{d}_f(i, j)}{\sum_{i,j \in G(f), i \neq j} \hat{d}(i, j)}$$

where $n_f = |G(f)|$. Due to assumption 2, the denominator is constant, and thus $\hat{\tau}_f$ follows a normal distribution with variance

$$\sigma^2(\hat{\tau}_f) = \frac{\sum_{i,j,k,l \in f, i \neq j, k \neq l} cov(\hat{d}_f(i, j), \hat{d}_f(k, l))}{(\sum_{i,j \in f, i \neq j} \hat{d}(i, j))^2}$$

Lateral Gene Transfer

Definition (lateral gene transfer). *In the present work, a lateral gene transfer (LGT) event is the transfer of a gene from a donor species d (or an ancestor thereof) to a recipient species r (or an ancestor thereof).*

Assumption 3. *Since the divergence of d and r, at most one LGT event per family of orthologs took place between the two lineages.*

Assumption 4. *The rate of evolution (the branch length on the phylogenetic tree) of a sequence after LGT is homogeneous among all donor and recipient lineages.*

Definition (δ). *Given a LGT event in family f between lineages of d and r, the evolutionary distance between the transfered sequence and the current sequences in r or d is expressed by δ (Fig. 1). The distance since LGT is the same for both species due to assumption 4.*

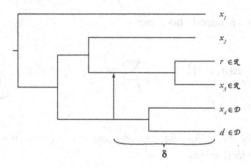

Fig. 1. Distance to LGT event as captured by the parameter δ. The LGT event is represented by the arrow.

Consequently, the expected distance between sequences in f of d and r is 2δ. For instance, if $\delta = 0$, the two proteins have not diverged since the LGT event, and thus the LGT is very recent.

2.2   Algorithm

DLIGHT identifies LGT events by considering, in all families of orthologs, all potential pairs of donor and recipient species. For each configuration, a likelihood ratio test is performed between the hypothesis of a LGT (alternative hypothesis) and the hypothesis of no LGT (null hypothesis). Formally, the set of significant LGT events is given by:

$$LGT = \left\{ (f,d,r) \mid f \in F; \ d, r \in G(f); \ \underset{\delta \geq 0 \in \mathbb{R}}{\operatorname{argmax}} \left(2\ln \frac{l(f,d,r,\delta)}{l(f,d,r,\delta = \infty)} \right) > \chi^2(\alpha, 1) \right\}$$

where F is the set of all families of orthologs, d a potential donor species, r a potential recipient species and $l(f,d,r,\delta)$ is the likelihood of an LGT in f from lineages of d and r at distance δ in the past. $l(f,d,r,\delta = \infty)$ is the likelihood under the null hypothesis (in which δ is fixed to ∞, see below), and $\chi^2(\alpha, 1)$ is the critical value of the chi-square distribution with significance level α and one degree of freedom. This test is known as the likelihood ratio test (see e.g. [17]). The ratio follows a chi-square distribution if the two models are nested, which is the case here, as we shall see below.

Below, we show how the likelihood of a LGT event $l(f, d, r, \delta)$ can be computed. The process can be split in three parts: first, given (f, d, r, δ), we infer which species of $G(f)$ belong to the set of donor species $\mathcal{D}$ and of recipient species $\mathcal{R}$. From these sets, we show how to compute the expected values of all $2|f| - 3$ evolutionary distances of pairs in f that involve r and/or d, as well as their variances and covariances. Finally, we compute the likelihood of the event, which is based on the deviation of the observed distances from the expected distances.

Step 1 – Assignment of Species to Sets of Donors ($\mathcal{D}$) and Recipients ($\mathcal{R}$). First, given a quartet (f, d, r, δ), we infer members of $G(f)$ belonging to the donor and recipient lineages, that is, the set of species that directly descend from the donor (set $\mathcal{D}$) and recipient species (set $\mathcal{R}$). These subsets of $G(f)$ can be defined as follows:

$$\mathcal{D} = \{j \in G(f) \mid \tau_f \cdot d(j, d) \leq 2\delta\}$$

$$\mathcal{R} = \{j \in G(f) \mid \tau_f \cdot d(j, r) \leq 2\delta\}$$

We shall now justify these definitions (illustrated in Fig. 1). First, note that as could be reasonably expected, $d \in \mathcal{D}, r \in \mathcal{R}$, because in both cases the distance to themselves is 0, and δ being a distance is non-negative. As for the other species of $G(f)$, the definitions use assumption 4 (we focus on the definition of $\mathcal{D}$; the rational for $\mathcal{R}$ is similar): if all sequences from the donor lineage in f evolve at the same rate, they will all be δ away from the LGT. Further, by definition, members of the donor lineage have speciated after the LGT event, and therefore, their sequences in f are separated by a distance of at most 2δ.

To build these sets, we must rely on the estimators $\hat{\tau}_f$ and $\hat{d}(j, d)$ (or $\hat{d}(j, r)$ in the case of $\mathcal{R}$). Since the interspecies distances are point estimates (assumption 2), we only need to consider the distribution of $\hat{\tau}_f$ (see Sect. 2.1): the sets of donors and recipients differ depending on the value of the estimator $\hat{\tau}_f$. Fig. 2 depicts the distribution with the critical values of $\hat{\tau}_f$ for the assignment of a species j to $\mathcal{D}$ and $\mathcal{R}$.

Thus, if we consider the two critical values for all species j in $G(f)$, the distribution of $\hat{\tau}_f$ will be partitioned into $2|f| + 1$ ranges. Each of these ranges map to particular $\mathcal{D}_i$ and $\mathcal{R}_i$, whose probability is the area of the density function $\text{pdf}(\hat{\tau}_f)$ in that particular range. We refer to the probability of the ith range as p_i. We will compute for each of these sets of donors and recipients the corresponding likelihood, and then average them according to their probability. The next step is therefore repeated for all $2|f| + 1$ possible assignments of $\mathcal{D}, \mathcal{R}$.

Step 2 – Pairwise Distance Statistics. Given a sextet $(f, d, r, \delta, \mathcal{D}_i, \mathcal{R}_i)$, the computation of the likelihood of a particular LGT event is based on the $2|f| - 3$ pairwise distances in f that involve d or r. These distances are of interest because

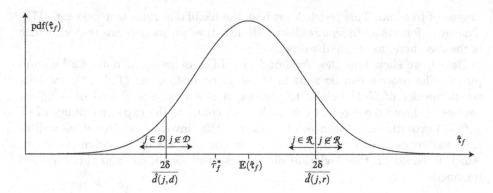

Fig. 2. The assignment of sequence j to sets $\mathcal{D}, \mathcal{R}$ depends on $\hat{\tau}_f$. For instance, at the point $\hat{\tau}_f^*$, j is in $\mathcal{R}$, but not in $\mathcal{D}$.

they are particularly altered by the LGT event, but the procedure could trivially be extended to all $\binom{|f|}{2}$ pairs in f.

The *observed* distances are simply the ML estimators for the relevant pairs of sequences of f. Estimators for the *expected* distances are provided in Table 1. Most distances involving the donor species d are unaffected by the LGT event, i.e. they are expected to follow the interspecies distances scaled by the family rate. Distances to the recipient species r however are mostly expected to follow the scaled interspecies distance to the donor d, because the sequence originated from the donor lineage, and after the LGT event, they evolved at the same rate as in the donor lineage (assumption 4). The special cases are: (i) distances between two recipients: they are unaffected by the LGT because the transfer happened before they speciated; (ii) distances between recipient and donor species: they are expected to be 2δ per definition; (iii) distances involving *inconsistent* species: the estimators and parameters can be such that a species is in both $\mathcal{D}_i$ and $\mathcal{R}_i$, for instance if δ is particularly large. In those cases, we treat the distance the same way as under the null hypothesis (no LGT transfer) and assign it an expected value that corresponds to the scaled interspecies distance. In terms of the model, this also has the advantage that the null hypothesis of no LGT is equivalent to the special case of a LGT with parameter $\delta = \infty$. This means that the models are nested, and therefore that the likelihood ratio follows a chi-square distribution with number of degree of freedom given by the difference in free parameters (one in our case).

Note that in our model, both observed and expected pairwise distances are normally distributed random variables, which can be expressed using two $2|f|-3$ dimensional vectors x and y. In both case, we have estimators for their variance-covariance matrices Σ_x and Σ_y : for observed distances, the diagonal entries can be obtained by ML theory, and the covariances can be computed as described in [16]. As for the expected distances, the variance is either that of $\hat{\tau}_f$ scaled appropriately, or else null when $\hat{\tau}_f$ does not appear in the expression. The expected distances do not covary, and thus all off-diagonal entries are null.

Table 1. LGT event: expected distances to r and d in f. Note that the last row (*inconsistent*) can occur in our model if δ is large; the adverse impact of such inherently inconsistent case is limited by using the same expected distances as under the null hypothesis (no LGT event).

label	$\in \mathcal{D}_i$	$\in \mathcal{R}_i$	$\hat{\mathbb{E}}(\hat{d}_f(j,d))$	$\hat{\mathbb{E}}(\hat{d}_f(j,r))$
outgroup	no	no	$\hat{\tau}_f \cdot \hat{d}(j,d)$	$\hat{\tau}_f \cdot \hat{d}(j,d)$
donor	yes	no	$\hat{\tau}_f \cdot \hat{d}(j,d)$	2δ
recipient	no	yes	2δ	$\hat{\tau}_f \cdot \hat{d}(j,r)$
inconsistent	yes	yes	$\hat{\tau}_f \cdot \hat{d}(j,d)$	$\hat{\tau}_f \cdot \hat{d}(j,r)$

Let $z = x - y$. The vector z is normally distributed, with expected value $\mathbb{E}(z) = 0$. If we now assume that x, y are independent, $\Sigma_z = \Sigma_x + \Sigma_y$. In reality, they are not strictly independent, because x is a component (albeit a minor one) of $\hat{\tau}_f$, which itself is used in the computation of y.

Step 3 – Computation of the Likelihoods, and Estimation of δ. The likelihood of the LGT event $(f, d, r, \delta, \mathcal{D}_i, \mathcal{R}_i)$ can be computed from the multivariate normal probability density function of the vector z and covariance matrix Σ_z from the previous section:

$$l(f, d, r, \delta, \mathcal{D}_i, \mathcal{R}_i) = \frac{exp(-\frac{1}{2}z^T \Sigma_z^{-1} z)}{\sqrt{(2\pi)^{2|f|-3}|\Sigma_z|}}$$

We can now marginalize over the $2|f|+1$ different sets of donors and recipients (see step 1) to compute the likelihood of the LGT event (f, d, r, δ):

$$l(f, d, r, \delta) = \sum_i p_i \cdot l(f, d, r, \delta, \mathcal{D}_i, \mathcal{R}_i)$$

Furthermore, the parameter δ can be estimated by maximizing the likelihood. As mentioned above, the likelihood for the null hypothesis of no LGT event is obtained by the special case with parameter $\delta = \infty$.

2.3   Model Violations and Test of Multivariate Normality

DLIGHT is based on assumptions that do not always hold, in particular when dealing with biological sequences whose evolution strongly deviates from the markovian model. To limit the adverse effect of such model violations, we test the multivariate normality of the data by computing a p-value based on the squared Mahalanobis distance $z^T \Sigma_z^{-1} z$, which is known to be chi-square distributed if z is multivariate normal. Data falling in extreme quantiles are considered dubious. In experiments reported here, predictions with data falling in the $(1 - 10^{-10})$ quantile were considered artifacts due to model violation, and were disregarded.

Furthermore, in case of poorly estimated variances or covariances, the matrix Σ_z may not be positive definite, or it may be singular if the sequences of two species are identical. In our implementation, we still try to identify LGT events by working with a subset of the family that constitute a well-posed problem (the problematic sequences are excluded on the basis of a simple greedy approach).

2.4 Combination of Results and Correction for Multiple Testing

As we presented above, DLIGHT computes a likelihood ratio test in all families of orthologs, for all different possible pairs of potential donor and recipient species. This raises the issues of combining results and correcting for multiple testing. Currently, we take the conservative approach of combining results that are consistent, for instance when a LGT event happened before speciation of the recipient species into two species g_1 and g_2: the algorithm may detect a transfer when run with both species as recipient, but if in both cases the estimated δ suggests a transfer prior to their speciation, the prediction is consistent and can be combined. Another common case for combination are pairs of results that report LGT between consistent sets of donor and recipient genomes, but with reverse direction. The direction of some LGT events, such as transfers between close species, is inherently difficult to assess. Nevertheless, if one direction has a significantly higher probability, and provided that the estimated parameter δ is consistent, the direction of the LGT can be infered.

We address the issue of multiple testing by using the Bonferroni adjustment, a common approach that discounts the significance by a factor corresponding to the total number of tests. If the tests are not independent from each other, which is the case here, the correction is excessive and some sensitivity is wasted.

3   Validation and Results

DLIGHT was tested using four different approaches: simulation, artifical LGT events, real biological data and comparison with previous results from the literature. The results of simulation are also reported for three simple LGT detection methods that serve as benchmark: methods based on GC-content, best-hits, and perturbed-distances. They are described in the *Appendix*.

3.1   *In Silico* Evolution Scenarios

Although a simulation will never fully capture the complexity and diversity of natural evolutionary processes, it allows the evaluation of algorithms with knowledge of the history of events, and therefore constitutes a tracable baseline. Synthetic genomes were generated using the software *EVA* (manuscript in preparation). EVA starts from a single organism and simulates the following evolutionary mechanisms: codon mutation based on empirical substitution probabilities [18], with biased genome-specific GC contents and gene-specific mutation rates, codon insertion and deletion, gene duplication, gene loss, LGT (both orthologous replacement and novel gene acquisition), and speciation. The probabilities of LGTs, gene duplications and gene loss were set to a proportion of 1:2:3, thereby keeping the expected number of genes constant (as suggested in [19]). The two types of LGT events, novel gene acquisition and orthologous replacement, were set to have an equal probability of occurence. Table 2 details the remaining parameters of the two different evolutionary scenarios investigated here. Genes from the resulting genomes were grouped in orthologous familes using the OMA algorithm [20].

Table 2. Overview of the simulation parameters. In *simulation 1* closely related organisms are used while in *simulation 2* more distantly related organisms are analysed.

Name	# of species	Avg. # of genes	Avg. genome distance (expect. identity)	# LGT	# of families
simulation 1	9	197	16 PAM (85.4%)	50	241
simulation 2	9	202	74 PAM (50.7%)	42	295

The different algorithms were run on the two datasets and the performances were analysed in terms of both sensitivity and specifity, at three levels of precision: first, the ability to report families of orthologs that contain at least one laterally transferred gene; second, the ability to identity the protein involved in a LGT event, that is, either report a *donor* or a *recipient* species; and third, the ability to correctly identify the direction of the LGT, in addition to the species involved. The six resulting ROC curves are presented in Fig. 3. Overall, DLIGHT showed significantly higher sensitivity and specificity than the other methods. It also performed more consistently than the other methods, with curves of similar shape across all experiments. The significance threshold is rather conservative (a consequence of the stringent Bonferoni correction) and led to 100% specifity in most cases. In the case in which the direction of LGTs was required, in distantly related species, the GC content and the perturbed distance approach outperformed DLIGHT. This may be due to the difficulties in estimating distances and variances when organisms are so far apart. In those cases, simpler methods may prove to be more robust.

3.2 Artificial LGT Events in Real Data

LGT events between real biological genomes can be simulated by introducing a gene from one species into another, either as substitute for its ortholog ("orthologous replacement") or as additional sequence. Such *artificially introduced* LGT event allows the testing of the algorithm on real biological data while having a positive control. However, only the specific case of very recently introduced genes can be simulated. Furthermore, real occurences of LGTs may already be present in the dataset and their signals may conflict with the artificially introduced ones.

The biological data consisted of 15 archaea with 2273 gene families, of which 727 families had at least 6 genes. 200 cases of LGT events from random donors to random recipients were introduced, as orthologous replacement, in families with at least 6 genes. Fig. 4 presents the results of the tests. The 200 top scoring

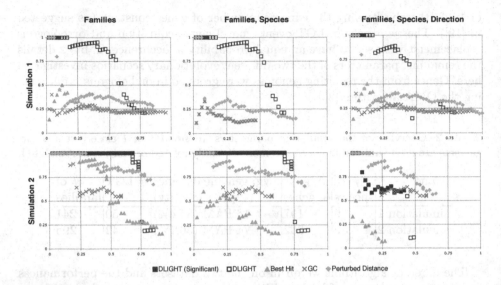

Fig. 3. ROC analyses. Sensitivity is ploted along the X axis, specifity along the Y axis. Plots on the first line were obtained from a simulation with closer species, plots on the second line from more distantly related ones. The left column shows results of identifying families with LGT events. The middle column shows results of identifying families with LGT events and the involved species. The right column shows results of identifying families with LGT events, the involved species and the direction of the transfers.

predictions were compared to the set of artificially introduced LGTs. Of all four methods, our performed best. Given the relatively good results obtained with the perturbed-distance approach in the previous test, its performance here is surprisingly poor, with only 7 artificial LGTs recovered. Note also that being recent, transfers introduced here constitue ideal conditions for both the GC method (the composition has not had time to adapt to the new host) and the best-hit approach (transfer after all speciations).

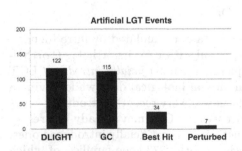

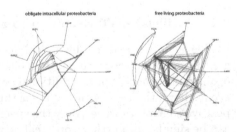

Fig. 4. Artificially introduced LGT. The number of such LGTs among the top 200 predictions is given.

Fig. 5. LGT flow among proteobacteria. LGTs are drawn with arrows indicating the direction of the transfer. DLIGHT was run with the same parameters on both datasets individually.

3.3   Real Biological Data

LGT events are believed to happen throughout the prokaryotes, but not uniformly so. Some organisms are considered to be little affected by LGT while others are thought to have acquired many genes from distant species. Endosymbionts and endoparasites are micro-organisms that spend most of their life inside a host cell. As a consequence, for an LGT event to happen, foreign DNA would need to cross the membrane and defensive system of both the organism and its host. Therefore, such organisms are expected to have very few genes aquired through LGT compared to free living micro-organisms [21].

Our algorithm was verified against these observations by comparing predictions of two different datasets. We inferred LGTs for 9 endosymbionts[2] and for 9 free living pathogenic proteobacteria[3]. The organisms were classified according to HAMAP [22].

DLIGHT detected between 1 and 22 foreign genes (6.3 in average) in endosymbionts, and between 2 and 70 genes (40.7 in average) in free living bacteria. Normalized with the genome sizes, this gives between 0.15% and 0.89% percent of foreign genes in endosymbionts, versus 0.12% to 2.43% in free living. Thus, endosymbionts appear indeed to have lower LGT rates than their free living counterparts. In figure 5 the LGT events are indicated in both trees as thin lines and there too, the difference in LGT occurences is clearly visible.

The detected percentages of foreign genes is much lower than the values of 2% to 60% found in previous reports [23,24,25]. However, these higher numbers represent all genes received by any organism outside the vertical genealogy, while our data reflect only gene transfer among 9 bacteria.

A larger set with 15 archaea[4] consisting of 2273 orthologous families was analysed in a similar way. The average LGTs per gene was at 1.07%, with 292 detected LGT events in all 15 archaea. The number of acquired genes varies from 1 for *Nanoarchaeum equitans* to 37 for *Methanosarcina mazei* . Looking at the relative gene uptake with regard to the genome size, *Nanoarchaeum equitans* still recieved the fewest genes with 0.19%. *Thermoplasma volcanium* received the most genes with 2.4%. It has been proposed previously that LGT is common between Thermoplasmatales and Sulfolobales [1]. In our dataset, *Thermoplasma*

---

[2] Candidatus Blochmannia floridanus, Blochmannia pennsylvanicus (strain BPEN), Buchnera aphidicola (subsp. Schizaphis graminum), Lawsonia intracellularis (strain PHE/MN1-00), Sodalis glossinidius (strain morsitans), Vibrio fischeri (strain ATCC 700601 / ES114), Wigglesworthia glossinidia brevipalpis, Wolbachia pipientis wMel, Wolbachia sp. (subsp. Brugia malayi) (strain TRS)

[3] Campylobacter jejuni, Escherichia coli O6, Escherichia coli, Haemophilus influenzae (strain 86-028NP), Neisseria meningitidis serogroup A, Pasteurella multocida, Pseudomonas aeruginosa, Shigella flexneri, Vibrio cholerae

[4] Methanocaldococcus jannaschii, Methanosarcina mazei, Pyrobaculum aerophilum, Sulfolobus solfataricus, Methanosarcina acetivorans, Aeropyrum pernix, Archaeoglobus fulgidus, Halobacterium salinarium, Methanobacterium thermoautotrophicum, Methanopyrus kandleri, Pyrococcus horikoshii, Thermoplasma volcanium, Nanoarchaeum equitans, Thermoplasma acidophilum, Methanococcus maripaludis

volcanium exchanged 14 genes with *Sulfolobus solfataricus* and *Thermoplasma acidophilum* also 14 genes with *Sulfolobus solfataricus*. This is significantly more than the 3.6 average LGTs between archaea.

In addition to these tests, DLIGHT was applied to a dataset of 10 mammals[5]. Although LGT between higher eukaryotes and bacteria are found by some authors, we are not aware of any case of LGT between two mammals. Mammals serve therefore as negative control for our LGT detection method. Indeed, DLIGHT did not detect any LGT among the 10 mammals.

3.4   Comparision with Previous Results

Results from different LGT inference approaches can be very inconsistent, with overlaps at times smaller than expected by random [26]. This is particularily true when comparing the results of parametric and phylogenetic methods. Thus, the results of DLIGHT were compared with two studies based on phylogenetic approaches.

Comparison with Zhaxybayeva *et al.* (2006). In [27], the authors used an embedded quartet decomposition analysis to search events of LGT in 11 completey sequenced cyanobacteria. Orthologs were grouped via reciprocal top-scoring blast hits, resulting in families with few paralogs. A set of 1128 ortholgous genes was found to be present in at least nine of the 11 cyanobacterial genomes and taken as input for the LGT search. Within the group of cyanobacteria, 135 LGTs were detected, mostly between *Gloeobacter violaceus* and *Synechococcus elongatus* (45) and *Prochlorococcus marinus* SS120 and *Prochlorococcus marinus* (strain MIT 9313) (28).

We tried to confirm the predictions of LGT in these 135 families using DLIGHT. In 54 families (40%), significant LGTs were reported. In 32 of them, the species predicted to be involved were either the same, or in agreement with the trees constructed by [27]. The 22 other predictions were conflicting with their trees. Additionally, it should be noted that the interspecies distances estimated by DLIGHT were computed on the basis of these 135 families, none of which is congruent to the species tree according to [27]; this suggests that DLIGHT is relatively robust with respect to perturbations in the data.

Comparison with Beiko *et al.* (2005). DLIGHT was compared with results from [10], a large scale LGT inference study using an explicit phylogenetic method. For 22,437 families of proteins in 144 genomes, they constructed gene trees and compared in each tree all bifurcations to a reference species tree. They reported bifurcations with significant posterior probability (PP), classified in either consistent or conflicting with the species tree.

A subset of their 8,315 protein families of size up to 15 sequences was randomly selected. Based on their bifurcation analysis, these familes were partitioned in

---

[5] Homo sapiens, Mus musculus, Canis familiaris, Rattus norvegicus, Bos taurus, Pan troglodytes, Monodelphis domestica, Macaca mulatta, Loxodonta africana, Oryctolagus cuniculus

four categories: i. 28.5% families with strong support of no LGT (all bifurcations consistent with species tree with $PP \geq 0.95$), ii. 38.4% families with mild support of no LGT (no conflicting bifurcation with $PP \geq 0.5$), iii. 15.2% families with mild support of LGT (at least one conflicting bifurcation with $PP \geq 0.5$, none with $PP \geq 0.95$), and iv. 17.8% families with strong support of LGT ($PP \geq 0.95$).

DLIGHT was run on this dataset, with, as sole input, the protein sequences labeled with family and species identifiers. The computation of all pairwise evolutionary distances within families required about 2 days on a single AMD Opteron 1.8 GHz. DLIGHT used another day to predict significant LGT events, which were found in 634 families. The distribution of inferred LGT events among the four categories defined from their predictions was as follows: i. 7.1%, ii. 13.1%, iii. 19.2%, and iv. 60.6%. As almost 80% of the predictions are the same, the level of agreement between the two methods is quite high, especially considering the large differences in methodologies.

4 Conclusion

In this article, we introduce a new implicit phylogenetic method for LGT detection, based on pairwise evolutionary distances in a probabilistic framework. Validation shows that it compares favorably with existing parametric and implicit phylogenetic methods. Furthermore, its advantages over explicit phylogenetic methods include speed and lack of reliance on multiple sequence alignments and gene tree inference.

There are, though, a number of aspects that could be the object of further improvement: the sensitivity could be increased by the computation of the likelihoods using all pairwise distances within gene families, and not only the distances to the transfered genes; confidence intervals in the estimation of the interspecies distances. Instead of the approximation of multivariate normality, and at expense of increased time complexity, the distribution of the distances could possibly be estimated in an MCMC framework.

Acknowledgements

The authors thank Manual Gil, Adrian Schneider, Sukman Dessimoz, Gina Cannarozzi, Maria Anisimova, Susan Holmes, Dan Graur, and three anonymous reviewers for helpful comments and suggestions.

References

1. Philippe, H., Douady, C.J.: Horizontal gene transfer and phylogenetics. Curr. Opin. Microbiol. 6, 498–505 (2003)
2. Lawrence, J.G., Ochman, H.: Reconciling the many faces of lateral gene transfer. Trends Microbiol. 10, 1–4 (2002)

3. Lawrence, J.G., Ochman, H.: Amelioration of bacterial genomes: rates of change and exchange. J. Mol. Evol. 44, 383–397 (1997)
4. Lawrence, J.G., Ochman, H.: Molecular archaeology of the Escherichia coli genome. Proc. Natl. Acad. Sci. U S A 95, 9413–9417 (1998)
5. Karlin, S.: Global dinucleotide signatures and analysis of genomic heterogeneity. Curr. Opin. Microbiol. 1, 598–610 (1998)
6. Moszer, I., Rocha, E.P., Danchin, A.: Codon usage and lateral gene transfer in Bacillus subtilis. Curr. Opin. Microbiol. 2, 524–528 (1999)
7. Mrazek, J., Karlin, S.: Detecting alien genes in bacterial genomes. Ann. N. Y. Acad. Sci. 870, 314–329 (1999)
8. Medigue, C., Rouxel, T., Vigier, P., Henaut, A., Danchin, A.: Evidence for horizontal gene transfer in Escherichia coli speciation. J. Mol. Biol. 222, 851–856 (1991) (Comparative Study)
9. Hamady, M., Betterton, M.D., Knight, R.: Using the nucleotide substitution rate matrix to detect horizontal gene transfer. BMC Bioinformatics 7, 476 (2006)
10. Beiko, R.G., Harlow, T.J., Ragan, M.A.: Highways of gene sharing in prokaryotes. Proc. Natl. Acad. Sci. U S A 102, 14332–14337 (2005) (Comparative Study)
11. Gophna, U., Ron, E.Z., Graur, D.: Bacterial type III secretion systems are ancient and evolved by multiple horizontal-transfer events. Gene 312, 151–163 (2003)
12. Lawrence, J.G., Hartl, D.L.: Inference of horizontal genetic transfer from molecular data: an approach using the bootstrap. Genetics 131, 753–760 (1992)
13. Clarke, G.D.P., Beiko, R.G., Ragan, M.A., Charlebois, R.L.: Inferring genome trees by using a filter to eliminate phylogenetically discordant sequences and a distance matrix based on mean normalized BLASTP scores. J. Bacteriol. 184, 2072–2080 (2002)
14. Koski, L.B., Golding, G.B.: The closest BLAST hit is often not the nearest neighbor. J. Mol. Evol. 52, 540–542 (2001)
15. Pupko, T., Huchon, D., Cao, Y., Okada, N., Hasegawa, M.: Combining multiple data sets in a likelihood analysis: which models are the best?. Mol. Biol. Evol. 19, 2294–2307 (2002)
16. Susko, E.: Confidence regions and hypothesis tests for topologies using generalized least squares. Mol. Biol. Evol. 20, 862–868 (2003)
17. Felsenstein, J.: Inferring Phylogenies. Sinauer Associates Inc., Sunderland (2004)
18. Schneider, A., Cannarozzi, G.M., Gonnet, G.H.: Empirical codon substitution matrix. BMC Bioinformatics 6 (2005)
19. Kunin, V., Ouzounis, C.A.: The balance of driving forces during genome evolution in prokaryotes. Genome Res. 13, 1589–1594 (2003)
20. Dessimoz, C., Cannarozzi, G., Gil, M., Margadant, D., Roth, A., Schneider, A., Gonnet, G.: OMA, a comprehensive, automated project for the identification of orthologs from complete genome data: Introduction and first achievements. In: McLysaght, A., Huson, D.H. (eds.) RECOMB 2005. LNCS (LNBI), vol. 3678, pp. 61–72. Springer, Heidelberg (2005)
21. Lawrence, J.G., Hendrickson, H.: Lateral gene transfer: when will adolescence end? Mol. Microbiol. 50, 739–749 (2003)
22. Boeckmann, B., Bairoch, A., Apweiler, R., Blatter, M.C., Estreicher, A., Gasteiger, E., Martin, M.J., Michoud, K., O'Donovan, C., Phan, I., Pilbout, S., Schneider, M.: The swiss-prot protein knowledgebase and its supplement trembl in 2003. Nucleic Acids Res. 31, 365–370 (2003)

23. Lerat, E., Daubin, V., Ochman, H., Moran, N.A.: Evolutionary origins of genomic repertoires in bacteria. PLoS Biol. 3, e130 (2005)
24. Ge, F., Wang, L.S., Kim, J.: The cobweb of life revealed by genome-scale estimates of horizontal gene transfer. PLoS Biol. 3, e316 (2005)
25. Dagan, T., Martin, W.: Ancestral genome sizes specify the minimum rate of lateral gene transfer during prokaryote evolution. Proc. Natl. Acad. Sci. USA 104, 870–875 (2007)
26. Ragan, M.A.: On surrogate methods for detecting lateral gene transfer. FEMS Microbiol. Lett. 201, 187–191 (2001)
27. Zhaxybayeva, O., Gogarten, J.P., Charlebois, R.L., Doolittle, W.F., Papke, R.T.: Phylogenetic analyses of cyanobacterial genomes: quantification of horizontal gene transfer events. Genome Res. 16, 1099–1108 (2006)

Appendix

4.1 Benchmark Methods

The three benchmark methods used in the validation section are described here. All three consist of a scoring function which is used to rank all genes as potentially laterally transfered candidates.

GC Content. The GC method used in this paper is a basic implementation of this common parametric approach. A more advanced implementation can be found in [3]. The version used here considers the GC content on the first and third codon position, without performing a codon usage analysis. The score for a gene x in a species X is computed as follows:

$$S_{GC}(x) = \frac{(GC(x,1) - \mu_{GC}(X,1))^2}{\sigma_{GC}^2(X,1)} + \frac{(GC(x,3) - \mu_{GC}(X,3))^2}{\sigma_{GC}^2(X,3)}$$

where $GC(x,i)$ is the average GC content of the gene x at its ith codon position, and $\mu_{GC}(X,i), \sigma_{GC}^2(X,i)$ the average and variance of GC content among all ith codon position of genes in species X.

Best Hit Approach. The best hit method infers LGT when the highest scoring hit of a particular sequence is in a distant species [13]. Our implementation improves this idea by considering the shortest evolutionary distance rather than the top similarity score. More precisely, the score of a gene x from a species X and family of orthologs f is computed as follows:

$$S_{BH}(x) = \frac{Rank_f(T)}{|f|}$$

where T is the organism in which x has its closest homolog, $Rank_f(T)$ the rank of T among the species represented in f ordered by increasing average interspecies distance to X.

Perturbed-Distances Approach. The third method detects LGT using the same underlying idea as DLIGHT – the discrepancy between gene and inter-species pairwise distances that results from an LGT event – but in a much cruder way: the score of a gene x from an species X, in family f is

$$S_{PD}(x) = \frac{1}{|f| - 1} \sum_{y \in f, y \neq x} (d(x,y) - d(X,Y))$$

where $d(x,y)$ denotes the evolutionary distance between genes x and y, $d(X,Y)$ the interspecies distance between X and Y.

Computation of Median Gene Clusters

Sebastian Böcker[1], Katharina Jahn[2], Julia Mixtacki[2],
and Jens Stoye[3]

[1] Institut für Informatik, Friedrich-Schiller-Universität Jena, Germany
boecker@minet.uni-jena.de
[2] International NRW Graduate School in Bioinformatics and Genome Research,
Universität Bielefeld, Germany
{kjahn,mixtacki}@cebitec.uni-bielefeld.de
[3] Technische Fakultät, Universität Bielefeld, Germany
stoye@techfak.uni-bielefeld.de

Abstract. Whole genome comparison based on gene order has become
a popular approach in comparative genomics. An important task in this
field is the detection of gene clusters, i.e. sets of genes that occur co-
localized in several genomes. For most applications it is preferable to extend this definition to allow for small deviations in the gene content of the
cluster occurrences. However, relaxing the equality constraint increases
the computational complexity of gene cluster detection drastically. Existing approaches deal with this problem by using simplifying constraints
on the cluster definition and/or allowing only pairwise genome comparison. In this paper we introduce a cluster concept named *median gene
clusters* that improves over existing models and present efficient algorithms for their computation that allow for the detection of approximate
gene clusters in multiple genomes.

1 Introduction and Related Work

The increasing availability of completely sequenced and assembled genomes
opens the opportunity to compare whole genomes based on their gene order.
It is well known that, during the course of evolution, rearrangement events, gene
loss and gene duplications lead to a divergence of genomes that initially had
the same gene order and gene content. If no selective pressure was acting on
these processes, gene order and content would be randomized over time. Therefore, the existence of conserved regions is used as a source of information for
comparative genomics [5]. For that purpose genomes are modeled as strings or
permutations of integers so that genes belonging to the same gene family are encoded by the same integer. A recent approach in this context is the computation
of *gene clusters*, which are sets of genes that occur as single contiguous blocks in
several genomes. Variable gene order and multiple occurrences of the same gene
within the blocks are usually allowed. Gene clusters of this type are known as
common intervals and there exist efficient algorithms for their computation, for
example [2,6,10,11,14,15].

M. Vingron and L. Wong (Eds.): RECOMB 2008, LNBI 4955, pp. 331–345, 2008.
© Springer-Verlag Berlin Heidelberg 2008

However, for most applications the requirement of exact occurrences of gene clusters in the genomes turned out to be too strict. Hence, the concept of *approximate gene clusters* arose recently, which allows for small deviations in the gene content of cluster locations. The problem of this model extension is that the search space of approximate gene cluster detection increases exponentially - depending on the cluster concept - either with the number of allowed deviations [4] or the number of compared sequences [9].

One approach to handle deviations of the gene content is by imposing constraints on the cluster locations: For example, *max-gap clusters* [3,9] allow for an arbitrary number of gaps in the cluster locations, each up to a certain length, but find no approximate locations that have lost some genes of the cluster. Despite these restrictions the complexity of this problem increases exponentially with the number of sequences, but is in $O(n^2)$ for two sequences, where n is the length of the longest sequence.

Another approach with a constrained cluster definition is an algorithm presented in [1] that computes gene clusters with a perfect location (reference interval) in one genome and an approximate occurrence in another sequence in $O(n^3 + occ)$ time using $O(n^3)$ space. Computation of gene clusters restricted in this way is a subproblem of our approach to median gene cluster computation. We introduce an algorithm that solves this problem in $O(n^2(1 + \delta)^2)$ time and $O(n^2)$ space, where $\delta \ll n$.

A less constrained model was presented in [13], resulting in a very general gene cluster model, including most other existing ones. In their approach the authors solve the approximate gene cluster problem by an integer linear program.

In this paper we introduce a new cluster concept, named *median gene clusters*, that constrains only the sum of errors that may occur in the approximate occurrences of a gene cluster. This means that we take from each genome the best location of a gene cluster and sum over the missing and interrupting genes in these locations. In the main part of this paper (Sections 3–6) we present an approach for the efficient computation of all median gene clusters in an arbitrary number of genomes, in Section 7 we apply our method to different genomic datasets, compare it to the approaches presented in [9] and [13] and show its applicability to multiple genomes.

2   Basic Definitions

In our context a genome is a string of integers over a finite alphabet $\Sigma = \{1, \ldots, \sigma\}$. Genes belonging to the same gene family are represented by the same integer value. Given a string S, $|S|$ denotes the length of the string and $S[i]$ refers to its ith character. By $S[i, j]$ we refer to the substring of S that starts with its ith and ends with its jth character, $1 \leq i \leq j \leq |S|$. We define the *character set* of a substring $S[i, j]$ of S as

$$CS(S[i, j]) = \{S[m] \mid i \leq m \leq j\}.$$

Inversely, a substring $S[i, j]$ is called a *location* of a character set $C \subseteq \Sigma$ if and only if $C = CS(S[i, j])$. Substrings $S_1[i_1, j_1], \ldots, S_k[i_k, j_k]$ of two or more strings

$S_1, \ldots, S_k$ of equal character content $CS(S_1[i_1, j_1]) = \ldots = CS(S_k[i_k, j_k])$ are called *common intervals* of $S_1, \ldots, S_k$.

To simplify the notation of the following definitions we assume that a sequence S of length n is extended by a terminal character $S[0] = S[n+1] \notin \Sigma$. A substring $S[i, j]$ is *left-maximal* if $S[i-1] \notin CS(S[i, j])$, *right-maximal* if $S[j+1] \notin CS(S[i, j])$ and *maximal* if it is both left- and right-maximal.

We define the following metric on two character sets $C, C' \subseteq \Sigma$, called the *symmetric set distance*:

$$D(C, C') = |C \setminus C'| + |C' \setminus C|.$$

A *d-location* of a character set C in a string S is a substring $S[i, j]$ such that $D(C, CS(S[i, j])) \leq d$.

A character set $C \subseteq \Sigma$ is a *median* of a set of k character sets $C_1, \ldots, C_k \subseteq \Sigma$ if and only if $\sum_{l=1}^{k} D(C, C_l) \leq \sum_{l=1}^{k} D(C', C_l)$ for all $C' \subseteq \Sigma$. Note that a median in this context is not necessarily unique. This is due to the fact that for even k a character occurring in the median can occur in exactly half of the k character sets. When removing this character from the median, the total distance to the character sets stays unchanged and the remaining characters form an alternative median.

The problem considered in this paper is the following.

Problem 1. Given k sequences $S_1, \ldots, S_k$, a minimum cluster size s and a distance threshold δ, we want to compute all sets $C \subseteq \Sigma$ with $|C| \geq s$ for which there exist $S_1[i_1, j_1], \ldots, S_k[i_k, j_k]$ with pairwise intersecting character sets and C is a median of $CS(S_1[i_1, j_1]), \ldots, CS(S_k[i_k, j_k])$ with

$$\sum_{l=1}^{k} D(C, CS(S[i_l, j_l])) \leq \delta. \tag{1}$$

Such a set C is called a *median gene cluster* of $S_1, \ldots, S_k$.

Defining gene cluster properties that are biologically meaningful and algorithmically feasible is a delicate task (a survey of different cluster properties can be found in [12]). Therefore, variants of the above problem formulation and additional cluster properties will also be discussed in the Appendix.

3 A Three Step Approach to Median Gene Clusters

Our strategy for finding all median gene clusters is based on the observation that whenever inequality (1) holds, the distances between the character sets of the involved substrings are limited by the following upper bound:

Lemma 1. *Let* $S_1, \ldots, S_k$ *be sequences with substrings* $S_1[i_1, j_1], \ldots, S_k[i_k, j_k]$ *such that for a given* $\delta \geq 0$ *there exists a* $C \subseteq \Sigma$ *with* $\sum_{l=1}^{k} D(C, CS(S[i_l, j_l])) \leq \delta$. *Then,*

there is at least one substring $S_m[i_m, j_m]$, $1 \leq m \leq k$, *with* $C' = CS(S_m[i_m, j_m])$
and

$$\sum_{l=1}^{k} D(C', CS(S_l[i_l, j_l])) \leq 2\frac{k-1}{k}\delta. \tag{2}$$

Proof. Among the substrings $S_1[i_1, j_1], \ldots, S_k[i_k, j_k]$ chose $S_m[i_m, j_m]$, $1 \leq m \leq k$, such that $D(C, CS(S_m[i_m, j_m])) \leq \frac{\delta}{k}$. Let $C' = CS(S_m[i_m, j_m])$. From the triangle inequality we infer:

$$\sum_{l=1}^{k} D(C', CS(S_l[i_l, j_l])) \leq \sum_{l \neq m} \left(D(C', C) + D(C, CS(S_l[i_l, j_l])) \right)$$

$$\leq (k-2)\frac{\delta}{k} + \sum_{l=1}^{k} D(C, CS(S_l[i_l, j_l]))$$

$$\leq (k-2)\frac{\delta}{k} + \delta \leq 2\frac{k-1}{k}\delta.$$

$\square$

Character sets such as the above C' are used to filter the search space of potential median gene clusters and are therefore named *cluster filters*.

Lemma 1 gives rise to the following approach, consisting of three steps:

1. First, we compute the set of all cluster filters C' for $S_1, \ldots, S_k$. For that purpose we test for all substrings of the k sequences whether their corresponding character sets meet the conditions given by lemma 1.
2. In the second step, for each cluster filter C' we compute k-tuples of the form $(S_1[i_1, j_1], \ldots, S_k[i_k, j_k])$ where at least one of the elements is a location of C' and inequality (2) holds.
3. Finally we compute for each k-tuple from Step 2 the median(s) of the corresponding character sets. Medians that comply with the distance threshold of inequality (1) are reported as median gene clusters.

4   Computation of Cluster Filters (Step 1)

In k sequences of length at most n there are $O(kn^2)$ substrings. A naive algorithm can determine the cluster filters in $O(k^2 n^4)$ time by computing the pairwise distances between all pairs of substrings. In this section we present two better approaches that are based on the algorithm *Connecting Intervals* (CI) [14] for the computation of common intervals in a pair of sequences.

For simplicity we give the detailed description of our algorithm for just two sequences. The extension to multiple sequences is straightforward and will be briefly addressed in Section 4.4. In the following let $d = 2\frac{k-1}{k}\delta$. For $k = 2$ sequences this cancels out to $d = \delta$. At first, we will review the basic concepts of the original algorithm CI, before we show in Sections 4.2 and 4.3 how it can be adapted to find cluster filters.

4.1 The Connecting Intervals Algorithm

Algorithm CI, presented in [14], finds all common intervals of two sequences S_1 and S_2 of length at most n in $O(n^2)$ time and space.

In a preprocessing step an array called POS and a table called NUM are computed. POS is of length $|\Sigma|$ and lists for each character $c \in \Sigma$ all positions where it occurs in S_2. NUM is a $|S_2| \times |S_2|$ table such that entry $NUM[i,j]$ contains the number of different characters that occur in the substring $S_2[i,j]$. For an example, see Figure 1.

$POS[1] = 1$

$POS[2] = 4, 7, 11$

$POS[3] = 2, 9$

$POS[4] = 5, 8$

$POS[5] = 3, 6, 12$

$POS[6] = 10$

$NUM[i,j]$:

$i \backslash j$	1	2	3	4	5	6	7	8	9	10	11	12
1	1	2	3	4	5	5	5	5	5	6	6	6
2		1	2	3	4	4	4	4	4	5	5	5
3			1	2	3	3	3	3	4	5	5	5
4				1	2	3	3	3	4	5	5	5
5					1	2	3	3	4	5	5	5
6						1	2	3	4	5	5	5
7							1	2	3	4	4	5
8								1	2	3	4	5
9									1	2	3	4
10										1	2	3
11											1	2
12												1

Fig. 1. For $S_2 = (1,3,5,2,4,5,2,4,3,6,2,5)$ with $\Sigma = \{1,\ldots,6\}$, the positions of each occurrence of a character c are stored in $POS[c]$. The entries of the table $NUM[i,j]$ equal $|\mathcal{CS}(S_2[i,j])|$.

The basic idea of the main algorithm is that while going systematically through all maximal substrings $S_1[i,j]$ of the first sequence, using the array POS one generates and iteratively extends marked intervals in the second sequence that consist only of characters occurring in the current interval $S_1[i,j]$.

Common intervals are detected by comparing the character content of $S_1[i,j]$ and the marked intervals in S_2. Since by construction the character sets of the marked intervals are subsets of $\mathcal{CS}(S_1[i,j])$, this can be tested by comparing their size, using the table NUM, and keeping track of the current size of $\mathcal{CS}(S_1[i,j])$. Only those intervals in S_2 that were extended by the latest character of the current $S_1[i,j]$ need to be considered for this test. (Other intervals do not contain this character and thus have a different character set.) Because of the systematic traversal of the maximal substrings of S_1, where for a fixed i the maximal substrings starting at i are processed one after the other for increasing values of j, each character is at most $|S_1|$ times the latest character of a substring of S_1. Hence, each position in S_2 becomes marked at most $|S_1|$ times and each time extends one marked interval, or merges two intervals or constitutes a new marked interval if its neighbors are not yet marked. Thus there are at most $|S_1| \cdot |S_2|$ interval extensions and the same number of character set comparisons. In total this algorithm takes $O(n^2)$ time and $O(n^2)$ space.

Algorithm 1. Connecting Intervals with Errors (CIE)

1: build data structures POS and NUM for S_2
2: $resultSet \leftarrow \emptyset$
3: **for** $i = 1, \ldots, |S_1|$ **do**
4:    **for each** $c \in \Sigma$ let $OCC[c] \leftarrow 0$
5:    $|OCC| \leftarrow 0$
6:    $minDist \leftarrow 0$
7:    $j = i$
8:    **while** $j \leq |S_1|$ **and** $S_1(i, j)$ is left-maximal **do**
9:       $c \leftarrow S_1[j]$
10:       $OCC[c] \leftarrow 1$
11:       $|OCC| \leftarrow |OCC| + 1$
12:       **while** $S_1[i, j]$ is not right-maximal **do**
13:          $j \leftarrow j + 1$
14:       **end while**
15:       $minDist \leftarrow minDist + 1$
16:       **for each** position p in $POS[c]$ **do**
17:          mark position p in S_2
18:          find positions $l_1, \ldots, l_{\delta+1}$ and $r_1, \ldots, r_{\delta+1}$
19:          **for each** pair (l_x, r_y) with $1 \leq x, y \leq \delta + 1$ **do**
20:             $z \leftarrow$ the number of different unmarked characters in $S_2[l_x + 1, r_y - 1]$
21:             $dist \leftarrow |OCC| - NUM[l_x + 1, r_y - 1] + 2z$
22:             **if** $dist < minDist$ **then**
23:                $minDist \leftarrow dist$
24:             **end if**
25:          **end for**
26:       **end for**
27:       **if** $minDist \leq d$ **then**
28:          $resultSet \leftarrow resultSet \cup (i, j))$
29:       **end if**
30:       $j \leftarrow j + 1$
31:    **end while**
32: **end for**

4.2 An $O(n^2(n + \delta^2))$ Time Algorithm for Cluster Filter Detection

Our first algorithm for cluster filter detection is a straightforward extension of Algorithm CI that we call *Connecting Intervals with Errors* (CIE). Pseudocode is given in Algorithm 1. It uses the same preprocessing tables NUM and POS for S_2 as described above.

In the main part of the algorithm we iterate through all maximal substrings $S_1[i, j]$ of S_1. We refer to the current $S_1[i, j]$ as *reference interval*. With array OCC and counter $|OCC|$ we keep track of the characters occurring in the current reference interval. In variable $minDist$ we store the minimal distance found so far between $CS(S_1[i, j])$ and S_2. Like in the Connecting Intervals algorithm for each latest character c in $S_1[i, j]$ we mark each position p where this character occurs in the other sequence (lines 16, 17 of Algorithm 1). But then we have to do some extra work: While marking a position p in S_2, there is no need to keep track of maximal intervals of marked positions. Instead, positions to the left and right of p with increasing numbers $x, y \geq 1$ of unmarked characters are computed:

$$l_x(p) = \max(\{l \mid S_2[l, p] \text{ contains } x \text{ different unmarked characters}\} \cup \{0\})$$
$$r_y(p) = \min(\{r \mid S_2[p, r] \text{ contains } y \text{ different unmarked characters}\} \cup \{|S_2| + 1\})$$

By definition, the intervals $S_2[l_x + 1, r_y - 1]$ then contain at most $x + y - 2$ characters not occurring in $S_1[i, j]$ and are maximal. Hence, in order to find all occurrences of $S_1[i, j]$ around p with up to δ errors, it suffices to consider intervals $S_2[l_x + 1, r_y - 1]$ with $1 \leq x, y \leq \delta + 1$. An example is illustrated in Fig 2.

	1	2	3	4	5	6	7	8	9	10	11	12
$S_2 =$	(1	3	5	2	4	5	2	4	3	6	2	5)
	l_2						l_1	p		r_1		r_2

Fig. 2. For a substring of S_1 with character set $\{2, 3, 4\}$, its characters are marked in S_2. For $c = 2$ being the latest marked character and position $p = 7$, we have $l_1(7) = 6$, $l_2(7) = 1$ and $r_1(7) = 10$, $r_2(7) = 12$. Occurrences around p with up to $\delta = 1$ errors that need to be checked are $S_2[2, 9]$, $S_2[2, 11]$, $S_2[7, 9]$ and $S_2[7, 11]$.

In line 21 we compute the distance of each of these $(\delta + 1)^2$ intervals to $CS(S_1[i, j])$, i.e. $D(CS(S_1[i, j], CS(S_2[l_x + 1, r_y - 1]))$. This is equal to the value of $|OCC| - NUM[l_x + 1, r_y - 1]$ plus twice the number of different unmarked characters in $S_2[l_x + 1, r_y - 1]$. In case this value is smaller than the current value of $minDist$ we update $minDist$.

Since we are also interested in intervals with missing characters, we need to consider intervals that do not contain c at all. But for these we know that their distance to the current substring $S_1[i, j]$ equals the distance to the previous $S_1[i, j']$ plus 1, with $j' < j$. We account for this in line 15 by increasing the value of $minDist$ by 1 after each extension of the reference interval. When we have finished all occurrences of c we check the value of $minDist$ to decide whether the current $S_1[i, j]$ qualifies as a cluster filter.

The crucial part in the analysis of Algorithm CIE is the **for** loop in line 16. From the analysis of Algorithm CI it follows immediately that each position p in S_2 is marked $O(n)$ times so that in total we mark $O(n^2)$ times a position. For each such position we search for the positions $l_1, \ldots, l_{\delta+1}$ and $r_1, \ldots, r_{\delta+1}$ (line 18). Performing this search in single steps takes $O(n)$ time. Then we test for each of the $(\delta + 1)^2$ pairs whether it fulfills the distance constraints (line 22), which can be done in constant time if we keep track of the number of unmarked characters in the substrings $S_2[l_x + 1, r_y - 1]$ while going through the **for** loop in line 19. In total we thus have an $O(n^2(n + \delta^2))$ time algorithm, using $O(n^2)$ space for table NUM.

Remark 1. If we assume an upper bound b for the number of repetitions of each character in sequence S_2, the number of steps to locate the positions $l_1, \ldots, l_{\delta+1}$ and $r_1, \ldots, r_{\delta+1}$ for a position p is bounded by $O(\min\{n, b\delta\})$. Hence, the overall runtime decreases to $O(n^2(1 + \min\{n, b\delta\} + \delta^2))$. This is especially relevant for genetic sequences, where the value of b is usually very small as it refers to the number of copies of a single gene in a genome.

4.3   An $O(n^2(1 + \delta^2))$ Time Algorithm for Cluster Filter Detection

The runtime of the algorithm introduced in the previous section can be reduced to $O(n^2(1 + \delta^2))$ when additional space of size $O(n\delta)$ is available. The speed-up is based on the observation that for each position p in sequence S_2 the values l_x and r_y are the same for all reference intervals $S_1[i, j]$ with a common left border. In a preprocessing step we compute for the left-most left border in S_1, i.e. $i = 1$, for each position p in S_2 the values $l_1, \ldots, l_{\delta+1}$ and $r_1, \ldots, r_{\delta+1}$. These are stored in two tables L and R of size $\delta \times |S_2|$ each. The values of these arrays need to be updated each time the left border i in S_1 is moved to the right which happens $O(n)$ times.

The details of the initialization and update of the arrays L and R are given in the following. For simplicity, as in [6] we re-name the characters in the sequences S_1 and S_2 by the rank of their first occurrence in the concatenated string $S_1[i, |S_1|]S_2$, initially for $i = 1$, and after each shift of the left border i. This re-naming is a bijection RANK $: \Sigma \to \{1, \ldots, |\Sigma|\}$. The consequence of the re-naming is that at the time the positions of a character c in S_2 are marked, the remaining unmarked characters c' are such that RANK$(c') >$ RANK(c).

The initialization of tables L (and R) is as follows: For each position p in S_2, we go to its left (and right) and look for the first $\delta + 1$ different characters with a rank greater than RANK$(S_2(p))$. We store as $l_1, \ldots, l_{\delta+1}$ (and $r_1, \ldots, r_{\delta+1}$) the positions where a new different character is found. An example for RANK and the tables L and R is given in Fig. 3.

(a)	(b)		1	2	3	4	5	6	7	8	9	10	11	12
RANK[1] = 4	$L \backslash S_2'$		4	3	5	1	2	5	1	2	3	6	1	5
RANK[2] = 1	l_1		0	1	0	3	3	0	6	6	6	0	10	10
RANK[3] = 3	l_2		0	0	0	2	2	0	5	2	1	0	9	0
RANK[4] = 2	l_3		0	0	0	1	1	0	2	1	0	0	8	0
RANK[5] = 5														
RANK[6] = 6	$R \backslash S_2'$		4	3	5	1	2	5	1	2	3	6	1	5
	r_1		3	3	10	5	6	10	8	9	10	13	12	13
	r_2		10	10	13	6	9	13	9	10	12	13	13	13
	r_3		13	13	13	9	10	13	10	12	13	13	13	13

Fig. 3. Initialization of (a) the rank for all characters and (b) the tables L and R. The characters of $S_1 = (2, 4, 2, 3, 4, 1, 4, 5, 4, 3, 6)$ and $S_2 = (1, 3, 5, 2, 4, 5, 2, 4, 3, 6, 2, 5)$ are re-named by the bijection RANK, defined by their first occurrence in the concatenated string $S_1[1, 11]S_2$. The tables L and R are computed for the re-named sequence S_2'.

When the left border in the substring of S_1 is shifted from i to $i+1$, the rank for all characters occurring between i and the next occurrence of the character $S_1[i]$ decreases by one while the rank of $c_{old} = S_1[i]$ increases by the number of different characters between the two occurrences. The tables L and R change in the following way. At positions belonging to occurrences of c_{old} in S_2 the table entries can change completely due to a possibly large change in the character number. We compute these entries anew by going through S_2 once from left to right and once from right to left and remembering the positions of the $\delta+1$ last read different characters with a rank greater than the new number of c_{old}. If a character is read more than once we only remember its latest occurrence. Once we reach a position of c_{old} in S_2 we fill the corresponding entries in L (respectively R) with the remembered positions. For positions in S_2 with a character different from c_{old} the entries in L and R can only change if the rank of the character is smaller than the new value of c_{old}. For these positions we need to check whether an occurrence of c_{old} is close enough to become an entry in L and/or R. We test this by going through S_2 once from left to right and once from right to left and remembering the latest position of the character c_{old} in S_2. Once we reach a position with a character of smaller rank than the new value of c_{old}, we go through its entries in L (respectively R) and insert the remembered position of c_{old} at the right position in the field.

The initialization takes $O(n^2\delta)$ time and the update $O(n\delta)$ time for each increment of i. Combined with the unmodified rest of Algorithm CIE, the overall runtime becomes $O(n^2\delta + n^2(1+\delta^2)) = O(n^2(1+\delta^2))$. The space consumption is $O(n\delta + n^2) = O(n^2)$.

4.4   Extension to Multiple Genomes

In this section we show how the computation of cluster filters can be generalized to more than two genomes. First note that in order not to miss any possible cluster filter C' we have to consider all substrings of *any* of the strings $S_1, \ldots, S_k$ as reference intervals, and not just substrings of S_1.

A reference interval $S_l[i,j]$ qualifies as a cluster filter if the sum of the minimal distances to the other $k-1$ sequences does not exceed $d = 2\frac{k-1}{k}\delta$. The threshold for pairwise distances is still δ, otherwise the total distance to the median exceeds δ due to the triangle inequality of symmetric set distance. Hence, we need to examine for the most recently added character $S_l[j]$ all its occurrences in the other $k-1$ sequences and compute for each occurrence the distance of the corresponding $(\delta+1)^2$ intervals to $S_l[i,j]$. While doing so, we keep track of the minimum distances found in each of the $k-1$ sequences separately. If in the end they sum up to a value smaller or equal to d we have found a new cluster filter. Due to this approach we have to store the data structures POS and NUM and, if required, also L and R for $k-1$ sequences at a time.

From these modifications it follows that the runtime multiplies by $O(k^2)$ for each of the presented algorithms while space requirements increase to $O(kn^2)$.

5   Collection of δ-Locations of Cluster Filters (Step 2)

In the second step of the overall algorithm, for each cluster filter C' its maximal δ-locations in each of the sequences $S_1, \ldots, S_k$ are searched in order to form k-tuples $(S_1[i_1, j_1], \ldots, S_k[i_k, j_k])$ with pairwise intersecting character sets that satisfy

$$\sum_{l=1}^{k} D(C', \mathcal{CS}(S_l[i_l, j_l])) \leq d.$$

Maximal δ-locations of C' can be found efficiently by a modified version of Algorithm 1 that iterates through a location of C' and generates uniquely all maximal δ-locations. Details of the algorithm are left to the reader.

While the number of maximal δ-locations is in $O(kn^2)$, the number of k-tuples can be exponential in k even for small δ as the following example shows: For $\delta = 0$, $s = 3$ and k sequences of the form $S_l = (abcx_l)^n$, $1 \leq l \leq k$ and $x_i \neq x_j$ for $i \neq j$, there are $O(n^k)$ k-tuples. However, for gene sequences where $|\Sigma|$ is in $\Theta(n)$ our experience shows that this approach is feasible for reasonable values of δ.

6   Computation of Median Gene Clusters from k-Tuples (Step 3)

The computation of the median of a k-tuple consists of a simple majority vote of the characters occurring as its elements, i.e. a gene occurring in at least half of the tuple elements becomes an element of the median. The median of each k-tuple is checked whether it fulfills inequality (1) and in case it fulfills the distance constraint it is reported as a median gene cluster.

Note that there can be several medians that have to be tested: If k is even, there may be ties when some character occurs in exactly $k/2$ of the elements. However, since each tie adds $k/2$ to the sum of distances, a median exists only if the number of ties is less than or equal to $\frac{2\delta}{k}$. In this case, there exist $2^{\frac{2\delta}{k}}$ different medians as the example of Fig. 4 illustrates.

									1	2	3	4	5
$S_1 =$	(1	2	1	3	1	4	1	5)	1	1	0	0	0
$S_2 =$	(1	2	4	1	2	1	3)		1	1	0	1	0
$S_3 =$	(1	3	3	1	2	1	2)		1	1	0	0	0
$S_4 =$	(1	4	1	1)					1	0	0	1	0
									1	1	0	T	0

Fig. 4. For $\delta = 3$ and the cluster filter $C' = \{1, 2\}$, one of its 3-locations in each of the sequences is given by the underlined substrings. The tie of the fourth character (denoted by T) yields the two medians $\{1, 2\}$ and $\{1, 2, 4\}$.

Moreover it can happen that the same character set is generated more than once either by duplicate k-tuples if more than one of the k-tuple elements is a cluster filter or by different k-tuples that have the same median by chance. In our implementation we filter away such multiple occurrences.

7 Experimental Results

In an initial test we compared the performance of our two algorithms on several datasets. Surprisingly, we found that running times are highly similar between these algorithms in practice (data not shown). The following results were achieved using the second of the two algorithms.

To demonstrate the ability of our method we applied it to approximate gene cluster detection in various genomic datasets. We compared it to previous approaches for gene cluster detection in two sequences and additionally show its applicability to multiple genomes. All computations reported in this section were performed with a 1.66 GHz Intel®Core Duo T2300 processor with 520 Mb of main memory running under the Suse Linux operating system.

7.1 Comparison to HomologyTeams

We reproduced the gene clusters reported in [9] with our program. The dataset consisting of the genomes of *E. coli* and *B. subtilis* annotated with COG numbers was downloaded from http://euler.slu.edu/~goldwasser/homologyteams/. Setting the parameters of our method to $s = 4$ and $\delta = 1$ we detected 1070 median gene clusters in this dataset, among them the ten operons studied in [9]. These findings show that our method finds a superset of the gene clusters detected by the HomologyTeams software. A biological evaluation of the additional gene clusters is currently in progress.

7.2 Comparison to ILP Approach

We downloaded the genome datasets from http://gi.cebitec.uni-bielefeld.de/comet used in [13]. The dataset consists of the annotated genomes of *C. glutamicum* and *M. tuberculosis* where labeling of genes according to gene family membership already took place.

Our program found the gene cluster reported in [13] in 17 seconds using appropriate values for the parameters δ and s while the ILP using CPLEX 9.03 took more than one hour on a superior processor. In order to detect this cluster, an approach based on *max-gap clusters* needs to set its gap-size threshold as big as twelve such that the longest gap of unmatched genes can be bridged.

To evaluate our method on a broader basis, we conducted a similar series of experiments as reported in [13] to find optimal gene clusters for each size between 5 and 150. Since our method finds gene clusters based on a distance threshold and not for a certain size, we had to run our algorithm several times for different minimal cluster sizes and distance thresholds. Despite this overhead our method was able to find all optimal gene clusters in this size range within 3 hrs and 4 min.

7.3 Experimental Results on Multiple Genomes

Although both of the approaches above are in general applicable to multiple genomes, no experimental results on the comparison of more than two genomes were shown in the respective publications. To show the applicability of our method to multiple genomes, we searched for approximate gene clusters in three bacterial genomes: *Bacillus subtilis*, *Buchnera aphidicola* and *Escherichia coli* with different combinations of s and δ. Results are shown in Table 1.

Table 1. The number of distinct median gene clusters found for different combinations of s and δ in three bacterial genomes and the corresponding computation times

	$s=10, \delta=0$	$s=20, \delta=5$	$s=25, \delta=10$	$s=30, \delta=15$	$s=35, \delta=25$
distinct medians	91	152	101	21	1
computation time in sec.	4.7	7.2	12.0	26.1	186.2

Many of the gene clusters found in this experiment belong to well-conserved ribosomal protein operons. In order to find other gene clusters, genes associated with ribosomal proteins were masked in the genomes for an additional test. Distance thresholds needed to be chosen larger for a fixed s in this setting in order to find gene clusters. For example, with $s = 13$ and $\delta = 10$, we found five distinct gene clusters, among them the following gene cluster involved in flagellar biosynthesis:

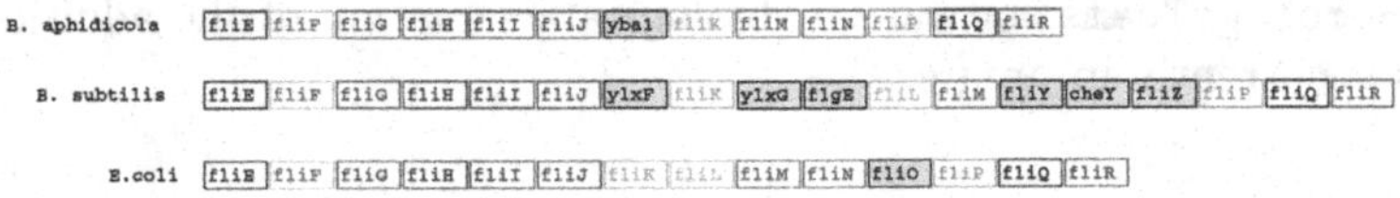

```
B. aphidicola  [fliE][fliF][fliG][fliH][fliI][fliJ][ybal][fliK][fliN][fliN][fliP][fliQ][fliR]

B. subtilis    [fliE][fliF][fliG][fliH][fliI][fliJ][ylxF][fliK][ylxG][flgE][fliM][fliY][cheY][fliZ][fliP][fliQ][fliR]

E.coli         [fliE][fliF][fliG][fliH][fliI][fliJ][fliK][fliL][fliM][fliN][fliO][fliP][fliQ][fliR]
```

Fig. 5. A gene cluster involved in flagellar biosythesis, detected by our method with parameters set to $\delta = 10$ and $s = 13$

8 Conclusion

In this paper we introduced the concept of median gene clusters for the detection of approximate gene clusters in a set of k genomes based on gene order. We applied a filter method to narrow down the search space of potential clusters efficiently, allowing for fast detection of gene clusters in multiple genomes.

Our cluster model improves over *max-gap clusters* [3,9] in two ways: The problem of low global cluster density reported in [12] does not arise as no fixed gap length needs to be specified. Unlike *max-gap clusters* our method is capable of finding approximate clusters that contain genes that are missing in some cluster occurrences. This becomes important in particular for multiple genome comparison.

We also compared our method to an approach using an ILP program for approximate gene cluster detection. While the underlying cluster models are similar, gene cluster computation was shown to be more efficient with our approach.

We believe that the main advantage of our method is its applicability to multiple genomes. Initial results show that the detection of gene clusters in multiple genomes is feasible, supporting our conjecture that the combinatorial explosion in Step 2 of our method does not occur with real-world data when parameters are chosen reasonably. A broader analysis of the influence of s and δ on sensitivity, specificity, and running time of our method is currently in progress. As the method is fastest when δ is small, we propose for practical applications to iteratively increase δ for some fixed s until clusters are detected that are potentially biologically meaningful.

In the future, we want to extend our method to detect median gene clusters that occur only in a subset of the input genomes. We also want to provide a statistical analysis of the detected clusters to rank the reported clusters according to their significance.

References

1. Amir, A., Gasieniec, L., Shalom, R.: Improved approximate common interval. Inf. Process. Lett. 103(4), 142–149 (2007)
2. Bergeron, A., Chauve, C., de Mongolfier, F., Raffinot, M.: Computing common intervals of k permutations, with applications to modular decomposition of graphs. In: Brodal, G.S., Leonardi, S. (eds.) ESA 2005. LNCS, vol. 3669, pp. 779–790. Springer, Heidelberg (2005)
3. Bergeron, A., Corteel, S., Raffinot, M.: The algorithmic of gene teams. In: Guigó, R., Gusfield, D. (eds.) WABI 2002. LNCS, vol. 2452, pp. 464–476. Springer, Heidelberg (2002)
4. Chauve, C., Diekmann, Y., Heber, S., Mixtacki, J., Rahmann, S., Stoye, J.: On common intervals with errors. Report 2006-02, Technische Fakultät der Universität Bielefeld, Abteilung Informationstechnik (2006)
5. Dandekar, T., Snel, B., Huynen, M., Bork, P.: Conservation of gene order: a fingerprint of proteins that physically interact. Trends Biochem. Sci. 23(9), 324–328 (1998)
6. Didier, G., Schmidt, T., Stoye, J., Tsur, D.: Character sets of strings. J. Discr. Alg. 5, 330–340 (2007)
7. Frances, M., Litman, A.: On covering problems of codes. Theor. Comput. Sci. 30, 119–133 (1997)
8. Gramm, J., Niedermeier, R., Rossmanith, P.: Fixed-parameter algorithms for closest string and related problems. Algorithmica 37, 25–42 (2003)
9. He, X., Goldwasser, M.H.: Identifying conserved gene clusters in the presence of homology families. J. Comp. Biol. 12, 638–656 (2005)
10. Heber, S., Stoye, J.: Algorithms for finding gene clusters. In: Gascuel, O., Moret, B.M.E. (eds.) WABI 2001. LNCS, vol. 2149, pp. 252–263. Springer, Heidelberg (2001)
11. Heber, S., Stoye, J.: Finding all common intervals of k permutations. In: Amir, A., Landau, G.M. (eds.) CPM 2001. LNCS, vol. 2089, pp. 207–218. Springer, Heidelberg (2001)

12. Hoberman, R., Durand, D.: The incompatible desiderata of gene cluster properties. In: McLysaght, A., Huson, D.H. (eds.) RECOMB 2005. LNCS (LNBI), vol. 3678, pp. 73–87. Springer, Heidelberg (2005)
13. Rahmann, S., Klau, G.W.: Integer linear programs for discovering approximate gene clusters. In: Bücher, P., Moret, B.M.E. (eds.) WABI 2006. LNCS (LNBI), vol. 4175, pp. 298–309. Springer, Heidelberg (2006)
14. Schmidt, T., Stoye, J.: Quadratic time algorithms for finding common intervals in two and more sequences. In: Sahinalp, S.C., Muthukrishnan, S.M., Dogrusoz, U. (eds.) CPM 2004. LNCS, vol. 3109, pp. 347–358. Springer, Heidelberg (2004)
15. Uno, T., Yagiura, M.: Fast algorithms to enumerate all common intervals of two permutations. Algorithmica 26, 290–309 (2000)

Appendix: Alternatives

Some variations of the model described in the main part of this paper are discussed in the following.

Transformation Set Distance

We can define a set distance based on the maximal set difference instead of the symmetric set difference:

$$D_T(C, C') = \max\{|C \setminus C'|, |C' \setminus C|\}.$$

This distance measure is called the *transformation set distance* between C and C' and is also a metric. It is easy to derive a simple linear time algorithm that finds for a given character set C and a sequence S all starting positions of substrings in S that have a transformation set distance that is smaller or equal to a given distance threshold d. Therefore, we can compute gene cluster candidates for the transformation set distance in time $O(k^2 n^3)$. But the problem with respect to application in gene cluster detection is that we lack efficient methods to compute the median of character sets under transformation set distance.

Center Representative

While being computationally tractable, selection of median representatives is probably not the best approach for gene cluster computation. The problem with median representatives is that the distances between the median and single objects (in our case sequences) are not directly restricted, but only via the sum of all distances. Hence, a rather large distance to a single sequence can be compensated by less than average distances to other sequences. Apparently, this effect can be the stronger the larger the number of sequences becomes. In an evolutionary context it makes possibly more sense to limit the distance between each of the sequences and their common ancestor:

$$\max_{1 \leq l \leq k} \{D'(C, S_l)\} \leq \delta.$$

Such a set C is called a *center representative*.

The approach described in Sections 4 and 5 is compatible with this new distance threshold. Step 1 is modified such that we search for substrings with distance at most 2δ to each other sequence, and in Step 2 we compute the k-tuples according to this distance threshold. This threshold is stronger than the one for median gene clusters since the value of δ will be chosen relatively small compared to the one for the median representative because it refers to a single distance and not to the sum of k distances.

However, the crucial point is that in Step 3 median computation needs to be replaced by the computation of the center sequence, which is known to be NP-hard [7]. There exist fixed-parameter algorithms that run in polynomial time for a fixed distance [8], but these are of limited use for this application.

BCL-2: From Translocation to Therapy

Suzanne Cory

Walter and Eliza Hall Institute of Medical Research, Melbourne
cory@wehi.edu.au

Abstract. Impaired apoptosis is a critical step in the development of cancer and a major impediment to effective therapy. Bcl-2, the oncoprotein activated by chromosome translocation in human follicular lymphoma, inhibits cells from undergoing apoptosis in response to many cytotoxic agents. Exciting recent developments suggest that small molecules which neutralise Bcl-2 and its anti-apoptotic homologues will be effective cancer therapeutics.

About the Keynote Speaker. Professor Suzanne Cory is one of Australia's most distinguished molecular biologists. She was born in Melbourne, Australia and graduated in biochemistry from The University of Melbourne. She gained her PhD from the University of Cambridge, England and then continued studies at the University of Geneva before returning to Melbourne in 1971, to a research position at The Walter and Eliza Hall Institute of Medical Research. She is currently Director of The Walter and Eliza Hall Institute and Professor of Medical Biology of The University of Melbourne. Her research has had a major impact in the fields of immunology and cancer and her scientific achievements have attracted numerous honours and awards.

M. Vingron and L. Wong (Eds.): RECOMB 2008, LNBI 4955, p. 346, 2008.
© Springer-Verlag Berlin Heidelberg 2008

Detecting Disease-Specific Dysregulated Pathways Via Analysis of Clinical Expression Profiles

Igor Ulitsky[1], Richard M. Karp[2], and Ron Shamir[1]

[1] School of Computer Science, Tel-Aviv University, Tel-Aviv 69978, Israel
{ulitskyi,rshamir}@post.tau.ac.il.
[2] International Computer Science Institute, 1947 Center St., Berkeley, CA 94704
karp@icsi.berkeley.edu

Abstract. We present a method for identifying connected gene subnetworks significantly enriched for genes that are dysregulated in specimens of a disease. These subnetworks provide a signature of the disease potentially useful for diagnosis, pinpoint possible pathways affected by the disease, and suggest targets for drug intervention. Our method uses microarray gene expression profiles derived in clinical case-control studies to identify genes significantly dysregulated in disease specimens, combined with protein interaction data to identify connected sets of genes. Our core algorithm searches for minimal connected subnetworks in which the number of dysregulated genes in each diseased sample exceeds a given threshold. We have applied the method in a study of Huntington's disease caudate nucleus expression profiles and in a meta-analysis of breast cancer studies. In both cases the results were statistically significant and appeared to home in on compact pathways enriched with hallmarks of the diseases.

1 Introduction

Systems biology has the potential to revolutionize the diagnosis and treatment of complex disease by offering a comprehensive view of the molecular mechanisms underlying the pathology. To achieve these goals, a computational analysis extracting mechanistic understanding from the masses of available data is needed. To date, such data include mainly microarray measurements of genome-wide expression profiles, with over 160,000 profiles stored in GEO alone as of August 2007. A wide variety of approaches for elucidating molecular mechanisms from expression data have been suggested [1]. However, most of these methods are effective only when using expression profiles obtained under diverse conditions and perturbations, while the bulk of data currently available from clinical studies are expression profiles of groups of diseased individuals and matched controls. The standard "pipeline" for analysis of such datasets involves the application of statistical and machine learning methods for identification of the genes that best predict the pathological status of the samples [2]. While these methods are successful in identifying potent signatures for classification purposes, the insights that can be obtained from examining the gene lists they produce are frequently limited [3].

It is thus desirable to develop computational tools that can extract more knowledge from clinical case-control gene expression studies. A challenge of particular interest is to identify the pathways involved in the disease, as such knowledge can expedite

M. Vingron and L. Wong (Eds.): RECOMB 2008, LNBI 4955, pp. 347–359, 2008.

development of directed drug treatments. One strategy of solution to this problem uses predefined gene sets describing pathways and quantifies the change in their expression levels [4]. The drawback of this approach is that pathway boundaries are often difficult to assign, and in many cases only part of the pathway is altered during disease. To overcome these problems, the use of gene networks has been suggested [5]. The appeal of using network information increases as the quality and scale of experimental data on such interaction networks improve.

Several approaches for integrating microarray measurements with network knowledge were described in the literature. Some (including us) proposed computational methods for detection of subnetworks that show correlated expression [6,7]. A successful method for detection of 'active subnetworks' was proposed by Ideker et al. and extended by other groups [8,9,10,11,12]. These methods are based on assigning a significance score to every gene in every sample and looking for subnetworks with statistically significant combined scores. Breitling et al. proposed a simple method named GiGA which receives a list of genes ordered by their differential expression significance and extracts subnetworks corresponding to the most differentially expressed genes [13]. Other tools use network and expression information together for classification purposes [5,14].

Methods based on correlated expression patterns do not use the sample labels, and thus their applicability for case-control data is limited, as correlation between transcript levels can stem from numerous confounding factors not directly related to the disease (e.g., age or gender). The extant methods that do use the sample labels rely on the assumption that the same genes in the pathway are differentially expressed in all the samples (an exception is jActiveModules which can identify a subset of the conditions in which the subnetwork is active [8]). This assumption may hold in simple organisms (e.g., yeast or bacteria) or in cell line studies. However, in human disease studies, the samples are expected to exhibit intrinsic differences due to genetic background, environmental effects, tissue heterogeneity, disease grade and other confounding factors. Here we propose a new viewpoint for analysis of clinical gene expression samples in the context of interaction networks, which avoids the above assumption.

Our approach aims to detect subnetworks in which multiple genes are dysregulated in the diseased specimens, while allowing for distinct affected gene sets in each patient. We call such modules *dysregulated pathways* (DPs). Specifically, we look for minimal connected subnetworks in which the number of dysregulated genes in each diseased sample exceeds a given threshold. By comparing to statistics of randomized networks, we can identify statistically significant DPs. As finding such modules is NP-hard, we propose heuristics and algorithms with provable approximation ratios and study their performance on real and simulated data. Our approach has several important advantages over the existing methods: (a) the dysregulated genes in a DP can vary between patients; (b) the method is robust to outliers (i.e., patients with unusual profiles); (c) the DPs can contain relevant genes based on their interaction pattern, even if they are not dysregulated; (d) it has only two parameters, both of which have an intuitive biological interpretation; (e) while not guaranteeing optimality, the algorithmic backbone of the method has a provable performance guarantee.

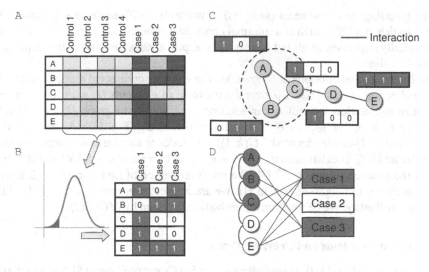

Fig. 1. From case-control profiles to dysregulated pathways. (A) The first input to our method is the gene expression matrix where the columns correspond to samples taken from case/control subjects and rows correspond to genes. (B) In a preprocessing step, differential expression is analyzed and, for each gene, the set of cases in which it is differentially expressed (up-regulated, down-regulated or both) is extracted. (C) A second input is a protein interaction network with nodes corresponding to genes and edges to interactions. The row next to each gene is its dysregulation pattern (its row from B). The goal is to find a smallest possible subnetwork in which, in all but l cases, at least k genes are differentially expressed. In this example, the circled subnetwork satisfies the condition with $k = 2, l = 1$: (i) A and C are dysregulated in case 1; (ii) A and B are dysregulated in case 3. (D) The bipartite graph representation of the data. Genes (left) are connected to the cases (right) in which they are differentially expressed. Edges between genes constitute the protein interaction network. The genes of the minimal cover and the samples covered by them are in green.

We first tested the performance of our method on simulated data. We then used it to dissect the gene expression profiles of samples taken from the caudate nucleus of Huntington's Disease (HD) patients. We reveal specific subnetworks that are up and down regulated in cases in comparison to controls, and show that they are significantly enriched with known HD-related genes. Finally, we performed a network-based meta-analysis of six breast cancer datasets, extracting DPs associated with good and poor outcome of the disease. In all cases, the DPs are significantly enriched with genes from relevant pathways and contain both known and novel potential drug targets.

For lack of space, some details and proofs are not included in this manuscript.

2 Methods

2.1 Problem Formulation

In this section we describe the theoretical foundations of our methodology (**Fig. 1**). The known gene network is presented as an undirected graph, where each node (gene) has a corresponding set of elements (samples) in which it is differentially expressed. Our goal is to detect a DP, which is a minimal connected subnetwork with at least k nodes differentially expressed in all but l analyzed samples (l thus denotes of the number of allowed 'outliers').

We formalize these notions as follows. We are given an undirected graph $G = (V, E)$ and a collection of sets $\{S_v\}_{v \in V}$ over the universe of elements U, with $|U| = n$. For ease of representation, we will use, in addition to G, a bipartite graph $B = (V, U, E^B)$ where $(v, u) \in E^B, v \in V, u \in U$ if and only if $u \in S_v$ (**Fig. 1D**). A set $C \subseteq V$ is a *connected* (k, l)-*cover* (denoted $CC(k, l)$) if C induces a connected component in G and a subset $U' \subseteq U$ exists such that $|U'| = n - l$ and for all $u' \in U', |N(u') \cap C| \geq k$, i.e., in the induced subgraph (C, U') the minimal degree of nodes in U' is at least k ($N(x)$ is the set of neighbors of x in B). We are interested in finding a $CC(k, l)$ of the smallest cardinality. We denote this minimization problem by *MCC(k,l)*.

2.2 Similar Problems and Previous Work

If G is a clique, $MCC(1, 0)$ is equivalent to the Set Cover problem [15]. For this classical NP-hard problem, Johnson proposed a simple greedy algorithm with approximation ratio $O(ln(n))$ [15]. If $k > 1$ and G is a clique, the $MCC(k, 0)$ problem is equivalent to the *set multicover problem*, also known as the *set k-cover problem*, a variant of the Set Cover problem in which every element has to be covered k times. The set multicover problem can be approximated to factor of $O(p)$, where p is the number of sets covering the element that appears in the largest number of sets [15]. The greedy algorithm for set multicover was shown to achieve an approximation ratio of $O(\log(n))$ [16]. See [15] for a comprehensive review of the available approximation results on set cover and set multicover problems.

For a general G, $MCC(1, 0)$ is the *Connected Set Cover problem*, which has been recently studied in the context of wavelength assignment of broadcast connections in optical networks [17]. It was shown to be NP-Hard even if at most one vertex of G has degree greater than two, and approximation algorithms were suggested for the cases where G is a line graph or a spider graph. Both of these special cases are not applicable in our biological context.

2.3 Greedy Algorithms for $MCC(k, l)$

We tested two variants of the classical greedy approximation for Set Cover. For simplicity we will describe them for $MCC(1, 0)$. The first algorithm, *ExpandingGreedy* works as follows: Given a partial cover $W \subseteq V$ and the set of corresponding covered elements $X \subseteq U$, the algorithm picks a node $v \in V$ that is adjacent to W and that covers the largest number of elements of $U \setminus X$, adds v to the cover and adds $N(v) \cap U$ to X. Initially $W = \emptyset$, $X = \emptyset$ and the first node is picked without connectivity constraints. Unfortunately, *ExpandingGreedy* can be shown to give a solution that is $O(|V|)$ times the optimal solution. Specifically, it runs into difficulties in cases where all the nodes in the immediate neighborhood of the current solution have equal benefit, and the next addition to the cover is difficult to pick. The second algorithm, *ConnectingGreedy*, first uses the simple greedy algorithm [15] to find a set cover that ignores the connectivity constraints and then augments it with additional nodes in order to obtain a proper cover. The *diameter* of a graph is the maximum length of a shortest path between a pair of nodes in V. It can be shown that *ConnectingGreedy* guarantees an approximation ratio of $O(D \log n)$ for $MCC(1, 0)$, where D is the diameter of G.

2.4   The CUSP Algorithm

We next describe an algorithm called Covering Using Shortest Paths (*CUSP*). Let $d(v, w)$ be the distance in edges between v and w in G. For each *root* node r and for each element $u \in U$ the algorithm computes distances $(M[r, u]_1, ..., M[r, u]_k)$ and pointers $(P[r, u]_1, ..., P[r, u]_k)$ to the k nodes closest to r that cover u. This can be done by computing the distances from r to all the nodes in V that cover u, and then retrieving the k closest nodes, which is an instance of the *selection problem* and can be solved in expected linear time [18]. Now take X_r, the union of the paths to the nodes covering the $n - l$ elements for which $max_q\{d(r, P[r, u]_q), 1 \le q \le k\}$ are the smallest. X_r is a proper $CC(k, l)$: (a) it is a subtree of T and thus induces a connected component in G; (b) $n - l$ elements of U are covered k times by the corresponding $\{P[r, u]_i\}$. The final solution is $X = \arg\min |Y_v|$. This algorithm can be proved to give a $k(n - l)$ approximation for $MCC(k, l)$.

In terms of computational complexity, the total amount of work for each choice of r is $O(|V| + |E| + |E^D|)$ and the overall complexity is $O(|V|(|V| + |E| + |E^B|))$. Note that it is not necessary to execute the algorithm from every root node, but only from the $l + 1$ nodes that cover elements from $U' \subseteq U$ for which $max_{u' \in U'} |N(u')|$ is minimal.

2.5   Practical Heuristics and Implementation Details

In order to improve the performance of the proposed algorithms, we implemented several practical heuristics.

CUSP* - starting from high coverage cores: A drawback of CUSP is that it ignores the number of elements covered by each node, and treats the coverage of every element separately. We therefore also implemented the CUSP* heuristic: For each root, it uses dynamic programming to identify a subnetwork of k nodes that offers a good coverage of the elements, and then extends it to a proper $CC(k, l)$ as in CUSP.

Clean-up: The DPs produced by all the described algorithms may contain superfluous nodes that are not necessary neither for the cover requirements nor for subnetwork connectivity. In all algorithms we therefore perform a clean-up step that iteratively removes such nodes until no further reduction is possible.

Shortest path tree construction: While the approximation bound of CUSP holds regardless of the shortest paths used, some sets of such paths may eventually give rise to smaller covers than others. We used the following heuristic in the BFS algorithm: at each level of the constructed BFS tree, we sort the nodes in descending order based on the added coverage they offer. The nodes are then scanned in this order and the next level of the tree is built.

Starting points: The performance of the algorithms depends on the number of starting points/seeds used. In all the results described here we executed all algorithms starting from the 30 nodes that had the highest degrees in B.

Assessment of DP significance: CUSP produces a set of DPs for a range of k values. To select the most significant DP, 200 random networks were generated by degree-preserving randomization [19]. CUSP was executed on each network, for a range of k values, and an empirical p-value was computed. The k value for which the size of the DP was most significant was subsequently used. In case of a tie, a normal distribution was fitted to the random scores, and k yielding the subnetwork with the most significant z-score was selected.

Finding multiple DPs: After recovering the first DP V_1, we seek additional DPs by removing all the edges adjacent to V_1 from E^B and reapplying the search procedure. This is repeated until no significant DP is found.

Our algorithms were implemented in Java, and source code of the implementation is available upon request. A user-friendly graphical interface for the algorithms described here is currently in development.

3   Results

Human Protein Interaction Network: We compiled a human protein-protein interaction network encompassing 7,384 nodes corresponding to Entrez Gene identifiers and 23,462 interactions. The interactions are based mostly on small-scale experiments and were obtained from several interaction databases. The network and the sources information are available at our website http://acgt.cs.tau.ac.il/clean.

3.1   Simulation

We first evaluated the algorithms on simulated data in which a single DP is planted. We used the human protein interaction network as G, created a biclique between a connected subgraph of G and a specified number of elements in U and added noise to B by randomly removing and inserting edges. In the simulations (results not shown) *ExpandingGreedy* generally found the smallest covers. The results produced by CUSP and CUSP* were only slightly inferior. However, the covers produced by CUSP and CUSP* were much more compact, giving a much lower mean shortest path length between nodes in the cover.

3.2   Analysis of Huntington's Disease Caudate Nucleus Expression Profiles

Huntington's disease (HD) is a devastating autosomal dominant neurological disorder caused by an expansion of glutamine repeats in the ubiquitously expressed huntingtin (*htt*) protein. HD pathology is well understood at a histological level but its effect on the molecular level in the human brain is poorly understood. Recent studies have shown that mutant huntingtin interferes with the function of widely expressed transcription factors, suggesting that gene expression may be altered in a variety of tissues in HD. Hodges

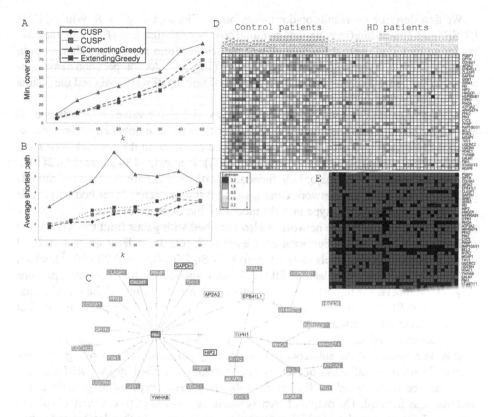

Fig. 2. Subnetwork identified by the CUSP algorithm as down-regulated in the caudate nucleus of HD patients. (A) Comparison of the minimal cover size obtained by the greedy and the CUSP algorithms. (B) Comparison of the average shortest path length between nodes in the minimal cover obtained by the greedy and the CUSP algorithms. (C) The subnetwork obtained for $k = 25$ and $l = 8$. HD modifiers described in [21] are in yellow. KEGG HD pathway genes are drawn with thick border. Note that HD is the official name of huntingtin (*htt*). (D) Heat map of the normalized expression values of the subnetwork genes in the control and HD groups. (E) The subnetwork genes and their differential expression in each HD samples. Red cells correspond to significantly down-regulated genes.

et al. reported gene expression profiles in grade 0-2 HD brains obtained using oligonucleotide arrays [20]. We focused our analysis on 38 patient samples and 32 unaffected control samples from that study, all taken from the caudate nucleus region of the brain, as this is the region where the disease is manifested the most. For every sample (patient), differentially expressed genes were selected based on comparison to the controls. The expression pattern of each gene was first standardized to mean 0 and standard deviation of 1. For every gene v, a normal distribution was fitted to its expression values in the control group, and for every HD sample u, a one-tailed p-value p_v^u was computed. We then introduced an edge (v, u) to E^B if and only if $p_v^u < 0.05$. At this significance level, 1,073 (1,696) genes were selected as down (up) regulated in a sample on average.

We first describe the results on down-regulation (**Fig. 2**), using $l = 8$. While CUSP, CUSP* and *ExpandingGreedy* found minimal covers of similar size (**Fig. 2A**), the covers found by CUSP were the most compact, as evident from the average shortest path length between a pair of nodes in the subnetwork (**Fig. 2B**). As compact and dense subnetworks are more likely to correspond to real biological pathways, we used the results of CUSP in further analysis.

Our significance evaluation of the results showed that for values of k between 10 and 40 the cover found was significantly smaller than the one obtained at random, indicating that genes dysregulated in HD are indeed clustered in the network. The most significant DP was obtained for $k = 25$ ($p < 0.005$). It contained 34 genes (**Fig. 2C-E**), with the *htt* protein as the major hub. Indeed, mutations in *htt* are the cause of the HD pathology. Moreover, the network contains six additional genes identified as genetic modifiers of the HD phenotype in a fly model of the disease [21] (the modifiers are highlighted in **Fig. 2C**). The network is also enriched with genes from the KEGG HD pathway ($p = 7.95 \cdot 10^{-7}$). Furthermore, the network contains at least six genes related to regulation of calcium levels (data taken from MSigDB [4], $p = 9.23 \cdot 10^{-7}$), which is known to be intimately related to HD [22]. An inspection of the expression patterns (**Fig. 2D**) indicates the importance of the outlier parameter l. A few of the samples (patients 16,103,86) have profiles that differ from those of the other patients, but this fact does not affect the algorithm.

A comparison of the DP we identified with gene sets identified using other methods (**Table 1**) reveals that the subnetwork produced by our method is more significantly enriched with most hallmarks of HD. The subnetwork identified by jActiveModules is also enriched for these hallmarks, but this subnetwork is an order of magnitude larger, and thus less focused. The output of jActiveModules consists of (i) the 'active' subnetwork; and (ii) the samples in which the subnetwork is active. In this dataset, the active subnetwork produced by this algorithm was based on a single sample, and thus it does not reflect common dysregulation across most patients in the study.

The running time on this dataset, for $k = 25$, was 10.6 seconds on a PC with two 2.67GHz processors and 4GB of memory. A search for additional down-regulated DPs (see Methods) did not produce significant networks.

Similar analysis of genes up-regulated in HD samples identified a marginally significant subnetwork ($k = 10, p = 0.11$) of 14 nodes centered at BRCA1 and p53, which are master regulators of DNA damage response, and are known to be hyperactive in HD affected cells [24]. Interestingly, p53 and BRCA1 are not differentially expressed in most HD samples, and the functional category 'DNA damage response' is not enriched in the 100 genes most significantly up-regulated in the HD samples (as obtained by a *t*-test). This further underlines the ability of our method to extract relevant pathways even if only part of the pathway is differentially expressed in diseased specimens. Another hub in this focused subnetwork is HDAC1, a histone deacetylase known to be elevated in HD neurons [25]. Sodium phenylbutyrate, a histone deacetylase inhibitor, is currently tested as a potent drug for HD [26], and was shown to revert HD transcriptional dysregulation in mouse and human brain and blood tissues [27,28]. Hence, the inclusion of HDAC1 in a focused subnetwork identified as up-regulated in diseased caudate nuclei demonstrates the ability of our method to detect potential therapeutic targets.

Table 1. Comparison of gene sets identified as down-regulated in HD caudate nucleus using different methods. GiGA was implemented as described in [13] and used to produce a subnetwork of 34 nodes. jActiveModules [8] was executed from Cytoscape and yielded five subnetworks. The reported results are for the highest scoring subnetwork. 't-test top' refers to the 34 down regulated genes with the most significant t-scores. HD modifiers are taken from [21]. HD relevant genes are taken from [23]. Calcium signalling genes are taken from MSigDB [4].

	CUSP	GiGA	jActiveModules	t-test top	t-test FDR < 0.05
Number of genes	34	34	282	34	1762
Contains Huntingtin?	**Yes**	No	No	No	**Yes**
HD modifiers	$6\,(7.7 \cdot 10^{-10})$	$3\,(1.55 \cdot 10^{-4})$	$12\,(\mathbf{3.15 \cdot 10^{-11}})$	$2\,(0.001)$	$16\,(3.47 \cdot 10^{-5})$
HD relevant	$7\,(\mathbf{4.29 \cdot 10^{-11}})$	$2\,(0.008)$	$14\,(1.42 \cdot 10^{-9})$	$1\,(0.124)$	$18\,(6.06 \cdot 10^{-5})$
KEGG HD pathway	$4\,(\mathbf{7.95 \cdot 10^{-7}})$	0	$4\,(0.003)$	0	$8\,(0.03)$
Calcium signaling	$6\,(9.23 \cdot 10^{-7})$	$5\,(1.99 \cdot 10^{-5})$	$10\,(5.68 \cdot 10^{-4})$	$3\,(0.005)$	$49\,(\mathbf{2.97 \cdot 10^{-12}})$

3.3 Meta-analysis of Breast Cancer Studies

In order to test our methodology on other diseases and on inter-study comparisons we performed meta-analysis of six breast cancer studies, spanning together expression profiles of 1,004 patients. Full details on the studies are available at our website. These studies compared breast cancer tumor samples, for which follow-up outcome information was available. We focused on comparison of tumors with good and poor prognosis (defined as development of distant metastases within five years [2]). In each study, using a one-tailed t-test, we extracted a set of differentially expressed genes between good and poor prognosis patients ($p = 0.05$ was used as a threshold). Here we applied CUSP to the genes vs. studies matrix. The most significant DP up-regulated in poor prognosis cancers is shown in **Fig. 3A** ($k = 40, l = 2, p < 0.005$). This network is highly enriched with cell-cycle genes (28 out of 51 genes are associated with cell-cycle in GO, $p = 2.44 \cdot 10^{-26}$). Cell cycle and proliferation genes are known to be associated with higher grade, poor prognosis tumors in numerous studies (see [29] and the references therein). In addition, this DP contains 15 genes shown to be regulated by YY1 (as found in [30], $p = 2.42 \cdot 10^{-16}$), known to be associated with overexpression of the ERBB2 oncogene and with poor prognosis of breast cancer [31]. We recovered an additional significant DP which is described on our website.

The most significant DP down-regulated in poor prognosis cancers ($k = 25, p < 0.005$, **Fig. 3B**) is enriched with genes associated with drug resistance and metabolism (Source:MSigDB, $p = 3.54 \cdot 10^{-9}$), p53 signalling ($p = 3.54 \cdot 10^{-9}$) and the JAK-STAT signalling pathway ($p = 3.68 \cdot 10^{-4}$). The latter pathway mediates the signals of a wide range of cytokines, growth factors and hormones, making its aberrant activation prone to lead to malignancy. This pathway was also linked specifically to breast cancer [32]. Our results indicate the down-regulation of this pathway on the expression level is associated with cancers with poor prognosis. Interestingly, this subnetwork, but not the up-regulated one, was enriched with genes that are frequently mutated in cancer in general ($p = 1.14 \cdot 10^{-7}$) and in breast cancer in particular ($p = 3.2 \cdot 10^{-4}$, both sets taken from [33]). A search for additional DPs did not yield significant results.

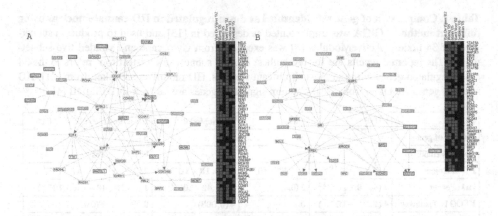

Fig. 3. DPs identified in breast cancer meta-analysis. In the differential expression maps (right) red cells correspond to differentially expressed genes. (A) a DP up-regulated in poor prognosis breast cancers ($k = 40, p < 0.005$). Cell cycle genes (from GO) are in yellow. YY1 regulated genes are drawn with thick border. (B) DP with a lower expression in poor prognosis breast cancers ($k = 25$). Drug resistance pathway genes appear in pink. JAK-STAT signalling pathway genes are drawn with thick border.

4 Discussion

We have developed a novel computational technique for network-based analysis of clinical gene expression data. The method is aimed at identifying pathways in the interaction network that exhibit ample evidence of disruption of transcription that is specific to diseased patients. Application of the method to a large-scale human protein-protein interaction network and a Huntington's disease study as well as meta-analysis of six breast cancer studies has shown its potential in outlining subnetworks with a high relevance to the mechanisms of pathogenesis. Comparison to extant techniques for analysis of gene expression data highlights the advantages of our approach in identifying clinically sound pathways.

While the results presented here are encouraging, there is certainly room for further development of these methods. Currently, we look for multiple subnetworks by iteratively finding and removing the most significant DP from the network. Better methods are needed to detect overlapping DPs. Furthermore, one can obtain significance scores for individual nodes in the DPs using established statistical methods such as bootstrapping [34].

Our problem formulation used a fixed k value, thus requiring that the same least number of genes is altered in all patients (or studies). All the algorithms and proofs presented are generalizable to the scenario where different samples have different thresholds. This case can be attractive if, for example, the number of differentially expressed genes varies significantly among patients or studies, and the goal is to detect subnetworks covering a fixed percentage of the differentially expressed genes. The value of l used in the examples presented here was set to 20% of the elements (cases or studies) in the dataset. While we observed that our method is rather robust to l values in the range of 15-40% of the cases, the methodology for a more rigorous selection of the l value is also an interesting subject for further research.

One of the main goals of case-control studies using microarrays is the detection of biomarkers, leading to an improved characterization of the pathologies of each patient. We believe that the fact that the subnetworks that we identified for HD and breast cancer contain numerous established therapeutic targets carries the promise that an integrative analysis of such studies with complementary molecular datasets can also indicate specific points for medical intervention.

Acknowledgements

We thank David Burstein, Yonit Halperin and Chaim Linhart for helpful discussions. IU is a fellow of the Edmond J. Safra Bioinformatics Program at Tel-Aviv University. This research was supported by the GENEPARK project which is funded by the European Commission within its FP6 Programme (contract number EU-LSHB-CT-2006-037544).

References

1. Bansal, M., Belcastro, V., Ambesi-Impiombato, A., di Bernardo, D.: How to infer gene networks from expression profiles. Molecular Systems Biology 3, 78 (2007)
2. van't Veer, L., Dai, H., van de Vijver, M., He, Y., Hart, A., Mao, M., Peterse, H., van der Kooy, K., Marton, M., Witteveen, A., et al.: Gene expression profiling predicts clinical outcome of breast cancer. Nature 415, 530–536 (2002)
3. Segal, E., Friedman, N., Kaminski, N., Regev, A., Koller, D.: From signatures to models: understanding cancer using microarrays. Nat Genet 37(suppl.), S38–S45 (2005)
4. Subramanian, A., Tamayo, P., Mootha, V., Mukherjee, S., Ebert, B., Gillette, M., Paulovich, A., Pomeroy, S., Golub, T., Lander, E., Mesirov, J.: Gene set enrichment analysis: A knowledge-based approach for interpreting genome-wide expression profiles. PNAS 102, 15545–15550 (2005)
5. Rapaport, F., Zinovyev, A., Dutreix, M., Barillot, E., Vert, J.: Classification of microarray data using gene networks. BMC Bioinformatics 8, 35 (2007)
6. Ulitsky, I., Shamir, R.: Identification of functional modules using network topology and high-throughput data. BMC Systems Biology 1 (2007)
7. Segal, E., Wang, H., Koller, D.: Discovering molecular pathways from protein interaction and gene expression data. Bioinformatics 19, I264–I272 (2003)
8. Ideker, T., Ozier, O., Schwikowski, B., Siegel, A.F.: Discovering regulatory and signalling circuits in molecular interaction networks. Bioinformatics 18, S233–S240 (2002)
9. Rajagopalan, D., Agarwal, P.: Inferring pathways from gene lists using a literature-derived network of biological relationships. Bioinformatics 21, 788–793 (2005)
10. Cabusora, L., Sutton, E., Fulmer, A., Forst, C.: Differential network expression during drug and stress response. Bioinformatics 21, 2898–2905 (2005)
11. Nacu, S., Critchley-Thorne, R., Lee, P., Holmes, S.: Gene expression network analysis and applications to immunology. Bioinformatics 23, 850 (2007)
12. Liu, M., Liberzon, A., Kong, S., Lai, W., Park, P., Kohane, I., Kasif, S.: Network-based analysis of affected biological processes in type 2 diabetes models. PLoS Genetics 3, e96+ (2007)

13. Breitling, R., Amtmann, A., Herzyk, P.: Graph-based iterative group analysis enhances microarray interpretation. BMC Bioinformatics 5, 100 (2004)
14. Chuang, H., Lee, E., Liu, Y., Lee, D., Ideker, T.: Network-based classification of breast cancer metastasis. Mol. Syst. Biol. 3 (2007)
15. Hochbaum, D.S.: Approximating covering and packing problems: set cover, vertex cover, independent set, and related problems. In: Hochbaum, D.S. (ed.) Approximation algorithms for NP-hard problems, PWS, Boston, pp. 94–143 (1997)
16. Dobson, G.: Worst-case analysis of greedy heuristics for integer programming with non-negative data. Mathematics of Operations Research 7, 515–531 (1982)
17. Shuai, T., Hu, X.: Connected set cover problem and its applications. In: Cheng, S.-W., Poon, C.K. (eds.) AAIM 2006. LNCS, vol. 4041, pp. 243–254. Springer, Heidelberg (2006)
18. Cormen, T.H., Leiserson, C.E., Rivest, R.L.: Introduction to Algorithms. MIT Press, Cambridge (1990)
19. Milo, R., Shen-Orr, S., Itzkovitz, S., Kashtan, N., Chklovskii, D., Alon, U.: Network motifs: simple building blocks of complex networks. Science 298, 824–827 (2002)
20. Hodges, A., Strand, A., Aragaki, A., Kuhn, A., Sengstag, T., Hughes, G., Elliston, L., Hartog, C., Goldstein, D., Thu, D., et al.: Regional and cellular gene expression changes in human Huntington's disease brain. Human Molecular Genetics 15, 965 (2006)
21. Kaltenbach, L., Romero, E., et al.: Huntingtin interacting proteins are genetic modifiers of neurodegeneration. PLoS Genet 3, e82 (2007)
22. Rockabrand, E., Slepko, N., Pantalone, A., Nukala, V., Kazantsev, A., Marsh, J., Sullivan, P., Steffan, J., Sensi, S., Thompson, L.: The first 17 amino acids of Huntingtin modulate its sub-cellular localization, aggregation and effects on calcium homeostasis. Human Molecular Genetics 16, 61 (2007)
23. Borrell-Pagès, M., Zala, D., Humbert, S., Saudou, F.: Huntington's disease: from huntingtin function and dysfunction to therapeutic strategies. Cellular and Molecular Life Sciences (CMLS) 63, 2642–2660 (2006)
24. Giuliano, P., De Cristofaro, T., et al.: DNA damage induced by polyglutamine-expanded proteins. Human Molecular Genetics 12, 2301–2309 (2003)
25. Hoshino, M., Tagawa, K., et al.: Histone deacetylase activity is retained in primary neurons expressing mutant huntingtin protein. J. Neurochem. 87, 257–267 (2003)
26. Butler, R., Bates, G.: Histone deacetylase inhibitors as therapeutics for polyglutamine disorders. Nat. Rev. Neurosci. 7, 784–796 (2006)
27. Ferrante, R., Kubilus, J., Lee, J., Ryu, H., Beesen, A., Zucker, B., Smith, K., Kowall, N., Ratan, R., Luthi-Carter, R., et al.: Histone deacetylase inhibition by sodium butyrate chemotherapy ameliorates the neurodegenerative phenotype in Huntington's disease mice. Journal of Neuroscience 23, 9418–9427 (2003)
28. Borovecki, F., Lovrecic, L., Zhou, J., Jeong, H., Then, F., Rosas, H.D., Hersch, S.M., Hogarth, P., Bouzou, B., Jensen, R.V., Krainc, D.: Genome-wide expression profiling of human blood reveals biomarkers for Huntington's disease. Proc. Natl. Acad. Sci. USA 102, 11023–11028 (2005)
29. Sotiriou, C., Wirapati, P., Loi, S., Harris, A., Fox, S., Smeds, J., Nordgren, H., Farmer, P., Praz, V., Haibe-Kains, B., et al.: Gene expression profiling in breast cancer: understanding the molecular basis of histologic grade to improve prognosis. J. Natl. Cancer Inst. 98, 262–272 (2006)
30. Affar, E., Gay, F., Shi, Y., Liu, H., Huarte, M., Wu, S., Collins, T., Li, E., Shi, Y.: Essential Dosage-Dependent Functions of the Transcription Factor Yin Yang 1 in Late Embryonic Development and Cell Cycle Progression. Molecular and Cellular Biology 26, 3565–3581 (2006)

31. Begon, D., Delacroix, L., Vernimmen, D., Jackers, P., Winkler, R.: Yin Yang 1 Cooperates with Activator Protein 2 to Stimulate ERBB2 Gene Expression in Mammary Cancer Cells. Journal of Biological Chemistry 280, 24428–24434 (2005)
32. Li, L., Shaw, P.: Autocrine-mediated activation of STAT3 correlates with cell proliferation in breast carcinoma lines. Journal of Biological Chemistry 277, 17397–17405 (2002)
33. Futreal, P., Coin, L., Marshall, M., Down, T., Hubbard, T., Wooster, R., Rahman, N., Stratton, M.: A census of human cancer genes. Nature Reviews Cancer 4, 177–183 (2004)
34. Efron, B., Tibshirani, R.: An introduction to the bootstrap. Chapman & Hall, New York (1993)

Constructing Treatment Portfolios
Using Affinity Propagation

Delbert Dueck[1], Brendan J. Frey[1,2], Nebojsa Jojic[3], Vladimir Jojic[1,3,4],
Guri Giaever[2], Andrew Emili[2], Gabe Musso[2], and Robert Hegele[5]

[1] Electrical and Computer Engineering, University of Toronto, Canada
[2] Center for Cellular and Biomolecular Research, University of Toronto, Canada
[3] Machine Learning and Statistics, Microsoft Research, Redmond, USA
[4] Computer Science, Stanford University, USA
[5] Cardiovascular Genetics Laboratory, Robarts Research Institute, London, Canada

Abstract. A key problem of interest to biologists and medical researchers is the
selection of a subset of queries or treatments that provide maximum utility for
a population of targets. For example, when studying how gene deletion mutants
respond to each of thousands of drugs, it is desirable to identify a small subset
of genes that nearly uniquely define a drug 'footprint' that provides maximum
predictability about the organism's response to the drugs. As another example,
when designing a cocktail of HIV genome sequences to be used as a vaccine, it
is desirable to identify a small number of sequences that provide maximum immunological protection to a specified population of recipients. We refer to this
task as 'treatment portfolio design' and formalize it as a facility location problem. Finding a treatment portfolio is NP-hard in the size of portfolio and number
of targets, but a variety of greedy algorithms can be applied. We introduce a new
algorithm for treatment portfolio design based on similar insights that made the
recently-published affinity propagation algorithm work quite well for clustering
tasks. We demonstrate this method using the two problems described above: selecting a subset of yeast genes that act as a drug-response footprint, and selecting
a subset of vaccine sequences that provide maximum epitope coverage for an
HIV genome population.

1 Treatment Portfolio Design (TPD)

A central question for any computational research collaborating with a biologist or
medical researcher is in what form computational analyses should be handed over to
the experimentalist or clinician. While application-specific predictions are often most
appropriate, we have found that in many cases what is needed is a selection of potential
options available to the biologist/medical researcher, so as to maximize the amount of
information gleaned from an experiment, which often can be viewed as consisting of
independently assayed targets. If the number of options is not too large, these can be
discussed and selected by hand. On the other hand, if the number of possibilities is
large, a computational approach may be needed to select the appropriate options. This
paper describes the framework and approaches that emerged while trying to address
problems of this type with our collaborators. In particular, we show how the affinity
propagation algorithm [1] can be used to effectively to approach this task.

M. Vingron and L. Wong (Eds.): RECOMB 2008, LNBI 4955, pp. 360–371, 2008.
© Springer-Verlag Berlin Heidelberg 2008

For concreteness, we will refer to the possible set of options as 'treatments' and the assays used to measure the suitability of the treatments as 'targets'. Each treatment has a utility for each target and the goal of what we will refer to as treatment portfolio design (TPD) is to select a subset of treatments (the portfolio) so as to maximize the net utility of the targets. The terms 'treatment', 'target' and 'utility' can take on quite different meanings, depending on the application. Treatments might correspond to queries, probes or experimental procedures, while targets might correspond to disease conditions, genes or DNA binding events.

Example 1: The treatments are a set of potential yeast gene deletion strains used to query drug response, the targets are all ~6000 yeast gene deletion strains, the utility is the number of gene-drug interactions in all strains that are predicted by the selected portfolio of strains.

Example 2: The treatments are a large set of potential vaccines derived from HIV genomes, the targets are a population of HIV epitopes likely to be present in a demographic with high infection risk, the utility is the level of immunological protection, *i.e.*, number of epitopes present in the selected portfolio of HIV vaccines.

Example 3: The treatments are a set of baseline demographic, anthropometric, bio chemical and DNA SNP variables thought to be predictive of cardiovascular end-points and postulated to form a clinical set of risk factors, the targets are 1,000,000 disease end-point targets comprising ~20,000 patients and ~200 conditions, the utility is the predictability of disease end-points, including risk.

Example 4: The treatments are a set of laboratory procedures used to synthesize biologically active compounds, the targets are a list of desired compounds to be synthesized, the utility is the negative financial cost needed to synthesize all target compounds using the selected portfolio of laboratory procedures.

Example 5: The treatments are a large set of microRNAs potentially involved in regulating the expression of disease-associated genes, the targets are a list of gene-disease pairs, the utility is the net corrected correlation between gene expression and expression of microRNAs in portfolio for all disease conditions.

The input to TPD is a set of potential treatments or queries T, a representative population of targets $\mathcal{R}$ and a utility function $u : T \times \mathcal{R} \to \mathbb{R}$, where $u(T, R)$ is the utility of applying treatment $T \in T$ to target $R \in \mathcal{R}$. This utility may be based on a variety of factors, including the benefit of the treatment, the cost, the time to application, the time to response, the estimated risk, *etc.* The goal of computational TPD is to select a subset of treatments $\mathcal{P} \subseteq T$ (called the 'portfolio') so as to maximize their net utility for the target population. A defining aspect of the utility function is that it is additive; for portfolio $\mathcal{P}$, the net utility is

$$\sum_{R \in \mathcal{R}} \max_{T \in \mathcal{P}} u(T, R).$$

To account for the fact that some treatments are preferable to others regardless of their efficacy for the targets (*e.g.*, different setup costs), we use a treatment-specific cost function $c : T \to \mathbb{R}$. The net utility, including the treatment cost is

$$U(\mathcal{P}) = \sum_{R \in \mathcal{R}} \max_{T \in \mathcal{P}} u(T, R) - \sum_{T \in \mathcal{P}} c(T). \tag{1}$$

Provided with T, $\mathcal{R}$, u and c, the computational task is to find $\max_{\mathcal{P} \subseteq T} U(\mathcal{P})$. Note that the number of treatments in the portfolio will be determined by balancing the utility with the treatment cost.

1.1   Relationship to K-Medians Clustering and Facility Location

Under certain conditions, TPD can be viewed as a K-medians clustering problem or facility location problem (a.k.a. the p-median model) [3, 4, 5]. Given a set of points $\mathcal{X}$ and a pairwise distance measure $d : \mathcal{X} \times \mathcal{X} \rightarrow \mathbb{R}$, the goal of K-medians clustering is to select a subset of points as centers and assign every other point to its nearest center, so as to minimize the sum of distances. To control the number of identified centers, either the number of centers is pre-specified or a cost is associated with each center. Because of the combinatorics involved in selecting K centers, the K-medians clustering problem is NP-hard (c.f. [5]). TPD can be viewed as K-medians clustering, if the treatment set equals the target set ($T = \mathcal{R}$). The application of K-medians clustering algorithms to TPD is discussed below.

In general, the treatment set does not equal the target set. Then, TPD can be viewed as a facility location problem, which is framed as opening up facilities or warehouses to service customers. Given a set of facilities that can potentially be opened $\mathcal{F}$, a set of customers $\mathcal{C}$, a distance function $d : \mathcal{C} \times \mathcal{F} \rightarrow \mathbb{R}$ and a facility opening cost function $c : \mathcal{F} \rightarrow \mathbb{R}$, the facility location problem [5] consists of identifying a subset of facilities and assigning every customer to a facility so as to minimize the net facility opening cost and distance of customers to facilities. Because the number of possible combinations of facilities to choose from is exponential in the number of potential and chosen facilities, the facility location problem is NP-hard.

Most work on approximation algorithms for K-medians clustering and facility location relies on d being metric (convex) (c.f. [5] for a review), but this is not necessary. Both problems can be formulated as binary-valued integer programs and then relaxed to linear programs. If the linear program solution is non-integer, it can be rounded, giving rise to various approximations. Unfortunately, these approximation algorithms are of limited practical value. We have experimented extensively with a data set of 400 Euclidean points derived from images of faces and found that CPLEX 7.1 takes several hours and gigabytes of memory to find solutions that can be found in less than one minute using a couple megabytes of memory via the affinity propagation algorithm [1] (data available at www.psi.toronto.edu/affinitypropagation).

2   Standard Algorithms Adapted to TPD

A variety of non-iterative and iterative algorithms for optimizing the objective functions can be formulated in a straight-forward fashion. A simple method is to start with an empty portfolio and add the single treatment T_1 that maximizes the net utility (including the treatment cost). All targets are assigned to that treatment and the current utility $v(R)$ for target R is set to $u(T_1, R)$. Next, another treatment is added to the portfolio and this treatment is chosen so as to maximize the net utility. This involves examining every treatment T not currently in the portfolio, computing a net utility gain

$\sum_{R \in \mathcal{R}} \max(u(T, R) - v(R), 0) - c(T)$, and selecting the treatment with maximum utility gain. This treatment, denoted T_2, is added to the portfolio. This procedure is repeated until another treatment cannot be added to the portfolio without decreasing the net utility. Note that in the absence of a treatment cost (*i.e.*, $c = 0$, assuming that all utilities are non-negative), all treatments would be added to the portfolio. Then, the number of treatments K may instead be specified, in which case the algorithm is terminated when K treatments have been added to the portfolio.

ALGORITHM 1: K-TREATMENTS CLUSTERING

Input: $T, \mathcal{R}, u, c, K, M$

Repeat M times:

    $\mathcal{P}' \leftarrow$ random subset of T of size K

    $\forall R \in \mathcal{R}, \; p(R) \leftarrow \mathrm{argmax}_{T \in \mathcal{P}'} u(T, R)$

    Repeat until convergence:

        Select $T \in \mathcal{P}'$ (in order)

        $\forall T' \in T \setminus \mathcal{P}'$, compute

            $g(T') \leftarrow \sum_{R : p(R) = T} u(T', R) - u(T, R) - c(T') + c(T)$

        $T^{\mathrm{alt}} \leftarrow \mathrm{argmax}_{T' \in T \setminus \mathcal{P}'} g(T')$

        If $g(T^{\mathrm{alt}}) > 0$ then set $\mathcal{P}' \leftarrow (\mathcal{P}' \setminus \{T\}) \cup \{T^{\mathrm{alt}}\}$

        $\forall R \in \mathcal{R}, \; p(R) \leftarrow \mathrm{argmax}_{T \in \mathcal{P}'} u(T, R)$

    If first repetition, then $\mathcal{P} \leftarrow \mathcal{P}'$; else if $U(\mathcal{P}') > U(\mathcal{P})$, then $\mathcal{P} \leftarrow \mathcal{P}'$

Output: $\mathcal{P}$

One problem with the above method is that after the subsequent addition of a treatment, previously-added treatments may no longer be the ones that maximize the utility. A natural extension is to initially find K treatments and then iteratively revisit treatments and consider replacing them with other treatments not in the current portfolio, until the portfolio converges. Alternatively, instead of deterministically initializing the portfolio, it can be randomly initialized to K treatments and then iteratively refined. The advantage of this approach is that a large number, M, of random initializations can be tried and the refined portfolio with highest net utility can be selected. To reflect the similarity of this approach to the standard K-means clustering and K-medians clustering algorithms [3], we will refer to it as 'K-treatments clustering'. Note that K-treatments clustering is not identical to K-means clustering or K-medians clustering, because treatments are neither means nor medians – in fact, they generally lie in a different space than the targets that are assigned to them. See Alg. 1 for details.

3 Modified Affinity Propagation for TPD

The recently-introduced affinity propagation algorithm is an exemplar-based clustering algorithm that operates by exchanging messages between data points until a subset of data points emerge as the cluster centers (exemplars) [1]. Unlike most other clustering algorithms, affinity propagation does not store and refine a fixed number of potential cluster centers, but instead simultaneously considers all data points as potential cluster centers. Data points exchange two kinds of message: the responsibility sent from point i to point k indicates how well-suited point k is as the exemplar for point i in contrast to other potential exemplars; the availability sent from point k to point i indicates how much support point k has received from other points for being an exemplar. As the message-passing procedure proceeds, responsibilities and availabilities become more extreme until a clear set of exemplars and clusters emerges.

In [1], affinity propagation was shown to find better solutions than other frequently-used methods, including K-medians (K-centers) clustering and hierarchical agglomerative clustering. It should be kept in mind that for small problems, *e.g.* < 500 points, linear programming [5] can often be used to find an exact solution. Also, when the number of sought-after exemplars is quite low (*e.g.*, < 10), methods that use random initialization (*e.g.*, K-centers clustering) can work quite well. One randomly-initialized method that works quite well is the vertex substitution heuristic. Recently in [2], affinity propagation was shown to be significantly faster than the vertex substitution heuristic for moderately large problems. For example, 20 randomly-initialized runs of the vertex substitution heuristic took ~10 days to find 454 clusters in 17,770 Netflix movies, whereas affinity propagation took ~2 hours and achieved lower error.

The input to the affinity propagation algorithm is a set of similarities $\{s(i, k)\}$, where $s(i, k)$ is the similarity of point i to k, and a set of preferences $\{p(k)\}$, where $p(k)$ is the *a priori* preference that point k be chosen as an exemplar. After exchanging messages, affinity propagation identifies a set of exemplars $\mathcal{K}$ so as to maximize the net similarity $\sum_{i \notin \mathcal{K}} \max_{k \in \mathcal{K}} s(i, k) + \sum_{k \in \mathcal{K}} p(k)$. Viewing treatments as potential exemplars, we can adapt affinity propagation to TPD: if point i is a target and point k is a treatment, we can set $s(i, k)$ to the utility of that treatment for that target; if point k is a treatment, we can set $p(k)$ to the negative cost for that treatment.

However, one important difference between the problem statements for exemplar-based clustering and TPD is the distinction between treatments and targets. The original affinity propagation algorithm treats all points as potential exemplars and every non-exemplar point must be assigned to an exemplar. In TPD, only treatments can be selected as exemplars, and only targets have utilities for being assigned to exemplars (treatments). Treatments that are not selected for the portfolio (exemplar set) are neither exemplars nor assigned to another exemplar (treatment).

To allow some treatments to not be selected for the portfolio and also not be assigned to any other points, we introduce a special 'garbage collector' point and set the similarities of treatments to this point to zero. So, unless there is a net benefit in utility minus cost when including a treatment in the portfolio (exemplar set), it will be assigned to the garbage collector point. In summary, the following similarity constraints account for the *bipartite* structure of TPD:

$$s(\text{target}, \text{treatment}) = u(\text{treatment}, \text{target}) \text{ and } s(\text{target}, \text{target}') = s(\text{target}, \text{garbage}) = -\infty$$
$$s(\text{treatment}, \text{garbage}) = 0 \text{ and } s(\text{treatment}, \text{target}) = s(\text{treatment}, \text{treatment}') = -\infty$$
$$s(\text{garbage}, \text{target}) = s(\text{garbage}, \text{treatment}) = -\infty$$
$$p(\text{treatment}) = -c(\text{treatment}) \text{ and } p(\text{target}) = p(\text{garbage}) = \infty$$

The last constraints ensure that targets cannot be selected as exemplars and that the garbage collection point is always available as an exemplar.

It turns out that when the above constraints are inserted into the original affinity propagation updates, certain simplifications occur and the garbage collection point need not be explicitly represented. In fact, many messages need not be computed and the algorithm reduces to messages exchanged on a bipartite graph connecting treatments and targets. The resulting algorithm is provided in Alg. 2 – note that messages need only be exchanged between a treatment and target if the utility is not $-\infty$. The input $\lambda \in (0, 1)$ is a damping factor that is used to improve convergence.

ALGORITHM 2: BIPARTITE AFFINITY PROPAGATION

Input: $\mathcal{T}, \mathcal{R}, u, c, \lambda$

Initialization: $\forall T, R : u(T, R) > -\infty$, set $a(T, R) \leftarrow 0$, $r(T, R) \leftarrow 0$

Repeat until convergence:

Update responsibilities: $\forall T, R : u(T, R) > -\infty$, set

$$r(T, R) \leftarrow \lambda r(T, R) + (1-\lambda) \cdot \left(u(T, R) - \max_{\substack{T' \in \mathcal{T}: \\ u(T', R) > -\infty}} \{ u(T', R) + a(T', R) \} \right) \tag{2}$$

Update availabilities: $\forall T, R : u(T, R) > -\infty$, set

$$a(T, R) \leftarrow \lambda a(T, R) + (1-\lambda) \cdot \min \{0, \sum_{\substack{R' \in \mathcal{R} \setminus \{R\}: \\ u(T, R') > -\infty}} \max(0, r(T, R')) - c(T)\} \tag{3}$$

Output:

$$\mathcal{P} \leftarrow \{ T : \sum_{\substack{R \in \mathcal{R}: \\ u(T, R) > -\infty}} \max(0, r(T, R)) > c(T) \} \tag{4}$$

This algorithm can be derived as an instance of the max-product algorithm in a factor graph describing Eq. 1. Here, we provide some intuition for why the updates make sense. Initially, the availabilities $a(\cdot, \cdot)$ are zero and Eq. 2 indicates that the responsibility of treatment T for target R is set to its utility minus the largest competing utility of another treatment for the same target. In subsequent iterations, the utilities of competing treatments are modulated by their availabilities. As indicated in Eq. 3, the availability of treatment T for target R is set to the sum of its responsibilities for other treatments minus its cost. Only positive responsibilities are included in this sum, because for a treatment to be deemed useful, it is only necessary that some targets yield high utility, not that all targets yield high utility. The availability is not allowed to rise above zero; this acts to prevent a treatment that accounts for a large number of targets from dominating other potential treatments. After convergence, Eq. 4 compares the net responsibility of target T with its cost, and includes it in the portfolio if the benefit outweighs the cost.

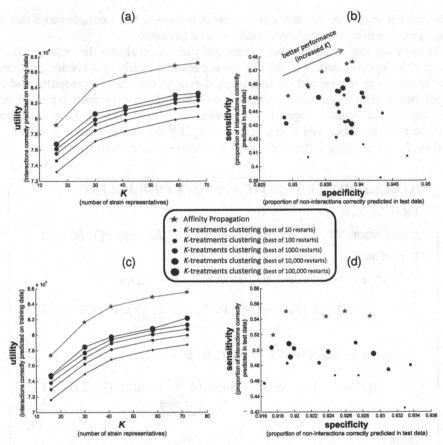

Fig. 1. Performance with affinity propagation and K-treatments clustering on finding yeast strains that are representative of gene-drug interaction profiles. (a) and (b) show training and test results, for a held-out test set of randomly-selected drugs; (c) and (d) show similar results, but where the held-out test drugs consisted of a small number of drugs with different dosage levels and exposure times. (a) and (c) show that affinity propagation maximizes the net utility of the training set better than the best of hundreds of thousands of random restarts of K-treatments clustering. (b) and (d) show that on the test drug set, affinity propagation has better sensitivity at a given specificity than K-treatments clustering.

4   Application 1: Selecting Yeast Gene Deletion Queries for Drug Profiling

We applied TPD to a data set showing the interaction of 1259 different drug-dosage-exposure combinations (called just drugs hereon in for brevity) with yeast gene-deletion strains. Our goal was to find a small query set of genes (a subset of the 5985 yeast genes) on which new drugs could be tested for the purpose of predicting interactions between the drug and genes not in the query set. The selection of a query set would avoid needing to test all genes on new drugs, which can be costly, especially when expanding tests, *e.g.*, to include proteomic profiling. In this application, the treatment set equals the target set ($T = R$); in the next section, we describe results on an application where the treatment set and target set are disjoint.

Gene-drug interactions were measured using TAG3 molecular barcode arrays [6]. z-scores were calculated by standardizing the rank-normalized intensity for each strain under the given drug against the intensity from a set of untreated controls. We imposed a threshold score of 5 for there to be an interaction between the strain and drug, which yielded roughly $150,000$ interactions of a possible 7.5 million combinations (2%). The utility of a potential query gene deletion strain (treatment) for representing another gene deletion strain (target) was set to the number of drug responses that were active for both. To test the predictive power of TPD, this utility was computed using a training set of drugs consisting of only 90% of the original; results are reported for both uniformly sampling the 10% test set of drugs and using a non-random test set consisting of all different dosage levels and exposure times for a smaller set of drugs (still 10% of the total number of conditions).

TPD was performed using both affinity propagation and K-treatments clustering (using a huge number of random restarts), with results shown in Fig. 1. All runs of affinity propagation took less than five hours whereas the many runs of K treatments clustering took over 3 days. Results for the uniformly-sampled test set are shown in Fig. 1(a) and (b), whereas similar results are shown in Fig. 1(c) and (d) for the test set with fewer drugs with varying dosages and exposure times. Fig. 1(a) and (c) plot the net utility (total number of drug-gene interactions in the training set accounted for by the portfolio) versus the number of treatments. Affinity propagation finds better representative yeast strains than K-treatments clustering, for both training sets.

Additionally, Fig. 1(c) and (d) plot sensitivity vs. specificity curves for both algorithms, using the held-out test sets. At any given specificity (proportion of non-interactions correctly predicted in the test data), affinity propagation achieves a higher sensitivity (proportion of interactions correctly predicted in the test data) than K-treatments clustering. Note that as the size of the portfolio (K) increases, both specificity and sensitivity experience corresponding increases.

It is evident from Fig. 1 that many thousands of random restarts of K-treatments clustering are needed to find good solutions. For example, for all data points shown in both plots, we found that the portfolios found by affinity propagation never represented fewer than 7 target genes, whereas in all but one case for K-treatments clustering (the lowest, $K = 16$), the portfolios included singleton genes, leading to less-accurate predictions.

Exemplar genes found by both affinity propagation and the K-treatments algorithm were also analyzed for functional enrichment [7]. In nearly all cases, the list of exemplars was not over-enriched for any functional category, indicating that these representatives were well dispersed in terms of biological function.

5   Application 2: Selecting HIV Strains that Maximize Immune Target Coverage

Next, we pose the problem of HIV vaccine cocktail design as a TPD. The idea here is to find a set of optimal HIV strains for the purpose of priming the immune systems of many patients. The treatments T are thousands of HIV strain sequences (available at www.hiv.lanl.gov). The targets $\mathcal{R}$ are a set of short sequences (patches, fragments) that correspond to the epitopes that immune systems respond to (we use all nonamers or 9-mers). The utility $u(T, R)$ of a strain T for a fragment R would ideally be set to its potential for immunological protection, but following [8, 9, 10, 12] we set it to the frequency of the fragment in the database of HIV sequences, if fragment R is present in strain T and 0 otherwise. The net utility is also referred to as 'coverage'.[1]

Fig. 2 shows aligned pieces of Gag protein from several different strains, with two variable sites marked by arrows as well as known or predicted T-cell epitopes for the MHC molecules of five different patients taken from the WA cohort [11]. Epitopes recognizable by a single patient are shown in a single color, and each patient is assigned a different color. The white patient could be immunized against three forms of the same epitope: KKYKLKHIV, KKYQLKHIV, KKYRLKHIV. In this small example, we can design a vaccine consisting of the following segments which epitomizes in an immunological sense the seven strains shown in the figure: VLSGGKLDKWEKIRLRPGGKKKYK-LKHIVWASRELERF LSGGKLDRWEKIRLR KKKYQLKHIVW KKKYRLKHIVW.

A lot of discussion among HIV vaccine experts has been focused on the need for constraining vaccine constructs optimized for coverage to resemble naturally occurring strains [10, 9]. This is motivated by several pieces of evidence that deviation from naturally occurring strains often reduces efficacy in animal models as well as in vaccine trials, both in terms of the cellular and antibody responses. Thus, [9] proposes enrichment of the vaccine with a sequence that sits in the center of the HIV phylogenetic tree, so that this single native-like (but still artificially derived) strain is used to provide high coverage of immune targets in as natural way as possible, while the additional coverage is achieved with an epitome fragment or fragments. In contrast, in their recent paper [10] Fischer *et. al.* avoid the use of fragments altogether, and propose building the entire vaccine out of multiple strain-like constructs optimized by simulated strain recombination, dubbed 'mosaics'. A mosaic vaccine is therefore a cocktail of artificially-derived strains, not existent among the observed strains of the virus, but achievable by recombining many times the existing strains. These vaccine components resemble natural strains, but have higher nonamer coverage than what would be expected from a cocktail of natural strains. Mosaics can always achieve higher coverage

---

[1] Despite its simplicity, this problem set-up is quite biologically relevant. The immune system recognizes pathogens by short protein segments called epitopes. These targets are recognized both on the surface of the free viral particles, as well as on the surface of the infected cells, where the peptide targets are presented by the cell's own MHC molecules in charge of signaling about the normal or abnormal protein expression in the cell. Due to a need for efficiency, the immune system takes longer to recognize a new pathogen the first time it is encountered than in subsequent infections. To prime such an adaptive system against foreign intruders, various vaccination strategies have been developed, all essentially with the same goal — to expose the patient to a harmless vaccine that shares similarities with a targeted pathogen, so that in the immune system gets prepared for the true infection. For many pathogens, with HIV being a prime example, genetic diversity poses a significant problem for vaccine design. The goal of vaccine design is to load the vaccine with targets that would work for multiple strains and multiple patients [8, 9, 10, 12].

Fig. 2. Fragments of Gag protein with epitopes recognized by several HIV-infected patients. Epitopes recognizable by a single patient are shown in a single color; mutations marked by red arrows escape MHC I binding.

than natural strains, so while they may not be viable as vaccines, they provide an upper bound on potential coverage.

As the data set of known HIV sequences is constantly growing, the potential for achieving high coverage with a cocktail of true natural strains is growing as well. Newly discovered strains differ from existing ones mostly by the combination of previously seen mutations, rather than by the presence of completely new nonamers. In fact, Fischer *et. al.* have increased the Gag vaccine coverage by their use of mosaic by some 4–5% in comparison to natural strain cocktails for Gag and Nef protein vaccines consisting of 3–5 components. This is not much larger than the differences of around 2% of all nonamers that they report among coverage scores of mosaics optimized on different datasets (even within the same HIV clade). Furthermore, as the problem is NP-hard, the natural strain cocktails (treatment portfolios) in their paper are found by a greedy technique (a combination of K-treatments clustering and the vertex substitution heuristic), which may further decrease the perceived potential of natural strain cocktails, especially for a larger number of components. For a large M-clade dataset consisting of 1755 Gag proteins from the LANL database, a Gag sequence consisting of the best 4 natural strains we could find had only 3% lower coverage than the mosaic of the same size optimized on the same data (69% vs 66%); the differences in the Pol gene were even lower. Obviously, as the dataset grows, the computational burden for finding the optimal cocktail grows exponentially, as is the case for the general TPD problem.

Furthermore, while potentially important for the cellular arm of the immune system, a vaccine components' closeness to natural strains is even more important for properly presenting potential targets of the humoral (antibody) arm of the immune system. As opposed to the T-cell epitopes, antibody epitopes are found on the surface of the folded proteins. It has been shown that slight changes in the HIV Env protein can cause it to mis-fold, and so naturally occurring HIV strains are more likely to function properly than artificially derived Env proteins.

In our experiments, we solve the TPD problem for the Gag vaccine cocktail optimization for larger cocktails, where the coverage approaches 80% or more, and where exhaustive search is computationally out of the question. Instead, we used the affinity propagation algorithm described above, and compare its achieved utility (coverage) with that of the greedy method and the mosaic upper bound [10]. Table 1 summarizes our results on 1755 strains from M-clade (combination of all clades, and thus the most diverse).

Table 1. The utility ("epitope coverage") of vaccine portfolios found by affinity propagation and a greedy method, including an upper bound on utility (found using mosaics)

Problem	Natural strains		Artificial mosaic strains (upper bound)
	Affinity propagation	Greedy	
Gag, $K = 20$	77.54%	77.34%	80.84%
Gag, $K = 30$	80.92%	80.14%	82.74%
Gag, $K = 38$	82.13%	81.62%	83.64%
Gag, $K = 52$	84.19%	83.53%	84.83%

These results show that affinity propagation achieves higher coverage than the greedy method. Importantly, these results also suggest that the sacrifice in coverage necessary to satisfy the vaccine community's often-emphasized need for natural components, may in fact be bearable if large datasets and appropriate algorithms are used to optimize coverage.

6   Summary

We introduced a computation problem called 'treatment portfolio design', which is a key problem for biologists and medical researchers who need select a set of options useful for extracting maximum information or utility from a set of targets. This problem is equivalent to the non-metric K-medians problem or facility location problem, but while these problems are not new, practical algorithms that produce good solutions are still elusive. We showed how the recently-introduced affinity propagation algorithm can be modified to perform treatment portfolio design. We demonstrated a greedy algorithm for TPD and affinity propagation on the problem of identifying a yeast gene-deletion query set for the purpose of drug profiling and identifying strains of HIV that together maximize coverage of epitopes. Both methods were useful for identifying treatment sets, but we found that affinity propagation achieved significantly higher utility values and better test set performance (in terms of sensitivity and specificity), even when the greedy method was re-run hundreds of thousands of times using different random initializations.

References

1. Frey, B.J., Dueck, D.: Clustering by passing messages between data points. Science 315, 972–976 (2007)
2. Frey, B.J., Dueck, D.: Affinity propagation and the vertex substitution heuristic. Science (in press)
3. MacQueen, J.: Some methods for classification and analysis of multivariate observations. In: Proc. 5th Berkeley Symp. on Mathematical Statistics and Probability, pp. 281–297. Univ. of California Press (1967)
4. Balinksi, M.L.: On finding integer solutions to linear programs. In: Proc. IBM Scientific Computing Symp. on Combinatorial Problems, pp. 225–248 (1966)
5. Charikar, M., Guha, S., Tardos, A., Shmoys, D.B.: A constant-factor approximation algorithm for the k-median problem. J. Comp. and Sys. Sci. 65(1), 129–149 (2002)
6. Pierce, S.E., Fung, E.L., Jaramillo, D.F., Chu, A.M., Davis, R.W., Nislow, C., Giaever, G.: A unique and universal molecular barcode array. Nature Methods 3(8), 601–603 (2006)
7. Maere, S., Heymans, K., Kuiper, M.: BiNGO: a Cytoscape plugin to assess overrepresentation of gene ontology categories in biological networks. Bioinformatics 21, 3448–3449 (2005)
8. Jojic, N., Jojic, V., Frey, B., Meek, C., Heckerman, D.: Using epitomes to model genetic diversity: Rational design of HIV vaccine cocktails. NIPS 18, 587–594 (2005)
9. Nickle, D.C., et al.: Coping with Viral Diversity in HIV Vaccine Design. PLoS Computational Biology 3(4), e75 (2007)
10. Fischer, W., Perkins, S., et al.: Polyvalent vaccines for optimal coverage of potential T-cell epitopes in global HIV-1 variants. Nature Medicine 13, 100–106 (2006)
11. Mallal, S.: The Western Australian HIV Cohort Study, Perth, Australia. Journal of Acquired Immune Deficiency Syndromes and Human Retrovirology 17(Suppl. 1), 23–27 (1998)
12. Jojic, V.: Algorithms for rational vaccine design. Ph.D. Thesis, University of Toronto (2007)

Bubbles: Alternative Splicing Events of Arbitrary Dimension in Splicing Graphs

Michael Sammeth[1], Gabriel Valiente[2], and Roderic Guigó[1]

[1] Center for Genomic Regulation, Genome Bioinformatics Lab, E-08003 Barcelona
micha@sammeth.net
[2] Algorithms, Bioinformatics, Complexity and Formal Methods Research Group,
Technical University of Catalonia, E-08034 Barcelona

Abstract. Eukaryotic splicing structures are known to involve a high degree of alternative forms derived from a premature transcript by alternative splicing (AS). With the advent of new sequencing technologies, evidence for new splice forms becomes more and more easily available—bit by bit revealing that the true splicing diversity of "AS events" often comprises more than two alternatives and therefore cannot be sufficiently described by pairwise comparisons as conducted in analyzes hitherto. Further challenges emerge from the richness of data (millions of transcripts) and artifacts introduced during the technical process of obtaining transcript sequences (noise)—especially when dealing with single-read sequences known as expressed sequence tags (ESTs). We describe a novel method to efficiently predict AS events in different resolutions (i.e., dimensions) from transcript annotations that allows for combination of fragmented EST data with full-length cDNAs and can cope with large datasets containing noise. Applying this method to estimate the real complexity of alternative splicing, we found in human thousands of novel AS events that either have been disregarded or mischaracterized in earlier works. In fact, the majority of exons that are observed as "mutually exclusive" in pairwise comparisons truly involve at least one other alternative splice form that disagrees with their mutual exclusion. We identified four major classes that contain such "optional" neighboring exons and show that they clearly differ from each other in characteristics, especially in the length distribution of the middle intron.

General Terms: Alternative Splicing, ESTs, New Sequencing Technologies, Algorithms, Graph Theory.

Keywords: exon-intron structure, splicing variation, alternative splicing event, expressed sequence tags, high-throughput sequencing, parallel sequencing, directed acyclic graph, galled network, blob, bubble.

1 Introduction

Alternative splicing (AS), a fundamental molecular process regulating eukaryotic gene expression, generates a substantial part of the human proteome diversity [1]

M. Vingron and L. Wong (Eds.): RECOMB 2008, LNBI 4955, pp. 372–395, 2008.

and is involved in numerous human diseases [2,3,4]. Over the recent years splicing variations for several organisms have been collected in various databases and several attempts to analyze the complexity of AS throughout different genomes have been undertaken [17,18,19,20]. Usually, splicing diversity is classified according to the observed pattern of exon-intron variation into structurally different events. However, all works hitherto focused on alternative splicing in a limited context by considering only pairwise comparisons—either of a reference transcript to other splice forms, or by comparing the transcript data in an all-against-all fashion. By focusing exclusively on pairs of transcripts one may—although describing the atomary elements of a pattern—separate alternatives that actually form more complex splicing structures into different events. Imagine for instance the effort to reconstruct the mutual exclusion of 3 (or more) neighboring exons from a set of pairwise events. It has already been noticed that comparing transcripts one by one is not satisfactory for describing such complex variations and novel ways to deal with this shortcoming have been postulated [21].

In this work, we describe a technique to exhaustively and efficiently describe arbitrarily large splicing variations in *annotations*, i.e. transcript data aligned to the genome. W.r.t. the quality of the sequenced transcript data, such alignments are usually not free of artifacts, i.o., sequencing errors can lead to misalignments of the transcripts to the genome. Consequently, gaps introduced in the transcript sequence during alignment and misaligning nucleotides that are arbitrarily distributed to the left or right of an intron lead to the observation of wrong or shifted introns and artificially suggest variations of the exon-intron structure (so-called *noise*). Another source of noise is due to technical difficulties in obtaining 5'- and 3'-complete transcript sequences, and especially single read sequences—typically not longer than one half kbase—without subsequent assembly usually represent fragments of transcribed genes called *expressed sequence tags* (ESTs). In contrast, ESTs that have been assembled to larger transcript sequences and contain part of an (hypothetical) open reading frame (i.e., a subsequence of the transcript that does not exhibit a stop codon in one of the 3 possible frames) are usually considered as *messenger RNAs* (mRNAs) although they often are still truncated parts of the real mRNA molecules. Based on these EST and mRNA data, curated reference transcripts are built (e.g., the NCBI mRNA reference sequence collection [25]) which are considered to be *full-length* but normally do not comprise all evidence for differential splicing. Naturally there is much more EST evidence available than mRNA sequences and full-length transcripts are the minority, e.g., Genbank [26] currently contains ~ 8 million human ESTs, $\sim 260,000$ mRNAs and $\sim 25,000$ RefSeq transcripts. Considering these—continuously growing—numbers and the advent of new sequencing technologies that already have been applied to explore transcript diversity in a new dimension [27,28], the need for efficient methods to analyze huge annotation datasets becomes evident.

The rest of the paper is organized as follows: in Section 2 we define AS events that involve two or more alternatives (Section 2.1) and the terminology that we subsequently use for splicing graphs inferred from annotations (Section 2.2). In these graphs, we describe the properties of *bubbles*, specific graph substructures that imply potential AS events (Section 2.3). We show how AS events are obtained from the bubble subgraphs (Section 2.4) and we propose an efficient algorithm to exhaustively and non-redundantly extract all AS events reflected by a splicing graph—w.r.t. possible artifacts from the upstream annotation pipeline (Section 2.5). In Section 3, we exemplarily demonstrate that the time complexity of our method is dominated by the linear dependence on the number of transcripts in the annotation (Section 3.1) s.t. even the biggest datasets can be analyzed within few hours. From the number of AS events that we find to have more than two alternatives, we derive the fraction of AS events that have either been disregarded or misjudged in analyzes hitherto (Section 3.2). This fraction turns out to be extremely high for the "mutually exclusive exons" observed in pairwise transcript comparisons and further analysis of the different groups that cluster these events reveals attributes that clearly distinguish between the groups and that may allow for conclusions on different underlying molecular mechanisms (Section 3.3).

2   Methods

2.1   AS Events of Arbitrary Dimension d

An RNA sequence aligned to the genome (here after called *transcript*) can be described by a sequence of exon boundaries (i.e., *sites*) $rna = \langle s_i \rangle_{i=1}^n, n \geqslant 2$ ordered by their position $pos(s_i)$ from 5' to 3' (Fig. 1a). Genomic coordinates of sites that align to the negative strand are inverted $-pos(s_i)$ to preserve the 5'→3' directionality when ordering them. Furthermore, a site $s = (pos(s), transcripts(s), class(s), type(s))$ is characterized by its functional role $type(s) \in \{\text{root, start, donor, acceptor, end, leaf}\}$, the set of supporting transcripts $transcripts(s)$ and their respective category $class(s) \in \{\text{RefSeq, mRNA, EST}\}$. The category $class(s)$ is assigned in a hierarchical manner, i.e., it is RefSeq if at least one RefSeq transcript is in $transcripts(s)$, if not it is mRNA if at least one of the mRNAs is supporting s, else $class(s) = \text{EST}$. When investigating AS, we compare all sets of k transcripts $\{rna_i\}_{i=1}^k$ that overlap in the genomic region they align to on the same strand (a *locus*) and we distinguish differences observed in the exonic structure as *variants*.

Definition 1. *A "variant" is a sequence of sites $\langle s_i \rangle_{i=1}^m$ shared by a non-empty set of transcripts $X = \bigcap_{i=1}^m transcripts(s_i)$. We refer to X as the "partition" of supporting transcripts because different variants between $pos(s_1)$ and $pos(s_m)$ split the k transcripts of the corresponding locus into disjoint sets. A variant $p = (s_1, s_m, X_p)$ can equivalently be described by the delimiting sites (s_1, s_m) and its partition X_p because for each $rna_j \in X_p$ holds $\langle s_i \rangle_{i=1}^m = \langle rna_j[x_j], \ldots, rna_j[y_j] \rangle, 1 \leqslant x_j < y_j \leqslant n_j$.*

AS events involve $d \geqslant 2$ variants, where d is said the *dimension* of the event. To delineate AS events of dimension d in a locus, we extend an earlier described definition for *pairwise* AS events ($d = 2$, Appendix A.1) as follows:

Definition 2. *An AS event of dimension $d \geqslant 2$ comprises $\{p_i\}_{i=1}^d$ different variants, such that (i) the first and the last site of each variant p_i coincide ("common sites") or the first/last site of each variant is the first/last site of the supporting transcripts, (ii) any site besides common first/last sites of a variant p_i is not used in at least one other variant ("variable sites"), and (iii) amongst these variable sites there is at least one splice site overlapping the genomic region of all variants ("alternative splice site").*

In other words, Def. 2 characterizes an AS event as a sequence of variable sites (ii), delimited upstream and downstream either by a common site of all variants (i) or by the start/end of the respective transcripts, disregarding in the latter case variations in the exon-intron structure caused by alternative transcription start/termination rather than by different choices of the splicing machinery (iii). Parameter $d \geqslant 2$ is determined *a priori*, hitherto to our knowledge only pairwise events have been considered in AS analyzes. If d includes all variants shown by the annotation in the corresponding genomic region, we call the event "complete" w.r.t. the given annotation.

2.2 Splicing Graphs

As described previously in [29], we define a *splicing graph* $G(V, E)$ on a genomic locus as a directed acyclic graph (DAG) with each vertex $s \in V$ describing non-redundantly a site of the transcripts in a locus. Now each edge $s \rightarrow t \in E$ consequently represents an exon ($type(s) \in \{\text{start}, \text{acceptor}\}$) or an intron ($type(s) = \text{donor}$) with transcript support $transcripts(s \rightarrow t) = transcripts(s) \cap transcripts(t)$. In order to deal with alignment errors, introns with unusual splice site sequences are marked by $valid(s \rightarrow t) = \textbf{false}$. In this work, introns have been considered as "trustworthy" ($valid(s \rightarrow t) = \textbf{true}$) if they matched the sequence combinations (donor/acceptor) GT/AG, GC/AG, ATATC/AG, ATATC/AC, ATATC/AT, GTATC/AT or ATATC/AA as these constitute over 90% of the human introns—including introns spliced by the U12 spliceosome. However, other criteria may be applied to distinguish introns from alignment errors, for instance involving their length or $class(s)$ and $class(t)$ of the delimiting sites.

Subsequently, V is completed by inserting two virtual sites, $root = (-\infty, T, \text{RefSeq}, \text{root})$ and $leaf = (+\infty, T, \text{RefSeq}, \text{leaf})$, which are connected to/from each transcription start/end site: $E \cup (root \rightarrow s) \cup (t \rightarrow leaf)$ for all $s, t \in V$ with $type(s) = \text{start}$ and $type(t) = \text{end}$. Furthermore, w.r.t. truncated transcripts, all edges $s_i \rightarrow t$ at the transcript extremities with $type(s_i) = \text{start}, class(s_i) = \text{EST}$ are coalesced into one edge $s \rightarrow t, s = minarg_{s_i}(pos(s_i))$, and correspondingly all edges $s \rightarrow t_j$ with $type(t_j) = \text{end}, class(t_j) = \text{EST}$ into one edge $s \rightarrow t, t = maxarg_{t_j}(pos(t_j))$. This removes variants that only differ from each other in the truncation point of the first/last exon. Additionally,

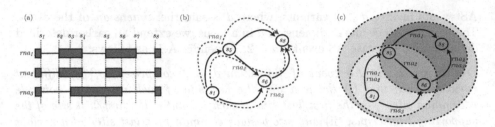

Fig. 1. (a) A cutoff showing 8 sites $\langle s_1, \ldots, s_8 \rangle$ from a locus with $k = 3$ transcripts (rna_1, rna_2 and rna_3). The exon-intron structure is shown schematically, i.e., exons (green boxes) and introns (lines) are not drawn to scale. Different variants can be observed, for instance $(s_1, s_5, \{rna_1, rna_2\})$. (b) The corresponding splicing graph structure after contracting uninformative vertices. Dotted lines indicate the paths supported by single transcripts rna_1 (red), rna_2 (green) and rna_3 (blue). (c) Ovals highlight all 3 bubbles, that is $(s_1, s_6, \{\{rna_2\}, \{rna_3\}\})$ (yellow), $(s_5, s_8, \{\{rna_1\}, \{rna_2\}\})$ (red) and $(s_1, s_8, \{\{rna_1\}, \{rna_2\}, \{rna_3\}\})$ (blue). The orange area in contrast does not represent a bubble because between s_5 and s_6 only exists one variant (i.e., rna_2).

in order to compensate for truncations at exon boundaries, transcription starts (respectively ends) s are replaced by acceptors (donors) if there is evidence for such in other transcript data at $pos(s)$. Finally, adopting the technique described in [30], vertices with $outdeg(s) = indeg(s) = 1$ are collapsed because they are uninformative w.r.t. the subsequently described technique (Appendix A.2). On the remaining vertices $s \in V$ (Fig. 1b) we define a preorder $\preceq$ by extending the natural total order of their genomic position $pos(s)$ as follows:

Definition 3. *The preorder $\preceq$ on the sites $s \in V$ orders them by the total order on their genomic position $pos(s)$ from 5' to 3' and their type, s.t. 5' exon boundaries precede 3' exon boundaries at the same genomic position $s \prec t$: $pos(s) < pos(t) \vee (pos(s) = pos(t) \wedge type(s) \in \{start, acceptor\})$.*

Note that by the properties of the graph construction, there exist no two vertices $s, t \in V, pos(s) = pos(t)$ with $type(s) = $ start, $type(t) = $ acceptor nor with $type(s) = $ end, $type(t) = $ donor. Obviously, for all $s \in V, s \notin \{root, leaf\}$, it holds $root \prec s \prec leaf$.

2.3 Subgraphs Describing AS Events: Bubbles

In G, a variant $p = (s, t, X_p)$ is a path $(s, \ldots, t)$ with a non-empty set of transcript support X_p (Def. 1), which excludes all paths in G describing splicing structures that have not been observed in nature. Subsequently, the variants between two vertices $s, t \in V$ are described by paths with s as common tail vertex, t as common head vertex and a set of non-empty partitions $\mathcal{X}_{s,t}$.

Observation 1. *For every pair of vertices with $transcripts(s) \cap transcripts(t) \neq \emptyset$, where $s \prec t$ there exists at least one variant $p = (s, t, X_p)$*

with $X_p \subseteq transcripts(s) \cap transcripts(t)$. An edge $u \rightarrow v, s \preccurlyeq u \prec v \preccurlyeq t$ with $(transcripts(u \rightarrow v) \cap X_p) \subsetneq X_p$ splits X_p into the partitions $X_{p'} = transcripts$ $(u \rightarrow v) \cap X_p$ and $X_{p''} = X_p \setminus X_{p'}$. Recursive splitting of partitions across all edges between s and t results in the "partition set" $\mathcal{X}_{s,t}$ describing all variants. For the union of transcripts in the partitions of $\mathcal{X}_{s,t}$ it naturally holds:

$$\bigcup_{rna \in X, X \in \mathcal{X}_{s,t}} rna = transcripts(s) \cap transcripts(t).$$

Consequently, if $|\mathcal{X}_{s,t}| > 1$ there exist different variants between s and t (Obs. 1). The subgraphs of G constituted by these variants are called *bubbles* (Fig. 1c).

Definition 4. *A "bubble" $(s, t, \mathcal{X}_{s,t})$ is a subgraph of G delimited by the vertices $s \prec t$ that comprises the maximal set of variants between s and t defined by the partition set $\mathcal{X}_{s,t}, |\mathcal{X}_{s,t}| \geqslant 2$ (maximality criterion for the number of variants) s.t. there exists no $\mathcal{X}_{u,v}$ with $\mathcal{X}_{s,t} \subseteq \mathcal{X}_{u,v}$ for any $s \preccurlyeq u \prec v \preccurlyeq t$ with $(s, t) \neq (u, v)$ (minimality criterion for the boundaries). We say s to be the "source" and t to be the "sink" of the bubble.*

By Def. 4, bubbles are subgraphs that involve cycles in the undirected graph underlying G. However, to our knowledge no graph structure described earlier on DAGs matches the requirements for a bubble.[1] Note that $(s, t, \mathcal{X}_{s,t})$ is equivalent to the notation of an AS event $\{p_i\}_{i=1}^d$ for $p_i = (s, t, X_{p_i}), \mathcal{X}_{s,t} = \bigcup_{p_i} (X_{p_i})$. It is straightforward to show that each d-dimensional AS event according to Def. 2 is reflected by a combination of variants $\{X_{p_i}\}_{i=1}^d \subseteq \mathcal{X}_{s,t}$ in a bubble of G (Lemma 1, proof in Appendix A.3).

Lemma 1. *For each AS event $\{p_i\}_{i=1}^d$ there exists a bubble $(s, t, \mathcal{X}_{s,t})$ with a combination of partitions $\{X_{p_i}\}_{i=1}^d \subseteq \mathcal{X}_{s,t}$ that describe the different variants of the event.*

Moreover, the following can directly be deduced from Def. 4:

Corollary 1. *Bubbles can intersect in vertices and edges.*

Corollary 1 (proof in Appendix A.4) outlines the complexity of overlaps between bubbles. Fig. 1c for instance shows 3 edge-intersecting bubbles. Theorem 1 however shows that there is a unique set of bubbles in G.

Theorem 1. *The set of all bubbles contained in G is unique.*

Proof. Def. 4 implies that—given a pair of vertices $(s, t), s \prec t$—there exists none $(|\mathcal{X}_{s,t}| < 2)$ or exactly one bubble (containing all $|\mathcal{X}_{s,t}|$ variants) between them. The complete set of bubbles as obtained from all tuples $(s, t) \in V$ is therefore unique and can be obtained by any arbitrary iteration over all $s \prec t$.   □

---

[1] *Blobs* described earlier [31] in phylogenetic networks with recombination cycles are defined as subgraphs involving all edge-intersecting cycles of the underlying undirected graph—thus a blob may coincide with an isolated bubble, but comprises multiple edge-intersecting bubbles. Whereas, a single recombination cycle—a *gall* [32]—does not necessarily describe the complete substructure of a bubble as by the maximality criterion for the number of variants in Def. 4 a bubble contains more than one cycle *iff* there are >2 variants between its source and sink.

2.4 AS Events Reflected by Bubbles

In the preceding section we show that each d-dimensional AS event is reflected by a variant combination in a bubble (Lemma 1). However, the mapping from AS events to d-tuple combinations of variants is *not injective*, hence not necessarily every combination of d variants in a bubble results in an AS event according to Def. 2. A bubble $(s, t, \mathcal{X}_{s,t})$ maximally harbors $\binom{|\mathcal{X}_{s,t}|}{d}$ events of dimension d, i.e., combinations $\{X_{p_i}\}_{i=1}^{d}$. Obviously, for $d > |\mathcal{X}_{s,t}|$ there exists none and for $d = |\mathcal{X}_{s,t}|$ there exists exactly one AS event. However, in the case of $d < |\mathcal{X}_{s,t}|$ variant combinations may occur that are all intersecting in one or more sites $u \notin \{s, t\}$ additional to the source/sink (Lemma 2, proof in Appendix A.5).

Lemma 2. *Some variants in a bubble* $(s, t, \mathcal{X}_{s,t}), |\mathcal{X}_{s,t}| > 2$ *may be intersecting in vertices (in addition to* $\{s, t\}$*) or in edges. Such variants imply the presence of at least one other bubble* $(u, v, \mathcal{X}_{u,v})$ *with* $s \preccurlyeq u \prec v \preccurlyeq t, (s, t) \neq (u, v)$*. In this specific geometry of overlapping bubbles* $(s, t, \mathcal{X}_{s,t})$ *is said the "outer bubble" that contains the "inner bubble"* $(u, v, \mathcal{X}_{u,v})$*.*

In Fig. 1c, the red and the yellow bubble are inner bubbles of the blue bubble. Further examples of outer and inner bubbles are shown in Fig. 2a and b. Note that inner bubbles comprise variants that are sub-paths of variants in the outer bubble and neither vertex-intersection (e.g., the yellow and the red bubble in Fig. 2b) nor edge-intersection (e.g., the yellow and the red bubble in Fig. 1c) nor the order of the source/sink vertices $s \preccurlyeq u \prec v \preccurlyeq t$ (e.g., the blue bubble in Fig. 2c) alone is sufficient for the geometry described in Lemma 2. Subsequently, Theorem 2 shows that for a combination of variants in an outer bubble that violates condition (ii) of Def. 2 there exists a combination of variants in an inner bubble containing the corresponding partitions that does not.

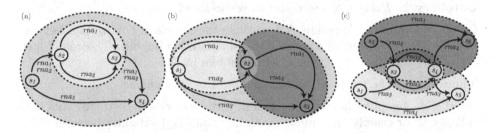

Fig. 2. Subgraphs showing different constellations of edge-intersecting bubbles with sites s_i numbered according to $pos(s_i)$. (a) The outer bubble $(s_1, s_4, \{\{rna_1\}, \{rna_2\}, \{rna_3\}\})$ (blue) contains the inner bubble $(s_2, s_3, \{\{rna_1\}, \{rna_2\}\})$ (yellow). (b) The inner bubbles $(s_1, s_2, \{\{rna_1\}, \{rna_2\}\})$ (yellow) and $(s_2, s_3, \{\{rna_1\}, \{rna_2\}\})$ (red) are both contained in the outer bubble $(s_1, s_3, \{\{rna_1\}, \{rna_2\}, \{rna_3\}\})$ (blue). (c) Although bubble $(s_3, s_4, \{\{rna_2\}, \{rna_3\}\})$ (blue) edge-intersects with $(s_2, s_6, \{\{rna_1\}, \{rna_2\}\})$ (red) and $(s_1, s_5, \{\{rna_3\}, \{rna_4\}\})$ (yellow), it is not an inner bubble because neither the partitions in $\mathcal{X}_{s_2,s_6}$ nor in $\mathcal{X}_{s_1,s_5}$ are all subsets of partitions in $\mathcal{X}_{s_3,s_4}$.

Theorem 2. *Given an outer bubble* $(s, t, \mathcal{X}_{s,t})$, *for any combination* $\{X_{p_i}\}_{i=1}^d$, $X_{p_i} \in \mathcal{X}_{s,t}$ *that corresponds to a set of variants* $\{p_1, \ldots, p_d\}$ *intersecting in more vertices than* s *and* t, *there exists a combination of partitions* $\{X_{p'_i}\}_{i=1}^d$, $X_{p'_i} \in \mathcal{X}_{u,v}$ *of an inner bubble* $(u, v, \mathcal{X}_{u,v})$ *s.t.* $X_{p_i} \subseteq X_{p'_i}$ *and* $\bigcap p'_i = \{u, v\}$.

Proof. By Lemma 2, an outer bubble $(s, t, \mathcal{X}_{s,t})$ contains $i \geqslant 2$ variants p_i that in G are super-paths of variants p'_i of its inner bubble $(u, v, \mathcal{X}_{u,v})$. As the partition X_p of a variant p is given by the intersection of the transcript support of all sites in p, it holds $X_{p_i} \subseteq X_{p'_i}$. □

For instance, for $d = 2$ the blue bubble in Fig. 1c gives rise to three partition combinations. Variants $(s_1, s_8, \{rna_1\})$ and $(s_1, s_8, \{rna_2\})$ contradict condition (ii) in Def. 2 because they intersect in $\{s_1, s_5, s_8\}$, but the variants $(s_5, s_8, \{rna_1\})$ and $(s_5, s_8, \{rna_2\})$ of the red inner bubble describe an AS event. Similarly the variants $(s_1, s_8, \{rna_2\})$ and $(s_1, s_8, \{rna_3\})$ are super-paths of variants $(s_1, s_6, \{rna_2\})$ respectively $(s_1, s_6, \{rna_3\})$ (yellow bubble) that represent an AS event. The partition combination $\{\{rna_1\}, \{rna_3\}\}$ in the blue bubble finally represent an AS event as there is no inner bubble $(u, v, \mathcal{X}_{u,v})$ with $\{rna_1\} \subsetneq X_i, \{rna_3\} \subseteq X_j, (X_i, X_j) \in \mathcal{X}_{u,v}$. Correspondingly, variants with the partition combination $\{\{rna_1\}, \{rna_2\}\}$ in the blue bubbles in Fig. 2a and 2b do not describe AS events.

2.5   An Exact Method for the Exhaustive Extraction of AS Events with Arbitrary Dimension

We now present an algorithm to extract all events of dimension d from a splicing graph G. Algorithm 1 iterates for all possible sinks of bubbles t over preceding edges $s \to u$, $s \prec t$ according to the order in Def. 3, until the 5'most source vertex with $transcripts(s) \supseteq transcripts(t)$ is reached. During the inner iteration, $\mathcal{X}$ stores the current set of partitions which is constantly subdivided by INTER-SECT() as new vertices s with $outdeg(s) \geqslant 2$ are iterated (Obs. 1). Initially, $\mathcal{X}$ consists of one partition containing all transcripts supporting t. $\mathcal{C}$ administrates d-tuples of partitions that are excluded from generating AS events in a hierarchical data-structure which ensures to also exclude tuples comprising subsets of these partitions as the splitting of $\mathcal{X}$ proceeds (consider the relation of partitions between variants of outer and inner bubbles in Theorem 2).

For any vertex s, transcript support of valid outedges $s \to u$ is intersected with the earlier found partitions in $\mathcal{X}$ to collect the partition set $\mathcal{X}_{s,t}$ (Obs. 1). Subsequently, EXTRACTEVENTS() outputs AS events as implied by all d-tuples in $\mathcal{X}_{s,t}$ that are not yet contained in $\mathcal{C}$ and for which the corresponding variants describe an alternative splice site (condition (iii) in Def. 2). In this subroutine, the partition d-tuples of all successfully extracted AS events are added to $\mathcal{C}$ in order to exclude them from generating AS events in further iterations of the inner loop.

Lemma 3. *The partition set* $\mathcal{X}_{s,t}$ *between two vertices* s *and* t *generated by Algorithm 1 is complete and corresponds to the partition set of all different variants from* s *to* t.

Input : A DAG $G(V, E)$ and the dimension of the AS events d.
Output: All AS events of dimension d reflected by G.

for *all vertices* $t \in V, indeg(t) \geqslant d$ *(in genomic order $\prec$)* **do**
  $\mathcal{X} \leftarrow \{transcripts(t)\}$
  $\mathcal{C} \leftarrow \emptyset$
  for *all inedges* $v \rightarrow t$ **do**
    $\lfloor \; \mathcal{C} \leftarrow transcripts(v \rightarrow t)$
  for *all vertices* $s \prec t, outdeg(s) \geqslant 2$ *(in reverse genomic order $\succ$)* **do**
    $\mathcal{X}_{s,t} \leftarrow \emptyset$
    for *all* $s \rightarrow u \in E$ **do**
      if $valid(s \rightarrow u)$ **then**
        $\lfloor \; \mathcal{X}_{s,t} \leftarrow \mathcal{X}_{s,t} \cup \text{INTERSECT}(transcripts(s \rightarrow u), \mathcal{X})$
      $\text{REMOVE}(transcripts(s \rightarrow u), \mathcal{X})$
    if $|\mathcal{X}_{s,t}| \geqslant d$ **then**
      $\lfloor \; \text{EXTRACTEVENTS}(s, t, \mathcal{X}_{s,t}, d, \mathcal{C})$
    $\mathcal{X} \leftarrow \mathcal{X} \cup \mathcal{X}_{s,t}$
    if $transcripts(t) \subseteq transcripts(s)$ **then**
      $\lfloor \;$ **break**

Algorithm 1. RETRIEVEASEVENTS(V, E, d)

By Theorem 2 partitions of outer bubbles that are included in partitions of inner bubbles (i.e., super-paths in G) do not give rise to AS events. To prove the correctness of Algorithm 1, it remains to be shown how such combinations are prevented:

Lemma 4. *Algorithm 1 does not consider variant combinations of outer bubbles with sub-paths that are part of an inner bubble.*

Given Lemma 3 (proof in Appendix A.6) and Lemma 4 (proof in Appendix A.7), it can be shown that the set of bubbles found by Algorithm 1 is complete and non-redundant:

Theorem 3. *Algorithm 1 finds all bubbles in G with at least d partitions and extracts AS events that comply with Def. 2.*

Proof. The main double loop considers all possible boundaries $(s \prec t)$ of bubbles with dimension d. Every boundary pair (s, t) is iterated exactly once, hence no bubble is found twice by the procedure. Given Lemma 3, it can be concluded that Algorithm 1 iterates all bubbles in G (Theorem 1). For each bubble $(s, t, \mathcal{X}_{s,t})$, all $\binom{|\mathcal{X}_{s,t}|}{d}$ variant combinations satisfying condition (i) of Def. 2 are considered. Lemma 4 shows how tuples contradicting condition (ii) are excluded. The check for the presence of at least one alternative splice site in EXTRACTEVENTS() ensures that the extracted AS events fulfill condition (iii) of Def. 2 and concludes the proof. □

To estimate the complexity of Algorithm 1, we consider that in each of the $\mathcal{O}(|V|)$ iterations of the outer loop, the transcripts of at most $\mathcal{O}(|E|)$ edges have to be compared to the transcripts in the current partition-set $\mathcal{X}$. Let $outdeg(s)$ be the average out-degree of a vertex $s \in V$, $|transcripts(s \to t)|$ be the average number of transcript support for an edge, and $|\mathcal{X}|$ be the average size of the partition set between two vertices. Assuming the comparison of transcript sets to be possible in (sub-)linear time (Appendix B), time complexity for retrieving all bubbles can be estimated by $\mathcal{O}(|V| \cdot |transcripts(s \to t)| \cdot |\mathcal{X}|)$ which, considering the reciprocal relation $|transcripts(s \to t)| \sim \frac{k}{|\mathcal{X}|}$, can be approximated by $\mathcal{O}(|V|k)$. Disregarding the update time of $\mathcal{C}$ and $\mathcal{X}$, and assuming the check for valid partition combinations in $\mathcal{C}$ to be feasible in constant time, the complexity of EXTRACTEVENTS() is determined by the output size $\sum_{(s,t) \in V} (|\mathcal{X}_{s,t}|^d)$ plus some additional iterations for variants at the transcript extremities that are lacking an alternative splice. Space requirements for storing the splicing graph as given in Algorithm 1 are $\mathcal{O}(k(|V| + |E|))$, but can technically be reduced to $max(\mathcal{O}(k|V|), \mathcal{O}(k|E|))$. Additionally, another $\mathcal{O}(|\mathcal{X}| \cdot |transcripts(s \to t)|) \sim \mathcal{O}(k)$ is required to store the current partition set in the double main loop of Algorithm 1.

3 Results

For our analysis, we used human and mouse annotations as downloaded from the UCSC genome browser [33] that contain transcripts of the NCBI reference sequence database (RefSeq) [25], the mRNAs in GenBank [26] and ESTs from the GenBank subset called dbEST [34] aligned to the genomes (reference sequence hg18 as generated by the centers of the Human Genome Sequencing Consortium [35], respectively mm8 by the Mouse Genome Sequencing Consortium [36]) using the program blat [23]. These two organisms have the largest amount of EST data available in dbEST (8,134,045 ESTs for human, respectively 4,850,243 for mouse), of which we took the subset of ESTs that show signs of splicing – as specified by UCSC (i.e., "intronESTs"—the "spliced ESTs" track). Subsequently, Algorithm 1 has been applied iteratively for each locus in the datasets.

3.1 The Information of mRNAs and ESTs about New Splice Forms Is Highly Redundant

As a proof of principle, we applied our program to transcript annotations of different sizes (i.e., RefSeq, RefSeq+mRNAs and RefSeq+mRNAs+ESTs in human and mouse) and observed the time needed to explore the AS diversity found for different dimensions $d \in \{2, 3, 4\}$. Detailed characteristics of these test runs are shown in Appendix B Tab. 2, from which we summarize here the dependency of the running time on the number of transcripts in the input (Fig. 3).

Obviously, the time effort grows linear with the number of transcripts k in the dataset—independent of different values for d. These characteristics indicate the following attributes for the input: first, there is a high degree of redundancy

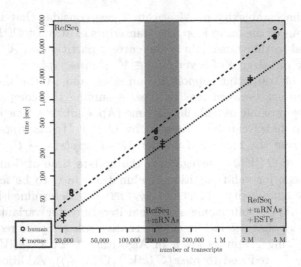

Fig. 3. Relation between the number of transcripts in the input and the running time of our implementation of Algorithm 1 for the datasets RefSeq (yellow), RefSeq+mRNAs (red) and RefSeq+mRNAs+ESTs (blue). The datapoints (circles for human and crosses for mouse) are taken from the extraction of events with dimension $d = 2$, $d = 3$ and $d = 4$ in the respective dataset. Clearly, the running time shows a linear dependency on the number of transcripts in the input (dashed line for human and dotted line for mouse). Packages from the R software [37] have been used when creating the plot.

in new splice forms because the comparatively big amount of transcript data in mRNAs and the even bigger one in ESTs does not add many more informative sites $|V|$. Second, the time complexity does not depend on dimension d of the extracted events which suggests that—when retrieving AS events of higher dimension—the time needed for additional combinations is about compensated by the time gained by disregarding all bubbles with $|\mathcal{X}_{s,t}| < d$. Indeed, the number of complete AS events with d variants decreases in an exponential manner when considering higher dimensions d (Appendix B Fig. 5).

3.2 More than a Quarter of the AS Events in Human Involve More than 2 Variants

From the results of Section 3.1 we observe in human 24,904 new events—in 6,945 different structures—that describe splicing variations comprising more than 2 alternatives and constituting >27% of all events. Motivated by the large fraction of events with a true dimension $d > 2$, we set off to explore up to which degree pairwise transcript comparison conducted usually provides an adequate picture of the true splicing complexity. To this end, we compare for the human annotation the number of the 5 hitherto analyzed AS patterns found in complete events with the number obtained by projecting all splicing variations to events of dimension $d = 2$ (Tab. 1).

Table 1. For each of the 5 types of AS events usually considered in literature, the number of corresponding structures found in pairwise events ($d = 2$) is shown in comparison to the fraction of them obtained when considering complete events. As can be seen, the fraction of exons that are erroneously observed as "mutually exclusive" in the pairwise projection is especially high.

structure	$d = 2$ events	complete events	fraction
skipped exon	42,054	24,547	58.4%
alt. acceptors	19,382	14,315	73.9%
alt. donors	17,727	13,647	77.0%
retained intron	7,939	5,990	75.5%
mutually exclusive exons	5,567	327	5.9%

As can be seen from Tab. 1, the fraction of splicing variations that are correctly described by a single pairwise event varies substantially amongst the different AS patterns. Whereas $\sim \frac{1}{4}$ of the pairwise AS events that describe retained introns or alternative donors/acceptors are part of variations between more than two transcripts, this holds $> 40\%$ of the skipped exons. Strikingly, most ($\sim$94%) of the splicing variations that in pairwise transcript comparisons lead to the observation of "mutually exclusive exons" are in reality part of events with $d > 2$. Fig. 6 (Appendix C) summarizes the most abundant complete AS events with $d > 2$. Obviously, the majority of alleged exon skipping events in reality is part of a variations where another exon upstream or downstream is also missing in some transcript evidence. These events are located in the red area of Fig. 6 and constitute in total 4,928 ($\sim$28%) of the "wrong" pairwise events. Another substantial part of $d = 2$ skipped exons ($\sim$26%$=$ 4,425 events) indeed co-occurs with splice donor/acceptor variations in the upstream/downstream exon (orange area in Fig. 6). The latter events also make up about half of the erroneously observed alt. donors/acceptors of which the rest mainly involves structures exhibiting more than one option for the donor/acceptor site (3,646 events, yellow area in Fig. 6). Retained introns in events with $d > 2$ often co-occur with variable donors/acceptors (886, that is nearly half of the retained introns observed when projecting to $d = 2$, are in the green and cyan areas of Fig. 6). 2,032 presumptive mutually exclusive exons are in AS events that show optional inclusion of 2 or 3 neighboring exons (violet area in Fig. 6). In the next section we elaborate on interesting differences in properties between these groups.

3.3 Characteristics of Events with Mutually Exclusive Exons Distinguish from Events Describing 2 Optionally Included Exons

Amongst the variations that involve 2 neighboring exons that are optionally included, 4 major groups that differ in structure can be distinguished: group 1 – events that include strictly one of the exons (327 cases, $\sim$6%), group 2 – events that show inclusion of none or exclusively one of the alternative exons (849 cases, 15%), group 3 – events that include one or both of them (460 cases, $\sim$8%) and group 4 – events that include one, both or none of the alternative exons (377 cases, $\sim$7%).

Usually, the picture of "mutually exclusive exons" fits the structure of group 1, however, one may still term the exons of events in group 2 as "mutually exclusive" since there is no transcript evidence including both of them.

Fig. 4 shows the distribution of intron lengths separately for the 3 introns of events in group 1 to 4 (in coding regions of the human genome)—which clearly differs: events of group 1 show significantly shorter 2nd introns (median $1,100$ nt, p-value$\sim 5e^{-7}$ two-sample Kolgomorov-Smirnov test) than the two flanking introns (median $1,929$ nt). In contrast, middle introns in group 4 are only slightly shorter (median $2,059$ vs $2,246$ nt, p-value> 0.69). Events of group 2 exhibit 2nd introns that are a bit, but not significantly, longer (median $2,897$ vs $2,342$ nt, p-value> 0.13), whereas group 3 contains more longer such introns ($3,288$ vs $2,305$ nt, p-value$\sim 1.3^{-4}$).

To get a deeper resolution on the differences, we analyzed the histograms of the length distributions for the middle introns in each of the 4 groups (Appendix C Fig. 7). Group 1 shows $> 50\%$ of events (91) with short 2nd introns ($< 1,500$ nt, the median intron length in human) and $\sim \frac{1}{3}$ (59 introns) very short introns

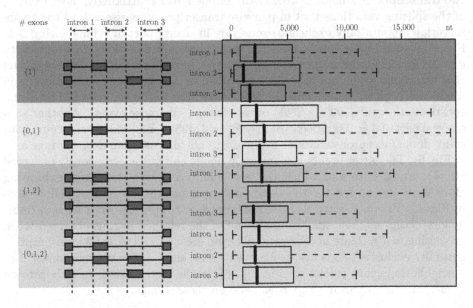

Fig. 4. Distribution of intron lengths occurring in AS events with two neighboring exons that are optionally included in coding regions of the human genome: mutually exclusive exons (exactly one exon included, red), events that include one of the two exons or none (number of included exons=$\{0,1\}$, yellow), events that include one of the two exons or both (green) and events that include one of the two exons, none or both (blue). Distributions are shown separately for the intron upstream of the first (i.e., the most 5') exon ("intron 1"), the intron between the two optional exons ("intron 2") and the intron downstream of the second exon ("intron 3"). The figure has been generated involving the boxplot() function of the R package [37] where the boxes represent the lengths for the 2nd and 3rd quartile of the distribution, separated by the median (bold vertical line). The 1st respectively the 4th quartile are shown as dashed lines.

($<$ 500 nt). Also group 2 exhibits several events with short (124) and very short (67) middle introns, still about twice as many as in group 3 and 4 (66 short and 33 very short introns). These observations support the hypothesis that events in group 1 and 2 (in contrast to events of group 3 and 4) may involve common molecular mechanisms to ensure mutual exclusion of the exons.

4   Discussion

We have developed an algorithm that efficiently retrieves arbitrarily complex AS events non-redundantly from large datasets, combining full-length cDNAs with highly fragmented ESTs. Due to an intrinsic clustering of transcripts into loci, no preliminary EST-clustering is required. Furthermore, by disregarding introns that appear to stem from technical artifacts in the transcript annotation, variations in the exon-intron structure are reduced to presumptive AS events. The method—comprising the steps of parsing the file with the annotated transcripts, clustering them into loci, retrieving the splice site sequence for all introns and extracting the events—copes with the to our knowledge biggest dataset of $\sim$ 4 million spliced transcripts in human in about $2\frac{1}{2}$ hours and retrieves more than 90,000 bona fide AS events. With the area of new sequencing technologies having just begun, ESTs can be produced rapidly (about a week from tissue harvesting to completion of DNA sequencing) and at very low costs (e.g., $<$\$0.03 per EST using pyrosequencing). Therefore, we see a high potential of our method in high-throughput systems to explore rapidly splicing variations reflected by the transcriptome of different organisms, tissues, cells or cell states.

In human, our method describes a plethora ($>$ 24, 000) AS events that involve actually more than two alternatives and that have hitherto not been described in their real complexity. Consequently, we estimated that at least $\sim \frac{1}{4}$ of the splicing variation has been ignored or miscategorized in previous works. Especially when regarding the alternative inclusion of two neighboring exons, we find that only the minority of these are what fits the common understanding of "mutually exclusive exons". We show that each of the most frequent of those patterns exhibits unique characteristics in the length distribution of the introns located between the two optional exons: many events that include exclusively one exon comprise (very) short such middle introns in contrast to events that alternatively include both exons. One could hypothesize that the common property of the former events could stem from a common molecular mechanism as described by steric blocking effects in the splicing process of mutually exclusive exons [38]. However, additional analyzes have to be conducted to support this hypothesis as literature describes a wide spectrum of mechanisms that can lead to mutual exclusion of exons, e.g., the relative strengths of 5' and 3'-splice sites [39,40,41], the pyrimidine content of a 3'-splice site [42,41], the location and number of branchpoints [43,44,45,46,38,47,46,48], branchpoint sequences [49,50,41], intron sequences between branchpoint and 3'-splice site [48,51,47], exon sequences [52,53,54,55,56,57,58,59,60] and trans-acting factors such as ASF or SF2 [61,62].

Acknowledgment

MS thanks the whole genome bioinformatics laboratory group of the CRG, especially Sylvain Foissac, Sarah Djebali, Christoforos Nikolau, Hagen Tilgner and Julien Lagarde for many discussions that contributed to this work, Tyler Alioto for elucidating the intron-based perspective of splicing and Oscar Gonzales for assistance in hard- and software problems.

References

1. The human sequencing consortium. Initial sequencing and analysis of the human genome. Nature 409, 860–921 (2001)
2. Smith, C.W., Valcarcel, J.: Alternative pre-mrna splicing: the logic of combinatorial control. Annu. Rev. Genet. 25, 381–388 (2000)
3. Lopez, A.J.: Alternative splicing of pre-mrna: developmental consequences and mechanisms of regulation. Annu. Rev. Genet. 32, 279–305 (1998)
4. Kuyumcu-Martinez, N.M., Cooper, T.A.: Mis-regulation of alternative splicing causes pathogenesis in myotonic dystrophy. Prog. Mol. Subcell. Biol. 44, 133–159 (2006)
5. Stamm, S., Riethoven, J.J., Le Texier, V., Gopalakrishnan, C., Kumanduri, V., Tang, Y., Barbosa-Morais, N.L., Thanaraj, T.A.: ASD: A bioinformatics resource on alternative splicing. Nucleic Acids Res. 34, D46–55 (2006)
6. Le Texier, V., Riethoven, J.J., Kumanduri, V., Gopalakrishnan, C., Lopez, F., Gautheret, D., Thanaraj, T.A.: AltTrans: Transcript pattern variants annotated for both alternative splicing and alternative polyadenylation. BMC Bioinformatics 7, 169 (2006)
7. Dralyuk, I., Brudno, M., Gelfand, M.S., Zorn, M., Dubchak, I.: ASDB: Database of alternatively spliced genes. BMC Bioinformatics 28, 296–297 (2000)
8. Holste, D., Huo, G., Tung, V., Burge, C.B.: HOLLYWOOD: a comparative relational database of alternative splicing. Nucleic Acids Res. 34, D56–62 (2006)
9. Zhou, Y., Zhou, C., Ye, L., Dong, J., Xu, H., Cai, L., Zhang, L., Wei, L.: Database and analyses of known alternatively spliced genes in plants. Genomics 82, 584–595 (2003)
10. Coward, E., Haas, S., Vingron, M.: SpliceNest: visualizing gene structure and alternative splicing based on EST clusters. Trends in Genetics 18, 53–55 (2002)
11. Huang, Y.H., Chen, Y.T., Lai, J.J., Yang, S.T., Yang, U.C.: PALS dbç: Putative alternative splicing database. Nucleic Acids Res. 30, 186–190 (2002)
12. Burset, M., Seledtsov, I.A., Solovyev, V.V.: SpliceDB: database of canonical and non-canonical mammalian splice sites. Nucleic Acids Res. 29, 255–259 (2001)
13. Ji, H., Zhou, Q., Wen, F., Xia, H., Lu, X., Li, Y.: AsMamDB: An alternative splice database of mammals. Nucleic Acids Res. 29, 260–263 (2001)
14. Modrek, B., Resch, A., Grasso, C., Lee, C.: Genome-wide analysis of alternative splicing using human expressed sequence data. Nucleic Acids Res. 29, 2850–2859 (2001)
15. Huang, H.D., Horng, J.T., Lee, C.C., Liu, B.J.: Prosplicer: A database of putative alternative splicing information derived from protein, mrna and expressed sequence tag sequence data. Genome Biol. 4, R29 (2003)
16. Bhasi, A., Pandey, R.V., Utharasamy, S.P., Senapathy, P.: ASD: a bioinformatics resource on alternative splicing. Boinformatics 23, 1815–1823 (2007)

17. Nagasaki, H., Arita, M., Nishizawa, T., Suwa, M., Gotoh, O.: Species-specific variation of alternative splicing and transcriptional initiation in six eukaryotes. Gene 364, 53–62 (2005)
18. Kim, E., Magen, A., Ast, G.: Different levels of alternative splicing among eukaryotes. Nucleic Acids Res. 35, 125–131 (2007)
19. Yandell, M., Mungall, C.J., Smith, C., Prochnik, S., Kaminker, J., Hartzell, G., Lewis, G.M., Rubin, S.: Large-scale trends in the evolution of gene structures within 11 animal genomes. PLoS Comput. Biol., vol. 2, p. 15 (2006)
20. Grasso, C., Modrek, B., Xing, Y., Lee, C.: Genome-wide detection of alternative splicing in expressed sequences using partial order multiple sequence alignment graphs. In: Pac. Symp. Biocomput., pp. 29–41 (2004)
21. Zavolan, M., van Nimwegen, E.: The types and prevalence of alternative splice forms. Curr. Opin. Struct. Biol. 16, 1–6 (2006)
22. Florea, L., Hartzell, G., Zhang, Z., Rubin, G.M., Miller, W.: A computer program for aligning a cDNA sequence with a genomic DNA sequence. Genome Res. 8, 967–974 (1998)
23. Kent, W.J.: BLAT - the blast-like alignment tool. Genome Res. 12, 656–664 (2002)
24. Bonizzoni, P., Rizzi, R., Pesole, G.: ASPIC: a novel method to predict the exon-intron structure of a gene that is optimally compatible to a set of transcript sequences. BMC Bioinformatics 6, 244 (2005)
25. Pruitt, K.D., Tatusova, T., Maglott, D.R.: NCBI reference sequences (RefSeq). a curated non-redundant sequence database of genomes, transcripts and proteins. Nucleic Acids Res. 35, Db1–D65 (2007)
26. Benson, D.A., Karsch-Mizrachi, I., Lipman, D.J., Ostell, J., Wheeler, D.L.: GenBank. Nucleic Acids Res. 35, D21–D25 (2007)
27. Weber, A.P., Weber, K.L., Carr, K., C.,, Wilkerson, O.J.B.: Sampling the arabidopsis transcriptome with massively parallel pyrosequencing. Plant Physiol. 144, 32–42 (2007)
28. Ruan, Y., Ooi, H.S., Choo, S.W., Chiu, K.P., Zhao, X.D., Srinivasan, K.G., Yao, F., Choo, C.Y., Liu, J., Ariyaratne, P., Bin, W.G.W., Kuznetsov, V.A., Shahab, A., Sung, W.-K., Bourque, G., Palanisamy, N., Weil, C.-L.: Fusion transcripts and transcribed retrotransposed loci discovered through comprehensive transcriptome analysis using paired-end ditags (pets). Genome Res. 17, 828–838 (2007)
29. Sugnet, C.W., Kent, W.J., Ares, M., Haussler, D.: Transcriptome and genome conservation of alternative splicing events in humans and mice. In: Pac. Symp. Biocomput., pp. 66–77 (2004)
30. Heber, S., Alekseyev, M., Sing-Hoi, S., Pevzner, P.: Splicng graphs and EST assembly problem. Bioinformatics 18, 181–188 (2002)
31. Gusfield, D., Bansal, V.: A fundamental decomposition theorem for phylogenetic networks and incompatible characters. In: Miyano, S., Mesirov, J., Kasif, S., Istrail, S., Pevzner, P.A., Waterman, M. (eds.) RECOMB 2005. LNCS (LNBI), vol. 3500, pp. 217–232. Springer, Heidelberg (2005)
32. Gusfield, D., Eddhu, S., Langley, C.: Optimal, efficient reconstruction of phylogenetic networks with constrained recombination. J. Bioinformatics and Computational Biology 2, 173–213 (2004)
33. University of California Santa Cruz (UCSC) Genome Browser, http://genome.ucsc.edu
34. Boguski, M.S., Lowe, T.M., Tolstoshev, C.M.: dbEST–database for "expressed sequence tags. Nat. Genet. 4, 332–333 (1993)
35. Human Genome Sequencing Consortium, http://genome.ucsc.edu/goldenPath/labs.html

36. Mouse Genome Sequencing Consortium,
    http://www.ensembl.org/Mus_musculus/credits.html
37. R Development Core Team. R: A Language and Environment for Statistical Computing. R Foundation for Statistical Computing, Vienna, Austria (2007) ISBN 3-900051-07-0
38. Smith, C.W., Nadal-Ginard, B.: Mutually exclusive splicing of alpha-tropomyosin exons enforced by an unusual lariat branch point location: implications for constitutive splicing. Cell 56, 749–758 (1989)
39. Zhuang, Y., Leung, H., Weiner, A.M.: The natural 5' splice site of simian virus 40 large t antigen can be improved by increasing the base complementarity to u1 rna. Mol. Cell Biol. 7, 3018–3020 (1987)
40. Kuo, H.C., Nasim, F.H., Grabowski, P.J.: Control of alternative splicing by the differential binding of u1 small nuclear ribonucleoprotein particle. Science 251, 1045–1050 (1991)
41. Mullen, M.P., Smith, C.W.J., Patton, J.G., Nadal-Girnard, B.: α-tropomyosin mutually exclusive exon selection: competition between branchpoint/polypyrimidine tracts determines default exon choice. Genes Dev. 5, 642–655 (1991)
42. Fu, X.Y., Ge, H., Manley, J.L.: In vitro splicing of mutually exclusive exons from the chicken β-tropomyosin gene: role of the branch point location and very long pyrimidine stretch. EMBO J. 7, 809–817 (1988)
43. Noble, J.C., Pan, Z.Q., Prives, C., Manley, J.L.: Splicing of sv40 early pre-mrna to large t and small t mrnas utilizes different patterns of lariat branch sites. Cell 27, 227–236 (1987)
44. Noble, J.C., Prives, C., Manley, J.L.: Alternative splicing of sv40 early pre-mrna is determined by branch site selection. Genes Dev. 2, 1460–1475 (1988)
45. Gattoni, R., Schmitt, P., Stevenin, J.: In vitro splicing of adenovirus e1a transcripts: characterization of novel reactions and of multiple branch points abnormally far from the 3' splice site. Nucleic Acids Res. 16, 2389–2409 (1988)
46. Helfman, D.M., Ricci, W.M.: Branch point selection in alternative splicing of tropomyosin pre-mrnas. Nucleic Acids Res. 17, 5633–5650 (1989)
47. Goux-Pelletan, M., Libri, D., d'Aubenton-Carafa, Y., Fiszman, M., Brody, E., Marie, J.: In vitro splicing of mutually exclusive exons from the chicken β-tropomyosin gene: role of the branch point location and very long pyrimidine stretch. EMBO J. 9, 241–249 (1990)
48. Helfman, D.M., Roscigno, R.F., Mulligan, G.J., Finn, L.A., Weber, K.S.: Identification of two distinct intron elements involved in alternative splicing of the β-tropomyosin pre-mRNA. Genes Dev. 4, 98–110 (1990)
49. Reed, R., Maniatis, T.: The role of the mammalian branchpoint sequence in pre-mrna splicing. Genes Dev. 2, 1268–1276 (1988)
50. Zhuang, Y.A., Goldstein, A.M., Weiner, A.M.: Uacuaac is the preferred branch site for mammalian mrna splicing. Proc. Natl. Acad. Sci. USA 86, 2752–2756 (1989)
51. Libri, D., Goux-Pelletan, M., Brody, E., Fiszman, M.Y.: Exon as well as intron sequences are cis-regulating elements for the mutually exclusive alternative splicing of the β tropomyosin gene. Mol. Cell Biol. 10, 5036–5046 (1990)
52. Reed, R., Maniatis, T.: A role for exon sequences and splice-site proximity in splice-site selection. Cell 46, 681–690 (1986)
53. Mardon, H.J., Sebastio, G., Baralle, F.E.: A role for exon sequences in alternative splicing of the human fibronectin gene. Nucleic Acids Res. 15, 7725–7733 (1987)
54. Somasekhar, M.B., Mertz, J.E.: Exon mutations that affect the choice of splice sites used in processing the sv40 late transcripts. Nucleic Acids Res. 13, 5591–5609 (1985)

55. Helfman, D.M., Ricci, W.M., Finn, L.A.: Alternative splicing of tropomyosin pre-mrnas in vitro and in vivo. Genes Dev. 2, 1627–1638 (1988)
56. Cooper, T.A., Ordahl, C.P.: Nucleotide substitutions within the cardiac troponin t alternative exon disrupt pre-mrna alternative splicing. Nucleic Acids Res. 17, 7905–7921 (1989)
57. Hampson, R.K., La Follette, L., Rottman, F.M.: Alternative processing of bovine growth hormone mRNA is influenced by downstream exon sequences. Mol. Cell Biol. 9, 1604–1610 (1989)
58. Streuli, M., Saito, H.: Regulation of tissue-specific alternative splicing: exon-specific cis-elements govern the splicing of leukocyte common antigen pre-mRNA. EMBO J. 8, 787–796 (1989)
59. Black, D.L.: Does steric interference between splice sites block the splicing of a short c-src neuron-specific exon in non-neuronal cells? Genes Dev. 5, 389–402 (1991)
60. Libri, D., Piseri, A., Fiszman, M.Y.: Exon as well as intron sequences are cis-regulating elements for the mutually exclusive alternative splicing of the β tropomyosin gene. Science 252, 1842–1845 (1991)
61. Ge, H., Manley, J.L.: A protein factor, asf, controls cell-specific alternative splicing of sv40 early pre-mrna in vitro. cell 13, 25–34 (1990)
62. Krainer, A.R., Conway, G.C., Kozak, D.: The essential pre-mrna splicing factor sf2 influences 5' splice site selection by activating proximal sites. Cell 13, 35–42 (1990)
63. Foissac, S., Sammeth, M., Astalavista: dynamic and flexible analysis of alternative splicing events in custom gene datasets. Nucleic Acids Res. 35, W297–W299 (2007)

A   Complementary Definitions and Proofs

A.1   A General Definition for Pairwise AS Events

Without restricting the molecular mechanism acting during the splicing process on a premature transcript *a priori* to exon or intron definition and—by this—to allow for possible interactions of parts of the splicing machinery across all exons and introns when delimiting AS events in exon-intron variations, we earlier have proposed a definition for pairwise ($d = 2$) AS events as stated in Def. 5 (Sammeth *et al.*, to appear).

Definition 5. *Comparing two transcripts* (rna_1, rna_2), *an AS event is a sequence of sites* $A = \langle s_i^A \rangle_{i=1}^g$ *satisfying the following conditions: (i) (consecutiveness of sites) all sites* s_i^A *that are supported by* $rna_1 = \langle s_i^{rna_1} \rangle_{i=1}^{n_1}$ *form a consecutive subsequence* $\bigcup_{s_i} \{s_i : rna_1 \in transcripts(s_i^A)\} = \langle s_j^{rna_1} \rangle_{j=a}^{b}$ *with* $1 \leqslant a < b \leqslant n$ *(and correspondingly all sites of A that are in* rna_2). *(ii) (minimality of common flanks) with the exception of the common sites at the flanks of the event, all sites are variable:* $\{rna_1, rna_2\} \not\subset transcripts(s_i^A)$ *for all* $1 < i < g$. *(iii) (prerequisite of an alternative splice site) A contains an alternative splice site* $s_i, 1 < i < g$ *(Def. 6).*

Note that by Def. 5 there can exist different pairwise AS events with the same set of variable splice sites but different common sites at their flanks, which are not to be treated identical as the flanking splice sites delimit the first/last variable exon/intron. Furthermore, an AS event requires the presence of an alternative splice site in the pair of transcripts (Def. 6).

Definition 6. *Comparing any two transcripts* $rna_1 = \langle s_i^{rna_1} \rangle_{i=1}^{n_1}$ *and* $rna_2 = \langle s_i^{rna_2} \rangle_{i=1}^{n_2}$, *an alternative splice site* s *is a variable site (i.e.,* $| transcripts(s) \cap \{rna_1, rna_2\}| = 1$*) that (i) is a splice site* $(type(s) \in \{donor, acceptor\})$*, and (ii) is contained within the common genomic region of both transcripts.*

A.2   Uninformative Vertices

When constructing the splicing graph $G = (V, E)$, we collapse vertices that are *uninformative* for delimiting bubbles in G as demonstrated by Lemma 5.

Lemma 5. *Vertices* v *with* $indeg(v) = outdeg(v) = 1$ *are uninformative for delimiting bubbles in splicing graphs and can be collapsed without loss of generality of Algorithm 1.*

Proof. From the minimality criterion for the boundaries in Def. 4 can directly be deduced that for a bubble $(s, t, \mathcal{X}_{s,t})$ holds $outdeg(s) > 1$ and $indeg(t) > 1$. Vertex v therefore can neither be source nor sink of a bubble. Let $s \preccurlyeq u \prec v \prec w \preccurlyeq t$ be contained in the bubble and $u \to v$ be the inedge and $v \to w$ be the outedge of v. Then naturally $transcripts(u \to v) = transcripts(v) = transcripts(v \to w) = transcripts(u) \cap transcripts(w)$ holds and the corresponding partition is equivalently described by a single edge $v \to w$. □

A.3   Proof of Lemma 1

Proof. Clearly, there exists a set of paths in G according to the variants of an AS event $\{p_i\}_{i=1}^{d}$. As by Def. 2, an AS event of dimension d is a set of sites that are not common to all of the d compared transcripts, flanked by splice sites (s, t) contained in all of them or the transcript start/end. Since the completion of G by $(root, leaf)$ ensures common vertices also in AS events including variable transcript extremities, $\bigcap_{p_i} (X_{p_i}) = \{s, t\}$ holds for all AS events according to Def. 2 and consequently there exists a bubble $(s, t, \mathcal{X}_{s,t})$ with $\{X_{p_i}\}_{i=1}^{d} \subseteq \mathcal{X}_{s,t}$. □

A.4   Proof of Corollary 1

Proof. As different splicing structures may involve common splice sites, exons or introns, two variants (p_1, p_2) with $p_1[1] \neq p_2[1] \lor p_1[m_1] \neq p_2[m_2]$ can intersect in vertices or edges. Given additionally the variants p_3 and p_4 with $p_3 \cap p_1 = \{p_1[1], p_1[m_1]\}$ and $p_4 \cap p_2 = \{p_2[1], p_2[m_2]\}$, p_1, p_3 and on the other hand p_2, p_4 are part of different bubbles. □

A.5   Proof of Corollary 2

Proof. Let the bubble $(s, t, \{X_{p_1}, X_{p_2}, X_{p_3}\})$ contain 3 variants s.t. $p_1 \cap p_2 \cap p_3 = \{s, t\}$ and without violating Def. 4 $p_1 \cap p_2 = \{s, t, u\}$. By $p_1 \neq p_2$ and $s \prec u \prec t$ (Def. 3), p_1 and p_2 differ between s and u or between u and t (or in both parts). Correspondingly, there exists a bubble $(s, u, \mathcal{X}_{s,u}), \{X_{p_1}, X_{p_2}\} \subset \mathcal{X}_{s,u}$ and/or a bubble $(u, t, \mathcal{X}_{u,t}), \{X_{p_1}, X_{p_2}\} \subset \mathcal{X}_{u,t}$. The argumentation can straightforwardly be extended if p_1 and p_2 intersect in more than one vertex and hence differ at least between two vertices u and v. □

A.6 Proof of Lemma 3

Proof. Algorithm 1 initializes the partition sets with $transcripts(t)$ and subsequently subdivides them by INTERSECT() with the transcript sets $X_{s,u}$ along out-edges of s. Thus, all variants are found based on Obs. 1.   □

A.7 Proof of Lemma 4

Proof. Given an outer bubble $(s, t, \mathcal{X}_{s,t})$ and an inner bubble $(u, v, \mathcal{X}_{u,v})$, we distinguish two cases. (i) $v = t$ (e.g., the red and the blue bubble in Fig. 2b): by the iteration order over sink vertices in genomic order $\prec$ and over source vertices in reverse genomic order $\succ$, $(u, v, \mathcal{X}_{u,v})$ is iterated before $(s, t, \mathcal{X}_{s,t})$ and partition combinations are stored in $\mathcal{C}$ accordingly. (ii) $v \prec t$: if $v \rightarrow t \in E$, the initialization of $\mathcal{C}$ with the transcript support of all in-edges of t prevents from variant combinations that are super-paths of variants in the inner bubble (e.g., the yellow and the blue bubble in Fig. 2a); otherwise (e.g., the yellow and the blue bubble in Fig. 2b) there exists a bubble $(v, t, \mathcal{X}_{v,t})$ (the red bubble in Fig. 2b) as inner bubble of $(s, t, \mathcal{X}_{s,t})$ and the problem is reduced to case (i).   □

B   Implementation and Benchmark Details

Algorithm 1 has been implemented as described in JAVA (compliance level 1.5). Technical optimizations include the separation of operations in Algorithm 1 from program parts that handle the I/O by different JAVA threads. Parsing the input data and performing the clustering of genes into loci may take a considerable amount of time—given that the input file including EST data in human is nearly as big as the complete sequence of the human genome—as does the writing of the found AS events to disk if output size is big. Time benchmarks with this parallelized program therefore reflect about the algorithmic time effort. Furthermore, partition comparisons (e.g., intersection operations) are optimized by encoding the respective transcript support in 64-bit number arrays where each bit indicates whether a corresponding transcript of the locus is contained in the partition or not. Memory requirements of storing a bit for every transcript in the input are negligible compared to the speed up gained in partition comparisons. In this way, the current implementation runs optimally on a 64-bit system architecture with two or more CPUs.

Tab. 2 shows characteristics of the running time for different input annotations and varying projection level $d \in \{2, 3, 4\}$, or when describing every variation in its true dimension as reflected in the annotation (complete events). To give a comparison on the times measured, we extracted pairwise events using the Astalavista web server [63] (reference). The time complexity of this technique is quadratic with the number of transcripts in the input (since an all-against-all comparison is performed on them) as well as quadratic with the number of AS events found in each locus (as they are compared 1-by-1 for non-redundancy). Note that the number of events found by this reference method differs as (i) events are filtered for introns with non-canonical (GT/AG) introns in the *variable* part of an AS event, and (ii) the method does not perform the confidence check for the transcription start/end sites of mRNAs or ESTs and therefore finds a magnitude of variations that stem from truncated transcript data. Thereforetimes for runs of the reference method are not directly comparable to our method, but it is clear that—although performing faster in the RefSeq data set of human and mouse, the reference is much slower then the here described technique when including mRNA data, and inapplicable to EST data.

Table 2. From the human and the mouse annotation the RefSeq transcripts (yellow), a dataset containing RefSeq transcripts and mRNAs ("+mRNA", red) and a dataset containing RefSeq transcripts, mRNAs and ESTs ("+ESTs", blue) have been applied in order to benchmark our method in the extraction of AS events with varying dimension d. The number of transcripts and the time (hh:mm:ss) of a reference method for pairwise AS event extraction is additionally shown.

dataset	human			mouse		
	RefSeq	+mRNA	+ESTs	RefSeq	+mRNA	+ESTs
transcripts	25,161	206,779	4,093,918	20,618	244,882	2,219,200
time (reference)	0:00:30	1:10:24^a	$-^b$	0:00:14	1:19:58^c	$-^b$
events $d = 2$	9,477	34,603	134,875	2,638	19,127	52,508
time	0:01:04	0:05:30	2:28:36	0:00:30	0:04:23	0:32:02
events $d = 3$	7,173	17,294	97,193	2,669	4,839	15,896
time	0:00:58	0:05:10	1:51:34	0:00:26	0:04:00	0:30:54
events $d = 4$	21,620	43,476	153,292	9,321	10,570	17,025
time	0:01:02	0:05:12	1:57:22	0:00:26	0:04:21	00:27:30
complete events	6,493	26,412	91,117	1,787	16,184	42,198
time	0:00:58	0:05:10	1:55:12	0:00:32	0:04:42	0:29:53

---

^{a} Skipping 5 loci with > 1000 transcripts, i.e., a locus on chr2 (88,937,526–89,411,301) encoding parts of antibodies—mostly variable regions (2,645 transcripts), the locus on chr14 (21,180,949–22,090,938) encoding the T-cell receptor α chain (1,493 transcripts), a locus on chr14 (104,896,270–106,354,328) containing genes for several immunoglobulin heavy α, β and γ chains (9,766 transcripts), the locus on chr7 (141,647,256–142,210,559) coding for the T-cell receptor β chain (2,934 transcripts) and the locus on chr22 (20,710,462–21,595,078) encoding the immunoglobulin λ chain (1,931 transcripts).

^{b} Test run exceeded memory (>16GB) or time limits (5 days).

^{c} Skipping 9 loci with > 500 transcripts, i.e., 5 loci on chr6 (with 629, 866, 878, 1,167 respectively 1,198 transcripts) encoding parts of antibodies (IgG *kappa* chain, variable regions, etc. ...), the locus of the Trpm1 gene on chr7 (510 transcripts), the locus of the Eef2 gene on chr10 (548 transcripts), an immunoglobulin locus on chr12 (10,275 transcripts), and a locus with several olfactory receptor genes on chr14 (1,196 transcripts).

As can be observed, running times increase mostly linear with the number of transcripts in the input. The expected exponential increase of the output size (and therefore, of the running time) is compensated by an exponential decrease of the number of events found for increasing dimension size (Fig. 5).

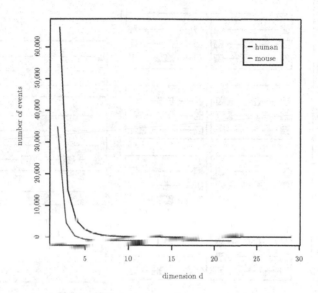

Fig. 5. Exponentially decreasing relation between the dimension d of an event and the number of complete AS events with d variants in the human (black curve) and the mouse (red curve) transcriptome

C   AS Events of Arbitrary Dimension

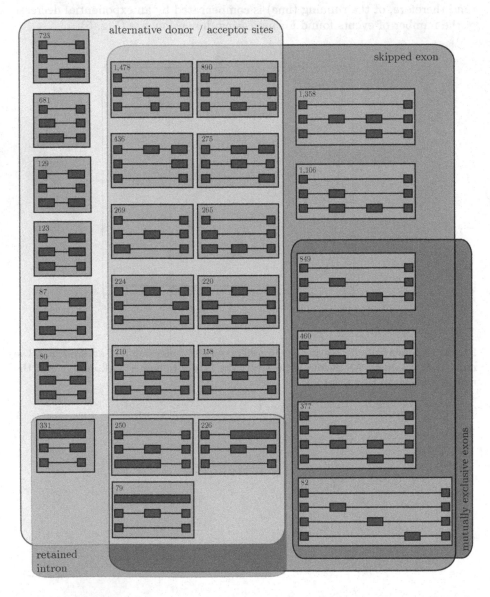

Fig. 6. AS events with $d > 2$ found in the 50 most abundant complete events of the human transcriptome. In pairwise projections these examples show multiple instances of alternative donors/accecptors (yellow area), skipped exons (red area), alternative donors/acceptors and skipped exons (orange area), skipped exons and mutually exclusive exons (violet area), alternate donors/acceptors and retained introns (green area) and of alternate donors/acceptors, skipped exons and retained introns (cyan area). Above each pictogram the number of events with the corresponding structure is given.

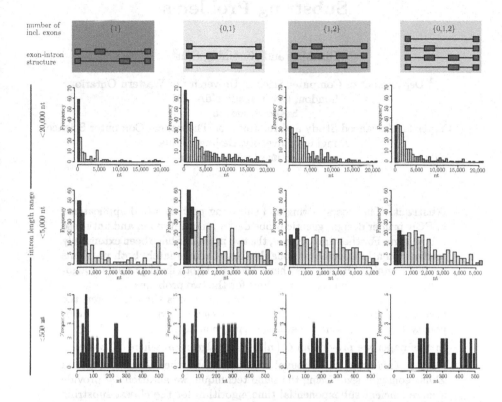

Fig. 7. Distribution of lengths exhibited by the 2nd intron of events that involve alternative neighboring exons, of which either strictly one is included (mutually exclusive exons, red), or one of them or none (number of included exons= $\{0,1\}$, yellow), one or both are included (number of included exons= $\{1,2\}$, green), or, none, one or both are included (number of included exons= $\{0,1,2\}$, blue). Red bars mark introns between 1nt and 500nt, yellow bars mark introns between 501nt and 5,000nt, and green bars show introns between 5,000nt and 20,000nt. Mutually exclusive exons show a large number of small and some very small middle introns that is partially present in events where none or one exon is included, but missing in the other variations.

More Efficient Algorithms for Closest String and Substring Problems

Bin Ma[1] and Xiaoming Sun[2]

[1] Department of Computer Science, University of Western Ontario
London, ON, Canada N6A 5B7
bma@csd.uwo.ca
[2] Center for Advanced Study and Institute for Theoretical Computer Science
Tsinghua University, Beijing, China
xiaomings@tsinghua.edu.cn

Abstract. The closest string and substring problems find applications in PCR primer design, genetic probe design, motif finding, and antisense drug design. For their importance, the two problems have been extensively studied recently in computational biology. Unfortunately both problems are NP-complete. Researchers have developed both fixed-parameter algorithms and approximation algorithms for the two problems.

In terms of fixed-parameter, when the radius d is the parameter, the best-known fixed-parameter algorithm for closest string has time complexity $O(nd^{d+1})$, which is still superpolynomial even if $d = O(\log n)$. In this paper we provide an $O\left(n|\Sigma|^{O(d)}\right)$ algorithm where Σ is the alphabet. This gives a polynomial time algorithm when $d = O(\log n)$ and Σ has constant size. Using the same technique, we additionally provide a more efficient subexponential time algorithm for the closest substring problem.

In terms of approximation, both closest string and closest substring problems admit polynomial time approximation schemes (PTAS). The best known time complexity of the PTAS is $O(n^{O(\epsilon^{-2} \log \frac{1}{\epsilon})})$. In this paper we present a PTAS with time complexity $O(n^{O(\epsilon^{-2})})$.

At last, we prove that a restricted version of the closest substring has the same parameterized complexity as closest substring, answering an open question in the literature.

1 Introduction

The closest string and substring problems have been recently studied extensively in computational biology [16,18,22,13,24,12,23,15,7,11,29,4,26,30]. The two problems have a variety of applications in bioinformatics, such as universal PCR primer design [20,16,5,27,12,31], genetic probe design [16], antisense drug design [16,4], finding unbiased consensus of a protein family [2], and motif finding [16,12,30,3,9]. In all these applications, a common task is to design a new DNA or protein sequence that is very similar to (a substring of) each of the given sequences. In the first three applications, the designed DNA sequence can bind

M. Vingron and L. Wong (Eds.): RECOMB 2008, LNBI 4955, pp. 396–409, 2008.
© Springer-Verlag Berlin Heidelberg 2008

to each of the given DNA sequences in order to perform its designated functions. In the last two applications, the designed sequence acts as an unbiased representative of all the given sequences. The common task has been formulated as the closest string problem and the closest substring problem.

Given n length-m strings s_1, s_2, ..., s_n, and a radius d, the *closest string* problem seeks for a new length-m string s such that $d(s, s_i) \leq d$. Here $d(\cdot, \cdot)$ is the Hamming distance. The *closest substring* problem seeks for a length L ($L \leq m$) string t such that for every $i = 1, 2, \ldots, n$, there is a substring t_i of s_i with length L such that $d(t, t_i) \leq d$. The problems may also be described as optimization problems where the objectives are to minimize the radius d.

Unfortunately, both of these two problems are NP-complete [10,16]. In addition to many heuristic algorithms without any performance guarantee (for example [19,23,24]), researchers have developed approximation algorithms and fixed-parameter algorithms for the two problems. Approximation algorithms sacrifice the quality of the solution in order to achieve polynomial time [14]. A polynomial time approximation scheme (PTAS) achieves ratio $1 + \epsilon$ in polynomial time for any fixed $\epsilon > 0$. Fixed-parameter algorithms find optimal solutions with time complexity $f(k) \cdot n^c$ for a constant c and any function f [6]. Here k is a parameter naturally associated to the input instance.

For fixed-parameter algorithms, Stojanovic et al. [29] provided a linear time algorithm for $d = 1$. Gramm et al. [13] provided the first fixed-parameter algorithm for closest string with running time $O(nm + nd^{d+1})$. Therefore, for small values of d their algorithm can solve closest string in acceptable time. In this paper we present a novel algorithm that finds the optimal solution of closest string problem with running time $O\left(nm + nd \cdot (16|\Sigma|)^d\right)$. This is exponentially better than the previous fixed-parameter algorithm when the alphabet has constant size.

The closest substring problem appeared to be harder than closest string in terms of parameterized complexity. For unbounded alphabet size, it has been shown that the problem is $W[1]$-hard even if all d, n, L are parameters [8,9]. The $W[1]$-hardness indicates that the problem unlikely has a fixed-parameter polynomial time algorithm [6]. For $|\Sigma|$ being a constant or a parameter, the problem is $W[1]$-hard even if both d and n are parameters [22]. For a more complete review of the parameterized complexities of the closest substring problem, we refer the readers to [9,22,25]. Marx [22] gave a $|\Sigma|^{d(\log d + 2)}(nm)^{\log d + O(1)}$ algorithm for the closest substring problem. In this paper we present a new algorithm for closest substring with improved time complexity $O\left((16|\Sigma|)^d \cdot nm^{\lceil \log d \rceil + 1}\right)$.

For approximation algorithms, Lanctot et al. [16] gave the first polynomial time approximation algorithm with approximation ratio $\frac{4}{3} + o(1)$. Li et al. [17] provided a PTAS for closest string with time complexity $O(mn^{O(\epsilon^{-5})})$. Ma [21] provided a PTAS for closest substring problem. These two PTAS results were collected in [18]. There have been many negative comments regarding the large exponent of the PTAS [9,3,11,13,22]. By using a lemma in [22] and an idea of [17], Andoni et al. [1] proposed a PTAS to obtain a much better time complexity $O(mn^{O(\epsilon^{-2} \log \frac{1}{\epsilon})})$. By combining our new fixed-parameter algorithm, in Section 5 we provide a simpler PTAS with further improved time complexity $O(mn^{O(\epsilon^{-2})})$.

Noticing the hardness of closest substring problem, Moan and Rusu [25] studied a more restricted version of closest substring. They put a diameter constraint on top of the original closest substring problem by further requiring the pairwise distances between substrings in the solution do not exceed a diameter D for some $D < 2d$. They hoped that such a constraint may reduce the parameterized complexity of closest substring when D is close enough to d. The condition for this to happen is left as a main open problem in [25]. In this paper we answer this question by proving that such condition does not exist. That is, for any given $\epsilon > 0$, all parameterized complexity results of closest string preserve in the constrained instances for $D < (1 + \epsilon)d$.

2   Preliminaries and Notations

Let Σ be an alphabet with constant size $|\Sigma|$. Suppose s is a string over Σ. $|s|$ denotes the length of s. $s[i]$ denotes the i-th letter of s. Therefore, $s = s[1]s[2]\ldots s[m]$, where m is the length of s. Let s and t be two strings with the same length m, $d(s,t)$ denotes the Hamming distance between s and t. Use $[1,m]$ to denote the set $\{1, 2, \ldots, m\}$. For $P = \{i_1, i_2, \ldots, i_k\} \subseteq [1, m]$, define $s|_P = s[i_1]s[i_2]\ldots s[i_k]$ and $d^P(s,t) = d(s|_P, t|_P)$. Let $Q = [1, m] \setminus P$. From the definition of Hamming distance, clearly $d(s,t) = d^P(s,t) + d^Q(s,t)$. Let $Q(s,t)$ denote the set of positions where s and t agree, i.e., $Q(s,t) = \{j \mid s[j] = t[j]\}$. Similarly, for k given strings $s_1, s_2, \ldots, s_k$ of same length, $Q(s_1, s_2, \ldots, s_k)$ denotes the position set where all strings agree. Let $P(s,t)$ denote the position set where s and t disagree.

Let s_1, s_2, ..., s_n be n strings of length m. The *closest string problem* asks for a string center s such that $d = \max_{i=1}^{n} d(s, t_i)$ is minimized. The minimum value of d is called the *radius* of the n input strings. $D = \max_{1 \leq i, j \leq n} d(s_i, s_j)$ is called the *diameter* of the n input strings. Let $L \leq m$. The *closest substring problem* asks for a length-L string center s and a length-L substring t_i of each s_i, such that $d = \max_{i=1}^{n} d(s, t_i)$ is minimized.

In this paper we will also study a more generalized version of closest string problem, the *neighbor string problem*: Given n strings $s_1, s_2, \ldots, s_n$ with length m, and n nonnegative integers $d_1, d_2, \ldots, d_n$, the neighbor string problem seeks for a length m string s such that $d(s, s_i) \leq d_i$ for every $1 \leq i \leq n$. An instance of the neighbor string problem is given by $\langle (s_1, d_1), (s_2, d_2), \ldots, (s_n, d_n) \rangle$.

3 $O\left(nm + nd \cdot (16|\Sigma|)^d\right)$ Algorithm for Closest String Problem

Parameterized complexity has been used to tackle NP-hard problems [6]. In principle, a fixed-parameter polynomial time algorithm is a well-structured super-polynomial algorithm such that the superpolynomial factor is only with respect to one parameter of the given instance. Many NP-hard problems have been found to be fixed-parameter tractable, which means that an algorithm with running time $f(k)\cdot n^c$ exists to solve the problem. Here k is a parameter naturally associated with the problem; n is the size of the input and c is a constant. Clearly $f(k)$ must be superpolynomial if $\mathbf{P} \neq \mathbf{NP}$. The hope is that this $f(k)$ will not grow too fast, and parameter k is small for practical instances; and hence the problem can be solved efficiently in practice.

Gramm et al. [13] provided a fixed-parameter polynomial time algorithm for closest string when the radius d is used as the fixed parameter. For a given instance $\{s_1, s_2, \ldots, s_n\}$ and a given value d, their algorithm finds a center string s such that $d(s, s_i) \leq d$ in $O(nm + nd^{d+1})$ time, if such a string exists.

In this section we provide a new algorithm for closest string problem with time complexity $O(nm + nd \cdot (16|\Sigma|)^d)$. When the alphabet size is a constant, our algorithm is exponentially faster than the previous algorithm. In order to design the algorithm for closest string, let us focus on the more generalized neighbor string problem.

Lemma 1.[1] *Let* $\langle (s_1, d_1), \ldots, (s_n, d_n) \rangle$ *be an instance of the neighbor string problem. If j satisfies $d(s_1, s_j) > d_j$, then for $Q = Q(s_1, s_j)$ and any solution s of the neighbor string problem, $d^Q(s, s_1) < \frac{d_1}{2}$.*

Proof. Let s be a solution, i.e. $d(s, s_i) \leq d_i$ for $i = 1, 2, \ldots, n$. Let $P = [1, m] \setminus Q$. Then

$$d^P(s, s_1) + d^P(s, s_j) \geq d^P(s_1, s_j) = d(s_1, s_j) > d_j. \tag{1}$$

On the other hand,

$$\begin{aligned}
&d^P(s, s_1) + d^P(s, s_j) \\
&= \big(d(s, s_1) - d^Q(s, s_1)\big) + \big(d(s, s_j) - d^Q(s, s_j)\big) \\
&= d(s, s_1) + d(s, s_j) - 2\,d^Q(s, s_1) \\
&\leq d_1 + d_j - 2\,d^Q(s, s_1)
\end{aligned}$$

The second equation in the above derivation is because $s_1|_Q = s_j|_Q$. Combining with (1), we get $d_1 + d_j - 2\,d^Q(s, s_i) > d_j$. Consequently, $d^Q(s, s_1) < \frac{d_1}{2}$. The lemma is proved. □

Theorem 1. *Let $d = \max_{1 \leq i \leq n} d_i$. If there is a solution s such that $d(s, s_i) \leq d_i$ ($1 \leq i \leq n$), then algorithm StringSearch in Fig. 1 outputs a solution s' such that $d(s', s_i) \leq d_i$ in time $O(mn + nd \cdot T(d, d_1))$, where the size of the search tree*

$$T(d, d_1) \leq \binom{d + d_1}{d_1} (|\Sigma| - 1)^{d_1} \cdot 2^{2d_1}.$$

Proof. First let us prove the correctness of the algorithm. It is easy to verify that when the algorithm returns a non-null string in either line 2 or line 4.3,

---

[1] Lemma 1 uses a similar idea as Lemma 2.2 in [22]. However the lemma in [22] cannot be directly used in our algorithms.

Algorithm StringSearch
Input: An instance of neighbor string $\langle (s_1, d_1), (s_2, d_2), \ldots, (s_n, d_n) \rangle$.
Output: String s such that $d(s, s_i) \leq d_i$ $(i = 1, \ldots, n)$, or NULL if there is no solution.
1. Try to find i_0 such that $d(s_1, s_{i_0}) > d_{i_0}$.
2. If step 1 fails, return s_1.
3. Let $Q = Q(s_1, s_{i_0})$, $P = [1, |s_1|] \setminus Q$.
4. For every possible string t of length $|P|$ such that $d(t, s_1|_P) \leq d_1$ and $d(t, s_{i_0}|_P) \leq d_{i_0}$
4.1   Let $e_i = d_i - d(t, s_i|_P)$ for $i \neq 1$, and $e_1 = \min\{d_1 - d(t, s_1|_P), \lceil d_1/2 \rceil - 1\}$;
4.2   Use **StringSearch** to find the solution u of $\langle (s_1|_Q, e_1), (s_2|_Q, e_2), \ldots, (s_n|_Q, e_n) \rangle$;
4.3   If $u \neq$ NULL then let $s|_P = t$ and $s|_Q = u$ and return s.
5. Return NULL.

Fig. 1. The algorithm **StringSearch**

the string is a solution of the input instance. Let us prove that when there is a solution of the input instance, then the algorithm can find it. We prove this by using induction on d_1. If $d_1 = 0$ then clearly the algorithm is correct. When $d_1 > 0$ and line 1 finds i_0 successfully, by Lemma 1, the Q and P defined in line 3 are such that there is a solution s satisfying $d(s|_Q, s_1|_Q) \leq e_1$. Therefore, this s is such that $d(s|_Q, s_i|_Q) \leq e_i$ for $1 \leq i \leq n$. As a result, when $t = s|_P$ in line 4, by induction, the recursive call to Algorithm **StringSearch** at line 4.2 will find u such that $d(u, s_i|_Q) \leq e_i$ for $1 \leq i \leq n$. Then it is easy to verify that the s returned in line 4.3 is a desired solution.

Next let us examine the time complexity of the algorithm **StringSearch**. We estimate the size (number of leaves) of the search tree first. In line 4, assume t is an eligible string and $d(t, s_1|_P) = k$. Then $|P| = d(s_1|_P, s_{i_0}|_P) \leq d(t|_P, s_1|_P) + d(t|_P, s_{i_0}|_P) \leq d_{i_0} + k \leq d + k$. Therefore, there are at most $\binom{|P|}{k}(|\Sigma| - 1)^k \leq \binom{d+k}{k}(|\Sigma| - 1)^k$ such strings t. For each of them, the size of the subtree rooted at t of the search tree is bounded by $T(d, \min\{d_1 - k, \lceil d_1/2 \rceil - 1\})$. k can take values from 0 to d_1. Therefore, the search tree size satisfies

$$T(d, d_1) \leq \sum_{k=\lfloor d_1/2 \rfloor + 1}^{d_1} \binom{d+k}{k}(|\Sigma| - 1)^k T(d, d_1 - k)$$

$$+ \sum_{k=0}^{\lfloor d_1/2 \rfloor} \binom{d+k}{k}(|\Sigma| - 1)^k T(d, \lceil d_1/2 \rceil - 1) \tag{2}$$

$$= I_1 + I_2 \tag{3}$$

Clearly $T(d, 0) = 1$ because in this case s_1 is the solution. We prove by induction that for $\tilde{d} \geq 1$,

$$T(d, \tilde{d}) \leq 2^{2\tilde{d}} \binom{d + \tilde{d}}{\tilde{d}} (|\Sigma| - 1)^{\tilde{d}}. \tag{4}$$

It is easy to verify that when $\tilde{d} = 1, T(d, 1) \leq (d+1)(|\Sigma|-1)+1$, the statement is true. When $\tilde{d} = 2$, because of (2), $T(d, 2) \leq \binom{d+2}{2}(|\Sigma| - 1)^2 + (d + 1)(|\Sigma| - 1) + 1 \leq 2\binom{d+2}{2}(|\Sigma| - 1)^2$, the statement is also true. Next we suppose $d_1 > 2$ and eq. (4) is true for $0 \leq \tilde{d} < d_1$. We bound the two terms of (3) separately. Let $k_0 = \lfloor d_1/2 \rfloor + 1$.

$$I_1 = \sum_{k=k_0}^{d_1} \binom{d+k}{k}(|\Sigma| - 1)^k T(d, d_1 - k)$$

$$\leq \sum_{k=k_0}^{d_1} \binom{d+d_1}{k}(|\Sigma| - 1)^k T(d, d_1 - k)$$

$$\leq \sum_{k=k_0}^{d_1} \binom{d+d_1}{k}(|\Sigma| - 1)^k \cdot \binom{d + d_1 - k}{d_1 - k}(|\Sigma| - 1)^{d_1-k} \cdot 2^{2(d_1-k)}$$

$$= \binom{d+d_1}{d_1}(|\Sigma| - 1)^{d_1} \sum_{k=k_0}^{d_1} \binom{d_1}{k} \cdot 2^{2(d_1-k)}$$

$$\leq \binom{d+d_1}{d_1}(|\Sigma| - 1)^{d_1} \cdot 2^{d_1 - 1} \sum_{k=k_0}^{d_1} \binom{d_1}{k}$$

$$\leq \binom{d+d_1}{d_1}(|\Sigma| - 1)^{d_1} \cdot 2^{2d_1-2}. \tag{5}$$

The rest is to bound I_2 by $3\binom{d+d_1}{d_1}(|\Sigma| - 1)^{d_1} \cdot 2^{2d_1-2}$.

$$I_2 = \sum_{k=0}^{k_0-1} \binom{d+k}{k}(|\Sigma| - 1)^{k_0} T(d, d_1 - k_0)$$

$$\leq \sum_{k=0}^{k_0-1} \binom{d+k}{k}(|\Sigma| - 1)^{k_0} \cdot \binom{d + d_1 - k_0}{d_1 - k_0}(|\Sigma| - 1)^{d_1-k_0} \cdot 2^{2(d_1-k_0)}$$

$$= \binom{d + d_1 - k_0}{d_1 - k_0}(|\Sigma| - 1)^{d_1} \cdot 2^{2(d_1-k_0)} \sum_{k=0}^{k_0-1} \binom{d+k}{k}$$

$$\leq \binom{d + d_1 - k_0}{d_1 - k_0}(|\Sigma| - 1)^{d_1} \cdot 2^{2(d_1-k_0)} \binom{d + k_0}{k_0}.$$

So we only need to prove

$$\binom{d + d_1 - k_0}{d_1 - k_0}\binom{d + k_0}{k_0} 2^{-2k_0} \leq \frac{3}{4} \cdot \binom{d + d_1}{d_1},$$

or equivalently,

$$\binom{d + d_1 - k_0}{d_1 - k_0}\binom{d_1}{k_0} \leq \frac{3}{4} \cdot 2^{2k_0}\binom{d + d_1}{d_1 - k_0}. \tag{6}$$

(6) is true because

$$\binom{d + d_1 - k_0}{d_1 - k_0} \leq \binom{d + d_1}{d_1 - k_0}, \quad \binom{d_1}{k_0} \leq \frac{1}{2} \cdot 2^{d_1+1} < \frac{3}{4} \cdot 2^{2k_0}.$$

Hence (4) is correct.

At each leaf, the time complexity of line 1 is $O(nm)$. By carefully remembering the previous distances and only update the $O(d)$ positions changed, this can be done in $O(nd)$ time. The total running time is dominated by the leaves. Therefore, the time complexity of the algorithm is $O(nm + nd \cdot T(d, d_1))$.   □

Corollary 1. *StringSearch solves the closest string problem in time*

$$O\left(nm + nd \cdot 2^{4d}(|\Sigma| - 1)^d\right).$$

4   More Efficient Algorithm For Closest Substring

In [22], a $|\Sigma|^{d(\log_2 d + 2)} N^{\log_2 d + O(1)}$ algorithm is given, where N is the total length of the input strings. In this section we improve it to $O\left(n|\Sigma|^{O(d)} m^{\lceil \log_2 d \rceil + 1}\right)$. That is, the $\log_2 d$ factor at the exponent of $|\Sigma|^{d(\log_2 d + 2)}$ is removed. Moreover, the total length N is replaced by the length m of the longest input string.

Again, in order to develop an algorithm for closest substring, we attempt to solve a more generalized version of closest substring. For convenience, we call the new problem *partial knowledge closest substring*. An instance of the partial knowledge closest substring problem is given by $\langle \{s_1, s_2, \ldots, s_n\}, d, d_1, L, O, \tilde{t} \rangle$, where $O \subset [1, L]$ and $\tilde{t}$ is a string of length $|O|$. The problem is to find a string t of length L, such that $t|_O = \tilde{t}$, $d^{[1,L] \setminus O}(t, s_1) \leq d_1$, and for every i, $d(t, t_i) \leq d$ for at least one substring t_i of s_i.

Theorem 2. *Algorithm SubstringTry in Fig. 2 finds a solution for closest substring with time complexity*

$$O\left((nL + nd \cdot 2^{4d}|\Sigma|^d \cdot m^{\lceil \log_2 d \rceil + 1}\right).$$

Sketch of Proof. When all the input strings have the same length L, a careful comparison between Algorithm **SubstringSearch** and the previous Algorithm **StringSearch** can see that the two algorithms are equivalent. The only difference is made when $|s_i| > L$. Then the "guess" operation in line 4 requires the algorithm to try all possible substrings of s_{i_0}. This expands the search tree size by a factor of at most m. Because of Lemma 1, the recursion of Algorithm **SubstringSearch** is at most $\lceil \log_2 d \rceil$ levels. This increases the search tree size by a factor of $m^{\lceil \log_2 d \rceil}$. Combining with Corollary 1, the theorem is proved.   □

5   More Efficient PTAS for Closest String

In [17,18], a PTAS for closest string problem was given. To achieve approximation ratio $1 + \epsilon$, the running time of the algorithm was $O\left(mn^{O(\epsilon^{-5})}\right)$. Apparently

---

Algorithm SubstringSearch
Input: $\langle\{s_1, s_2, \ldots, s_n\}, d, L, O, \tilde{t}\rangle$ such that $|s_1| = L$.
Output: A solution t of the partial knowledge closest substring, or NULL if there is no solution.
1. Let $O' = [1..L] \setminus O$. Let s be a string such that $s|_O = \tilde{t}$ and $s|_{O'} = s_1|_{O'}$.
2. Try to find i_0 such that $d(s, t_{i_0}) > d_{i_0}$ for every substring t_{i_0} of s_{i_0}.
3. If line 1 fails, return s.
4. **Guess** a substring t_{i_0} of s_{i_0}.
5. Let $P = P(s_1, t_{i_0}) \setminus O$.
6. For every possible string t of length $|P|$ such that $d(t, s_1|_P) \leq d_1$ and $d(t, t_{i_0}|_P) \leq d - d(\tilde{t}, t_{i_0}|_O)$
6.1   Let t' be a string such that $t'|_O = \tilde{t}$ and $t'|_P = t$.
6.2   Let $e_1 = \min\{d_1 - d(t, s_1|_P), \lceil d_1/2 \rceil - 1\}$.
6.3   Use **SubstringSearch** to find solution u of $\langle\{s_1, s_2, \ldots, s_n\}, d, e_1, L, O \cup P, t'|_{O \cup P}\rangle$.
6.4   If 6.3 is successful then return u.
7. Return NULL.

Algorithm SubstringTry
Input: $\langle\{s_1, s_2, \ldots, s_n\}, d, L\rangle$
1 for every length L substring t_1 of s_1,
1.1   call **SubstringSearch** with $\langle\{t_1, s_2, \ldots, s_n\}, d, d, L, \emptyset, e\rangle$.

Fig. 2. The algorithms **SubstringSearch** and **SubstringTry**

this PTAS has only theoretical value as the degree of the polynomial grows very fast when ϵ gets small. By using the Lemma 2.2 in [22] and an idea of [17,18], Andoni et al. [1] proposed a PTAS in [17] to get much better time complexity $O(mn^{O(\epsilon^{-2} \log \frac{1}{\epsilon})})$. The proof in [1] argued that when $d = \Omega(n/\epsilon^2)$, a standard linear programming relaxation method can solve the closest string problem with good approximation ratio. When $d = O(n/\epsilon^2)$, one can exhaustively enumerate all the possibilities of positions in the solution where r of the input strings do not agree. However, by using Lemma 2.2 of [22], r can be reduced from the original $O(\frac{1}{\epsilon})$ in [18] to $O(\log \frac{1}{\epsilon})$.

With our new fixed-parameter algorithm that runs $O\left(mn + nd \cdot (16|\Sigma|^d)\right)$ time, we can further reduce the time complexity by the following algorithm: Use the fixed-parameter algorithm to solve $d = O(n/\epsilon^2)$, and use the standard linear programming relaxation to solve the case $d = \Omega(n/\epsilon^2)$. It is easy to verify that this provides a simple $O(m \cdot n^{O(\epsilon^{-2})})$ PTAS.

Theorem 3. *Closest string has a PTAS that achieves approximation ratio $1 + \epsilon$ with time $O(m \cdot n^{O(\epsilon^{-2})})$.*

6   Hardness Result

Together with the development of fixed-parameter polynomial time algorithms, W-hierarchy has been developed to prove fixed-parameter intractability [6]. The

W[1]-hardness results reviewed in Section 1 indicate that the closest substring problem unlikely has fixed-parameter polynomial time algorithms even if both d and n are fixed-parameters. More parameterized complexity results about the closest substring problem can be found in [9,22,25].

Moan and Rusu [25] studied a variant of the closest substring problem by adding a constraint on the diameter of the solution, and hoped that the constraint can help reduce the parameterized complexity of the problem. The constraint is called the bounded Hamming distance (BHD) constraint in their paper. Then the *BHD-constrained closest substring (BCCS)* problem is defined as follows.

BCCS   Given a set of n strings $s_1, s_2, \ldots, s_n$, substring length L, radius d, and diameter D. Find length-L substring t_i of each s_i, $i = 1, 2, \ldots, n$, and a new length-L string t, such that $d(t_i, t_j) \leq D$, and $d(t, t_i) \leq d$.

Clearly, $d \leq D \leq 2d$. For any $c \geq \frac{4}{3}$, Moan and Rusu proved that the diameter constraint $D \leq c \cdot d$ does not reduce the complexity of closest substring problem. More precisely, with any $c \geq \frac{4}{3}$, all parameterized complexity results for closest substring preserve for BCCS when using any non-empty subset of the following values as parameters: the radius d, the alphabet size $|\Sigma|$, the number of input strings n, the length of desired substrings L.

However, Moan and Rusu pointed out that in computational biology, D is usually significantly smaller than $2d$. Therefore, they hoped that when $\frac{D}{d}$ is very close to 1, the BCCS problem might become easier than the original closest substring problem. If this is true, BCCS can be used to solve the practical closest substring problems. The finding of the necessary condition for that BCCS problem becomes easier is left as the "main open question" of the paper [25]. In this section, we answer this question negatively with the following theorem.

Theorem 4. *For any constant $\epsilon > 0$, with the diameter constraint $D \leq (1 + \epsilon)d$, all parameterized complexity (W[l]-hardness) results for closest substring preserve for BCCS when using any non-empty subset of the following values as parameters: the radius d, the alphabet size $|\Sigma|$, the number of input strings n, the length of desired substrings L.*

Proof. The proof is done in three steps: First, we construct an instance of closest string with radius $\tilde{d}$ and diameter $\tilde{D} = (1 + o(1))\tilde{d}$. Then, we show that an instance of closest substring with radius d and diameter D can be "merged" with an instance of the closest string with radius $\tilde{d}$ and diameter $\tilde{D}$, so that the new instance has radius $d + \tilde{d}$ and diameter $D + \tilde{D}$. Thirdly, by letting $\tilde{d} \gg d$ and $\tilde{D} \gg D$, we get an instance such that the diameter is very close to the radius. Thus, we reduce the closest substring problem to BCCS, and hence prove the theorem. The detailed proof can be found in the appendix.   □

7   Discussion

The closest string and closest substring are two problems motivated and well-studied in computational biology. We proposed a novel technique that leads to more efficient fixed-parameter algorithm for closest string. This is also the first polynomial algorithm for the problem when $d = O(\log n)$. The same technique is then used to give a more efficient algorithm for closest substring. As a consequence of the fixed-parameter algorithm, we presented a more efficient PTAS of the closest string problem. At last, we showed that a restricted version of the closest substring problem has the same parameterized complexity as the original closest substring problem. This answers an open question raised in [25].

An interesting observation is that the approximation and fixed-parameter strategies work complementarily for different d values. For smaller $d < \log_2 n$ and binary strings, our fixed-parameter algorithm has time complexity $O(nm + nd \cdot 2^{4d}) = O(nm + n^5 \log_2 n)$. For larger $d > c \ln n / \epsilon^2$ for some constant c, the linear programming relaxation's time complexity is dominated by the time to solve a linear programming of m variables and nm coefficients, which is again a low-degree polynomial. This scenario can be intuitively explained as follows. When d is small and n is large, each input string puts a strong constraint on the solution, and consequently removes a large portion of the search space in a fixed-parameter algorithm. Therefore, it is easier to design a fixed-parameter algorithm. Conversely, when d is large and n is small, the constraint superimposed by each input string is weaker and there are fewer constraints. Therefore, it is easier to find an approximate solution to roughly satisfy those constraints.

But when d falls in between $\log_2 n$ and $c \ln n / \epsilon^2$, the polynomial will have high degree for the fixed-parameter algorithm, and the approximation ratio of the linear programming relaxation may exceed $1 + \epsilon$. The instances with d in this range seem to be the "hardest" instances of the closest string problem. However, because the fixed-parameter algorithm has polynomial (although with high degree) running time on these instances, a proof for the "hardness" of these instances seem to be difficult too. We leave the finding of more efficient (approximation) algorithm for $\log_2 n < d < c \ln n / \epsilon^2$ as an open problem.

Acknowledgements

Bin Ma's work was partially done when he visited Professor Andrew Yao at ITCS at Tsinghua University, and was supported in part by China NSF 60553001, National Basic Research Program of China 2007CB807900,2007CB807901, NSERC and Canada Research Chair. Xiaoming Sun's was supported in part by National Natural Science Foundation of China Grant 60553001, 60603005, 60621062, and the National Basic Research Program of China Grant 2007CB807900, 2007CB807901.

References

1. Andoni, A., Indyk, P., Patrascu, M.: On the optimality of the dimensionality reduction method. In: Proceedings of the 47th Annual IEEE Symposium on Foundations of Computer Science, pp. 449–458 (2006)
2. Ben-Dor, A., Lancia, G., Perone, J., Ravi, R.: Banishing bias from consensus sequences. In: Hein, J., Apostolico, A. (eds.) CPM 1997. LNCS, vol. 1264, pp. 247–261. Springer, Heidelberg (1997)

3. Davila, J., Balla, S., Rajasekaran, S.: Space and time efficient algorithms for planted motif search. In: International Conference on Computational Science (2), pp. 822–829 (2006)
4. Deng, X., Li, G., Li, Z., Ma, B., Wang, L.: Genetic design of drugs without side-effects. SIAM Journal on Computing 32(4), 1073–1090 (2003)
5. Dopazo, J., Rodríguez, A., Sáiz, J.C., Sobrino, F.: Design of primers for PCR amplification of highly variable genomes. CABIOS 9, 123–125 (1993)
6. Downey, R.G., Fellows, M.R.: Parameterized complexity. In: Monographs in Computer Science, Springer, New York (1999)
7. Evans, P.A., Smith, A.D.: Complexity of approximating closest substring problems. In: Lingas, A., Nilsson, B.J. (eds.) FCT 2003. LNCS, vol. 2751, pp. 210–221. Springer, Heidelberg (2003)
8. Evans, P.A., Smith, A.D., Wareham, H.T.: On the complexity of finding common approximate substrings. Theoretical Computer Science 306(1-3), 407–430 (2003)
9. Fellows, M.R., Gramm, J., Niedermeier, R.: On the parameterized intractability of motif search problems. Combinatorica 26(2), 141–167 (2006)
10. Frances, M., Litman, A.: On covering problems of codes. Theoretical Computer Science 30, 113–119 (1997)
11. Gramm, J., Guo, J., Niedermeier, R.: On exact and approximation algorithms for distinguishing substring selection. In: Lingas, A., Nilsson, B.J. (eds.) FCT 2003. LNCS, vol. 2751, pp. 159–209. Springer, Heidelberg (2003)
12. Gramm, J., Hüffner, F., Niedermeier, R.: Closest strings, primer design, and motif search. In: Florea, L., et al. (eds.) Currents in Computational Molecular Biology, poster abstracts of RECOMB 2002, pp. 74–75 (2002)
13. Gramm, J., Niedermeier, R., Rossmanith, P.: Fixed-parameter algorithms for closest string and related problems. Algorithmica 37, 25–42 (2003)
14. Hochbaum, D.S. (ed.): Approximation Algorithms for NP-Hard Problems. PWS Publishing Company, Boston (1996)
15. Jiao, Y., Xu, J., Li, M.: On the k-closest substring and k-consensus pattern problems. In: Sahinalp, S.C., Muthukrishnan, S.M., Dogrusoz, U. (eds.) CPM 2004. LNCS, vol. 3109, pp. 130–144. Springer, Heidelberg (2004)
16. Lanctot, K., Li, M., Ma, B., Wang, S., Zhang, L.: Distinguishing string search problems. In: Proceedings of the 10th Annual ACM-SIAM Symposium on Discrete Algorithms (SODA), pp. 633–642 (1999)
17. Li, M., Ma, B., Wang, L.: Finding similar regions in many strings. In: Proceedings of the 31st ACM Symposium on Theory of Computing, pp. 473–482 (1999)
18. Li, M., Ma, B., Wang, L.: On the closest string and substring problems. Journal of the ACM 49(2), 157–171 (2002)
19. Liu, X., He, H., Sýkora, O.: Parallel genetic algorithm and parallel simulated annealing algorithm for the closest string problem. In: Li, X., Wang, S., Dong, Z.Y. (eds.) ADMA 2005. LNCS (LNAI), vol. 3584, pp. 591–597. Springer, Heidelberg (2005)
20. Lucas, K., Busch, M., Mössinger, S., Thompson, J.A.: An improved microcomputer program for finding gene- or gene family-specific oligonucleotides suitable as primers for polymerase chain reactions or as probes. CABIOS 7, 525–529 (1991)
21. Ma, B.: A polynomial time approximation scheme for the closest substring problem. In: Proceedings of the 11th Symposium on Combinatorial Pattern Matching, pp. 99–107 (2000)
22. Marx, D.: The closest substring problem with small distances. In: Proceedings of the 46th Annual IEEE Symposium on Foundations of Computer Science (FOCS), pp. 63–72 (2005)

23. Mauch, H., Melzer, M.J., Hu, J.S.: Genetic algorithm approach for the closest string problem. In: Proceedings of the 2nd IEEE Computer Society Bioinformatics Conference (CSB), pp. 560–561 (2003)
24. Meneses, C.N., Lu, Z., Oliveira, C.A.S., Pardalos, P.M.: Optimal solutions for the closest-string problem via integer programming. INFORMS Journal on Computing (2004)
25. Moan, C., Rusu, I.: Hard problems in similarity searching. Discrete Applied Mathematics 144, 213–227 (2004)
26. Nicolas, F., Rivals, E.: Complexities of the centre and median string problems. In: Proceedings of the 14th Annual Symposium on Combinatorial Pattern Matching, pp. 315–327 (2003)
27. Proutski, V., Holme, E.C.: Primer master: A new program for the design and analysis of PCR primers. CABIOS 12, 253–255 (1996)
28. Raghavan, P.: Probabilistic construction of deterministic algorithms: Approximating packing integer program. Journal of Computer and System Sciences 37, 130–143 (1988)
29. Stojanovic, N., Berman, P., Gumucio, D., Hardison, R., Miller, W.: A linear-time algorithm for the 1-mismatch problem. In: Proceedings of the 5th International Workshop on Algorithms and Data Structures, pp. 126–135 (1997)
30. Wang, L., Dong, L.: Randomized algorithms for motif detection. Journal of Bioinformatics and Computational Biology 3(5), 1039–1052 (2005)
31. Wang, Y., Chen, W., Li, X., Cheng, B.: Degenerated primer design to amplify the heavy chain variable region from immunoglobulin cDNA. BMC Bioinformatics 7(suppl. 4), S9 (2006)

Appendix - Proof of Theorem 4

STEP I

First let us construct an instance $\mathcal{I}_1$ of the closest string problem with very close radius and diameter. Let k be an even number. Examine the instance with k binary strings $x_1, x_2, \ldots, x_k$. Each x_i has length $\tilde{L} = \binom{k}{k/2}$. For each column j, exactly half of $x_1[j], x_2[j], \ldots, x_k[j]$ take value 0 and the other half take value 1. Hence there are in total $\binom{k}{k/2}$ ways to assign values to a column. Each of the $\binom{k}{k/2}$ columns takes a distinct way.

Claim 1. The radius of the constructed instance is $\tilde{d} = \tilde{L}/2$.

Proof. Because of the construction, each string has half of the $\tilde{L}$ letters as 0. Therefore, $d(0^{\tilde{L}}, x_i) = \tilde{L}/2$ for every x_i. Therefore, the radius is at most $\tilde{L}/2$.

On the other hand, for any center string x, at each column, the total number of differences between x_i ($i = 1, 2, \ldots, k$) and the center string is exactly $k/2$. Therefore, $\sum_{i=1}^{k} d(x, x_i) = k\tilde{L}/2$. Consequently, $\max_{i=1}^{k} d(x, x_i) \geq \tilde{L}/2$. The claim is proved. $\square$

Now let us examine the diameter of the constructed instance. For every two strings x_i and x_j, the Hamming distance is the number of columns such that x_i and x_j take different values. This is equivalent to the number of ways to split k elements into two equal-sized subsets, ensuring that elements i and j are separated. With simple combinatorics, this number is $2\binom{k-2}{\frac{k}{2}-1}$. Therefore,

$$\frac{\tilde{D}}{\tilde{d}} = \frac{2\binom{k-2}{\frac{k}{2}-1}}{\tilde{L}/2} = \frac{4\binom{k-2}{\frac{k}{2}-1}}{\binom{k}{k/2}} = \frac{k}{k-1}$$

In order to avoid the exponential growth of $\tilde{D}$ and $\tilde{d}$ with respect to k, we note that $\tilde{D}$ and $\tilde{d}$ can be enlarged while keeping the same ratio $\frac{\tilde{D}}{\tilde{d}}$ by replacing each x_i by $\underbrace{x_i x_i \ldots x_i}_{K}$, i.e., the concatenation of K copies of x_i. In the rest of the proof we consider $\mathcal{I}_1$ as such an enlarged instance, and the value K is to be determined later. The notations diameter $\tilde{D}$, radius $\tilde{d}$, input string x_i, and string length $\tilde{L}$ all correspond to the enlarged instance.

STEP II

Let $\mathcal{I} = \langle \{s_1, \ldots, s_n\}, L, d \rangle$ be an instance of the closest substring. We construct a new instance in the following.

Let X be an i.i.d. random binary string with length $7(L + \tilde{L})$. Then by using Chernoff's bound, it is easy to verify that with positive probability,

$$d(X[1..j], X[|X|-j+1, |X|]) + d(X[j+1..|X|], X[1..|X|-j]) > 3(L+\tilde{L}) \text{ for } j \neq 0 \text{ and } |X|. \tag{7}$$

By using the standard derandomization techniques such as conditional probability [28], we can easily design a polynomial time deterministic procedure to construct the binary string X satisfying (7). Here we omit the detail.

For each s_i $(i = 1, \ldots, n)$ and each x_j $(j = 1, \ldots, k)$, let

$$s_{ij} = X \; s_i[1..L]x_j \; XX \; s_i[2..L+1]x_j \; XX \; \ldots \; XX \; s_i[m-L+1..m]x_j \; X$$

The new instance is then

$$\mathcal{I}_2 = \langle \{s_{ij} | i = 1, \ldots, n, j = 1, \ldots, k\}, 15(L + \tilde{L}), d + \tilde{d} \rangle.$$

Claim 2. $\mathcal{I}$ has a solution with radius $\leq d$ and diameter $\leq D$ if and only if $\mathcal{I}_2$ has a solution with radius $\leq d + \tilde{d}$ and diameter $\leq D + \tilde{D}$.

Proof. Suppose $\mathcal{I}$ has a solution $s_i[l_i..l_i + L - 1]$, $i = 1, \ldots, n$ with radius d and diameter D. Then the substrings $X s_i[l_i..l_i+L-1]x_j X$, $i = 1, \ldots, n, j = 1, \ldots, k$ are a solution of $\mathcal{I}_2$. It is easy to verify that the radius and diameter are bounded by $d + \tilde{d}$ and $D + \tilde{D}$, respectively.

Now we prove the other direction. We first show that the solution of $\mathcal{I}_2$ is such that X from different strings are aligned exactly together. Otherwise, because of 7, the inexact overlaps between X from two input strings will cause at least $3(L + \tilde{L})$ minus two times of the length of $s_i[k..k + L]x_j$. This gives a diameter at least $L + \tilde{L} > D + \tilde{D}$, contradicting the condition.

Further, without making the solution worse, we can easily modify the solution by "sliding" so that every substring has the form $X s_i[l_i..l_i + L - 1]x_j X$ for some l_i. Next we show that $s_i[l_i..l_i + L - 1]$ $(i = 1, 2, \ldots, n)$ is the desired solution for $\mathcal{I}$.

Let $Xs\tilde{s}X$ be the center of $Xs_i[l_i..l_i + L - 1]x_jX$ with radius $d + \tilde{d}$. Because $\tilde{d}$ is the radius of $\mathcal{I}_1$, there is j_0 such that $d(\tilde{s}, x_{j_0}) = \tilde{d}$. Therefore, $d(Xs\tilde{s}X, Xs_i[l_i..l_i + L - 1]x_{j_0}X) \le d + \tilde{d}$ derives that $d(s, s_i[l_i..l_i + L - 1]) \le d$ for every i.

Similarly, there are j_0 and j_1 such that $d(x_{j_0}, x_{j_1}) = \tilde{D}$. Therefore, $d(Xs_i[l_i..l_i + L - 1]x_{j_0}X, Xs_{i'}[l_{i'}..l_{i'} + L - 1]x_{j_0}X) \le D + \tilde{D}$ derives that $d(s_i[l_i..l_i + L - 1], s_{i'}[l_{i'}..l_{i'} + L - 1]) \le D$ for every i and i'.

The claim is proved. $\square$

STEP III

For any $\epsilon > 0$, we let $k = \lceil \frac{2}{\epsilon} + 1 \rceil$ and $K = \left\lceil \frac{4D}{\binom{k}{k/2}\epsilon} \right\rceil$ in the construction of $\mathcal{I}_1$. Then $\tilde{d} = \frac{K}{2} \cdot \binom{k}{k/2} \ge \frac{2D}{\epsilon}$. Then in instance $\mathcal{I}_2$, the ratio between the diameter and radius is

$$\frac{D + \tilde{D}}{d + \tilde{d}} \le \frac{D}{\tilde{d}} + \frac{\tilde{D}}{\tilde{d}} \le \frac{\epsilon}{2} + \frac{k}{k-1} \le 1 + \epsilon$$

Thus, we successfully reduce the closest substring problem to the closest substring problem with the constraint that the diameter is within $1 + \epsilon$ times the radius. The number n, length m of the input strings are increased only by a constant factor determined by ϵ. The new length L and radius d of the substrings are polynomials of the old L and d. Therefore, all the W-complexities of the closest substring problem still hold with the diameter constraint $D \le (1 + \epsilon)d$ for any small $\epsilon > 0$. The theorem is proved. $\square$

Disruption of a Transcriptional Regulatory Pathway Contributes to Phenotypes in Carriers of Ataxia Telangiectasia

Denis Smirnov and Vivian G. Cheung

University of Pennsylvania, Philadelphia
vcheung@mail.med.upenn.edu

Abstract. Ataxia Telangiectasia, AT, is a recessive disorder caused by mutations in the ATM gene. Although it is a recessive disorder, population-based studies have shown that carriers of AT have increased risks of breast cancer and other diseases compared to non-carriers. The goal of this study is to characterize phenotypes in AT carriers. Since expression level of genes is a major determinant of cellular phenotypes, we studied gene expression in AT carriers and identified regulatory mechanisms that influence these expression phenotypes.

We found gene expression phenotypes that showed a recessive pattern, where AT carriers are similar to non-carriers but differ from AT patients. However, there are also expression phenotypes that showed a dominant pattern where AT carriers are similar to AT patients but differ from non-carriers. One of the dominant gene expression phenotypes is that of TNFSF4. We showed that ATM regulates TNFSF4 expression through a transcriptional regulatory pathway that includes transcription factors and miRNAs. In AT carriers and AT patients, this pathway is disrupted, resulting in higher expression of TNFSF4. In this presentation, I will describe this ATM-mediated pathway, and show that the disruption of this pathway leads to increased risk of breast cancer and cardiac death in AT carriers. The integration of molecular and computational analyses of gene and microRNA expression revealed the complex consequences of a human gene mutation.

About the keynote speaker. Vivian Cheung is an investigator of the Howard Hughes Medical Institute and an Associate Professor of Pediatrics and Genetics at the University of Pennsylvania. She received her bachelor degree from the University of California, Los Angeles, and her MD from Tufts University. She received clinical training in Pediatrics and Neurology at the University of California, Los Angeles and The Childrens Hospital of Philadelphia. Her research focuses on identifying the genetic determinants of human traits and developing tools that facilitate such studies. Her lab has contributed to understanding the genetic basis of variation in human gene expression and characterizing gene expression phenotypes in carriers of autosomal recessive diseases, such as ataxia telangiectasia. Her group is also studying the genetic basis of variation in the frequencies and locations of human meiotic recombination.

M. Vingron and L. Wong (Eds.): RECOMB 2008, LNBI 4955, p. 410, 2008.
© Springer-Verlag Berlin Heidelberg 2008

Accounting for Non-genetic Factors Improves the Power of eQTL Studies

Oliver Stegle[1], Anitha Kannan[2], Richard Durbin[3], and John Winn[2]

[1] University of Cambridge, UK
os252@cam.ac.uk
[2] Microsoft Research, Cambridge, UK
{ankannan,jwinn}@microsoft.com
[3] Wellcome Trust Sanger Institute, Hinxton, Cambridge, UK
rd@sanger.ac.uk

Abstract. The recent availability of large scale data sets profiling single nucleotide polymorphisms (SNPs) and gene expression across different human populations, has directed much attention towards discovering patterns of genetic variation and their association with gene regulation. The influence of environmental, developmental and other factors on gene expression can obscure such associations. We present a model that explicitly accounts for non-genetic factors so as to improve significantly the power of an expression Quantitative Trait Loci (eQTL) study. Our method also exploits the inherent block structure of haplotype data to further enhance its sensitivity. On data from the HapMap project, we find more than three times as many significant associations than a standard eQTL method.

1 Introduction

Discovering patterns of genetic variation that influence gene regulation has the potential to impact a broad range of biological endeavours, such as improving our understanding of genetic diseases. Recent advances in microarray and genotyping methods have made it feasible to investigate complex multi-gene associations on a genome-wide level, through expression Quantitative Trait Loci (eQTL) studies (see [1] and references therein). The vast number of potential associations and relatively small numbers of individuals in current data sets makes it challenging to discover statistically significant associations between genome and transcript. Methods for improving the sensitivity and accuracy of such studies are therefore of considerable interest.

In this paper, we describe a method to improve substantially the number of significant associations found in an eQTL study. The main insight is that much of the variation in gene expression is due to non-genetic factors, such as differing environmental conditions or developmental stages [2]. By explicitly accounting for non-genetic variation, we can greatly improve the statistical power of eQTL methods as most of the non-genetic variation is removed and real associations stand out to a greater extent.

M. Vingron and L. Wong (Eds.): RECOMB 2008, LNBI 4955, pp. 411–422, 2008.
© Springer-Verlag Berlin Heidelberg 2008

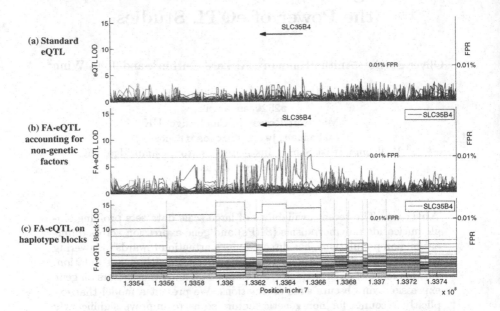

Fig. 1. Example results of **(a)** standard eQTL, **(b)** our proposed method FA-eQTL which accounts for non-genetic factors and **(c)** the same method applied to haplotype blocks. In this region, standard eQTL does not find any significant associations but our proposed methods finds a *cis* association for the gene *SLC35B4*. The significance of the association is improved when haplotype blocks are considered instead of individual SNPs.

Following [3], we also improve the accuracy of eQTL by exploiting the inherent block structure present in haplotype data. By jointly considering all SNPs in a haplotype block, it is possible to detect weaker associations than can be found using single SNPs. For example, if the relevant SNP lies between the measured marker SNPs, a haplotype block model can effectively perform imputation of this missing SNP value leading to a stronger detected association.

The contributions of this paper are best illustrated by the plots of Fig. 1 showing the results of different eQTL methods over the same region of chromosome 7. The top plot demonstrates that no associations have been found using a standard eQTL method, whilst the second plot shows a significant *cis* association which only becomes visible when non-genetic factors are accounted for. The bottom plot shows that, when haplotype blocks are used instead of individual SNPs, the significance of this association is further increased to well above the 0.01% False Positive Rate (FPR) level.

The structure of this paper is as follows. In Section 2, we compare several models of how non-genetic factors influence gene expression. The best of these models is incorporated into an eQTL method in Section 3 and their power demonstrated on data from the HapMap project [4]. Section 4 describes how this eQTL approach can be extended to exploit the block structure of haplotype data. Section 5 concludes with a discussion.

2   Modelling Non-genetic Factors

In addition to variation due to genomic differences, human gene expression levels vary because of differing developmental stages, environmental influences and other physiological and biological factors. In principle, when collecting gene expression data sets for eQTL, non-genetic factors should be controlled to be constant across all samples, but in practice this can only be achieved to a limited degree. Indeed, it is reasonable to expect that a substantial amount of the variation in gene expression is still due to non-genetic factors. Hence, eQTL studies face the challenge of distinguishing the expression variation due to genetic causes from the variation due to all non-genetic ones. Previous eQTL methods have addressed this issue by modelling non-genetic variation as independent noise [1], neglecting the fact that non-genetic causes can have widespread influence on large sets of genes, jointly promoting or inhibiting their expression. An alternative approach used by [2] is to ignore those genes whose measured expression level may be due to environmental factors (a heuristic score is used to represent the heritability component of a gene probe). However, this approach faces problems when non-genetic factors affect many of the gene expression levels since it would lead to discarding most of the data. We instead choose to model the non-genetic factors so as to account for their influence.

One of the difficulties in modelling non-genetic expression variation is that human gene expression data sets for eQTL currently include little or no information about the environmental, physiological or developmental factors that may have affected the expression measurements. Lacking this information, we treat non-genetic factors as unobserved latent variables and aim to estimate their influence on the gene expression values. Previously, linear Gaussian models [5] such as principal components analysis [6] have been used to describe the expression levels of genes as linear functions of hidden variables. Such models have been used to represent causes of variation such as cellular function [7], regulation of gene expression [8], co-expression of genes [9] or environmental conditions [10]. We use such a model to capture non-genetic variation so that it can be explained away, thereby significantly improve the power of our eQTL study.

Our model assumes the existence of K non-genetic factors $\mathbf{x} = \{x_1 \ldots x_K\}$ which linearly influence the observed gene expression levels $\mathbf{y} = \{y_1 \ldots y_G\}$ through a weight matrix $\mathbf{W}$:

$$\mathbf{y} = \mathbf{W}\mathbf{x} + \mathbf{v} \tag{1}$$

where $\mathbf{v}$ represents Gaussian-distributed observation noise. We considered three Bayesian variants of this model,

- **Principal Components Analysis (PCA)** where the prior on $\mathbf{x}$ is Gaussian and the prior noise variance is the same for each gene probe (each element of $\mathbf{v}$),
- **Factor Analysis (FA)** where the prior on $\mathbf{x}$ is also Gaussian but a separate prior noise variance is learned for each gene probe,
- **Independent Components Analysis (ICA)** is like PCA except that the prior on each component x_k is a mixture of two Gaussian distributions.

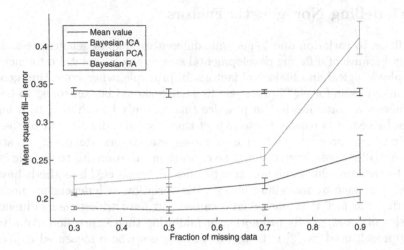

Fig. 2. Comparative performance of various linear Gaussian models for filling-in missing gene expression values for X chromosome genes. The plot shows the mean squared error in the fill-in predictions against the fraction of missing data ρ, averaged over four runs with different training/test splits. Error bars show one standard deviation. The factor analysis model gives the lowest fill-in error over a range of missing data rates.

For each method, we use an Automatic Relevance Determination (ARD) prior on the variance of each column of $\mathbf{W}$, so that the number of latent non-genetic factors is learned automatically [11].

2.1   Investigation on HapMap Expression Data

We investigated these three models on gene expression measurements of individuals from the HapMap project, consisting of the expression profiles for 47,294 gene probes profiled in EBV-transformed lymphoblastoid cell lines [12]. The parameters of each model were learned from the expression levels of 512 X chromosome gene probes from a randomly-selected 75% of the HapMap individuals, with the maximum number of non-genetic factors set to 40 (during learning several of these factors were switched off by ARD). Bayesian learning was achieved with a fully-factorised variational approximation using the VIBES software package [13]. For the 25% of individuals not used for training, we removed a fraction ρ of expression measurements and applied each learned model to fill-in these missing values. The idea behind this experiment is that models which better capture the latent causes of the observed gene expression levels, will better predict missing expression levels from partial observations. The accuracy of the fill-in predictions for each model was assessed in terms of mean squared error. The results are shown in Figure 2, along with a baseline prediction given by the mean expression across the training individuals. These results show that the factor analysis model gives the best fill-in performance, even when large fractions of the data are missing. Hence, we use factor analysis to model non-genetic effects.

3   Accounting for Non-genetic Factors in eQTL

Standard eQTL methods assess how well a particular gene expression level is modelled when genetic factors are taken into account, compared to how well it is modelled by a background model that ignores genetic factors [14]. The relevant quantity is the log-odds (LOD) score,

$$\log_{10}\left\{\prod_j \frac{P(y_{gj}\,|\,\mathbf{s}_j, \theta_g)}{P(y_{gj}\,|\,\theta_{\text{bck}})}\right\} \tag{2}$$

where $\mathbf{s}_j$ is a SNP measurement and y_{gj} the gene expression level of probe g, for the jth individual. The terms θ_g, θ_{bck} are parameters for probe g of the genetic and background models respectively. The LOD scores can then be plotted against the location of the SNP over a large genomic region to give an eQTL scan for each gene expression g.

To account for non-genetic factors, we modify this approach to use the factor analysis model of the previous section, denoting the new method FA-eQTL. In FA-eQTL, the LOD score compares a *full model* of both genetic and non-genetic factors to a *background model* which only includes non-genetic factors,

$$\log_{10}\left\{\prod_j \frac{P(y_{gj}\,|\,\mathbf{s}_j, \theta_g, \mathbf{w}_g, \mathbf{x}_j)}{P(y_{gj}\,|\,\mathbf{w}_g, \mathbf{x}_j)}\right\} \tag{3}$$

where $\mathbf{x}_j$ are the latent non-genetic causes for the jth individual and $\mathbf{w}_g$ is the gth row of the weight matrix $\mathbf{W}$ described in the previous section. For the full model, which incorporates both genetic and non-genetic factors, we model the expression value of gene g for the jth individual by

$$P(y_{gj}\,|\,\mathbf{s}_j, \theta_g, \mathbf{w}_g, \mathbf{x}_j) = \mathcal{N}\Big(\underbrace{\mathbf{s}_j.\theta_g}_{\text{genetic}} + \underbrace{\mathbf{w}_g\mathbf{x}_j}_{\text{non-genetic}}, \psi_g\Big), \tag{4}$$

where $\mathcal{N}(m, \tau)$ represents a Gaussian distribution with mean m and variance τ. The variable $\mathbf{s}_j$ encodes the state of a particular SNP whose relevance we want to assess, and θ_g captures the change in gene expression caused by this SNP. The SNP state $\mathbf{s}_j$ is the sum of two indicator vectors encoding the two alleles measured for this SNP. Each indicator vector has a one at the location corresponding to the measured allele and zeroes elsewhere. The noise is Gaussian with learned variance ψ_g. The full model is shown graphically in the Bayesian network of Figure 3.

For the background model, we use exactly the factor analysis model of the previous section. Hence, $P(y_{gj}\,|\,\mathbf{w}_g, \mathbf{x}_j)$ also has the Gaussian form of Eqn. 4, but where the mean consists of only the non-genetic term. For completeness, we also tested ICA and PCA as alternatives to FA but found that these led to weaker associations, showing that the results of the fill-in experiments of Section 2 do indeed indicate how well each model accounts for non-genetic effects.

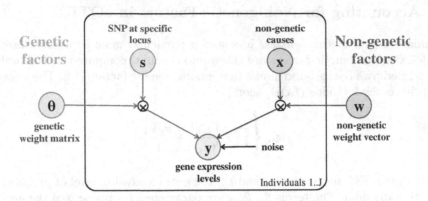

Fig. 3. The Bayesian network for the full model that includes both genetic (green) and non-genetic factors (red) when explaining gene expression levels. The rectangle indicates that contained variables are duplicated for each individual. See the text for a detailed explanation of this model.

When there is extra information about each individual, it can also be incorporated into the full and background models. For example, the HapMap individuals are divided into three distinct populations with African ancestry (YRI), European ancestry (CEU) and Asian ancestry (CHB,JPT). Certain SNP and probe measurements have differing statistics in each population, which can lead to false associations. To avoid such false associations, we introduce an additional three-valued 'virtual' SNP measurement encoding the population each individual belongs to, and extend the sum in the mean of Eqn. 4 to include a linear relation to this measurement. For similar reasons, we also include a binary measurement encoding each individual's gender. If desired, we can investigate the association of a probe to multiple SNPs jointly by extending Eqn. 4 to a sum over all SNPs in a region (or in multiple disjoint regions) as described in [3].

3.1 FA-eQTL on HapMap Data

We applied both the standard eQTL method and the FA-eQTL method to the HapMap Phase II genotype data [4] and corresponding gene expression measurements [12]. Both methods were applied to chromosomes 2, 7, 11 and X. For each chromosome, only probes for genes within that chromosome were included, so that only within-chromosome associations were tested.

An issue with using the FA-eQTL model for chromosome-wide scans is the very high computational cost of re-learning the factor analysis model at each locus. To avoid this, we learned each $\mathbf{w}_g$ and $\mathbf{x}_j$ once for the background model and kept them fixed when learning the full model. This approximation is accurate only if the genetic and non-genetic models are nearly orthogonal. To test this assumption, we estimated the contribution to the gene expression levels due to the non-genetic factors alone, given by $\mathbf{Wx}$, and treated it as expression data

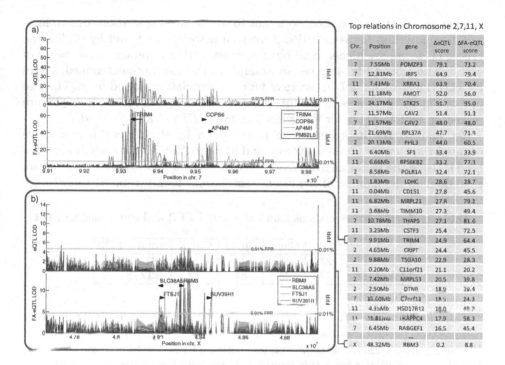

Top relations in Chromosome 2,7,11, X

Chr.	Position	gene	ΔeQTL score	ΔFA-eQTL score
7	7.55Mb	POMZP3	79.1	73.2
7	12.81Mb	IRF5	64.9	79.4
11	7.41Mb	XRRA1	63.9	70.4
X	11.18Mb	AMOT	52.0	56.0
2	24.17Mb	STK25	51.7	95.0
7	11.57Mb	CAV2	51.4	51.3
7	11.57Mb	CAV2	48.0	48.0
2	21.69Mb	RPL37A	47.7	71.9
2	20.13Mb	PHL3	44.0	60.5
11	6.40Mb	SF1	33.4	33.9
11	6.66Mb	RPS6KB2	33.2	77.3
2	8.58Mb	POLR1A	32.4	72.1
11	1.83Mb	LDHC	28.6	28.7
11	0.04Mb	CD1S1	27.8	45.6
11	6.82Mb	MRPL21	27.8	79.2
11	3.68Mb	TIMM10	27.3	49.4
7	10.78Mb	THAP5	27.1	81.6
11	3.29Mb	CSTF3	25.4	72.5
7	9.91Mb	TRIM4	24.9	64.4
2	4.65Mb	CRIPT	24.4	45.5
2	9.88Mb	TSGA10	22.9	28.3
11	0.20Mb	C11orf21	21.1	20.2
2	7.42Mb	MRPL53	20.5	39.8
2	2.50Mb	DTNB	18.9	19.4
7	16.00Mb	C7orf13	18.5	24.3
11	4.35Mb	HSD17B12	18.0	40.7
11	11.81Mb	RAPPC4	17.9	58.3
7	6.45Mb	RABGEF1	16.5	45.4
...				
X	48.32Mb	RBM3	0.2	8.8

Fig. 4. Examples of the improved association significance for FA-eQTL over standard eQTL. The plots a) and b) give two example regions where FA-eQTL increases the significance of *cis* associations found by standard eQTL and also finds additional *cis* associations. Horizontal lines indicate the threshold value corresponding to 0.01% FPR for relevant genes. Arrows indicate gene coding regions. The table shows associations ranked how much the eQTL score is above the FPR threshold (ΔeQTL) along with the corresponding score above threshold for FA-eQTL (ΔFA-eQTL). For the associations found by both methods, the FA-eQTL score is on average 21.8 higher than the eQTL score, demonstrating the advantage of accounting for non-genetic factors.

in standard eQTL. If the genetic and non-genetic models are nearly orthogonal, then we would expect that no significant association would be found between any SNP and these reconstructed expression levels. This was indeed the case, for example, the highest LOD score over the entirety of chromosome 2 was just 11.4 which is not statistically significant. Also for computational reasons, we apply maximum likelihood methods to estimate the parameters θ, rather than the variational approach which is Bayesian but much more expensive. Because maximum likelihood methods perform poorly with little data, we remove SNPs where two or fewer individuals have the minor allele.

Fig. 4 shows the results of FA-eQTL and standard eQTL applied to the HapMap data. The table lists associations ranked by the difference between the eQTL score and the 0.01% FPR threshold. For comparison the corresponding FA-eQTL score difference (computed from the highest score within 50 loci of the eQTL peak) is also listed. We consider all associations found in any 100kbp window as a single association, so as not to over count associations due to linkage disequilibrium between SNPs. Across all associations found by eQTL, the FA-eQTL scores are higher in all but five cases, with an average score change of +21.8. The plot of Fig. 4a gives an example of the improvement gained, where FA-eQTL increases the LOD score of four *cis* associations found by eQTL. The plot of Fig. 4b illustrates that some weaker *cis* associations missed by eQTL are picked up by FA-eQTL, for the genes *SLC38A5*, *FTSJ1* and *SUV39H1*.

To quantify the improvement in power given by the FA-eQTL model, we counted the number of associations found at a 0.01% FPR for each model (Table 1). Using the factor analysis model to explain away non-genetic effects more than doubles the number of significant associations found (from 81 to 222).

Table 1. Number of associations found at a 0.01% FPR and corresponding FDR

	Chr 2	Chr 7	Chr11	Chr X	Total	FDR
eQTL	24	13	39	5	**81**	2%
FA-eQTL	82	44	84	12	**222**	2%

False Positive Rates (FPRs) were estimated empirically for each chromosome using 30,000 permutations across randomly selected regions of length 500 SNPs. For each gene the threshold score was set to give a FPR of 0.01%. False Discovery Rates were calculated from this fixed FPR, the number of conducted tests and the number of associations found for a specific gene. Since almost all of the associations we find are *cis* each gene generally has either exactly one or no association, leading to the False Discovery Rate listed above for all *cis* associated genes. The reduction in irrelevant variation for FA-eQTL meant that a particular FPR was normally achieved at a lower LOD score than for standard eQTL and hence a higher number of the genes exceeded the significance threshold.

The majority of the discovered associations were *cis* associations, typically SNPs within 100kb of the interrogated probe. A small number of potential *trans* associations (ten in total) were found with SNPs further than 5MB from the expression probe. However, these were all weak associations with LOD scores close to the estimated 0.01% FPR threshold and so are most likely false associations.

4　Haplotype Block eQTL

The performance of our method can be further improved by relating expression values to haplotype blocks rather than to individual SNPs. A haplotype block, being a genomic region which has been inherited in its entirety from an ancestral genome, can be used as an intermediary for the values of missing SNP measurements. In addition, haplotype blocks are more correlated with population structure than individual SNPs. Hence, we would expect stronger evidence for association with blocks than SNPs if either:

- the true association is with an unmeasured 'missing' SNP,
- the association is only present in a particular sub-population of individuals,

— there is an epistatic interaction within the haplotype block where the ancestral genome captures the interaction better than any individual SNP.

In the case where the true association is with a measured marker SNP, the use of haplotype blocks could potentially reduce the evidence for the association. However, this reduction would typically be small since a SNP and its haplotype ancestor label are normally very strongly correlated.

We applied a method for discovering haplotype blocks that explains each individual's haplotype using pieces of a small number of learned 'ancestor' haplotypes, where the pieces break at block boundaries. The particular model we used is the recently proposed piSNP model [15] which accounts for population structure to give increased accuracy when learning the ancestral haplotypes. The learning process discovers the haplotype block boundaries and indicates which ancestral haplotype is used for each block. We define Block FA-eQTL to be the FA-eQTL method previously described but where the measurements s_j are over ancestors rather than alleles and are measured for each block rather than each locus. Also, as there are fewer blocks than SNPs, it is computationally affordable to enhance the noise model for Block FA-eQTL. To do this, we model v as a mixture of a Gaussian distribution and a uniform distribution. This heavy tailed noise model is more robust to outliers and considerably reduces the false positive rate, at the cost of around a fifty-fold increase in computation time.

4.1   Block FA-eQTL on HapMap Data

Due to the computational expense of Block FA-eQTL, it was only applied to 500 SNP regions within each chromosome, with the number of ancestors set to ten. A region was analysed if it was in the top 100 regions ranked by a soft-max criterion, $S = \frac{1}{N^{1/p}} \left(\sum_{n=1}^{N} (\text{LOD}_n)^p \right)^{1/p}$ with $p = 3$. This criterion identifies both regions with high single-locus peaks and regions where the LOD score is high over an extended area. The selected regions contained all the associations found with FA-eQTL in the previous section. Figure 5 illustrates the benefits of learning associations using haplotype blocks. The plot of Fig. 5a shows the LOD scores for FA-eQTL and Block FA-eQTL for a region containing a strong association. The form of this association is shown in Fig. 5b where the expression level of the *AMOT* gene plotted against the SNP allele and haplotype block at the locations marked with red arrows. The block model is able to pull out population-specific associations which are not apparent in the single SNP plot.

This difference is shown more clearly in the second example of Fig. 5 c,d where blocking causes a *cis* association to move from just above to well above the 0.01%

Table 2. Number of associations found at a 0.01% FPR and corresponding FDR

	Chr 2	Chr 7	Chr11	Chr X	Total	FDR
FA-eQTL	82	44	84	12	**222**	2%
Block FA-eQTL	117	57	86	14	**274**	2%

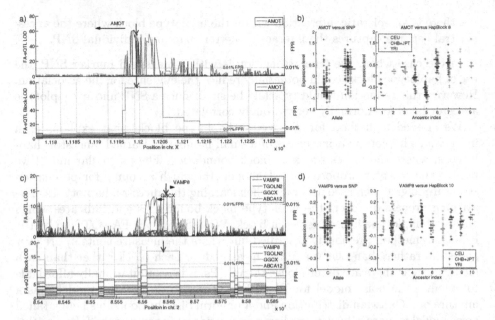

Fig. 5. Left: FA-eQTL scores for SNPs and haplotype blocks, showing the 0.01% FPR threshold for the gene corresponding to the strongest association. Blocking leads to improved association significance. **Right:** Scatter plots of mRNA levels against the SNP allele and block ancestor at the locations marked with red arrows in the left plots.

FPR significance threshold. Table 2 shows that for each of the four chromosomes tested we identify more significant associations using haplotype blocks compared to the single SNP approach. Out of the total of 274 associations at the 0.01% FPR level, only five are *trans*, all at sufficiently low significance levels to suggest that they are false associations.

Transcription factor study. We selected a subset of 960 gene probes listed as transcription factors in the BDB database ver 2.0 [16]. FA-eQTL was then applied genome-wide to search for *trans* associations to these probes. Only one significant *trans* association was found: between *DPF2* and a region in chromosome 12 (54.5Mb). Testing for associations of all expression levels to this genomic region, we identified an additional *cis* association with *RPS26* and a second *trans* association with *FLT1*. The expression profiles of *RPS26* and *FLT1* show strong correlation whilst *RPS26* and *DPF2* are strongly anti-correlated. A plausible biological explanation is that the ribosomal protein *RBS26* is mediating the expression of both *FLT1* (vascular endothelial growth factor) and *DPF2* (apoptosis response zinc finger protein).

5   Conclusion

In [17], Sen and Churchill described two effects that act to obscure QTL associations, "First is the environmental variation inherent in most quantitative phenotypes. Second is the incomplete nature of the genotype information, which can only be observed at the typed markers". By explicitly modelling non-genetic variation and by using a haplotype block model, we have accounted for both of these effects. Our results on HapMap data demonstrate that countering these effects leads to a more than threefold increase in the number of significant associations found. Given this performance of Block FA-eQTL, we now plan to scale it so that it can be applied exhaustively across all probes and SNPs.

Acknowledgments. The authors would like to thank Barbara Stranger and Manolis Dermitzakis for access to their gene expression data, Hetu Kamichetty for the use of his piSNP software and Leopold Parts for helpful discussions.

References

[1] Kendziorski, C.M., Chen, M., Yuan, M., Lan, H., Attie, A.D.: Statistical methods for expression quantitative trait loci (eQTL) mapping. Biometrics 62(1), 19–27 (2006)

[2] Brem, R.B., Kruglyak, L.: The landscape of genetic complexity across 5,700 gene expression traits in yeast. Proc. Natl. Acad. Sci. 102(5), 1572–1577 (2005)

[3] Huang, J., Kannan, A., Winn, J.: Bayesian association of haplotypes and non-genetic factors to regulatory and phenotypic variation in human populations. Bioinformatics 23(13), i212–i221 (2007)

[4] The International HapMap Consortium: A haplotype map of the human genome. Nature 437, 1299–1320 (2005)

[5] Roweis, S.T., Ghahramani, Z.: A unifying review of linear Gaussian models. Neural Computation 11(2), 305–345 (1999)

[6] Tipping, M.E., Bishop, C.M.: Probabilistic principal component analysis. Journal of the Royal Statistical Society, Series B 21(3), 611–622 (1999)

[7] Liebermeister, W.: Linear modes of gene expression determined by independent component analysis. Bioinformatics 18(1), 51–60 (2002)

[8] Iosifina, P., Lorenz, W.: Factor analysis for gene regulatory networks and transcription factor activity profiles. BMC Bioinformatics

[9] Lan, H., Stoehr, J.P., Nadler, S.T., Schueler, K., Yandell, B., Attie, A.D.: Dimension reduction for mapping mRNA abundance as quantitative traits. Genetics 121, 1607–1614 (2003)

[10] Hastie, T., Tibshirani, R., Eisen, A., Levy, R., Staudt, L., Chan, D., Brown, P.: Gene shaving as a method for identifying distinct sets of genes with similar expression patterns. Genome Biology (2000)

[11] Bishop, C.M.: Bayesian PCA. Advances in Neural Information Processing Systems 11, 382–388 (1999)

[12] Stranger, B., Forrest, M., Dunning, M., Ingle, C., Beazley, C., et al.: Relative impact of nucleotide and copy number variation on gene expression phenotypes. Science 315, 848–853 (2007)

[13] Bishop, C.M., Winn, J., Spiegelhalter, D.: VIBES: A variational inference engine for Bayesian networks. In: Advances in Neural Information Processing Systems, vol. 15, pp. 793–800 (2002)

[14] Lander, E., Botstein, D.: Mapping Mendelian Factors Underlying Quantitative Traits Using RFLP Linkage Maps. Genetics 121(1), 185–199 (1989)

[15] Kamisetty, H., Kannan, A., Winn, J.: A Bayesian model for population-stratified haplotype block inference, http://research.microsoft.com/mlp/bio/piSNP.html

[16] Kummerfeld, S., Teichmann, S.: DBD: a transcription factor prediction database. Nucleic Acids Res. 34(Database issue), D74–D81 (2006)

[17] Sen, S., Churchill, G.A.: A statistical framework for quantitative trait mapping. Genetics 159, 371–387 (2001)

Effects of Genetic Divergence in Identifying Ancestral Origin Using HAPAA

Andreas Sundquist[*], Eugene Fratkin[*], Chuong B. Do, and Serafim Batzoglou

Department of Computer Science, Stanford University, Stanford CA 94305, USA
{asundqui,fratkin,chuongdo,serafim}@cs.stanford.edu

Human population migration, adaptation, and admixture have a chaotic and mostly undocumented history, but we are at the cusp of an era where we will be able to unlock these records from our genomes. An admixed individual's genome with ancestors from isolated populations is a mosaic of chromosomal blocks, each following the statistical properties of variation seen in those populations. By analyzing polymorphisms in the admixed individual against those seen in representatives from the populations, we can infer the ancestral source of the individual's haploblocks. Several methods have emerged recently that use SNPs as a basis for variation to infer the ancestral population composition of admixed individuals.

In this paper we describe a novel approach for ancestry inference, HAPAA (HMM-based Analysis of Polymorphisms in Admixed Ancestries), that models the allelic and haplotypic variation in the populations and captures the signal of correlation due to linkage disequilibrium. We benchmarked HAPAA against the current best-performing method, the Markov-HMM-based SABER by Tang et al. on the HapMap dataset, consisting of three populations: North-Western Europeans, Yoruban-Africans, and East Asians. We first partitioned each population into a set used for training and a set used for test individual construction. Then, we constructed admixed test individual genotypes by simulating recombination and mating for up to 20 generations. Finally, we show that HAPAA significantly outperforms SABER with approximately half the mean-square-error.

Next, we introduce a methodology for studying the limitations of inference as a function of the genetic divergence between ancestral populations and time-to-admixture. Via simulation, we construct pairs of populations separated by 100…2000 generations of recombination, and partition them into individuals used for modeling and training and individuals for test set construction. For this dataset, we measure the ability of our inference algorithm to detect a single admixture event 1…20 generations ago. Our results show the connection between genetic divergence of populations, the number of generations of admixture, and the accuracy of inference, and indicate that with high probability we can detect admixture events of 10 or more generations using our methodology.

The full manuscript may be found in the Genome Research RECOMB issue.

---

[*] These authors contributed equally to the work.

M. Vingron and L. Wong (Eds.): RECOMB 2008, LNBI 4955, p. 423, 2008.
© Springer-Verlag Berlin Heidelberg 2008

On the Inference of Ancestries in Admixed Populations

Sriram Sankararaman[1,*], Gad Kimmel[1,2,*],
Eran Halperin[2], and Michael I. Jordan[1,3]

[1] Computer Science Division, University of California, Berkeley, CA 94720, USA
[2] International Computer Science Institute, Berkeley, CA 94704, USA
[3] Department of Statistics, University of California, Berkeley, CA 94720, USA

Abstract. Inference of ancestral information in recently admixed populations, in which every individual is composed of a mixed ancestry (e.g., African Americans in the US), is a challenging problem. Several previous model-based approaches have used hidden Markov models (HMM) to model the problem, however, the Markov Chain Monte Carlo (MCMC) algorithms underlying these models converge slowly on realistic datasets. While retaining the HMM as a model, we show that a combination of an accurate fast initialization and a local hill-climb in likelihood results in significantly improved estimates of ancestry. We studied this approach in two scenarios—the inference of locus-specific ancestries in a population that is assumed to originate from two unknown ancestral populations, and the inference of allele frequencies in one ancestral population given those in another.

1 Introduction

The recent advances in genotyping and sequencing technologies have resulted in exciting discoveries of links between genes and diseases via whole-genome association studies [1]. In these studies, cases and controls are collected and single nucleotide polymorphisms (SNPs) are genotyped across the entire genome of these two populations. A discrepancy in the allele distribution across the cases and the controls serves as evidence for an association between the SNP and the condition studied.

One of the main caveats of such association studies is their sensitivity to confounding effects. In particular, the ancestral background of the cases and the controls may affect the results. In order to overcome this problem, one could infer the ancestral background of each individual using the genotypes, and then apply a correction to the statistical tests based on this information [2].

The inference of ancestral information is a non-trivial problem, and the accuracy of existing methods on this task is currently limited. The problem is especially difficult when recently admixed populations are considered, in which every individual is composed of a mixed ancestry (e.g., African Americans in the

---

* These authors contributed equally to this paper.

M. Vingron and L. Wong (Eds.): RECOMB 2008, LNBI 4955, pp. 424–433, 2008.

US, Hispanic populations, and recently mixed populations in large metropolitan areas such as New York or the San Francisco bay area). These populations originate from two or more ancestral populations that were separated for a long time, and then started mixing a small number of generations ago (e.g., 10-20 generations ago). Due to recombination events, the genome of every such admixed individual is a mosaic of haplotypes that originated from the original ancestral populations. Thus, in order to describe their overall ancestry, we have to find the *locus specific ancestry* for each individual, or the ancestral origin of every locus in the genome of each of the individuals.

Given the genetic underpinnings of the ancestral origin problem it is natural to consider inference methods based on probabilistic models. Indeed, most previous work has made use of hidden Markov models (HMMs), where the states are ancestries, the transitions roughly correspond to historical recombination events and the emission matrix models population-specific allele frequencies [3,4,5,6]. Due to their assumption that alleles are independent once the ancestries are known, such models fail to capture population-specific background linkage disequilibrium (LD), but it is possible to augment the HMM to include additional Markovian dependencies among the observed alleles to attempt to account for this (short-range) form of LD [7].

The HMM is amenable to dynamic programming and therefore inference of ancestry is tractable under the HMM once its parameters are determined. But the need to estimate various hyperparameters of the admixture models has led researchers to Markov chain Monte Carlo (MCMC) procedures; these procedures have desirable properties in the limit of large numbers of samples, but in practice they can be overly slow for realistic data sets.

A rather different, non-model-based approach to inferring locus-specific ancestries has recently been proposed by Sankararaman, et al. [8]. This method (referred to as "LAMP") is based on running a window over the genome, computing the local ancestry of each individual within each window based on a local-likelihood model, and combining the results from the windows overlapping a given SNP using a majority vote. This method has been shown empirically to provide accurate estimates, improving on the HMM-based methods described above.

Practical applications of HMMs in other literatures, most notably the speech and signal processing literatures [9], emphasize the critical need for effective initialization of the parameter estimation procedure for HMMs. LAMP may yield an improvement over the HMM methods simply because the latter are being initialized randomly and the MCMC procedures are not mixing on a practical time scale. This suggests using the solution from LAMP to initialize the HMM. Hill-climbing in likelihood from the LAMP solution may provide an effective way to retain the advantages of a model-based method while not sacrificing performance. One of the goals of the current paper is to explore this possibility.

Another goal of the current paper is to consider an augmented form of the standard HMM for admixture [4] which includes explicit indicators of recombination events. Specifically, if a recombination event occurs between SNPs, then the ancestry of the SNPs are chosen independently; if recombination does not occur, then the ancestries are set equal. Treating these indicators as explicit latent variables in the model, we can attempt to infer historical recombinations.

Admixture models include a number of biologically-motivated parameters. In principle, all of these parameters can be estimated via an MCMC algorithm or an expectation-maximization (EM) algorithm, or some combination. It is also possible to integrate out parameters that are "nuisance parameters" for a particular inferential problem. Given, however, the biologically-interpretable of several of our parameters, it is also possible to use the model in settings in which estimates of some of the parameters are available from previous analyses. In particular, one of the scenarios we consider involves inferring locus-specific ancestries in a population that is assumed to originate from two *unknown* ancestral populations. This is the case for African-Americans, and other similar populations. This scenario has been investigated by earlier researchers (e.g., [8], [7]), and provides a point of comparison for the analyses that we present here.

In a second scenario, we assume that one of the ancestral populations is completely unknown, or its genotypes are not given. We show that it is possible to infer the allele frequencies in the unknown population. This scenario may be appropriate in situations in which it is difficult to obtain external estimates of the allele frequencies of one of the ancestral populations. This is the case, for example, in admixed populations that contain native Americans as one of the ancestries, such as the Puerto Rican population.

2   Methods

In this section, we describe the hidden Markov model that serves as the basis of our experiments. We then describe various forms of inference algorithms for this HMM, emphasizing the use of the EM procedure for parameter estimation.

2.1   Probabilistic Model

To simplify our presentation, let us assume that the number of populations that have been admixed is two (the notation is slightly more involved in the case of more than two populations but no new ideas are needed). Also, again for simplicity of presentation, we restrict our attention to haplotypes; genotypes can be handled in a straightforward manner

Let m denote the number of haploytpes, and let n denote the number of SNPs. Let X be the observed binary matrix of SNPs; i.e., $X_{i,j}$ is the j-th SNP of the i-th haplotype. Let p and q be the vectors of the allele frequencies in the ancestral populations. Hence, p_j is the probability to obtain '1' in the j-th SNP in the first population and q_j is the corresponding probability in the second population. The matrix Z denotes the ancestry information of each haplotype at each SNP: $Z_{i,j} \in \{0,1\}$ holds the ancestry of haplotype i of the j-th SNP (0 if SNP j of haplotype i originated from the first population and 1 if it originated from the second). We use the matrix W to denote recombination events. Specifically, $W_{i,j}$

equals '1' if a recombination event occurred during the history of the admixture process in the i-th haplotype between the $j - 1$-th SNP and the j-th SNP, and '0' otherwise. The $n - 1$-dimensional vector $\boldsymbol{\theta}$ denotes the probability for a recombination event, with θ_j corresponding to the $j - 1$-th and the j-th SNPs. The fraction of the first population in the ancestral population, which we call the *admixture fraction*, is denoted by α. Finally, g denotes the number of generations of the admixed process (in the sense that $\frac{1}{g-1}$ models the average length of ancestral chromosome blocks in the current admixed population).

Given the parameters g, α, $\boldsymbol{p}$, $\boldsymbol{q}$, and $\boldsymbol{\theta}$, a haplotype is generated as follows: recombination points are generated on each chromosome based on a Poisson process whose rate parameter depends on g and the recombination rate r. This process corresponds to setting some of the W's to 1. Then the ancestries of the resulting chromosomal blocks are determined independently for each block with α being the probability to choose the first ancestry. We assume that the mating is random across the populations. Given the ancestry at a particular position, an allele is generated using the corresponding ancestral allele frequency. We assume that the alleles are generated independently in a block.

We now describe the marginal and conditional distributions of the model. We assume a uniform prior over the interval $[0, 1]$ for each of the parameters α, $\boldsymbol{p}$, $\boldsymbol{q}$. g is assumed to be distributed uniformly over the interval $[g_{min}, g_{max}]$ for some $g_{max} > g_{min} > 1$. Given the ancestry and the allele frequencies of a specific SNP j in haplotype i, $X_{i,j}$ is a Bernoulli random variable with distribution:

$$\Pr(X_{i,j} = 1 | Z_{i,j}, p_j, q_j) = \begin{cases} p_j & Z_{i,j} = 0 \\ q_j & Z_{i,j} = 1 \end{cases}. \tag{1}$$

The distribution of the ancestry of a specific SNP depends on the occurrence of a recombination event. On the occurrence of a recombination between SNPs j and $j - 1$ of haplotype i, the ancestry $Z_{i,j}$ is chosen with probability α to be 0 and 1 otherwise. If there was no recombination, the ancestry stays the same as that at the previous SNP:

$$\Pr(Z_{i,j} | Z_{i,j-1}, W_{i,j}, \alpha) = \begin{cases} \delta(Z_{i,j}, Z_{i,j-1}) & W_{i,j} = 0 \\ (1 - \alpha)^{Z_{i,j}} \alpha^{(1-Z_{i,j})} & W_{i,j} = 1 \end{cases}.$$

where $\delta(x, y) = 1$, iff $x = y$.

Since we assume that the recombination process is a Poisson process, the $W_{i,j}$, $W_{i,k}, k \neq j$ are independent and the probability for a specific location between SNPs $j - 1$ and j to have at least one recombination depends solely on θ_j. For $j > 1$, we have $\Pr(W_{i,j} = 1 | \theta_j) = \theta_j$ and $\theta_j = 1 - e^{-(g-1)l_j r_j}$, where l_j is the distance between the $j - 1$-th SNP and the j-th SNP and r_j is the recombination rate in that region. In our specific problem, θ_j is a deterministic function of g. In other situations, it may be more appropriate for g to parameterize a prior over θ_j.

2.2 Inference Problems

In this section, we focus on two inferential problems that can be framed based on the HMM. In both the problems, we seek the *maximum a posteriori* (MAP) estimates of a subset of the variables in the model and we find parameter estimates via the EM algorithm. We assume that the number of generations g is constant and known, and therefore θ is known. This is often the case for admixed populations. The two problems that we consider are: (1) The admixture fraction is known, the allele frequencies are unknown, and the goal is to find the local ancestries for each SNP in each haplotype. The optimization problem is to find W, Z such that $\Pr(W, Z|X, \alpha, g)$ is maximized. We refer to this problem as the *local ancestries inference problem*. (2) The allele frequencies are known for one of the ancestral populations, and the goal is to find the allele frequencies of the other as well as the admixture fraction. Here, the local ancestries are missing variables. The optimization problem is to find q, α such that $\Pr(q, \alpha|X, p)$ is maximized. We refer to this problem as the *ancestral allele frequencies inference problem*.

Local Ancestries Problem. To compute the local ancestries, we would like to compute the MAP estimates of Z, W by solving the following optimization problem:

$$\arg\max_{Z,W} \log[\Pr(W, Z|X, \alpha, \theta)]. \tag{2}$$

In each iteration of EM, the updates to Z and W are computed by a Viterbi algorithm with the emission probabilities $\Pr(X_{i,j}|Z_{i,j}, p_j, q_j)$ replaced by an integral over p_j, q_j. The E-step involves computing the posterior probabilities of p_j, q_j; i.e., $\Pr(p_j, q_j|X_{.,j}, Z_{.,j}^{(t)})$. This can be done easily using Bayes' theorem. The M-step involves solving m separate optimization problems in $Z_i, W_i, i \in \{1, \dots, m\}$ where Z_i denotes the vector of ancestries for the i-th haplotype and W_i denotes the corresponding vector of recombination events:

$$\{\log[\Pr(Z_{i,1}|\alpha)] + I_{1,i}(Z_{i,1})\} + \sum_{j=2}^{n}\{I_{j,i}(Z_{i,j}) + f_{i,j-1,j}(Z_{i,j-1}, Z_{i,j}, W_{i,j})\} \tag{3}$$

where $f_{i,j-1,j}(Z_{i,j-1}, Z_{i,j}, W_{i,j}) \equiv \log[\Pr(Z_{i,j}|Z_{i,j-1}, W_{i,j}, \alpha)] + \log[\Pr(W_{i,j}|\theta_j)]$ correspond to log transition probabilities and $I_{j,i}(Z_{i,j}) \equiv \sum_{i=1}^{m} \sum_{j=1}^{n} \int \{\log[\Pr(X_{i,j}|Z_{i,j}, p_j, q_j)] \Pr(p_j, q_j|X_{.,j}, Z_{.,j}^{(t)}) dp_j dq_j\}$ are expectations of the log emission probabilities.

Generally, the values of $I_{j,i}(z)$ can be tabulated for each i, j, z by computing the integral over a grid on p_j, q_j. For our setting, we have a uniform prior over p_j, q_j which permits the integral to be evaluated analytically. We can maximize (3) by dynamic programming. The values obtained for Z, W are then used to recompute the integrals $I_{j,i}(Z_{i,j})$ and the procedure is iterated.

Ancestral Allele Frequencies Problem. To compute the ancestral allele frequencies, we compute the MAP estimates of q and α:

$$\arg\max_{q,\alpha} \log \Pr(q, \alpha|X, p, \theta) = \arg\max_{q,\alpha} \log \Pr(X|p, q, \alpha, \theta)$$

since we have a uniform prior on q and α. We assume g and p, to be known. Let $q^{(t)}$, $\alpha^{(t)}$ denote the current estimates of q, α. The EM algorithm produces new estimates $q^{(t+1)}$, $\alpha^{(t+1)}$ that improve the objective function.

$$q_j^{(t+1)} = \frac{\sum_{i=1}^m X_{i,j} d_{i,j}(z)}{\sum_{i=1}^m d_{i,j}(1)}, \alpha^{(t+1)} = \frac{\sum_{i=1}^m d_{i,1}(0) + \sum_{j=2}^n c_{i,j}(1,0)}{m + \sum_{i=1}^m \sum_{j=2}^n \sum_{z \in \{0,1\}} c_{i,j}(1,z)}$$

Here $c_{i,k}(w,z) \equiv \Pr(W_{i,k} = w, Z_{i,k} = z | X_i, q^{(t)}, \alpha^{(t)}, p, \theta)$ and $d_{i,j}(z) \equiv \Pr(Z_{i,j} = z | X_i, q^{(t)}, \alpha^{(t)}, p, \theta)$ are the posterior probabilities of (W, Z) and Z at the j-th SNP of haplotype i respectively and are computed by an application of the forward-backward algorithm in the E-step.

These updates have an intuitive interpretation. At each position j, the new value of q_j is the fraction of SNPs that are 1 out of all SNPs belonging to the second population (weighted by their posterior probabilities). The update for α is the fraction of ancestries chosen from the first population whenever a new haplotype is chosen (weighted by their posterior probabilities).

3 Results

The implementation of the HMM and the EM algorithm that we have described has been provided in a program that we term "SWITCH." In this section, we describe experiments aimed at evaluating SWITCH. These experiments were run on datasets generated from HapMap data [10]. We used SNPs found in the Affymetrix 500K GeneChip Assay® [11] from chromosome 1 for each of the HapMap populations; i.e., Yorubans (YRI), Japanese (JPT), Han Chinese (CHB), and western Europeans (CEU). For a pair of populations, we simulated admixture by picking individuals from two ancestral populations in the ratio $\alpha : 1 - \alpha$. In each generation, individuals mate randomly and produce offspring. The rate of the recombination process is set to 10^{-8} per base pair per generation [12]. The mixing process is repeated for g generations. We generated datasets consisting of admixtures of YRI-CEU, CEU-JPT and JPT-CHB populations. We set g to 7 and α to 0.20 since these roughly correspond to the admixing process in African-American populations [5,4,13]. To ensure that the SNPs are (roughly) independent, we greedily remove SNPs that have a high correlation coefficient, $r^2 > 0.1$, with any other SNPs. For each of the problems, we use only genotype data.

3.1 Local Ancestries Problem

We first compare the estimates of the ancestries obtained from SWITCH to the estimates obtained from SABER and LAMP. In these experiments, the methods are given g and α. We consider two settings depending on whether the ancestral frequencies, (p, q), are available. Even when the frequencies of the ancestral populations are available, it is still advantageous to use the data to update the frequency estimates, which may have drifted from the ancestral frequencies.

Table 1. Accuracies of ancestry estimates averaged over 100 datasets. The methods are compared under two settings. When the ancestral allele frequencies are known, the methods compared are LAMP-ANC, SWITCH-ANC, and SABER. When the ancestral allele frequencies are not known, the methods compared are SWITCH and LAMP.

Method	YRI-CEU	CEU-JPT	JPT-CHB
SWITCH-ANC	97.6±0.3	94.5±0.8	66.4±2.7
LAMP-ANC	94.9±0.6	93.7±0.7	69.9±2.1
SABER	89.4±0.8	85.2±1.2	68.2±1.9
SWITCH	96.0± 0.6	83.2±5.6	51.4±2.8
LAMP	94.0±0.8	82.9±5.5	50.6±2.5

When they are available, SWITCH uses a maximum-likelihood classification based on these frequencies as initialization. We refer to this variation of SWITCH as SWITCH-ANC. SABER also requires the ancestral allele frequencies. The version of LAMP that uses ancestral frequencies is termed LAMP-ANC.

When the ancestral allele frequencies are not known, LAMP can still be used. We use the resulting estimates of ancestries to initialize the EM algorithm.

For each individual i and SNP j, each method finds an estimate $\hat{a}^p_{ij} \in \{0, 0.5, 1\}$ for the true ancestry a^p_{ij}. We measure the accuracy of a method as the fraction of triplets (i, j, p) for which $a^p_{ij} = \hat{a}^p_{ij}$. The first half of Table 1 compares the accuracies of SABER, LAMP-ANC and SWITCH-ANC on 100 random datasets of YRI-CEU, CEU-JPT and JPT-CHB. SWITCH-ANC improves significantly over LAMP-ANC and SABER on the YRI-CEU dataset. On the CEU-JPT, SWITCH-ANC and LAMP-ANC have comparable performance, and both methods are more accurate than SABER. All methods perform poorly on the JPT-CHB dataset due to the closeness of the two populations. The second half of Table 1 compares the accuracies of SWITCH and LAMP. On the YRI-CEU data, SWITCH, with an accuracy of 96.0%, improves significantly over LAMP, which has an accuracy of 94.0% (Wilcoxon paired signed rank test p-value of 3.89×10^{-18}). Interestingly, SWITCH improves significantly over LAMP-ANC even though the latter uses the ancestral allele frequencies. On the CEU-JPT and the JPT-CHB datasets, SWITCH seems to have slightly higher accuracies than LAMP. We believe that using more informative priors on the variables p, q, should yield further improvements by improving the estimation of low-frequency alleles. These results indicate that the HMM is most useful when the mixing populations can be easily distinguished as is the case with the YRI-CEU admixture.

Although the versions of SWITCH have a factor of $5 - 10$ increase in running time compared to LAMP, they still run under an hour on large datasets making them feasible for genome-scale problems.

3.2   Role of the Inference Algorithm

To understand the impact of the inference algorithm and the initialization, we compared SWITCH to STRUCTURE. While the model used in SWITCH is the same as the model introduced in STRUCTURE (with additional switching variables W), the inference algorithms differ. SWITCH obtains the posterior mode of the ancestries Z using an EM algorithm with LAMP providing the initialization. STRUCTURE computes the posterior marginals of each $Z_{i,j}$ using an MCMC algorithm to integrate out the unknown parameters. We compared the ancestry estimates produced by the two methods on the YRI-CEU dataset. STRUCTURE was run for 10000 burn-in and 50000 MCMC iterations (see below for further discussion of this choice). The linkage model was used. STRUCTURE was run on non-overlapping sets of 4000 SNPs covering 36000 of the 38000 initial SNPs due to numerical instabilities when larger number of SNPs were used.

On the YRI-CEU dataset, SWITCH achieved an accuracy of 97% while STRUCTURE achieved an accuracy of 84%. To isolate the reason for this difference, we evaluated MCMC algorithms which differ from STRUCTURE in varying degrees. First, we ran MCMC from a random starting point for 1000 iterations with 100 iterations of burn-in and used the posterior mean as the ancestry estimates. This yielded estimates with an accuracy of 91.13%. When the LAMP estimates were used as a starting point, the accuracy was 94.9%. This suggests that the chain has not mixed. To test if the chain has converged, we simulated five such chains each from different random starting points. We then computed a multivariate potential scale reduction factor (PSRF) [14] for random sets of 100 p's and q's and found it to be consistently large (> 1.2). When the Markov chain is unable to converge quickly, the initialization influences the ancestry estimates. Given that the MCMC algorithms do not converge even after being run for several days (e.g., STRUCTURE was run for a little less than three days on a set of 4000 SNPs while the MCMC runs described took about a day to run), good initialization becomes essential.

Two other differences between STRUCTURE and the MCMC algorithm that we implemented are that the latter discards correlated SNPs and fixes the hyperparameters. We modified the MCMC to retain the correlated SNPs and the accuracy falls to 74.9%. We conclude that removing highly correlated SNPs has a large impact on the accuracy when the models do not account for such SNPs.

3.3 Predicting Recombinations

The switching variables W denote historic recombinations created by the mixing process after the initial admixture event. While a change in the ancestry between two SNPs implies a recombination event, many of the recombination events do not result in a change in the ancestry. When α is small, this happens quite often. Here we measured the accuracy of SWITCH in predicting such recombinations. If a predicted recombination falls within 30 Kbases of the SNPs flanking a true recombination, it is called a true positive. If multiple recombinations are predicted within this window, only one is counted as a true positive. False positives and false negatives are defined similarly. The precision and recall of the predictions are then computed as $Precision = \frac{TP}{TP+FP}$ and $Recall = \frac{TP}{TP+FN}$. As a baseline, we use a null model that predicts recombinations based on the exponentially-distributed lengths of the haplotypes. The total number of recombinations in the null model is set to the number of predicted recombinations and the precision

Table 2. Average L1 error in the estimates of q. The methods compared are SWITCH (which estimates q and α jointly) and two naive algorithms that are given the true $\alpha = 0.20$ and α estimated from the data respectively.

Method	YRI-CEU	CEU-JPT	JPT-CHB
SWITCH	7.7±0.5	8.5±0.6	11.7±1.3
Naive1	11.8±0.5	12.2±0.5	12.5±0.5
Naive2	11.8±1.2	12.3±1.2	12.6±1.2

and recall of the predictions are computed similarly. On the YRI-CEU dataset, SWITCH attains a precision of 83.1% with a recall of 57.9% while the null model attained a precision of 67.6% with a recall of 43.0%. SWITCH was found to be consistently more accurate than the null model on the CEU-JPT and JPT-CHB datasets as well (data not shown).

3.4   Ancestral Allele Frequencies Problem

We now turn to the problem of inferring ancestral allele frequencies. To obtain a benchmark, we implemented a naive algorithm. The naive algorithm is given the true value of α (which is *not* available to the model). The idea behind the naive algorithm is as follows. For a position j with minor allele frequency f_j, and allele frequencies p_j and q_j in the two populations, if the number of individuals is large, f_j can be written as $f_j = (1-\alpha)p_j + \alpha q_j$. So we compute the allele frequency q_j at position j as $q_j = \max\left(\min\left(\frac{f_j - (1-\alpha)p_j}{\alpha}, 1\right), 0\right)$. We used two different estimates of α, yielding algorithms that we refer to as Naive1 and Naive2. Naive1 uses the value of $\alpha = 0.20$ which is the admixture fraction in the first generation of admixture. Naive2 uses an α measured from each dataset. We calculated the L1 error between the estimated $\hat{q}$ and the true q. The L1 error averaged over 100 datasets of YRI-CEU, CEU-JPT and JPT-CHB is shown in Table 2. SWITCH reduces the L1 error by about 30% in the YRI-CEU and the CEU-JPT datasets while there is no significant difference for the JPT-CHB dataset.

4   Conclusions

While HMMs provide a reasonable approach to the modeling of admixture, the use of MCMC algorithms to estimate the model hyperparameters can be impractical on realistic datasets. By using an accurate initialization based on LAMP, and using an EM algorithm to hill-climb in likelihood from this initialization, we found that we were able to obtain highly accurate solutions in reasonable time. Thus we are able to retain the advantages of the model-based framework within a practical procedure. These advantages include the ability to infer location-specific ancestries, to predict historic recombinations, and to infer ancestral allele frequencies in an unknown population when one ancestral population is known.

It is interesting to note that SWITCH is more accurate than SABER which tries to model the background LD in the populations. This may be due in part to the fact that SABER has a larger number of parameters to estimate than the HMM and again there may be practical limitations on obtaining these estimates.

Acknowledgments. G.K. and E.H. were supported by NSF grant IIS-0513599 and NSF grant IIS-0713254. S.S was supported by NIH grant R33 HG003070-03.

References

1. Bonnen, P., Pe'er, I., Plenge, R., Salit, J., Lowe, J., Shapero, M., Lifton, R., Breslow, J., Daly, M., Reich, D., et al.: Evaluating potential for whole-genome studies in Kosrae, an isolated population in Micronesia. Nat. Genet. 38, 214–217 (2006)
2. Price, A., Patterson, N., Plenge, R., Weinblatt, M., Shadick, N., Reich, D.: Principal components analysis corrects for stratification in genome-wide association studies. Nature Genetics 38, 904–909 (2006)
3. Pritchard, J., Stephens, M., Donnelly, P.: Inference of population structure using multilocus genotype data. Genetics 155, 945–959 (2000)
4. Falush, D., Stephens, M., Pritchard, J.K.: Inference of population structure using multilocus genotype data: linked loci and correlated allele frequencies. Genetics 164, 1567–1587 (2003)
5. Patterson, N., Hattangadi, N., Lane, B., Lohmueller, K.F., Hafler, D.A., Oksenberg, J.R., Hauser, S.L., Smith, M.W., O'Brien, D.J., Altshuler, D., Daly, M.J., et al.: Methods for high-density admixture mapping of disease genes. Am. J. Hum. Genet. 74, 979–1000 (2004)
6. Hoggart, C., Shriver, M., Kittles, R., Clayton, D., McKeigue, P.: Design and analysis of admixture mapping studies. Am. J. Hum. Genet. 74, 965–978 (2004)
7. Tang, H., Coram, M., Wang, P., Zhu, X., Risch, N.: Reconstructing genetic ancestry blocks in admixed individuals. Am. J. Hum. Genet. 79, 1–12 (2006)
8. Sankararaman, S., Sridhar, S., Kimmel, G., Halperin, E.: Estimating local ancestry in admixed populations. American Journal of Human Genetics (to appear)
9. Huang, X., Acero, A., Hon, H.-W.: Spoken Language Processing. Prentice-Hall, Upper Saddle River (2001)
10. http://www.hapmap.org
11. http://www.affymetrix.com/products/arrays/specific/500k.affx
12. Nachman, M., Crowell, S.: Estimate of the mutation rate per nucleotide in humans. Genetics 156, 297–304 (2000)
13. Tian, C., Hinds, D.A., Shigeta, R., Kittles, R., Ballinger, D.G., Seldin, M.F.: A genomewide single-nucleotide-polymorphism panel with high ancestry information for African American admixture mapping. Am. J. Hum. Genet. 79, 640–649 (2006)
14. Brooks, S.P., Gelman, A.: General methods for monitoring convergence of iterative simulations. Journal of Computational and Graphical Statistics 7(4), 434–455 (1998)

Increasing Power in Association Studies by Using Linkage Disequilibrium Structure and Molecular Function as Prior Information

Eleazar Eskin

Department of Computer Science and Department of Human Genetics, University of
California Los Angeles, Los Angeles, CA, 90095

The availability of various types of genomic data provide an opportunity to incorporate this data as prior information in genetic association studies. This information includes knowledge of linkage disequilibrium structure as well as knowledge of which regions are likely to be involved in disease. In this paper, we present an approach for incorporating this information by revisiting how we perform multiple hypothesis correction. In a traditional association study, in order to correct for multiple hypothesis testing, the significance threshold at each marker, t, is set to control the total false positive rate. In our framework, we vary the threshold at each marker t_i and use these thresholds to incorporate prior information. We present a novel Multi-threshold Association Study Analysis (MASA) method for setting these threshold to maximize the statistical power of the study in the context of the additional information. Intuitively markers which are correlated with many polymorphisms will have higher thresholds than other markers. The simplest approach for encoding prior information is through assuming a causal probability distribution. In this setting, we assume that the causal polymorphism is chosen from this distribution and only one polymorphism is causal. We refer to the probability that the polymorphism i is causal as its causal probability, c_i. Given the causal probabilities, using the approach presented in this paper, we can numerically solve for the marker thresholds which maximize power. By taking advantage of this information, we show how our multi-threshold framework can significantly increase the power of association studies while still controlling the overall false positive rate, α, of the study as long as $\sum t_i = \alpha$. We present a numerical procedure for solving for thresholds that maximize association study power using prior information. We present the results of benchmark simulation experiments using the HapMap data which demonstrate a significant increase in association study power under this framework.

Our optimization algorithm is very efficient and we can obtain thresholds for whole genome associations in minutes. We also present an efficient permutation procedure for correctly adjusting the false positive rate for correlated markers and show how the this approach increases computational time only slightly relative to performing permutation tests for traditional association studies.

We provide a webserver for performing association studies using this method at http://masa.cs.ucla.edu/. On the website, we provide thresholds optimized for the the Affymetrix 500k and Illumina HumanHap 550 chips.

M. Vingron and L. Wong (Eds.): RECOMB 2008, LNBI 4955, p. 434, 2008.
© Springer-Verlag Berlin Heidelberg 2008

Panel Construction for Mapping in Admixed Populations Via Expected Mutual Information

Sivan Bercovici[1], Dan Geiger[1], Liran Shlush[2],
Karl Skorecki[2], and Alan Templeton[3]

[1] Computer Science Department
Technion, Haifa 32000, Israel
{sberco,dang}@cs.technion.ac.il
[2] Faculty of Medicine
Technion, Haifa 31096, Israel
{lshlush,skorecki}@tx.technion.ac.il
[3] Department of Biology
Washington University , St. Louis, Mo. 63130, USA
templeton@biology2.wustl.edu

Abstract. Mapping by Admixture Linkage Disequilibrium (MALD) is an economical and powerful approach for the identification of genomic regions harboring disease susceptibility genes in recently admixed populations. We develop an information-theory based measure, called EMI (expected mutual information), that computes the impact of a set of markers on the ability to infer ancestry at each chromosomal location. We then present a simple and effective algorithm for the selection of panels that strives to maximize the EMI score. Finally, we demonstrate via well established simulation tools that our panels provide considerably more power and accuracy for inferring disease gene loci via the MALD method in comparison to previous methods.

1 Introduction

Mapping by admixture linkage disequilibrium (MALD) is an economical and powerful approach for the identification of genomic regions harboring disease susceptibility genes in recently admixed populations [11,9]. For the method to be useful, the prevalence of the disease under study should be considerably different between the ancestral populations from which the admixed population was formed.

Myeloma, for example, is a type of cancer that is approximately three times more prevalent in Africans than in Europeans [11]. Hepatitis C clearance is approximately five times more prevalent in Europeans than in Africans. Stroke, lung cancer, prostate cancer, dementia, end-stage renal disease, multiple-sclerosis, hypertension and many more diseases all exhibit a higher morbidity in either Africans or Europeans, when the two ethnically-different populations are compared [11]. This difference in susceptibility to a specific disease is also evident in other ethnic populations. Native Americans suffer from a high prevalence of type 2 diabetes,

M. Vingron and L. Wong (Eds.): RECOMB 2008, LNBI 4955, pp. 435–449, 2008.

obesity and gallbladder disease, while showing a lower prevalence of asthma, relative to Europeans [8].

When examining an individual that originated from several ancestral populations, such as African-Americans, the likelihood that this individual will carry a given disease is influenced by the susceptibility to the disease in the ancestral populations. When such an admixed individual carries a hereditary disease, the chances are higher that the disease gene or genes are harbored in chromosomal segments that originated from the ancestral population with the higher risk.

The MALD method screens through the genome of either affected or both affected and healthy admixed individuals, looking for chromosomal segments with an unusually high representation of the high-risk ancestry population for the disease. MALD requires 200 to 500-fold fewer markers, in comparison to genome-wide association mapping, while offering the same power [9]. Consequently, the method has an economical advantage over alternative methods. Lately, successful results from admixture mapping have begun to emerge. For example, the usage of MALD led to the discovery of multiple risk alleles (gene variants) for prostate cancer [2].

In this paper, we develop an information-theory based measure, called *EMI* (expected mutual information), to select an effective panel of markers to be used in MALD. Our measure, presented in Section 4, computes the total impact of a set of markers on the ability to infer ancestry at each chromosomal location, averaged over all possible admixture related recombinations. This method improves previous measures such as the Shannon Information Content (SIC)[10], and Fisher Information Content (FIC)[6]. We then present, in Section 5, a simple and effective algorithm for the selection of panels that strives to maximize the EMI score. In Section 6, we demonstrate via well established simulation tools used in previous studies, that our panels provide considerably more power for inferring disease gene loci. For example, our simulations show that in the challenging case of a disease with an ethnicity risk ratio of 1.6 between two ancestral populations, the power increased from 50% to 68%, namely, an increase of about 36% in the ability to detect the loci of disease susceptibility genes. The detection accuracy has also significantly improved with the use of our new panels. The increase in power is particularly important in the detection of weak signals that underlie complex diseases. Section 7 concludes with extensions and discussion.

2   Background

The MALD method consists of three steps. First, an admixed population with a significantly higher risk for a specific disease in one of the ancestral populations is identified. Ancestry informative markers that effectively distinguish between the relevant ancestral populations are selected, and either cases or both cases and controls are genotyped. Second, the ancestry along the chromosomes of every individual is computed based on the sampled genotypes. Third, chromosomal regions with an elevated frequency of the ancestral population with the higher

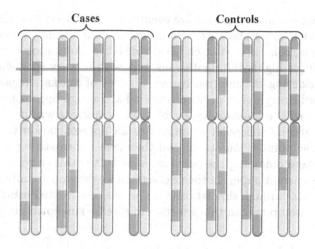

Fig. 1. Ancsetry informative markers are used to compute the ancestry across the chromosomes of cases and controls. The region indicated by the bar shows elevated frequency of one ancestral population in the cases versus the expected distribution of ancestry in the controls, suggesting a disease susceptibility locus.

disease prevalence are identified. Figure 1 shows the ancestral profile of eight individuals, of which half are cases and half controls. The ancestral profiles are indicated as dark and light segments along the chromosomes. The excess of one ancestral population in the cases at the locus marked by the bar suggests that the locus contains the disease susceptibility gene. In the controls, the ancestry at the same locus matches the expected distribution of ancestry, strengthening the hypothesis. The detection of suspected regions can be followed by methods such as high density SNP-based association studies, or a study of nearby candidate genomic regions.

Choosing ancestry informative markers (AIM) for the construction of MALD panels has been pursued in several studies. AIM panels were constructed for African-American [12,15], Mexican-American [14] and Hispanic/Latino [4,8] populations. The construction of such panels requires three ingredients: a database of markers, a measure for the informativeness of a set of markers regarding ancestry, and an algorithm that selects informative markers for the MALD panel.

The work of Rosenberg et. al. [10] introduced a measurement for the information multialleleic markers provide on ancestry, based on the Shannon Information Content. Pfaff et. al. [6] based their measurement on the Fisher Information Content.

The algorithms used for panel construction in the studies that followed were driven by two prime objectives: (1) choose markers with the highest ancestry-informativeness (2) choose evenly spread markers. These guidelines were set to provide informative panels for the estimation of ancestry at each point along the genome. Current panel construction algorithms are "greedy", attempting to locally maximize an informativeness criterion, whilst investing less effort in ensuring that the chosen markers are evenly spaced or that the informativeness along the genome is well balanced. Smith et. al. used a purely greedy algorithm for marker selection [12]. Tian et. al. divided the chromosome into windows, choosing multiple highly-informative markers within every such window [15].

When considering the informativeness of a set of markers regarding the ancestry at an arbitrary point, previous work offered rough approximations. Smith et. al. considered the informativeness of a set of markers within a constant-size window centered on the point examined as an approximation to the informativeness at that point. Tian et. al. used the mean informativeness between two adjacent markers bounding the point examined. It is this deficiency that is addressed in the current paper. In the next section we develop an improved measure and demonstrate through simulations that panels constructed using our measure provides increased power in the detection of disease susceptibility gene loci.

3   Admixed Individuals Model

The genome of a recently admixed individual is a mosaic of large chromosomal segments, where each segment originated from a single ancestral population. We use the following definitions to describe these segments in admixed individuals.

Definition 1. *An **admixed chromosome** is a chromosome that originated from more than one ancestral population.*

Definition 2. *A **Post Admixture Recombination** point (abbreviated PAR) is a recombination point in which either two chromosomes from different populations crossed, or two chromosomes crossed when at least one of the chromosomes is an admixed chromosome.*

Definition 3. *A **(PAR) block** is a chromosomal segment limited by two consecutive PAR points, or by a chromosome edge and its closest PAR point.*

An immediate implication of these definitions is that every PAR block originated from a single ancestral population, designated as the ancestry of the block, for otherwise the block would have been further divided.

Figure 2 illustrates the propagation of PAR points along three generations of admixture, and the PAR blocks they induce. In particular, Figure 2 shows a grandmother originating from one population, and a grandfather originating from two populations, yielding a parent with one admixed chromosome (with one PAR point) and one non-admixed chromosome. As the parent's chromosomes recombine to produce the child's admixed chromosome, a second PAR point is added. Hence, three recombination points reside on the child's chromosome, of which only two are PAR points (colored black). Three PAR blocks are defined rather than four as the leftmost recombination point is not a PAR point.

We denote the set of all observed markers by J, and the vector of an individual's PAR-blocks ancestries as Q. The set of an individual's PAR points defines a partition (denoted π) of the chromosomes into blocks. We use the random

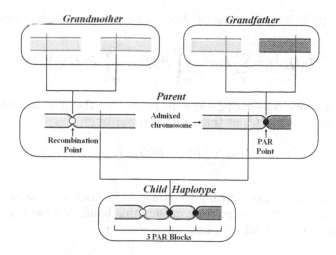

Fig. 2. Three generations admixture example. PAR blocks are limited only by PAR points and the chromosomes' ends.

available Q_π to denote the vector of ancestries corresponding to the PAR-blocks determined by π, $Q_{\pi,i}$ to denote the ancestry (out of K possible ancestral populations) of the ith PAR block in the given partition π, and the random vector $J_{\pi,i} = \{J_{\pi,i,1}, J_{\pi,i,2}, ..., J_{\pi,i,m_i}\}$ to denote the haplotype assignment of the set of $m_{\pi,i}$ observed markers within this block. Reference to subscript π will be omitted whenever π is clear from the context.

Markers within a PAR block are assigned according to the probability function of the corresponding ancestral population. We further assume that the ancestries of all PAR blocks of a given partition π are mutually independent. A graphical model showing these assumptions is given in Figure 3.

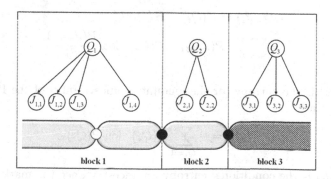

Fig. 3. Graphical model for $P(Q, J)$ assuming markers J_i within a PAR block are independent conditioned on ancestry Q_i. Ancestries of PAR blocks are mutually independent.

The joint probability distribution described via the graphical model is given by

$$P(Q, J) = \sum_\pi P(\pi) \cdot \prod_{i=1}^{|Q_\pi|} P(Q_{\pi,i}) \prod_{j=1}^{m_{\pi,i}} P(J_{\pi,i,j}|Q_{\pi,i}) \tag{1}$$

In particular, when considering a specific point x on the genome, the joint probability for the ancestry Q_x at that point is given by

$$P(Q_x, J) = \sum_\pi P(\pi) \cdot P(Q_x, J_{\pi,x}) \cdot P(\overline{J}_{\pi,x}) \tag{2}$$

where $J_{\pi,x}$ are the markers within the same PAR-block as location x, and $\overline{J}_{\pi,x}$ is the complementary set of markers outside this block. We use this joint distribution to derive our panel informativeness measure.

4   Informativeness of Panels

In this section we develop a measure for the contribution of a set of observed markers to the ability to infer the ancestry of a block conditioned on a partition π. We then extend this measure to account for the fact that π is unknown by computing the expectation over all possible partitions, while focusing on the inference of a single location x.

We start by exploring the relationship between observed markers and the ancestries of PAR blocks under the assumption that the partition is known. Using information-theory, we estimate the extent to which a set of markers contribute to the ability to infer ancestry by measuring the informativeness of a set of markers regarding ancestry. The information gain for ancestry due to observing a set of markers can be described by the well known Shannon Information Content (SIC)

$$I(Q_i; J_i) = H(Q_i) - H(Q_i|J_i) \tag{3}$$
$$= \sum_{Q_i} \sum_{J_i} P(J_i|Q_i) \cdot P(Q_i) \cdot \log \frac{P(J_i|Q_i)}{P(J_i)}$$

where $H(Q_i)$ is the entropy (or the amount of uncertainty) of the PAR block's ancestry, given by

$$H(Q_i) = - \sum_{Q_i=1}^{K} P(Q_i) \cdot \log P(Q_i)$$

and $H(Q_i|J_i)$ is the conditional entropy on ancestry once the markers observations are accounted for, given by

$$H(Q_i|J_i) = - \sum_{Q_i} \sum_{J_i} P(J_i, Q_i) \cdot \log P(Q_i|J_i).$$

In other words, the markers' informativeness is measured by the reduction in uncertainty regarding the ancestry of a given location due to observing these markers. This reduction in uncertainty originates from the fact that each ancestral population has a distinct distribution over the haplotype. The information gain in each PAR block is computed separately through Equation 3 due to our assumption of mutual independence.

The possible presence of linkage disequilibrium between markers within a block raises difficulties partially stemming from the need to estimate the joint probability of a haplotype J_i that contains multiple markers conditioned on the ancestry (i.e., $P(J_i|Q_i)$). To reduce computational cost, we assume conditional independence between all markers given ancestry, yielding a simpler form of mutual-information $I_{ind}(Q_i; J_i)$, explicated in Lemma 1. The relaxation of this assumption is pursued in Section 7.

Lemma 1. *For a given PAR Block, let Q_i be its ancestry, and $J_{i,j}$ be its jth marker (out of m_i markers). Under the assumption that the markers are conditionally independent given Q_i, the mutual information of Q_i and J_i is:*

$$I_{ind}(Q_i, J_i) = H(J_i) - \sum_{j=1}^{m_i} H(J_{i,j}|Q_i) \qquad (4)$$

Given a partition π, all PAR blocks are determined, and the informativeness of markers regarding ancestry Q_i, and in particular regarding ancestry Q_x of an arbitrary location x within the ith PAR block, is the informativeness of the markers in J_i alone. All other markers are not informative regarding Q_x. However, π is not known, and for every π a different block contains a location x. Hence, for each π, a different set of markers is informative. The expected informativeness of all markers regarding ancestry at location x is given, in principle, by

$$EMI(Q_x; J) = \sum_{\pi} P(\pi) \cdot I(Q_x; J|\pi) \qquad (5)$$

We call this measure EMI for Expected Mutual Information. Since summing over all possible partitions is not feasible, the rest of this section rewrites Equation 5 and explicates how to compute it.

Observe that for any two partitions π_1 and π_2 such that the PAR block that contains location x also contains the same set of markers $J_{\pi,x} \subseteq J$, the term $I(Q_x; J|\pi)$ in Equation 5 is equal. The probability for a partition π to contain a block that contains both location x and markers $J_{\pi,x}$ is defined by three events:

1. The minimal segment $[l, r]$ that spans over $J_{\pi,x}$ and x does not contain a PAR point.
2. The segment between l and the marker to its left (at l'), if such exists, contains a PAR point.
3. The segment between r and the marker to its right (at r'), if such exists, contains a PAR point.

Assuming PAR points are distributed independently, the aforementioned three events are independent as well. This holds because the corresponding three segments are mutually exclusive. Hence, the probability of a partition π to contain a PAR block containing location x and markers $J_{\pi,x}$ alone is given by the product

$$P_{(l,r)} = P(N_{[l',l]} \neq 0) \cdot P(N_{[l,r]} = 0) \cdot P(N_{[r,r']} \neq 0) \tag{6}$$

where $N_{[a,b]}$ is a random variable of the number of PAR points in segment $[a,b]$, and $[l,r]$ is the minimal segment containing location x and markers $J_{\pi,x}$.

The term $P(N_{[l',l]} \neq 0)$ depends on the existence of a marker at l', hence the term will not appear in Equation 6 in case there is no marker to the left of l. Similarly, $P(N_{[r,r']} \neq 0)$ will not appear in Equation 6 if there is no marker to the right of r.

Let $J_{[l,r]}$ denote a sequence of markers within a segment $[l,r]$, and $location(j)$ denote the location of a marker $j \in J$. To compute EMI, we weight the potential contribution $I(Q_x; J_{[l,r]})$ by the probability of such a contribution, namely the probability $P_{(l,r)}$ of a partition π to contain location x and markers $J_{[l,r]}$ within the same block.

Theorem 1. *Let Q_x be the ancestry at location x, and J the set of observed markers. The expected mutual information between Q_x and J is*

$$EMI(Q_x; J) = \sum_{l \in L} \sum_{r \in R} P_{(l,r)} \cdot I(Q_x; J_{[l,r]}) \tag{7}$$

where

$$L = \{location(j) \leq x | j \in J\} \cup \{x\},$$
$$R = \{location(j) \geq x | j \in J\} \cup \{x\}.$$

The common realization of the term $P_{(l,r)}$ in Equation 7 is via the exponential distribution. In particular $P(N_{[a,b]} = 0) = e^{-\lambda \cdot |b-a|}$, where λ is the rate of PAR points in admixed individuals, as derived from the admixture model being used. Consequently,

$$P_{(l,r)} = (1 - e^{-\lambda \cdot |l-l'|}) \cdot e^{-\lambda \cdot |r-l|} \cdot (1 - e^{-\lambda \cdot |r'-r|}). \tag{8}$$

Equation 7 defines the EMI at a specific location x. The average information gain regarding the entire chromosome is given by

$$EMI_{avg}(J) = \frac{1}{|N|} \cdot \sum_{x \in N} EMI(Q_x, J) \tag{9}$$

which measures the the average EMI along the chromosome. The set N consists of all locations x on an evenly spaced grid with a specific resolution. For example, for chromosome 1, a set N of 280 points means about one location per cM.

In the task of mapping disease genes in admixed populations using the MALD method, panels of high EMI_{avg} are shown in Section 6 to outperform previous panels.

5   Panel Construction

We employ a greedy algorithm that constructs panels of markers for which the EMI_{avg} is high. In principle, the algorithm iterates over the candidate markers, selecting the marker with the highest EMI_{avg} gain given the markers chosen so far. Namely, in each iteration the algorithm chooses a marker j that maximizes

$$\text{EMI}_{avg}(J \cup \{j\}) - \text{EMI}_{avg}(J) \tag{10}$$

where J is the set of markers selected so far.

The evaluation of EMI_{avg} is a computationally intensive task that is repeated with every iteration, and for every candidate marker. To reduce execution time, for each examined candidate we locally evaluate EMI_{avg} on a set of points located in a segment of length w centered on the candidate marker. Equation 11 evaluates the EMI_{avg} on a subset of points $w_j \subseteq N$

$$\text{EMI}_{avg}(J) = \frac{1}{|w_j|} \cdot \sum_{x \in w_j} \text{EMI}(Q_x, J) \tag{11}$$

where

$$w_j = \{p \in N \mid \quad location(j) - \frac{w}{2} \leq p \leq location(j) + \frac{w}{2}\}$$

rather than on the entire chromosome. Once a marker j is chosen, the EMI_{avg} gain in the next iteration is computed only for those markers that are within w_j, as the last chosen marker mostly affects their potential gain.

The most computationally dominant factor in EMI is the evaluation of $H(J)$ (Equation 4), as it is exponential in the number markers $|J|$. However, for a given PAR-block, a small number of ancestry informative markers suffice to nearly eliminate the uncertainty regarding its ancestry; the information gain regarding the ancestry of the PAR-block saturates rapidly as the number of markers within the PAR-block increases. Hence, limiting the number of markers used in the evaluation of Equation 4 yields an eligible approximation. In our implementation, we limited the number of markers in the evaluation of Equation 4 to a maximum of 17 markers, offering a plausible tradeoff between performance and approximation accuracy.

6   Evaluation

In this section we demonstrate the power of panels produced by our algorithm and EMI. We compare performance with the works of Smith et. al. [12], and Tian et. al. [15].

Similarly to the panels of Smith et. al. and Tian et. al., we constructed a panel for the African-American admixed population. The International HapMap Project [13] was used as the SNP allele frequencies source for the two ancestral populations, namely the west-African and European populations. HapMap has been shown to reflect these two distinct populations well [1].

We constructed a panel of 148 markers (denoted EMI-148) for chromosome 1, matching the number of corresponding markers in the screening panel of Tian et. al.. The panel of Smith et. al. contains 238 markers. We further constructed a more economical panel of 100 markers for chromosome 1 (denoted EMI-100) which is two thirds the number of markers in the panel constructed by Tian et. al. Based on the admixture-dynamics of African-Americans as described in [3,11,12], we used $\lambda = 6$ (Equation 8), a proportion of 0.8 African/0.2 European contribution to the admixed population, and $w = 45cM$ (Equation 11).

We first examine the performance of the four panels according to the EMI measure. The maximal EMI value is derived from the admixed-population characteristics, namely the number of admixed populations and the admixture co-efficient. In the case of African-Americans, the maximal value is approximately 0.5. As illustrated in Figure 4, the proportion of chromosome 1 above most EMI thresholds is higher in the panels constructed in the current work. Moreover, the EMI-148 panel constructed by our algorithm has a low EMI standard-deviation of 0.0041 in comparison to the screening panel of Tian et. al. (0.0142) and the panel of Smith et. al. (0.0178); indeed, our algorithm strives to balance the informativeness of markers across the chromosome. It is interesting to note that our lighter panel, EMI-100, has good performance as well, with a low EMI standard-deviation of 0.0056.

ANCESTRYMAP [5] is a tool we used for the estimation of the ancestral origin of a locus on the basis of sampled genotypes. Given genotypes of cases and controls, the tool can compute the likelihood of each point along the genome to be the disease locus, under a specified disease model. ANCESTRYMAP can also generate admixed-individual genotypes for cases and controls under a given

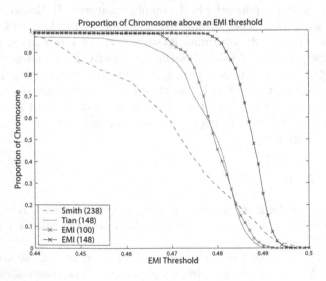

Fig. 4. Proportion of chromosome above an EMI threshold. For most levels of informativeness, our panel covers larger segments of the chromosome.

disease model. This software was used in [12] and [15] to evaluate the power of the Smith and Tian panels, respectively.

In the experiments conducted, we generated 576[1] admixed-individual cases per experiment through the use of ANCESTRYMAP. In each experiment, a single location on chromosome 1 was used as the disease predisposition locus. In order to evaluate the performance of the panel across the entire chromosome, a set of disease predisposition loci were chosen using a resolution of 4 points per cM; above 900 uniformly selected locations across chromosome 1 were used in the experiments. Multiple disease models were examined, all with higher prevalence in the European population. A range of ethnicity relative risk (ERR) factors, between 1.4 and 1.8, were set as the disease model parameters. We focused on this range as it captures diseases such as stroke and lung cancer [11], which are considered mild in their ERR, hence harder to detect. We proceeded by employing ANCESTRYMAP to locate the disease gene. Similar to the threshold used for the evaluation of Tian's panel [15], we used a LOD score above 4.0 as an indicator for successful detection. Figure 5 shows the power, namely, the detection success-rate, in a total of 5500 experiments per panel.

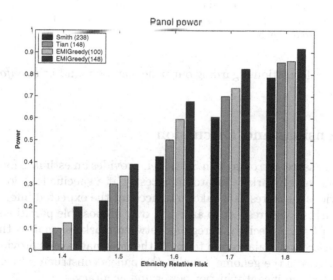

Fig. 5. EMI-148 achieves a significantly higher power in all tested disease models

Measuring the distance between the highest detection signal and the actual disease predisposition locus reveals that our panel also has a high detection accuracy, as illustrated in Figure 6. Approximately 55% of the experiments conducted on the entire range of ERR using our EMI-148 panel detected a signal within a 3 cM distance from the actual disease predisposition locus, whilst the other two panels achieved approximately 42% (Tian et. al.) and 37% (Smith et. al.). The EMI-100 panel achieved a 46% detection-rate within 3 cM.

---

[1] Commercial panel infrastructure requires sample size to be a multiplicative of 96.

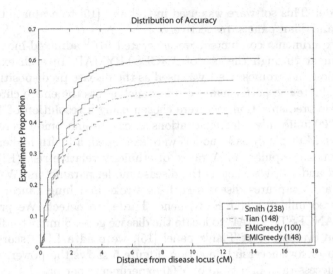

Fig. 6. Experiments percentage above accuracy threshold for each of the four panels. A higher percentage of the experiments yield higher accuracy for EMI-148, in comparison to the other three panels.

Detailed information regarding our panel can be found at *bioinfo.cs.technion. ac.il/MALD*.

7   Extensions and Discussion

The EMI measure, introduced in Section 4, provides an estimate for the informativeness of a set of markers regarding ancestry at a specific location. It improves upon previous measures as it takes into account the expected informativeness of a set of markers with respect to ancestry, over all possible partitions. The higher accuracy of EMI, especially in regions between markers, enables the creation of panels that are well balanced in terms of the informativeness provided by the set of markers across the genome. Finally, the panels constructed by our algorithm demonstrated significantly higher power and accuracy.

An immediate extension of EMI that we pursued addresses possible dependencies between markers given ancestry. Lemma 1 disregards LD within ancestral population in favor of a lower computational cost. We now use a first-order Markov-chain to model marker dependencies within ancestral populations in order to provide a more accurate model. Under this model, the transition probabilities are derived from the LD present between every two adjacent markers given the ancestry. Such a model still yields a computationally plausible form, as shown in the next lemma.

Lemma 2. *For a given PAR Block, let Q_i be the ancestry, and $J_{i,j}$ be the jth marker (out of m_i markers). Under the assumption that each marker is* *dependent on the preceding marker and conditionally independent of the rest of the markers given Q_i, the mutual information of Q_i and J_i is:*

$$I_{chain}(J_i; Q_i) = \quad H(J_i) - H(J_{i,1}|Q_i) - \sum_{j=2}^{m_i} H(J_{i,j}|Q_i, J_{i,j-1}) \quad (12)$$

However, we have not employed this extension in the panel constructed because the public data of LD categorized by ancestral population is still too sparse. In addition, as our algorithm inherently yields a balanced panel in terms of EMI_{avg}, the selected markers tend to be evenly spaced (hence with a large inter-distance), decreasing the probability for LD between two consecutive markers, given ancestry.

Another extension of EMI relaxes the assumption that the rate of PAR points, used in Equation 8, is constant across the chromosomes. Recombinational hot-spots can be taken into account by using a PAR point rate as a function of location $\lambda(x)$ instead of the constant rate λ. For example, assume that a chromosome is divided into regions of different PAR point rates $\lambda_1, \lambda_2, ..., \lambda_m$. For a segment $[l, r]$ that spans two consecutive regions with PAR rates λ_i and λ_{i+1}, the term $P(N_{[l,r]})$ in Equation 6 equals $e^{-(\lambda_i t + \lambda_{i+1} (1-t)) |r-l|}$, where t is the proportion of segment $[l, r]$ with PAR rate λ_i. Furthermore, the framework provided in this paper can address the effect on EMI of different admixture models, such as continuous-gene-flow [7], by considering a different realization of Equation 6.

We note that EMI assumes a model for haplotypes rather than genotypes, and that the allele frequencies $P(J|Q)$ are definite. In reality these frequencies are derived from a small set of samples (60, barring missing data, in the case of HapMap). In its current form, EMI lacks an appropriate treatment for the uncertainty involving allele frequencies. It is advisable to validate the allele frequencies by taking more samples for candidate markers, as done in [15].

The approach presented in this paper for panel construction also applies to the second phase of the MALD method. This second step currently employs a Markov chain model that assigns the most probable ancestry for each location, given the model and marker data [5,3]. By conditioning on possible partitions π, one can compute the expected ancestry $P(Q_x|J = j)$ of a point x given measurements $J = j$ via Equation 2, similarly to our computation of the expected informativeness. It would be interesting to see whether this approach yields higher accuracy in ancestry inference.

Finally, further research could determine the relative contribution of each of the following three ingredients to the reported increase in power: the richer set of SNP markers available in the current HapMap project, the validity of the EMI measure, and the success of our proposed algorithm. Nevertheless, we clearly showed that the panel produced using EMI has a well balanced high score in terms of informativeness of markers, yielding a significant improvement in both power and accuracy, compared to previous work.

Acknowledgements

The authors would like to thank Walter Wasser and Guennady Yudkovsky for the fruitful discussions, and Tamar Aizikowitz for her helpful comments. We also thank the Wolfson foundation for contributing a grid of 120 PCs that enabled us to conduct this research. This research is supported by Microsoft TCI grant.

References

1. Conrad, D.F., Jakobsson, M., Coop, G., Wen, X., Wall, J.D., Rosenberg, N.A., Pritchard, J.K.: A worldwide survey of haplotype variation and linkage disequilibrium in the human genome. Nature Genetics 38, 1251–1260 (2006)
2. Haiman, C.A., Patterson, N., Freedman, M.L., Myers, S.R., Pike, M.C., Waliszewska, A., Neubauer, J., Tandon, A., Schirmer, C., McDonald, G.J., Greenway, S.C., Stram, D.O., Marchand, L.L., Kolonel, L.N., Frasco, M., Wong, D., Pooler, L.C., Ardlie, K., Oakley-Girvan, I., Whittemore, A.S., Cooney, K.A., John, E.M., Ingles, S.A., Altshuler, D., Henderson, B.E., Reich, D.: Multiple regions within 8q24 independently affect risk for prostate cancer. Nature Genetics 39, 638–644 (2007)
3. Hoggart, C.J., Shriver, M.D., Kittles, R.A., Clayton, D.G., McKeigue, P.M.: Design and analysis of admixture mapping studies. The American Journal of Human Genetics 74, 965–978 (2004)
4. Mao, X., Bigham, A., Mei, R., Gutierrez, G., Weiss, K., Brutsaert, T., Leon-Velarde, F., Moore, L., Vargas, E., McKeigue, P., Shriver, M., Parra, E.: A genome-wide admixture mapping panel for hispanic/latino populations. The American Journal of Human Genetics 80, 1171–1178 (2007)
5. Patterson, N., Hattangadi, N., Lane, B., Lohmueller, K.E., Hafler, D.A., Oksenberg, J.R., Hauser, S.L., Smith, M.W., O'Brien, S.J., Altshuler, D., Daly, M.J., Reich, D.: Methods for high-density admixture mapping of disease genes. The American Journal of Human Genetics 74, 979–1000 (2004)
6. Pfaff, C.L., Barnholtz-Sloan, J., Wagner, J.K., Long, J.C.: Information on ancestry from genetic markers. Genetic Epidemiology 26, 305–315 (2004)
7. Pfaff, C.L., Parra, E.J., Bonilla, C., Hiester, K., McKeigue, P.M., Kamboh, M.I., Hutchinson, R.G., Ferrell, R.E., Boerwinkle, E., Shriver, M.D.: Population structure in admixed populations: Effect of admixture dynamics on the pattern of linkage disequilibrium. The American Journal of Human Genetics 68, 198–207 (2001)
8. Price, A.L., Patterson, N., Yu, F., Cox, D.R., Waliszewska, A., McDonald, G.J., Tandon, A., Schirmer, C., Neubauer, J., Bedoya, G., Duque, C., Villegas, A., Bortolini, M.C., Salzano, F.M., Gallo, C., Mazzotti, G., Tello-Ruiz, M., Riba, L., Aguilar-Salinas, C.A., Canizales-Quinteros, S., Menjivar, M., Klitz, W., Henderson, B., Haiman, C.A., Winkler, C., Tusie-Luna, T., Ruiz-Linares, A., Reich, D.: A genomewide admixture map for latino populations. The American Journal of Human Genetics 80, 1024–1036 (2007)
9. Reich, D., Patterson, N.: Will admixture mapping work to find disease genes? Philosophical Transcations of the Royal Society B 360, 1605–1607 (2005)
10. Rosenberg, N.A., Li, L.M., Ward, R., Pritchard, J.K.: Informativeness of genetic markers for inference of ancestry. The American Journal of Human Genetics 73, 1402–1422 (2003)

11. Smith, M.W., O'Brien, S.J.: Mapping by admixture linkage disequilibrium: advances, limitations and guidelines. Nature Review Genetics 6, 623–632 (2005)
12. Smith, M.W., Patterson, N., Lautenberger, J.A., Truelove, A.L., McDonald, G.J., Waliszewska, A., Kessing, B.D., Malasky, M.J., Scafe, C., Le, E., De Jager, P.L., Mignault, A.A., Yi, Z., De The, G., Essex, M., Sankale, J.L., Moore, J.H., Poku, K., Phair, J.P., Goedert, J.J., Vlahov, D., Williams, S.M., Tishkoff, S.A., Winkler, C.A., De La Vega, F.M., Woodage, T., Sninsky, J.J., Hafler, D.A., Altshuler, D., Gilbert, D.A., O'Brien, S.J., Reich, D.: A high-density admixture map for disease gene discovery in african americans. The American Journal of Human Genetics 74(5), 1001–1013 (2004)
13. The International HapMap Project. A haplotype map of the human genome. Nature Publishing Group 437, 1299–1319 (2005)
14. Tian, C., Hinds, D.A., Shigeta, R., Adler, S.G., Lee, A., Pahl, M.V., Silva, G., Belmont, J.W., Hanson, R.L., Knowler, W.C., Gregersen, P.K., Ballinger, D.G., Seldin, M.F.: A genomewide single-nucleotidepolymorphism panel for mexican american admixture mapping. The American Journal of Human Genetics 80, 1014–1023 (2007)
15. Tian, C., Hinds, D.A., Shigeta, R., Kittles, R., Ballinger, D.G., Seldin, M.F.: A genomewide single-nucleotide-polymorphism panel with high ancestry information for african american admixture mapping. The American Journal of Human Genetics 79, 640–649 (2006)

Constructing Level-2 Phylogenetic Networks from Triplets*

Leo van Iersel[1], Judith Keijsper[1], Steven Kelk[2], Leen Stougie[1,2],
Ferry Hagen[3], and Teun Boekhout[3]

[1] Department of Mathematics and Computer Science, Technische Universiteit
Eindhoven, Den Dolech 2, 5612 AX Eindhoven, The Netherlands
{l.j.j.v.iersel,j.c.m.keijsper,l.stougie}@tue.nl
[2] Centrum voor Wiskunde en Informatica (CWI), Kruislaan 413, 1098 SJ
Amsterdam, The Netherlands
{s.m.kelk,leen.stougie}@cwi.nl
[3] Centraalbureau voor Schimmelcultures (CBS), Fungal Biodiversity Center,
Uppsalalaan 8, 3584 CT Utrecht, The Netherlands
{f.hagen,t.boekhout}@cbs.knaw.nl

Abstract. Jansson and Sung showed that, given a dense set of input
triplets T (representing hypotheses about the local evolutionary relationships of triplets of taxa), it is possible to determine in polynomial
time whether there exists a *level-1 network* consistent with T, and if so
to construct such a network [18]. Here we extend this work by showing
that this problem is even polynomial-time solvable for the construction
of *level-2* networks. This shows that, assuming density, it is tractable to
construct plausible evolutionary histories from input triplets even when
such histories are heavily non-tree like. This further strengthens the case
for the use of triplet-based methods in the construction of phylogenetic
networks. We also implemented the algorithm and applied it to yeast
data.

1 Introduction

Phylogenetics is the field at the interface of biology, mathematics and computer-
science which studies the (re-)construction of plausible evolutionary scenarios
when confronted with incomplete and/or error-prone biological data. Until recently almost all research effort was directed at finding evolutionary trees. However, biologically, especially for lower order species, evolution does not necessarily
exhibit a tree structure. Thus, the quest for models and methods for more general evolutionary structures than trees has emerged naturally. This forms the
subject of this paper.

Many algorithmic strategies for constructing evolutionary trees have been
proposed in the literature. The most well-known techniques are Maximum Parsimony (MP), Maximum Likelihood (ML), Bayesian methods, Distance-based

---

* Part of this research has been funded by the Dutch BSIK/BRICKS project.

M. Vingron and L. Wong (Eds.): RECOMB 2008, LNBI 4955, pp. 450–462, 2008.
© Springer-Verlag Berlin Heidelberg 2008

methods (such as Neighbour Joining and UPMGA) and quartet-based methods, as well as various (meta-)combinations of these [3][10][19][24].

Quartet methods apply to the construction of unrooted evolutionary trees; less well studied is the problem of constructing *rooted* evolutionary trees, where the edges of the tree are directed to reflect the direction of evolution. The analogue of quartet methods in the case of rooted evolutionary trees are *triplet* methods: here we are given not unrooted trees on four leaves, but rooted binary trees on three leaves, as in Figure 1. The unique *rooted triplet* (*triplet* for short) on a leaf set $\{x, y, z\}$ in which the lowest common ancestor of x and y is a proper descendant of the lowest common ancestor of x and z is denoted by $xy|z$ (which is identical to $yx|z$). The triplet in the figure is $xy|z$.

Fig. 1. One of the three possible triplets on the leaves x, y and z. Note that, as with all figures in this article, all arcs are directed downwards, away from the root.

Aho et al. studied the problem of constructing a tree from a set of triplets. They showed that, given a set of triplets, it is possible to construct in polynomial time a rooted tree consistent with all the input triplets, or decide that no such tree exists [1]. This contrasts favourably with the corresponding quartet problem, which is NP-hard [25]. Various authors [2][6][14][15][26] have studied variations of the problem in cases where the algorithm of Aho et al. fails to return a tree. A well-studied one, albeit NP-hard [14], is to find a tree that maximises the number of input triplets it is consistent with.

In recent years attention has turned towards the construction of evolutionary scenarios that are not tree-like. This has been motivated by the fact that biological phenomena such as hybridisation, horizontal gene transfer, recombination, and gene duplication can cause lineages, which earlier in time diversified from a common ancestor, to once again intersect with each other later in time. These kind of evolutionary events are called *reticulation events* and lead to evolutionary scenarios where the underlying undirected graph potentially contains cycles. Rather than attempt to summarise this extremely varied area we refer the reader to [11], [22] and [23], all outstanding survey articles.

Informally, a level-k network is an evolutionary network where each biconnected component of the network contains at most k reticulation events. Jansson, Sung and Nguyen considered the problem of deciding whether, given a set of input triplets, it is possible to construct a *level-1* network (otherwise known as a *galled tree* [8] or a *galled network* [4][9][16]) consistent with all those triplets [16][18]. They showed that, in general, the level-1 problem is NP-hard. (In contrast, the algorithm of Aho et al. always runs in polynomial time.) However, they show that the problem can be solved in polynomial time when the input is *dense*, meaning that for each set of three taxa there is at least one triplet in the input. Density is a reasonable assumption if high-quality triplets can be constructed for all subsets of three taxa. The authors also give various upper bounds, lower bounds and approximation algorithms for the general case [16]. Related problems studied comprise the construction of level-1 networks from ultrametric distance matrices [4] and building level-1 networks where certain input triplets are *forbidden* [9]. Huson and Klöpper [12] consider a generalisation of level-1 networks that they call *galled networks* (defined differently than in [4][9][16]).

In this paper we extend considerably the work of Jansson and Sung by showing that, when the input set is dense, we can construct in polynomial time a *level-2* network consistent with all input triplets or decide that no such network exists. In case of a general (possibly non-dense) input triplet set, we claim that it is NP-complete to decide whether a level-2 network consistent with the input exists [13]. The proof of this claim, omitted here, is a nontrivial extension of the proof that this problem is NP-hard for level-1 networks [16]. In Section 3 we present our algorithm for constructing level-2 networks from dense triplet sets that runs in (slightly better than) cubic time in the number of input triplets. This significantly extends the power of triplet methods because it further extends the complexity of the evolutionary scenarios that can be constructed. For example, networks of the complexity shown in Figure 2 can be constructed by our algorithm. Our result and the way it is proved make it tempting to conjecture that, for fixed k, the level-k problem with dense triplet sets is polynomial-time solvable. However, it is not yet clear that the pivotal theorem in Section 3, Theorem 2, generalises easily to level-3 networks and higher.

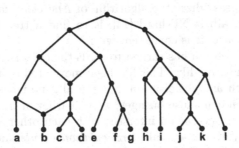

Fig. 2. An example of a level-2 network

An implementation of our algorithm in Java has been made publicly available [21]. We applied it to sequences from the yeast *Cryptococcus gattii*. We constructed a level-2 network for different isolates of this, potentially dangerous, yeast and are planning to use this network to find the origin of a *C. gattii* outbreak on the Westcoast of Canada. The results are reported in Section 4. In

Section 5 we discuss our conclusions and the many fascinating open problems that still remain in this area.

2 Preliminaries

A *phylogenetic tree* is a rooted binary tree with directed edges (arcs) and distinctly labelled leaves. A triplet $xy|z$ is thus a phylogenetic tree on three leaves. A *phylogenetic network* (*network* for short) is defined as a directed acyclic graph in which exactly one vertex has indegree 0 and outdegree 2 (the root) and all other vertices have either indegree 1 and outdegree 2 (*split vertices*), indegree 2 and outdegree 1 (*reticulation vertices*) or indegree 1 and outdegree 0 (*leaves*), where the leaves are distinctly labelled. In general directed acyclic graphs a reticulation vertex is a vertex with indegree 2. A directed acyclic graph is *connected* (also called "weakly connected") if there is an undirected path between any two vertices and *biconnected* if it contains no vertex whose removal disconnects the graph. A *biconnected component* of a network is a maximal biconnected subgraph.

Definition 1. *A network is said to be a level-k network if each biconnected component contains at most k reticulation vertices.*

A tree can thus be considered a level-0 network. A network that is level-k but not level-$(k - 1)$ is called a *strict level-k network*.

Denote the set of leaves in a network N by L_N. For any set T of triplets define $L(T) = \bigcup_{t \in T} L_t$ and let $n = |L(T)|$. A set T of triplets is called *dense* if for each $\{x, y, z\} \subseteq L(T)$ at least one of $xy|z$, $xz|y$ and $yz|x$ belongs to T. Furthermore, for a set of triplets T and a set of leaves $L' \subseteq L(T)$, we denote by $T|L'$ the triplets $t \in T$ with $L_t \subseteq L'$. We use L as shorthand for $L(T)$.

Definition 2. *A triplet $xy|z$ is* consistent *with a network N (interchangeably: N is consistent with $xy|z$) if N contains a subdivision of $xy|z$, i.e. if N contains vertices $u \neq v$ and pairwise internally vertex-disjoint paths $u \to x$, $u \to y$, $v \to u$ and $v \to z$.*

By extension, a set of triplets T is said to be *consistent* with N (interchangeably: N is consistent with T) if every triplet in T is consistent with N. To clarify triplet consistency we observe that the network in Figure 2 is consistent with (amongst others) $ab|c$, $bc|a$ and $dg|k$ but not consistent with (for example) $ah|f$ or $hk|i$.

We will now define SN-sets, introduced in [18], which will play a crucial role in the rest of the paper. For a triplet set T, let σ_T be the operation on subsets X of $L(T)$ defined by $\sigma_T(X) = X \cup \{c \in L(T) | \exists x, y \in X : xc|y \in T\}$. The set $SN_T(X)$ is defined as the closure of X w.r.t. the operation σ_T. Define an *SN-set* of T as a set of the form $SN_T(X)$ for some $X \subseteq L(T)$, i.e. SN-sets are the subsets of $L(T)$ that are closed under the operation σ_T. An SN-set X is *maximal* with respect to a triplet set T if $X \neq L(T)$ and $L(T)$ is the only SN-set that is a strict superset of X. In [18] it is shown that the maximal SN-sets of T partition $L(T)$ if T is dense.

We call an arc $a = (u, v)$ of a network N a *cut-arc* if its removal disconnects N and call it *trivial* if v is a leaf. A cut-arc is *highest* if there does not exist a cut-arc $a' = (u', v')$ such that u is reachable from v'. We say that a vertex is *below* (u, v) if it is reachable from v. A little thought should make it clear that for each cut-arc a in a network N consistent with a dense triplet set T, the set S of leaves below a is an SN-set of T and that the sets of leaves below highest cut-arcs partition L_N. The following lemma reveals a crucial characteristic that we exploit in our algorithm.

Lemma 1. *Let N be a network consistent with dense triplet set T. Each maximal SN-set S in T can be expressed as the union of the leaves below one or more highest cut-arcs in N.*

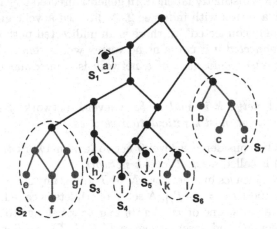

Fig. 3. Constructing a level-2 network by recursively constructing "simple" networks, given the partition $L = S_1 \cup \ldots \cup S_7$

3   Constructing Level-2 Networks from Dense Triplet Sets

This section describes our main result, a polynomial time algorithm that constructs a level-2 network from a dense triplet set T if such a network exists. The algorithm is recursive. The main idea is visualised in Figure 3. Suppose that we know the correct partition $L = S_1 \cup \ldots \cup S_7$ of the leaves. Then an algorithm can replace each set S_i by a single (meta-)leaf and start with constructing this "simple" level-2 network (in black). In Subsection 3.1 we formally introduce simple level-2 networks and show how they can be constructed. The complete level-2 network can then be obtained by replacing each meta-leaf S_i by a recursively created level-2 network. In spirit, this procedure resembles that for level-1 networks [18]. However, besides the fact that the simple level-2 networks are more complex, it also turns out that finding the right partition (and especially the proof of correctness) is far more involved than in the level-1 version of the problem. There does for example not always exist a level-2 network where the sets of leaves below highest cut-arcs correspond to the maximal SN-sets, as is the case for level-1 networks. How the recursion works in detail and how the correct partition can be found is explained in Subsection 3.2. Due to space constraints we have to defer all proofs to a full version of the paper.

3.1 Simple Level-2 Networks

We now introduce the class of level-2 networks that we name *simple* level-2 networks. Informally these are the basic building blocks of level-2 networks in the sense that each biconnected component of a level-2 network is in essence a simple level-2 network. We first introduce a *simple level-k generator*:

Definition 3. *A simple level-k generator is a directed acyclic biconnected multi-graph, which has a single root (indegree 0, outdegree 2), precisely k reticulation vertices (indegree 2, outdegree at most 1) and apart from that only split vertices (indegree 1, outdegree 2).*

In simple level-k generators, vertices with indegree 2 and outdegree 0 as well as all arcs are labelled and called *sides*. A simple case-analysis shows that there is only one simple level-1 generator and that there are four simple level-2 generators, depicted in Figure 4.

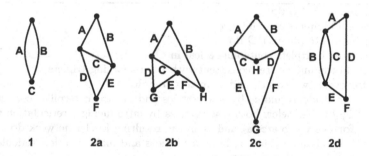

Fig. 4. The only simple level-1 generator and all four simple level-2 generators

Definition 4. *A simple level-k network N is a network obtained by applying the following transformation to some simple level-k generator such that the resulting graph is a valid network:*

1. *replace each arc X by a path and for each internal vertex v of the path add a new leaf x and an arc (v, x); and*
2. *for each vertex Y of indegree 2 and outdegree 0 add a new leaf y and an arc (Y, y).*

Note that in the above definition a path used to replace an edge possibly contains no internal vertices. An exception to this, however, is that whenever there are multiple arcs, we replace at least one of them by a path of at least three vertices. There is a nice and simple characterisation of simple level-k networks.

Lemma 2. *A strict level-k network is a simple level-k network if and only if it contains no nontrivial cut-arcs.*

All simple level-1 networks on dense triplet sets can be found by an algorithm by Jansson, Nguyen and Sung [16]. We designed Algorithm SL2 to find all simple level-2 networks consistent with a dense set of triplets T. Before describing the algorithm, we will first give an analysis of the structure of simple level-2 networks, whereon the algorithm is based. We use the notion *reticulation leaf* for a leaf whose parent is a reticulation vertex and say that a network is a *caterpillar* if the deletion of all leaves gives a directed path. We call a subset of the leaves L' a *caterpillarset* if there exists a network consistent with the input triplets that contains a caterpillar with leaves L' as subgraph. We prove that there are only $O(n)$ caterpillarsets and designed a routine to construct all of them from the triplets in $O(n^5)$ time. Now consider any simple level-2 network. If we remove a reticulation leaf and its parent, there is one reticulation vertex y left and below it is a caterpillar. If we now remove this last reticulation vertex and the caterpillar below it, we obtain a tree, which is unique [17] and can be constructed using the algorithm of Aho et al. [1]. The algorithm SL2 will first identify this tree and then reconstruct the simple level-2 network from it as follows.

Algorithm 1. SL2 (sketch)

1: **for** each leaf $x \in L$ **do**
2:     delete all triplets containing x
3:     **for** each caterpillarset Q **do**
4:         delete all triplets containing a leaf in Q
5:         build the unique tree consistent with the remaining triplets
6:         **for** every two arcs a_1 and a_2 in this tree **do**
7:             subdivide a_1 and a_2 by new vertices and put the caterpillar consistent with $T|Q$ below below both new vertices by introducing a reticulation vertex y
8:             **for** every two arcs a_3 and a_4 in the resulting level-1 network **do**
9:                 subdivide a_3 and a_4 by new vertices and make x a new reticulation leaf below both new vertices
10:                **if** the obtained network N is a simple level-2 network consistent with T **then**
11:                    output N.

Now suppose that triplet set T is consistent with some simple level-2 network N. At some iteration the algorithm will choose the right leaf and caterpillarset to remove and the right arcs to subdivide and the algorithm will construct the network N. Furthermore, for each constructed network the algorithm checks whether it is a simple level-2 network consistent with T. We conclude that the algorithm finds exactly all simple level-2 networks consistent with T.

For several reasons the actual algorithm is more complicated than the sketch above. Firstly, some special cases occur when there are no leaves on certain sides of the simple level-2 network. These cases can be dealt with by introducing dummy vertices. Secondly, we can improve the running time of the algorithm slightly if we loop through only a part of all combinations of arcs. Finally, we present an $O(n^3)$ routine that checks whether a given network N is a simple level-2 network consistent with a given triplet set T. This leads to an overall running time of $O(n^8)$.

Theorem 1. *Algorithm SL2 finds all simple level-2 networks consistent with a dense triplet set and can be implemented to run in time $O(n^8)$.* □

3.2   From Simple to General Level-2 Networks

This section explains how to build general level-2 networks by recursively building simple level-1 and -2 networks. The following theorem will be crucial.

Theorem 2. *Let T be a dense triplet set consistent with some level-2 network N. Then there exists a level-2 network N' consistent with T such that at most one maximal SN-set of T equals the union of the sets of leaves below two highest cut-arcs and each other maximal SN-set is equal to the set of leaves below just one highest cut-arc.*

For a collection $S = \{S_1, \ldots, S_q\}$ of SN-sets let $T\nabla S$ denote the *induced* set of triplets $S_i S_j | S_k$ such that there exist $x \in S_i$, $y \in S_j$, $z \in S_k$ with $xy|z \in T$ and i, j and k all distinct. The above theorem implies that, after possibly splitting one SN-set, we can replace each SN-set by a single leaf and the problem then essentially reduces to constructing a simple level-1 or -2 network for the induced triplet set. Given that there is at most one maximal SN-set that needs to be split into two subsets, we can simply try splitting each maximal SN-set of T in turn, as well as considering the case where no maximal SN-set of T is split. There are only $O(n)$ maximal SN-sets. The following lemma tells us how to split the chosen maximal SN-set into two subsets.

Lemma 3. *Let T be a dense set of triplets and N' a network with the properties described in Theorem 2. Suppose T contains a maximal SN-set X which occurs as the union of the sets S_1 and S_2 of leaves below two highest cut-arcs. Then $T|X$ contains precisely two maximal SN-sets and these are S_1 and S_2.*

The general outline of our algorithm LEVEL2, which constructs level-2 networks from dense triplet sets, is as follows. First we compute the maximal SN-sets. If there are precisely two maximal SN-sets then we recursively create two level-2 networks for the two maximal SN-sets and connect their roots to a new root. Otherwise, we try splitting each maximal SN-set in turn and we try the case where no maximal SN-set is split. If S is the obtained set of SN-sets then we compute the induced set of triplets $T\nabla S$ and try to construct a simple level-1 or -2 network N consistent with $T\nabla S$ using algorithm SL2. We recursively create level-2 networks for each SN-set in S and replace each leaf of N by the corresponding, recursively created, level-2 network.

It can furthermore be proven that if it is necessary to split an SN-set X then the simple level-2 network must be of type 2c and X must be below the two reticulation leaves. Exploiting this fact we can prove that LEVEL2 can be implemented to run in time $O(n^8)$, which is equal to $O(|T|^{\frac{8}{3}})$. A simplified version of the algorithm is displayed below.

Algorithm 2. LEVEL2 (sketch)

1: compute the set SN of maximal SN-sets of T
2: **if** $|SN| = 2$ **then**
3:     N consists of a root connected to two leaves: the elements of SN
4: **else**
5:     **if** $T \nabla SN$ is consistent with a simple level-1 network **then**
6:         let N be such a network
7:     **else if** $T \nabla SN$ is consistent with a simple level-2 network **then**
8:         let N be such a network
9:     **else**
10:         **for** $X \in SN$ **do**
11:             compute the set of maximal SN-sets SN' of $T|X$
12:             **if** $|SN'| = 2$ **then**
13:                 $S := SN \setminus \{X\} \cup SN'$
14:                 **if** $T \nabla S$ is consistent with a simple level-2 network of type 2c where the elements of SN' are the two reticulation leaves **then**
15:                     let N be such a network
16: replace each leaf X of N by a recursively created level-2 network for $T|X$.

Theorem 3. *Algorithm LEVEL2 constructs, in $O(|T|^{\frac{8}{3}})$ time, a level-2 network consistent with a dense set of triplets T if and only if such a network exists.*

4   Experimental Results

The algorithm from the previous section has been implemented in Java and applied to experimental data. The implementation was made publicly available [21]. The data of this application consists of sequences from different isolates of the yeast *Cryptococcus gattii*. This yeast is potentially dangerous and an ongoing outbreak on the Westcoast of Canada [20], which started in 1999, has caused many infections and even some fatalities. We have constructed a phylogenetic network for these isolates as a tool to find the origin of this *C. gattii* outbreak. We have blinded the names of the isolates here since the biological part of the research has not yet reached a conclusion.

Given the *C. gattii* sequences we have constructed a set of triplets as follows. Firstly, all identical sequences are combined into a single *sequence type (ST)*. One of the sequence types (that is only distantly related to the others) is used as an outgroup and we have applied the Maximum Likelihood method of PHYML [7] to each subset of four sequence types that includes the outgroup. Each output tree of PHYML gives us one input triplet for our algorithm LEVEL2. Running our algorithm for all ST's tells us that there exists no level-2 network consistent with all triplets. Therefore, we have applied our algorithm to a set containing as many

ST's as possible (where certain important ST's get priority over others) without destroying level-2-realisability. This set has been found by searching through all subsets. Given this subset and all triplets the execution of the algorithm LEVEL2 took 0.8 seconds on a Pentium IV, 3 GHz PC with 1GB memory. The resulting level-2 network is displayed in Figure 5. This network is consistent with all 1330 triplets that were generated over this set of taxa. The figure displays both a reticulate pattern and a dichotomous and tree-like structure. Our method is able to differentiate and visualise these.

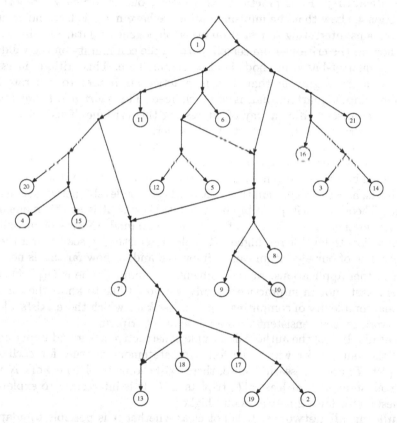

Fig. 5. The network constructed by the algorithm LEVEL2 for the triplets based on the yeast data set

A great advantage of our algorithm is that it is extremely fast. However, for data for which not all triplets can be found accurately or for which there are many reticulations that do not fit into a level-2 network, it might only be possible to find a phylogenetic network for a subset of the taxa. On the positive side, even in such cases, our level-2 networks could possibly give a better representation of reality than a tree or a level-1 network.

5   Conclusion and Open Questions

In this paper we have shown that in polynomial time we can decide whether a dense triplet set is consistent with a level-2 network, and if so construct such a network. In this way we have brought more complex, interwoven forms of evolution within reach of triplet methods. The described method has shown to be useful in practice. There remain, of course, many open questions and challenges, which we briefly list here.

1. **Applicability.** First practical experiments, one of which we reported in Section 4, show that the implementation we have made is fast and accurate. It remains interesting to test it on other phylogenetic data. We also wonder in how far the critique from certain parts of the community on the validity of many quartet-based methods is also relevant here. This critique in essence rests on the argument that it is in practice far harder to generate high-quality input quartets than is often claimed. The *short quartet method* [5] has been discussed as a way of addressing this critique. This debate needs to be addressed in the context of this paper.

2. **Complexity.** Is the non-dense level-k problem NP-hard for all fixed $k \geq 1$? Is the dense version polynomial-time solvable for all fixed k? In this regard it would be helpful to generalise Theorem 2, which captures the behaviour of SN-sets, and the algorithm that constructs simple level-2 networks. Generalising Theorem 2 will probably be difficult, because it is at this moment not clear whether the technique of "pushing" maximal SN-sets below cut-arcs generalises to level-3 and higher. It is also interesting to see how far the running time of our algorithm can be improved and/or how far this is necessary for further applications. At the moment the running time is $O(|T|^{\frac{8}{3}})$, which seems fast enough in practice. Finally, we would like to know the computational complexity of computing the smallest k for which there exists a level-k network that is consistent with a dense set of triplets.

3. **Bounds.** In [16] the authors determine constructive lower and upper bounds on the value τ for which the following statement is true: for each set of triplets T, not necessarily dense, there exists some level-1 network N which is consistent with at least $\tau|T|$ triplets in T. It is interesting to explore this question for level-2 networks and higher.

4. **Building all networks.** It is not clear whether it is possible to adapt our algorithm to generate *all* level-2 networks consistent with the input triplet set. If so, then such an adaptation could (even in the case that exponentially many networks are produced) be very useful for comparing the plausibility and/or relative similarity of the various solutions.

5. **Properties of constructed networks.** Under what conditions on the triplet set T is there only one network N consistent with T? Under what conditions does T permit some solution N such that the set of all triplets consistent with N, is exactly equal to T? These questions are also valid for level-1 networks.

6. **Different triplet restrictions.** Density is only one of very many possible restrictions on the input triplets. Interesting alternatives are what we have

named *minimal density* and *extreme density*. A minimally dense triplet set has exactly one triplet for every combination of three leaves. In the extreme dense case we assume that the set of input triplets is exactly equal to the set of triplets consistent with some network.

7. **Confidence.** At the moment all input triplets are assumed to be correct. Is there scope for attaching a confidence measure to each input triplet, and optimising on this basis? This is also related to the problem of ensuring that certain triplets are *excluded* from the output network, as explored in [9].

8. **Exponential-time exact algorithms.** It could be interesting, and useful, to develop exponential-time exact algorithms for solving the NP-hard problems for non-dense triplet sets.

Acknowledgements. We thank Katharina Huber for her useful ideas and many interesting discussions.

References

1. Aho, A.V., Sagiv, Y., Szymanski, T.G., Ullman, J.D.: Inferring a Tree from Lowest Common Ancestors with an Application to the Optimization of Relational Expressions. SIAM Journal on Computing 10(3), 405–421 (1981)
2. Bryant, D.: Building Trees, Hunting for Trees, and Comparing Trees: Theory and Methods in Phylogenetic Analysis, Ph.D. thesis, University of Canterbury, Christchurch, New Zealand (1997)
3. Bryant, D., Steel, M.: Constructing Optimal Trees from Quartets. Journal of Algorithms 38(1), 237–259 (2001)
4. Chan, H.-L., Jansson, J., Lam, T.W., Yiu, S.-M.: Reconstructing an Ultrametric Galled Phylogenetic Network from a Distance Matrix. Journal of Bioinformatics and Computational Biology 4(4), 807–832 (2006)
5. Erdös, P.L., Steel, M.A., Szekely, L.A., Warnow, T.: A few logs suffice to build (almost) all trees (Part II). Theoretical Computer Science 221(1), 77–118 (1999)
6. Gąsieniec, L., Jansson, J., Lingas, A., Östlin, A.: On the complexity of constructing evolutionary trees. Journal of Combinatorial Optimization 3, 183–197 (1999)
7. Guindon, S., Gascuel, O.: A simple, fast, and accurate algorithm to estimate large phylogenies by maximum likelihood. Systematic Biology 52(5), 696–704 (2003)
8. Gusfield, D., Eddhu, S., Langley, C.: Optimal, efficient reconstruction of phylogenetic networks with constrained recombination. Journal of Bioinformatics and Computational Biology 2, 173–213 (2004)
9. He, Y.-J., Huynh, T.N.D., Jansson, J., Sung, W.-K.: Inferring Phylogenetic Relationships Avoiding Forbidden Rooted Triplets. Journal of Bioinformatics and Computational Biology 4(1), 59–74 (2006)
10. Holder, M., Lewis, P.O.: Phylogeny estimation: Traditional and bayesian approaches. Nature Reviews Genetics 4, 275–284 (2003)
11. Huson, D.H., Bryant, D.: Application of Phylogenetic Networks in Evolutionary Studies. Molecular Biology and Evolution 23(2), 254–267 (2006)
12. Huson, D.H., Klöpper, T.H.: Beyond Galled Trees - Decomposition and Computation of Galled Networks. In: Speed, T., Huang, H. (eds.) RECOMB 2007. LNCS (LNBI), vol. 4453, pp. 211–225. Springer, Heidelberg (2007)

13. Iersel, L.J.J. van, Keijsper, J.C.M., Kelk, S.M., Stougie, L.: Constructing level-2 phylogenetic networks from triplets (preprint, 2007), http://arxiv.org/abs/0707.2890
14. Jansson, J.: On the complexity of inferring rooted evolutionary trees. In: proceedings of GRACO 2001, ENDM 7, pp. 121–125. Elsevier, Amsterdam (2001)
15. Jansson, J., Ng, J.H.-K., Sadakane, K., Sung, W.-K.: Rooted maximum agreement supertrees. Algorithmica 43, 293–307 (2005)
16. Jansson, J., Nguyen, N.B., Sung, W.-K.: Algorithms for Combining Rooted Triplets into a Galled Phylogenetic Network. SIAM Journal on Computing 35(5), 1098–1121 (2006)
17. Jansson, J., Sung, W.-K.: Inferring a Level-1 Phylogenetic Network from a Dense Set of Rooted Triplets. In: Chwa, K.-Y., Munro, J.I.J. (eds.) COCOON 2004. LNCS, vol. 3106, pp. 462–471. Springer, Heidelberg (2004)
18. Jansson, J., Sung, W.-K.: Inferring a Level-1 Phylogenetic Network from a Dense Set of Rooted Triplets. Theoretical Computer Science 363, 60–68 (2006)
19. Jiang, T., Kearney, P.E., Li, M.: A Polynomial Time Approximation Scheme for Inferring Evolutionary Trees from Quartet Topologies and Its Application. SIAM Journal on Computing 30(6), 1942–1961 (2000)
20. Kidd, S., Hagen, F., Tscharke, R., Huynh, M., Bartlett, K., Fyfe, M., MacDougall, L., Boekhout, T., Kwon-Chung, K.J., Meyer, W.: A rare genotype of Cryptococcus gattii caused the Cryptococcosis outbreak on Vancouver Island (British Columbia, Canada). In: Proceedings of the National Academy of Sciences of the United States of America, vol. 101, pp. 17258–17263 (2004)
21. LEVEL2: A fast method for constructing level-2 phylogenetic networks from dense sets of rooted triplets, http://homepages.cwi.nl/~kelk/level2triplets.html
22. Makarenkov, V., Kevorkov, D., Legendre, P.: Phylogenetic Network Reconstruction Approaches. In: Applied Mycology and Biotechnology. International Elsevier Series 6, Bioinformatics, vol. 6, pp. 61–97 (2006)
23. Moret, B.M.E., Nakhleh, L., Warnow, T., Linder, C.R., Tholse, A., Padolina, A., Sun, J., Timme, R.: Phylogenetic networks: modeling, reconstructibility, and accuracy. IEEE/ACM Transactions on Computational Biology and Bioinformatics 1(1), 13–23 (2004)
24. Semple, C., Steel, M.: Phylogenetics. Oxford University Press, Oxford (2003)
25. Steel, M.: The complexity of reconstructing trees from qualitative characters and subtrees. Journal of Classification 9, 91–116 (1992)
26. Wu, B.Y.: Constructing the maximum consensus tree from rooted triples. Journal of Combinatorial Optimization 8, 29–39 (2004)

Accurate Computation of Likelihoods in the Coalescent with Recombination Via Parsimony

Rune B. Lyngsø[1], Yun S. Song[2], and Jotun Hein[1]

[1] Department of Statistics, Oxford University, Oxford OX1 3TG, UK
[2] Computer Science Division and Department of Statistics, University of California, Berkeley, CA 94720, USA
lyngsoe@stats.ox.ac.uk, yss@cs.berkeley.edu, hein@stats.ox.ac.uk

Abstract. Understanding the variation of recombination rates across a given genome is crucial for disease gene mapping and for detecting signatures of selection, to name just a couple of applications. A widely-used method of estimating recombination rates is the maximum likelihood approach, and the problem of accurately computing likelihoods in the coalescent with recombination has received much attention in the past. A variety of sampling and approximation methods have been proposed, but no single method seems to perform consistently better than the rest, and there still is great value in developing better statistical methods for accurately computing likelihoods. So far, with the exception of some two-locus models, it has remained unknown how the true likelihood exactly behaves as a function of model parameters, or how close estimated likelihoods are to the true likelihood. In this paper, we develop a deterministic, parsimony-based method of accurately computing the likelihood for multi-locus input data of moderate size. We first find the set of all ancestral configurations (ACs) that occur in evolutionary histories with at most k crossover recombinations. Then, we compute the likelihood by summing over all evolutionary histories that can be constructed only using the ACs in that set. We allow for an arbitrary number of crossing over, coalescent and mutation events in a history, as long as the transitions stay within that restricted set of ACs. For given parameter values, by gradually increasing the bound k until the likelihood stabilizes, we can obtain an accurate estimate of the likelihood. At least for moderate crossover rates, the algorithm-based method described here opens up a new window of opportunities for testing and fine-tuning statistical methods for computing likelihoods.

1 Introduction

Estimating evolutionary parameters and making ancestral inference are an important part of molecular evolutionary genetics. Often, at the core of these studies is the problem of computing the probability of observing sample sequences for given parameter values. In the context of the coalescent model and its various extensions, closed-form formulas are generally not known for such likelihoods,

M. Vingron and L. Wong (Eds.): RECOMB 2008, LNBI 4955, pp. 463–477, 2008.
© Springer-Verlag Berlin Heidelberg 2008

and therefore several computationally intensive statistical methods have been proposed for approximating them. Most of these statistical approaches fall into one of two categories, one based on Markov chain Monte Carlo methods—for examples, see [3, 25, 26, 44]—and the other on importance sampling methods, some notable examples being [5, 6, 9, 12, 13, 14, 15, 41]. Both approaches involve sampling genealogies to estimate a sum over the genealogies consistent with the input data.

The problem of estimating recombination rates has received particular attention in the past and various methods have been proposed so far [9, 10, 11, 12, 23, 26, 28, 30, 42]. (Henceforward, when we say recombination, we will mean crossover recombination.) Since computing the full likelihood in the coalescent with recombination is difficult, several approximation methods have been proposed. Hudson's [23] composite likelihood method is a popular approximation method which treats pairs of loci as being independent and takes a product of two-locus full likelihoods over all pairs of loci. (Different versions of the composite likelihood idea have also been suggested. E.g, see [10, 11].) This method has been generalized to study the fine-scale crossover rate variation in the human genome [24, 32, 33].

On the algorithms side, much recent attention has focused on the problem of estimating the minimum number $R_{min}(D)$ of recombinations needed to derive a given set D of sequences, using some specified model of mutations. A commonly adopted model is the *infinite-sites model*, which implies that each site can mutate at most once in the entire evolutionary history of the sequences. Assuming that mutation model, it has been shown that computing $R_{min}(D)$ is NP-hard [4, 43], and previous algorithms that compute it exactly either work only on relatively small data sets [36, 38], or on problems with special structure [16, 17, 18]. Since there are no efficient algorithms to compute $R_{min}(D)$ for an arbitrary D, several papers have considered efficient computation of lower bounds on $R_{min}(D)$ [22, 19, 20, 34, 37, 1, 2, 17, 18, 16, 40], as well as practical upper bounds [40].

In a recent paper [29], we have made progress in making the exact computation of $R_{min}(D)$ more practical, significantly increasing the size of data sets that can be handled. Here, we extend some of the algorithmic ideas developed in that paper to address the aforementioned problem of computing likelihoods in the coalescent with recombination. To our knowledge, this is the first application of a parsimony-based algorithm to likelihood computations in the coalescent.

The main idea behind our approach goes as follows. Instead of attempting to sum over all genealogies, we sum only over a restricted subset of genealogies. To each genealogy, there corresponds a sequences of events, consisting of mutations, coalescences, and recombinations. When an event happens, going backwards in time, there is a change in ancestral configuration (AC) [39], defined as the set of all DNA sequences present at a particular point in time in the genealogy. Summing over all genealogies for D corresponds to summing of all paths of ACs consistent with D, i.e., with each path starting from the input data D and ending at an AC in which every site in the input data has found a common ancestor.

In our work, we first find the set of all ACs that occur in evolutionary histories with at most k recombinations. Then, we compute the likelihood by summing over all evolutionary histories that can be constructed only using the ancestral configurations in that set. We allow for an arbitrary number of recombination, coalescent and mutation events in the evolutionary history, as long as the transitions stay within that restricted set of ancestral configurations. By starting with $k = R_{min}(D)$ and incrementing the bound k gradually until the change in likelihood satisfies some stopping criteria, we can compute the likelihood accurately.

There exist well-defined recursions relating the probability of a given AC ψ to the probabilities of those ACs that can be reached from ψ using one event back in time [7, 8, 13, 14, 15, 12, 35]. Solving the system of recursion relations to evaluate the probability of $\psi = D$ effectively sums over all possible genealogies consistent with D. In our work, we systematically solve the system of recursion relations involving the probabilities of the ACs in the restricted set described above. Note that this effectively sums not only over genealogies with at most k recombinations, but over all genealogies that can be constructed using the ACs in the restricted set with an arbitrary number of recombination events.

Although our deterministic approach can currently handle only small data sets—say, with about ten sequences and half as many sites—the work described here should prove useful for evaluating the performance of Monte-Carlo-based methods. Further, some pseudo-likelihood methods [23, 32, 33] are based on accurate likelihood calculations for few (typically 2) sites, and the method presented here significantly extends this capability.

2   Methods

We use D to denote a set of single nucleotide polymorphisms (SNPs) with two alleles at each site. We assume that the ancestral allele type is known. (This assumption is only made for ease of exposition. The approach presented here has a straightforward generalization to the case in which the ancestral allele type is unknown, albeit with steeper time and space complexity. Our software handles both cases.) In what follows, the ancestral allele is denoted by 0, while the mutant allele type is denoted by 1. For given mutation and recombination rates, our goal is to compute the probability of observing D under the coalescent with recombination and the infinite sites model of mutation.

2.1   Possible Events Back in Time

We assume that D contains m segregating sites with positions $s_1, \ldots, s_m$. We rescale the region to a unit interval between 0 and 1 so that $0 = s_1 < s_2 < \cdots < s_m = 1$. We use θ and ρ to denote, respectively, the population-scaled mutation and recombination rates for the unit interval. We assume that both recombination and mutation rates are constant over the interval. For given θ and ρ, the probability of observing D is obtained by integrating over the probabilities of all evolutionary histories that derive D. Tracing an evolutionary history backwards in time gives a path of ancestral configurations, reached from D through the following types of events back in time:

Mutation. We assume the infinite sites model of mutation. So, for any particular site, if there is exactly one sequence carrying a 1 at that site, it may change to the ancestral type 0.

Recombination. A sequence x breaks up into two new sequences with a breakpoint between sites i and $i+1$. One new sequence carries the prefix of x up to site i, followed by a suffix of length $m-i$ carrying non-ancestral material, denoted by $*$s. The other new sequence carries the suffix of x starting from site $i+1$, preceded by a prefix of length i carrying non-ancestral material, again denoted by $*$s. Recombination events where there is no ancestral material (0 or 1) either to the left or to the right of the breakpoint are ignored.

Coalescent Type 1. Two identical sequences find a common ancestor.

Coalescent Type 2. Two distinct sequences find a common ancestor if there is no site in which one sequence carries a 1 and the other a 0. Suppose that two sequences x and y are replaced by a single sequence z via coalescence. Then, z contains a 1 (respectively, 0) at site i if either x or y contains a 1 (respectively, 0) at site i. Otherwise, z has a "$*$" at site i.

See [12] for a more detailed description of the coalescent with recombination.

2.2   The Full Recursion

Griffiths and Marjoram [12] constructed a system of recursion relations satisfied by ancestral configurations, assuming a continuous model of recombination. Obtaining a closed-form solution to the recursions is out of reach, so they proposed using an importance sampling method to obtain numerical solutions. More efficient importance sampling approaches now exist for computing coalescent likelihoods by sampling genealogies [41, 9, 5, 6], but the recursions found by Griffiths and Marjoram still provide a transparent framework for computation. In what follows, we devise a deterministic algorithm for numerically solving the recursions accurately. To make the problem tractable, we assume a discrete model of recombination such that breakpoints occur only at the midpoints between consecutive segregating sites. Such a discretized model of recombination has been adopted by others in the past [9, 27].

To describe the recursions in more detail, we first need to define some notation. An ancestral configuration is a multiset of strings from $X = \{0, 1, *\}^m \setminus \{*^m\}$. With a chosen ordering on X, we use $n \in \mathbb{Z}_{\geq 0}^{3^m - 1}$ to uniquely specify an AC by listing the multiplicity of each element in X. A subscript (respectively, superscript) on n denotes decreasing (respectively, increasing) the multiplicity of string i by 1. For example, n_i denotes changing the component n_i to $n_i + 1$,

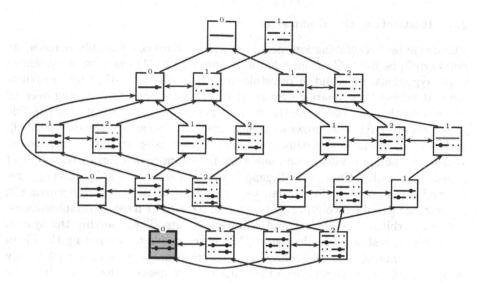

Fig. 1. Graph of all ACs for $D = \{01, 11\}$. Each node (box) corresponds to an AC, with "—" denoting an ancestral segment, " " a non-ancestral segment, and "•" a site carrying the mutant allele. The highlighted node at the bottom left corresponds to D. A directed edge joins an AC x to an AC y if there is an event (coalescence, mutation, or recombination) that transforms x to y. Each node is labeled with the minimum value of k for which that AC is in $\mathcal{C}_k(D)$, i.e. $\mathcal{C}_k(D)$ consists of all the nodes labeled k or less. ACs connected by horizontal bi-directional arrows form a strongly connected component of the graph. The probability of D for $\theta = 2$ and $\rho = 1$ is 0.125 when only ACs from $\mathcal{C}_0(D)$ are used, 0.193 when ACs from $\mathcal{C}_1(D)$ are used, and 0.202 when the full equation system is used. If cyclic structures of the recursions are eliminated by requiring that coalescing sequences have at least one site where they both carry ancestral material, as proposed in [31], the above probabilities reduce to 0.125, 0.172, and 0.174, respectively. This suggests that our parsimony-based approximation method of restricting the set of ACs is more accurate than forbidding certain classes of events.

while keeping n_j for $j \neq i$ unchanged. Then, the recursion relation satisfied by the probability $Q(n)$ of an AC n can be schematically written as

$$
Z(\boldsymbol{n}, \theta, \rho)\, Q(\boldsymbol{n}) = \underbrace{\sum c(\boldsymbol{n}, i)\, Q(\boldsymbol{n}_i)}_{\substack{\text{coalescent type 1} \\ x_i \text{ with } x_i}} + \underbrace{\sum c(\boldsymbol{n}, i, j, k)\, Q(\boldsymbol{n}_{ij}^k)}_{\substack{\text{coalescent type 2} \\ x_i \text{ with } x_j \to x_k}}
$$

$$
+ \theta \underbrace{\sum c(\boldsymbol{n}, k)\, Q(\boldsymbol{n}_i^k)}_{\substack{\text{mutation} \\ x_i \to x_k}} + \rho \underbrace{\sum c(\boldsymbol{n}, s_1, \ldots, s_m,, i, j, k)\, Q(\boldsymbol{n}_k^{ij})}_{\substack{\text{recombination} \\ x_k \to x_i \text{ and } x_j}}, \tag{1}
$$

where $c(\cdot)$ denote combinatorial coefficients that depend on the factors specified in the argument; $Z(\boldsymbol{n}, \theta, \rho)$ is a normalization constant; $x_i, x_j, x_k \in X$; and summations are performed over the events described in the previous subsection. Shown in Fig. 1 is an example of "unwrapping" the above recursion for an input data set D containing two length-2 sequences 01 and 11. In total there are 23 ACs for D, shown as rectangular boxes in Fig. 1, explained further below.

2.3   Restricting the Recursion

The discretized recombination model described above considerably reduces the number of possible ACs, from infinite to finite. Since (1) describes a system of linear equations, we could in principle find the probability of D by constructing and solving this equation system. It would correspond to summing over all genealogies that can derive D. However, although finite, the number of possible ACs for a given data set grows extremely fast with the size of the data set [39], and exact computation remains infeasible for practical purposes. As mentioned in Sect. 1, the main idea behind our work is to sum over a restricted subset of genealogies, rather than over all genealogies. We achieve this by solving a restricted system of recursions. First, we find the set of all ACs each occurring in at least one possible evolutionary history for D with at most k recombinations, but with arbitrary coalescent and mutation events. Then, solving the system of recursions restricted to that set of ACs corresponds to computing the likelihood by summing over all evolutionary histories that can be constructed only using the ACs in that set. Note that this is more general than summing over the genealogies with at most k recombinations. As long as transitions remain within the restricted set of ACs, our method allows for an arbitrary number of recombination events in a genealogy.

The method can be used either with a fixed value of k determined from $R_{min}(D)$, or increasing k until a stopping criteria is met. The simplest stopping criteria is to continue until the change in likelihood becomes less than some specified small number ϵ. From our experiment, we suggest using a stopping criteria based on diminishing returns, stopping when the change in likelihood begins to decrease.

Formally, we define the k-neighborhood $\mathcal{N}_k(n)$ of an AC n to be the set of all ACs reachable from n with no more than k recombinations. The inverse k-neighborhood of n is defined as $\mathcal{N}_k^{-1}(n) = \{n' \mid n \in \mathcal{N}_k(n')\}$. Finally, the k-configurations for D is defined as $\mathcal{C}_k(D) := \bigcup_{i=0}^k \left[\mathcal{N}_i(D) \cap \mathcal{N}_{k-i}^{-1}(\mathcal{A}) \right]$, where $\mathcal{A}$ denotes the set of ACs in which every site has found a common ancestor and $\mathcal{N}_i^{-1}(\mathcal{A}) := \bigcup_{a \in \mathcal{A}} \mathcal{N}_i^{-1}(a)$. Note that $\mathcal{C}_k(D)$ is the set of ACs that can occur in histories with at most k recombinations. Fig. 2 illustrates these concepts.

Our proposed method of computing the probability of the input data set D is to set $Q(n) = 0$ if $n \notin \mathcal{C}_k(D)$ and apply the recursion in (1) if $n \in \mathcal{C}_k(D)$. For a data set with n sequences and m segregating sites, $\mathcal{C}_{2n(m-1)}(D)$ will be equal to the set of all ACs for D, since any AC can be reached from D using at most $n(m-1)$ recombinations and an AC in $\mathcal{A}$ can then be reached using at most $n(m-1)$ additional recombinations. Therefore, for sufficiently large k, our method becomes equivalent to solving the full equation system.

2.4   Algorithmic and Implementation Details

The k-neighborhoods $\mathcal{N}_k(D)$ can be computed incrementally by increasing k one by one. However, in our work the entire k-neighborhood $\mathcal{N}_k(D)$ is not needed; only the k-configurations $\mathcal{C}_k(D) = \bigcup_{i=0}^k \left[\mathcal{N}_i(D) \cap \mathcal{N}_{k-i}^{-1}(\mathcal{A}) \right]$ are needed. First,

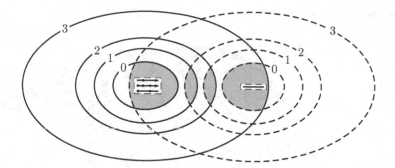

Fig. 2. Illustration of k-neighborhoods $\mathcal{N}_k(D)$ and k-configurations $\mathcal{C}_k(D)$. Neighborhoods around the input data D are shown as solid ellipses and inverse neighborhoods $\mathcal{N}_i^{-1}(a)$ around a particular most-recent-common-ancestor AC $a \in \mathcal{A}$ are shown with dashed ellipses. The minimum number of recombinations required for the data set is 3, and the regions corresponding to the set $\mathcal{C}_3(D)$ of 3-configurations are shaded in gray.

note that we can determine whether $n \in \mathcal{N}_{k-i}^{-1}(\mathcal{A})$ by checking whether the minimum number of recombinations needed to derive n is at most $k - i$. To compute that minimum number, we can employ the algorithm described in [29]. Second, $n \in \mathcal{N}_i(D) \cap \mathcal{N}_{k-i}^{-1}(\mathcal{A})$ only if there exists an AC $n' \in \mathcal{N}_{i-1}(D) \cap \mathcal{N}_{k-i+1}^{-1}(\mathcal{A})$ such that $n \in \mathcal{N}_1(n')$. Using these ideas, we can find $\mathcal{C}_k(D)$ without having to explore the entire set $\mathcal{N}_k(D)$, which can be significantly larger than $\mathcal{C}_k(D)$. Pictorially, in Fig. 2 we enumerate only the ACs in the shaded areas and their one-event neighbors, rather than the full k-neighborhoods of D. In this way, we can achieve a large reduction in both time and space requirement.

A dependency graph corresponding to the systems of recursions in (1) is a graph with one node for each AC and a directed edge from n to n' if n' appears on the right hand side of the recursion (1) for n. Once $\mathcal{C}_k(D)$ has been found, we determine the strongly connected components of the dependency graph and the directed acyclic graph connecting them. Then, subsystems of recursions corresponding to the strongly connected components are solved in reverse topological order. This reduces the time complexity from $O(|\mathcal{C}_k(D)|^3)$ to $O(|\mathcal{C}_k(D)| \times M^2)$, where M is the size of the largest strongly connected component.

In our implementation, a coarse grained *a priori* separation and sorting of connected components is obtained by sorting the ACs in $\mathcal{C}_k(D)$ by their total number of 0s and 1s. Going backwards in time in any evolutionary history, the total number of 0s and 1s will be non-increasing. This means that if the total number of 0s and 1s in n is larger than that in n', then $Q(n')$ does not depend on the value of $Q(n)$, thus allowing the recursions to be solved slice by slice in the order of increasing total number of 0s and 1s.

We have implemented our algorithm in C, using the UMFPack library. Our software is called cob, available at http://www.stats.ox.ac.uk/~lyngsoe/section26/ under the Lesser Gnu Public License.

Likelihood

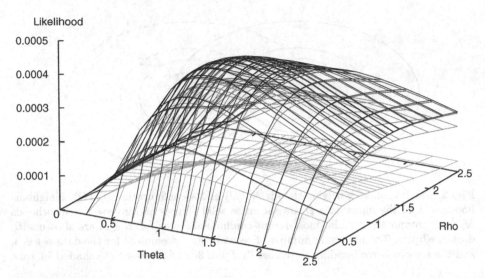

Fig. 3. Likelihood surfaces for $D = \{010, 010, 101, 101, 110\}$ computed using the ACs in $C_1(D)$ (light gray), $C_2(D)$ (medium gray), $C_3(D)$ (dark gray) and $C_{11}(D)$ (black). Both θ and ρ range from 0.0 to 2.5.

3   Results

When facing a hard computational problem, one usually needs to choose the right balance between accuracy and speed. In this section we explore these two aspects of our method. We will assess the quality of the approximation proposed in the previous section, by characterizing the behavior of the likelihood itself and also by studying the accuracy of the maximum likelihood estimates (MLEs) of the population-scaled mutation and recombination rates θ and ρ, respectively.

3.1   Comprehensive Analysis of Small Data Set

We first study a small data set $D = \{010, 010, 101, 101, 110\}$ with segregating sites at positions 0, 0.75 and 1. The minimum number $R_{min}(D)$ of recombinations for this data set is 1, and the size of $C_1(D)$ is 74. It turns out that all possible ACs can occur in evolutionary histories with at most 11 recombinations. The size of $C_{11}(D)$ is 400,820. This is sufficiently small that the full system of recursions can be solved in a reasonable time, allowing us to track the accuracy and resource requirement for the approximation based on $C_k(D)$, as k is varied from 1 to 11.

For a grid of θ and ρ values between 0.0 and 2.5, Fig. 3 shows four likelihood surfaces computed using four different levels of approximation: based on $C_1(D)$, $C_2(D)$, $C_3(D)$, and the full equation system (i.e., $C_{11}(D)$). The likelihood surfaces for the remaining $C_k(D)$ have been left out as they are sandwiched between the $C_3(D)$-surface and the $C_{11}(D)$-surface, and these are already very similar. Numerical values of the likelihood obtained from the $C_3(D)$-based equation

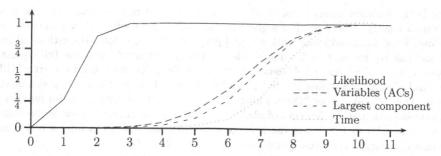

Fig. 4. The likelihood of $D = \{010, 010, 101, 101, 110\}$, the number of ACs (or variables), the size of the largest strongly connected component, and the running time, each plotted against the value of k in $C_k(D)$, as a fraction of the corresponding values for the full recursion system. The full recursion system produced a likelihood of 4.05×10^{-4} and had $400,820$ ACs in total. The largest strongly connected component contained $15,998$ ACs and the computation took 404.2 seconds.

system and that from the $C_{11}(D)$-based system differ by little, at least for the range of θ and ρ shown in Fig. 3. Consequently, MLEs θ^*, ρ^* for the two cases are very similar, with $\theta^* = 1.1426$, $\rho^* = 0.9631$ for $C_3(D)$ and $\theta^* = 1.1426$, $\rho^* = 0.9753$ for $C_{11}(D)$. Even the $C_2(D)$-surface is not far off from the $C_{11}(D)$-surface, with MLEs $\theta^* = 1.1523$ and $\rho^* = 0.8240$, which is beginning to show a trend of underestimating ρ. This trend becomes even more pronounced with the $C_1(D)$-based equation system, with estimates $\theta^* = 1.2427$ and $\rho^* = 0.4370$. This $C_1(D)$-based method also significantly underestimates the likelihood for most values of θ and ρ. As our heuristic is based on ignoring ACs that only contribute to the likelihood through evolutionary histories with many recombinations, for fixed θ, not surprisingly the difference between the approximated and the true likelihoods increases with increasing ρ. For fixed ρ and varying θ, the difference between the approximated and the true likelihoods tend to correlate more with the magnitude of the likelihood than with the value of θ.

As k in $C_k(D)$ varied from 1 to 11, the change in some key features of the computation is plotted in Fig. 4 where the likelihood was computed at the MLEs $\theta^* = 1.1426$, $\rho^* = 0.9753$ from the full equation system. All plots exhibit an S-curve behavior. A very encouraging feature of these plots is that the turning point for the likelihood plot occurs much earlier than the turning points for the time, the largest strongly connected component, and the equation system size plots. In fact, the likelihood all but coincides with the exact value before any of the other features shows any increasing tendency. Identifying the turning point of an assumed S-curve behavior of the likelihood is the basis of our diminishing-returns stopping criteria described before.

3.2 Average Behavior on Simulated Data Sets

The data set analyzed above is just one example. To study the average behavior, we used Hudson's [21] program **ms** to generate simulated data under the

Table 1. Average behavior of maximum likelihood estimation based on the near-minimal history restriction. For a given n and m, 100 data sets with n sequences and m sites were simulated using Hudson's [21] program ms. LR denotes the estimated likelihood relative to the true likelihood (denoted LH) computed using the full equation system, while $|\Delta\theta^*|$ and $|\Delta\rho^*|$ denote average absolute deviation from the true MLEs (θ^* and ρ^*) obtained using the full equation system. Running times are given in seconds. The columns under "Diminishing Returns" are for incrementing the number of recombinations until differences of likelihoods between increments no longer increases. The column labeled "k" lists the average value of k for which the final solution was obtained using $\mathcal{C}_{r+k}(D)$-configurations, where $r := R_{min}(D)$.

$n \times m$	$\mathcal{C}_r(D)$ LR	$|\Delta\theta^*|$	$|\Delta\rho^*|$	Time	$\mathcal{C}_{r+1}(D)$ LR	$|\Delta\theta^*|$	$|\Delta\rho^*|$	Time	$\mathcal{C}_{r+2}(D)$ LR	$|\Delta\theta^*|$	$|\Delta\rho^*|$	Time
2×2	1.00	0.00	0.00	0.00	1.00	0.00	0.00	0.00	1.00	0.0000	0.0000	0.02
3×2	0.99	0.01	0.06	0.00	1.00	0.01	0.03	0.10	1.00	0.0000	0.0003	0.12
4×2	1.00	0.01	0.09	0.00	1.00	0.01	0.05	0.16	1.00	0.0000	0.0014	0.19
5×2	0.98	0.02	0.12	0.02	1.00	0.01	0.06	0.21	1.00	0.0002	0.0018	0.36
6×2	0.98	0.02	0.17	0.09	0.99	0.01	0.08	0.33	1.00	0.0002	0.0034	0.56
2×3	1.00	0.00	0.00	0.00	1.00	0.00	0.00	0.10	1.00	0.0000	0.0000	0.10
3×3	0.99	0.02	0.05	0.00	1.00	0.01	0.03	0.24	1.00	0.0007	0.0031	0.31
4×3	0.98	0.03	0.10	0.05	1.00	0.01	0.05	0.45	1.00	0.0016	0.0057	0.88
5×3	0.99	0.01	0.07	0.09	1.00	0.01	0.04	0.59	1.00	0.0000	0.0036	0.90

$n \times m$	Diminishing Returns LR	$|\Delta\theta^*|$	$|\Delta\rho^*|$	Time	k	Full Equation System LH	θ^*	ρ^*	Time
2×2	1.00	0.0000	0.0000	0.002	1.00	0.15	5.08	0.00	0.10
3×2	1.00	0.0000	0.0000	0.15	1.27	0.06	3.88	0.07	0.19
4×2	1.00	0.0000	0.0000	0.28	1.29	0.05	4.16	0.09	0.50
5×2	1.00	0.0002	0.0002	0.70	1.76	0.03	3.41	0.14	1.12
6×2	1.00	0.0000	0.0014	0.94	1.57	0.02	3.45	0.21	2.34
2×3	1.00	0.0000	0.0000	0.10	1.00	0.16	5.86	0.00	1.20
3×3	1.00	0.0000	0.0000	0.62	1.16	0.04	4.55	0.05	18.7
4×3	1.00	0.0000	0.0004	2.11	1.28	0.02	3.91	0.13	326
5×3	1.00	0.0000	0.0000	2.28	1.18	0.01	3.10	0.07	11918

coalescent with recombination. We generated 100 data sets for a given number of sequences and a given number of sites. We considered two to six sequences with either two or three sites. Hudson's program actually uses a finite-sites model of recombination, requiring the user to specify the number of sites. In our study, all simulations were carried out with 10,000 sites in the recombination model. We set $\rho = 5$ and used -s option to fix the number of segregating sites. For each data set, we determined the MLE of θ and ρ by iterating eight times the likelihood computation on a five-by-five grid of θ and ρ, refining around the (θ, ρ) pair that yielded the highest likelihood. For each simulated data set, four different computations were done: using $\mathcal{C}_k(D)$-configurations, for $k = r, r+1, r+2$, where $r := R_{min}(D)$, or using the diminishing returns stopping criteria. Simulated data

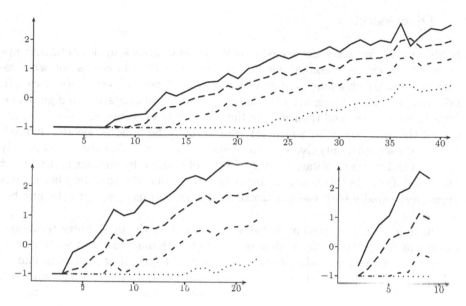

Fig. 5. Logarithm of running times in seconds as function of number of sequences for data sets with two segregating sites (top), three segregating sites (bottom left) and four segregating sites (bottom right). For each data set the likelihood at $\theta = 2$ and $\rho = 5$ was computed based on $C_r(D)$ (shown in "$\cdots$"), on $C_{r+1}(D)$ ("$- - -$"), on $C_{r+2}(D)$ ("$- -$"), and using the diminishing returns stopping criteria ("$\underline{\quad}$"). Average running times of less than 0.1 second were truncated to 0.1 second.

sets were sufficiently small that it was possible to solve the full equation system, thus allowing approximations to be compared with the true value.

Results are summarized in Table 1. Both the likelihood itself and MLE of θ are quite accurate even for the computation based on $C_r(D)$-configurations, while MLE of ρ becomes quite accurate when the equation system is expanded to $C_{r+2}(D)$-configurations. Applying the diminishing returns stopping criteria does slightly better than using $C_{r+2}(D)$-configurations, both in terms of accuracy and time. All in all, our method is quite accurate, while being substantially faster than using the full equation system.

Being able to compare results to the true values severely limits the data set sizes that can be investigated. Even for data sets with just five sequences and three segregating sites, we experienced an average running time of more than three hours to solve the full equation system at 200 (θ, ρ) grid points to obtain the MLEs. (In contrast, our method required only a few seconds on average to obtain very accurate estimates. See Table 1.) To investigate how large of a data set our method can handle, we simulated data sets with more number of sequences, while keeping the number of segregating sites to two, three or four. We again used Hudson's [21] program ms with $\rho = 5$. For each data set, we computed the likelihood for $\theta = 2$ and $\rho = 5$. Average computation times averaged over ten simulated data sets are plotted in Fig. 5.

4   Discussion

In this paper, we have developed a novel parsimony-based, deterministic approach for accurately computing the likelihood under the coalescent with recombination. Given enough computation time and memory, our method can, in principle, compute the exact likelihood, by finding all ancestral configurations for a given data set and then solving the full system of recursions. However, the size of the input data for which this can actually be done is severely limited. For a data set with only five sequences and three segregating sites, it currently takes several hours to obtain accurate MLEs of θ and ρ by computing the exact likelihood. Perhaps this is not so surprising, given that the total number of ACs grows very rapidly with the number of sequences and more so with the number of sites [39].

Our approximation method is based on restricting the probability recursions to certain ACs, namely those that occur in evolutionary histories with a near-minimal number of recombinations. The restricted system of recursions can be solved several orders of magnitude faster than the full recursion system, with no noticeable loss of accuracy. It dramatically increases the size of data sets for which one can compute the likelihood by solving the recursion system. For example, our approximation method takes only a few minutes to compute the probability of a data set with twenty sequences and three sites, while, in the same amount of time, one can only compute the probability of a data set with five sequences and three sites using the full equation system. However, even with the techniques introduced here, our method is limited to moderate-sized data sets. Despite the enormous reduction in time requirement of our method compared to the exact computation, the complexity of the problem grows so astronomically fast with data size that the speedup is dwarfed in comparison. For further details on this matter, we again refer to our previous work [39], where the growth of $C_{R_{min}}(D)$ as a function of data size was also investigated. We believe that new insights—e.g., regarding symmetries in the recursion structure, allowing ACs to be lumped together—are required for making this kind of algorithm-based approach applicable to large data sets.

Even so, the work presented here should be useful to the researchers in statistical genetics. For moderate-sized data sets, our method can be used to develop benchmarks with very well-characterized likelihoods. Such studies can be valuable for evaluating the performance of existing and new sampling-based approaches, and for fine-tuning them. Further, as some pseudo-likelihood methods [23, 32, 33] use likelihood calculations for few (typically 2) sites, the method developed here should be useful for improving such methods.

Acknowledgment

We would like to thank Thomas Mailund for useful discussions. This research is supported in part by BBSRC grants BB/D005418/1 and BB/D012139/1 (RBL and JH), and by NIH grants 1K99GM-080099 and 4R00-GM080099 (YSS).

References

1. Bafna, V., Bansal, V.: The number of recombination events in a sample history: conflict graph and lower bounds. IEEE/ACM Transactions on Computational Biology and Bioinformatics 1, 78–90 (2004)
2. Bafna, V., Bansal, V.: Improved Recombination Lower Bounds for Haplotype Data. In: Miyano, S., Mesirov, J., Kasif, S., Istrail, S., Pevzner, P.A., Waterman, M. (eds.) RECOMB 2005. LNCS (LNBI), vol. 3500, pp. 569–584. Springer, Heidelberg (2005)
3. Beaumont, M.: Detecting population expansion and decline using microsatellites. Genetics 153, 2013–2029 (1999)
4. Bordewich, M., Semple, C.: Computing the minimum number of hybridization events for a consistent evolutionary history. Discrete Applied Mathematics 155, 914–928 (2007)
5. De Iorio, M., Griffiths, R.C.: Importance sampling on coalescent histories. I. Adv. Appl. Prob. 36, 417–433 (2004)
6. De Iorio, M., Griffiths, R.C.: Importance sampling on coalescent histories. II: Subdivided population models. Adv. Appl. Prob. 36, 434–454 (2004)
7. Ethier, S.N., Griffiths, R.C.: The infinitely-many-sites model as a measure valued diffusion. Ann. Probab. 15, 515 545 (1987)
8. Ethier, S.N., Griffiths, R.C.: On the two-locus sampling distribution. J. Math. Biol. 20, 131 159 (1990)
9. Fearnhead, P., Donnelly, P.: Estimating recombination rates from population genetic data. Genetics 159, 1299–1318 (2001)
10. Fearnhead, P., Donnelly, P.: Approximate likelihood methods for estimating local recombination rates. J. R. Statist. Soc. B 64, 657–680 (2002)
11. Fearnhead, P., Smith, N.G.C.: A novel method with improved power to detect recombination hotspots from polymorphism data reveals multiple hotspots in human genes. Am. J. Hum. Genet. 77, 781–794 (2005)
12. Griffiths, R.C., Marjoram, P.: Ancestral inference from samples of DNA sequences with recombination. J. Comput. Biol. 3, 479–502 (1996)
13. Griffiths, R.C., Tavaré, S.: Ancestral inference in population genetics. Stat. Sci. 9, 307–319 (1994)
14. Griffiths, R.C., Tavaré, S.: Sampling theory for neutral alleles in a varying environment. Proc. R. Soc. London B. 344, 403–410 (1994)
15. Griffiths, R.C., Tavaré, S.: Simulating probability distributions in the coalescent. Theor. Popul. Biol. 46, 131–159 (1994)
16. Gusfield, D.: Optimal, efficient reconstruction of Root-Unknown phylogenetic networks with constrained recombination. J. Comput. Sys. Sci. 70, 381–398 (2005)
17. Gusfield, D., Eddhu, S., Langley, C.: The fine structure of galls in phylogenetic networks. INFORMS J. on Computing, special issue on Computational Biology 16, 459–469 (2004)
18. Gusfield, D., Eddhu, S., Langley, C.: Optimal, efficient reconstruction of phylogenetic networks with constrained recombination. J. Bioinf. Comput. Biol. 2, 173–213 (2004)
19. Hein, J.: Reconstructing evolution of sequences subject to recombination using parsimony. Math. Biosci. 98, 185–200 (1990)
20. Hein, J.: A heuristic method to reconstruct the history of sequences subject to recombination. J. Mol. Evol. 36, 396–405 (1993)
21. Hudson, R.R.: Generating Samples under the Wright-Fisher neutral model of genetic variation. Bioinformatics 18, 337–338 (2002)

22. Hudson, R., Kaplan, N.: Statistical properties of the number of recombination events in the history of a sample of DNA sequences. Genetics 111, 147–164 (1985)
23. Hudson, R.R.: Two-locus sampling distributions and their application. Genetics 159, 1805–1817 (2001)
24. International HapMap Consortium. A haplotype map of the human genome 437, 1299–1320 (2005)
25. Kuhner, M.K., Yamato, J., Felsenstein, J.: Estimating effective population size and mutation rate from sequence data using metropolis-hastings sampling. Genetics 140, 1421–1430 (1995)
26. Kuhner, M.K., Yamato, J., Felsenstein, J.: Maximum likelihood estimation of recombination rates from population data. Genetics 156, 1393–1401 (2000)
27. Larribe, F., Lessard, S., Schork, N.J.: Gene Mapping via the Ancestral Recombination Graph. Theor. Popul. Biol. 62, 2150–2229 (2002)
28. Li, N., Stephens, M.: Modeling linkage disequilibrium and identifying recombination hotspots using single-nucleotide polymorphism data. Genetics 165, 2213–2233 (2003)
29. Lyngsø, R.B., Song, Y.S., Hein, J.: Minimum recombination histories by branch and bound. In: Casadio, R., Myers, G. (eds.) WABI 2005. LNCS (LNBI), vol. 3692, pp. 239–250. Springer, Heidelberg (2005)
30. McVean, G., Awadalla, P., Fearnhead, P.: A coalescent-based method for detecting and estimating recombination from gene sequences. Genetics 160, 1231–1241 (2002)
31. McVean, G., Cardin, N.: Approximating the coalescent with recombination. Philos. Trans. R. Soc. Lond. B Biol. Sci. 360, 1387–1393 (2005)
32. McVean, G.A.T., Myers, S., Hunt, S., Deloukas, P., Bentley, D.R., Donnelly, P.: The fine-scale structure of recombination rate variation in the human genome. Science 304, 581–584 (2004)
33. Myers, S., Bottolo, L., Freeman, C., McVean, G., Donnelly, P.: A fine-scale map of recombination rates and hotspots across the human genome. Science 310, 321–324 (2005)
34. Myers, S.R., Griffiths, R.C.: Bounds on the minimum number of recombination events in a sample history. Genetics 163, 375–394 (2003)
35. Simonsen, K.L., Churchill, G.A.: A Markov chain model of coalescence with recombination. Theor. Popul. Biol. 52, 43–59 (1997)
36. Song, Y.S., Hein, J.: Parsimonious reconstruction of sequence evolution and haplotype blocks: Finding the minimum number of recombination events. In: Proc. of Workshop on Algorithms in Bioinformatics 2003, Berlin, Germany. LNCS, pp. 287–302. Springer, Berlin (2003)
37. Song, Y.S., Hein, J.: On the minimum number of recombination events in the evolutionary history of DNA sequences. J. Math. Biol. 48, 160–186 (2004)
38. Song, Y.S., Hein, J.: Constructing minimal ancestral recombination graphs. J. Comput. Biol. 12, 147–169 (2005)
39. Song, Y.S., Lyngsø, R.B., Hein, J.: Counting all possible ancestral configurations of sample sequences in population genetics. IEEE Transactions on Computational Biology and Bioinformatics 3(3), 239–251 (2006)
40. Song, Y.S., Wu, Y., Gusfield, D.: Efficient computation of close lower and upper bounds on the minimum number of needed recombinations in the evolution of biological sequences. In: Proc. of ISMB 2005, Bioinformatics, vol. 21, pp. 413–422 (2005)

41. Stephens, M., Donnelly, P.: Inference in molecular population genetics. J.R. Stat. Soc. Ser. B 62, 605–655 (2000)
42. Wall, J.D.: A comparison of estimators of the population recombination rate. Mol. Biol. Evol. 17, 156–163 (2000)
43. Wang, L., Zhang, K., Zhang, L.: Perfect phylogenetic networks with recombination. J. Comput. Biol. 8, 69–78 (2001)
44. Wilson, I.J., Balding, D.J.: Genealogical inference from microsatellite data. Genetics 150, 499–510 (1998)

Author Index

Lecture Notes in Bioinformatics

Vol. 4955: M. Vingron, L. Wong (Eds.), Research in Computational Molecular Biology. XVI, 480 pages. 2008.

Vol. 4780: C. Priami (Ed.), Transactions on Computational Systems Biology VIII. VII, 103 pages. 2007.

Vol. 4774: J.C. Rajapakse, B. Schmidt, G. Volkert (Eds.), Pattern Recognition in Bioinformatics. XVIII, 410 pages. 2007.

Vol. 4751: G. Tesler, D. Durand (Eds.), Comparative Genomics. IX, 193 pages. 2007.

Vol. 4695: M. Calder, S. Gilmore (Eds.), Computational Methods in Systems Biology. X, 249 pages. 2007.

Vol. 4689: K. Li, X. Li, G.W. Irwin, G. He (Eds.), Life System Modeling and Simulation. XIX, 561 pages. 2007.

Vol. 4645: R. Giancarlo, S. Hannenhalli (Eds.), Algorithms in Bioinformatics. XIII, 432 pages. 2007.

Vol. 4643: M.-F. Sagot, M.E.M.T. Walter (Eds.), Advances in Bioinformatics and Computational Biology. XII, 177 pages. 2007.

Vol. 4544: S. Cohen-Boulakia, V. Tannen (Eds.), Data Integration in the Life Sciences. XI, 282 pages. 2007.

Vol. 4532: T. Ideker, V. Bafna (Eds.), Systems Biology and Computational Proteomics. IX, 131 pages. 2007.

Vol. 4463: I.I. Măndoiu, A. Zelikovsky (Eds.), Bioinformatics Research and Applications. XV, 653 pages. 2007.

Vol. 4453: T. Speed, H. Huang (Eds.), Research in Computational Molecular Biology. XVI, 550 pages. 2007.

Vol. 4414: S. Hochreiter, R. Wagner (Eds.), Bioinformatics Research and Development. XVI, 482 pages. 2007.

Vol. 4366: K. Tuyls, R.L. Westra, Y. Saeys, A. Nowé (Eds.), Knowledge Discovery and Emergent Complexity in Bioinformatics. IX, 183 pages. 2007.

Vol. 4360: W. Dubitzky, A. Schuster, P.M.A. Sloot, M. Schröder, M. Romberg (Eds.), Distributed, High-Performance and Grid Computing in Computational Biology. X, 192 pages. 2007.

Vol. 4345: N. Maglaveras, I. Chouvarda, V. Koutkias, R. Brause (Eds.), Biological and Medical Data Analysis. XIII, 496 pages. 2006.

Vol. 4316: M.M. Dalkilic, S. Kim, J. Yang (Eds.), Data Mining and Bioinformatics. VIII, 197 pages. 2006.

Vol. 4230: C. Priami, A. Ingólfsdóttir, B. Mishra, H. Riis Nielson (Eds.), Transactions on Computational Systems Biology VII. VII, 185 pages. 2006.

Vol. 4220: C. Priami, G. Plotkin (Eds.), Transactions on Computational Systems Biology VI. VII, 247 pages. 2006.

Vol. 4216: M. R. Berthold, R.C. Glen, I. Fischer (Eds.), Computational Life Sciences II. XIII, 269 pages. 2006.

Vol. 4210: C. Priami (Ed.), Computational Methods in Systems Biology. X, 323 pages. 2006.

Vol. 4205: G. Bourque, N. El-Mabrouk (Eds.), Comparative Genomics. X, 231 pages. 2006.

Vol. 4175: P. Bücher, B.M.E. Moret (Eds.), Algorithms in Bioinformatics. XII, 402 pages. 2006.

Vol. 4146: J.C. Rajapakse, L. Wong, R. Acharya (Eds.), Pattern Recognition in Bioinformatics. XIV, 186 pages. 2006.

Vol. 4115: D.-S. Huang, K. Li, G.W. Irwin (Eds.), Computational Intelligence and Bioinformatics, Part III. XXI, 803 pages. 2006.

Vol. 4075: U. Leser, F. Naumann, B. Eckman (Eds.), Data Integration in the Life Sciences. XI, 298 pages. 2006.

Vol. 4070: C. Priami, X. Hu, Y. Pan, T.Y. Lin (Eds.), Transactions on Computational Systems Biology V. IX, 129 pages. 2006.

Vol. 4023: E. Eskin, T. Ideker, B. Raphael, C. Workman (Eds.), Systems Biology and Regulatory Genomics. X, 259 pages. 2007.

Vol. 3939: C. Priami, L. Cardelli, S. Emmott (Eds.), Transactions on Computational Systems Biology IV. VII, 141 pages. 2006.

Vol. 3916: J. Li, Q. Yang, A.-H. Tan (Eds.), Data Mining for Biomedical Applications. VIII, 155 pages. 2006.

Vol. 3909: A. Apostolico, C. Guerra, S. Istrail, P.A. Pevzner, M. Waterman (Eds.), Research in Computational Molecular Biology. XVII, 612 pages. 2006.

Vol. 3886: E.G. Bremer, J. Hakenberg, E.-H.(S.) Han, D. Berrar, W. Dubitzky (Eds.), Knowledge Discovery in Life Science Literature. XIV, 147 pages. 2006.

Vol. 3745: J.L. Oliveira, V. Maojo, F. Martín-Sánchez, A.S. Pereira (Eds.), Biological and Medical Data Analysis. XII, 422 pages. 2005.

Vol. 3737: C. Priami, E. Merelli, P. Gonzalez, A. Omicini (Eds.), Transactions on Computational Systems Biology III. VII, 169 pages. 2005.

Vol. 3695: M. R. Berthold, R.C. Glen, K. Diederichs, O. Kohlbacher, I. Fischer (Eds.), Computational Life Sciences. XI, 277 pages. 2005.

Vol. 3692: R. Casadio, G. Myers (Eds.), Algorithms in Bioinformatics. X, 436 pages. 2005.

Vol. 3680: C. Priami, A. Zelikovsky (Eds.), Transactions on Computational Systems Biology II. IX, 153 pages. 2005.

Vol. 3678: A. McLysaght, D.H. Huson (Eds.), Comparative Genomics. VIII, 167 pages. 2005.

Vol. 3615: B. Ludäscher, L. Raschid (Eds.), Data Integration in the Life Sciences. XII, 344 pages. 2005.

Vol. 3594: J.C. Setubal, S. Verjovski-Almeida (Eds.), Advances in Bioinformatics and Computational Biology. XIV, 258 pages. 2005.

Vol. 3500: S. Miyano, J. Mesirov, S. Kasif, S. Istrail, P.A. Pevzner, M. Waterman (Eds.), Research in Computational Molecular Biology. XVII, 632 pages. 2005.

Vol. 3388: J. Lagergren (Ed.), Comparative Genomics. VII, 133 pages. 2005.

Vol. 3380: C. Priami (Ed.), Transactions on Computational Systems Biology I. IX, 111 pages. 2005.

Vol. 3370: A. Konagaya, K. Satou (Eds.), Grid Computing in Life Science. X, 188 pages. 2005.

Vol. 3318: E. Eskin, C. Workman (Eds.), Regulatory Genomics. VII, 115 pages. 2005.

Vol. 3240: I. Jonassen, J. Kim (Eds.), Algorithms in Bioinformatics. IX, 476 pages. 2004.

Vol. 3082: V. Danos, V. Schachter (Eds.), Computational Methods in Systems Biology. IX, 280 pages. 2005.

Vol. 2994: E. Rahm (Ed.), Data Integration in the Life Sciences. X, 221 pages. 2004.

Vol. 2983: S. Istrail, M.S. Waterman, A. Clark (Eds.), Computational Methods for SNPs and Haplotype Inference. IX, 153 pages. 2004.

Vol. 2812: G. Benson, R.D.M. Page (Eds.), Algorithms in Bioinformatics. X, 528 pages. 2003.

Vol. 2666: C. Guerra, S. Istrail (Eds.), Mathematical Methods for Protein Structure Analysis and Design. XI, 157 pages. 2003.